Structural Design of Polymer Composites

Structural Design of Polymer Composites

EUROCOMP Design Code and Handbook

THE EUROPEAN STRUCTURAL POLYMERIC COMPOSITES GROUP

Edited by

JOHN L. CLARKE

Sir William Halcrow and Partners Ltd,
London, UK

E & FN SPON
An Imprint of Chapman & Hall
London · Weinheim · New York · Tokyo · Melbourne · Madras

Published by E & FN Spon, an imprint of Chapman & Hall, 2-6 Boundary Row, London SE1 8HN, UK

Chapman & Hall, 2-6 Boundary Row, London SE1 8HN, UK

Chapman & Hall, GmbH, Pappelallee 3, 69469 Weinheim, Germany

Chapman & Hall USA, 115 Fifth Avenue, New York, NY 10003, USA

Chapman & Hall Japan, ITP-Japan, Kyowa Building, 3F, 2-2-1 Hirakawacho, Chiyoda-ku, Tokyo 102, Japan

Chapman & Hall Australia, 102 Dodds Street, South Melbourne, Victoria 3205, Australia

Chapman & Hall India, R. Seshadri, 32 Second Main Road, CIT East, Madras 600 035

First edition 1996

Printed in Great Britain by St Edmundsbury Press, Bury St Edmunds, Suffolk

ISBN 0 419 19450 9

A catalogue record for this book is available from the British Library

Library of Congress Catalog Card Number: 96-67191

Publisher's Note This book has been prepared from camera ready copy provided by the Editor.

∞ Printed on acid-free paper, manufactured in accordance with ANSI/NISO Z39.48-1992 (Permanence of Paper)

CONTENTS

PREFACE

INTRODUCTION

In the autumn of 1989, a group of firms, having a common interest in the development of structural polymeric composites, met to discuss the formation of a consortium, which would have the objective of the production of a code of recommended practice for the design of structures made of polymeric composites. This initiative was prompted by the lack of available design information, methodology and history of performance in service, to designers or constructors, which consequently restricted the use of such materials.

The founding Group agreed to collaborate in this initiative and in view of the then approaching Single European Market, also agreed that the proposed Design Code should be applicable throughout Europe and could, if accepted as good practice, eventually form the basis for a future Eurocode.

ORGANISATION

The founding members of the Group set about identifying further funding partners and supporters throughout Europe and, in parallel, applied for governmental assistance through the EUREKA Initiative. The Project Plan, which was submitted to the EUREKA Secretariat, was accepted at the high-level meeting of ministers in Brussels in mid 1991 and allocated a EUREKA Number EU468.

The three-year plus Project commenced in July 1991 with financial assistance being given by the UK and Scandinavian EUREKA Secretariats and was completed by December 1994. The Project, executed under the terms and conditions of a joint Collaboration Agreement, was steered by a Project Steering Committee comprising a representative from each of the funding organisations, with overall management by Halcrow Polymerics Limited, a wholly-owned subsidiary of Sir William Halcrow and Partners, an international firm of Consulting Engineers and Architects who are independent of any commercial or manufacturing interests. Full details of the participating companies and Project Steering Committee Members are as follows:

UK

1.	Defence Research Agency (DRA)	- Mr D.J. Steel*
2.	Fibreforce Composites Ltd	- Mr P.C. White
3.	James Quinn Associates Ltd	- Mr J.A. Quinn*
4.	L G Mouchel & Partners Ltd,	- Mr D.B. Thompson/ Dr V. Peshkam*
5.	Halcrow Polymerics Limited	- Mr I.K. Lamb (Project Manager)
6.	Sir William Halcrow and Partners Ltd	- Mr R.W. Rothwell (Chairman,PSC) Mr P.S. Godfrey/ Dr J.L. Clarke*
7.	Taywood Engineering Ltd	- Dr P.B. Bamforth
8.	Vosper Thornycroft (UK) Ltd	- Dr R. Mableson
9.	W S Atkins Consultants Ltd	- Mr A.L. Gilbertson/ Mr L. Jones*

UK (Continued)

10. University of Lancaster	-	Dr G.J. Turvey
11. University of Strathclyde	-	Prof. W.M. Banks*
12. University of Warwick	-	Dr J.T. Mottram
13. Imperial College of Science, Technology and Medicine, London	-	Mr F.L. Matthews
14. University of Newcastle Upon Tyne	-	Professor A.G. Gibson
15. University of Glasgow	-	Professor M.J. Cowling
16. University of Surrey	-	Professor L. Hollaway
17. University of Southampton	-	Dr R.A. Shenoi

FINLAND

18. Neste OY	-	Mr J. Juselius/ Mr J. Hanka*
19. Helsinki University of Technology	-	Prof. T. Brander

SWEDEN

20. Royal Institute of Technology,	-	Dr I. Eriksson

FRANCE

21. CEBTP	-	Dr A. Morel
22. CEA	-	Mr P. Garcin
23. Electricite de France	-	Mr T. Le Courtois
24. SEPMA SA	-	Mr M. Baguet

* *Members of the Code Drafting Sub-Committee.*

EUREKA SECRETARIATS

25. London	-	Department of Trade and Industry (DTI)
26. Finland	-	Technology Develpment Centre of Finland (TEKES)
27. Sweden	-	Swedish National Board for Industrial and Technical Development (NUTEK)

The Project Steering Committee met for a total of 11 approximately quarterly Steering Committee Meetings. In addition, numerous meetings of various Task Area Sub-groups were held and informal discussions took place.

STRUCTURE AND STATUS

The objective of the Project has been the production of a practical Design Code for the Construction Industry, which would enable designers to consider the use of a broad range of polymeric composites for structural applications. The Design Code is intended for use by engineers familiar with design using conventional structural materials, such as steel and concrete. The scope is limited to glass FRP materials, components, connections and assemblies, but excludes entity structures, for example automotive and aerospace applications.

The resulting Document is in three parts; Part 1 is the EUROCOMP Design Code and Part 2 is the EUROCOMP Handbook which provides further background information. Part 3 consists of brief technical reports describing research carried out by participants during the course of the development of the EUROCOMP Design Code. In the absence of any authoritative Code, this Document is intended to set down recommended good practice. It is based on the best information available, but it has no legal status.

ACKNOWLEDGEMENTS

The Project Steering Committee acknowledges the valuable comment, assistance and support given by the many Regulatory Authorities, academic and other organisations and individuals listed below. No endorsement of the EUROCOMP Design Code is, however, implied by their involvement;

A. ASSOCIATE MEMBER PARTICIPANTS (REGULATORY BODIES)

1.	Lloyd's Register, Offshore Division	Mr J.C. Smith
2.	Department of the Environment	Mr Harding
3.	Swedish National Testing & Research Institute	Mr T. Gevert
4.	VTT Technical Research Centre of Finland	Professor A. Sarja
5.	Det Norske Veritas Research, Norway	Dr A.T. Echtermeyer

B. PROJECT SUPPORTERS

(i) Commercial/ Research Organisations:

1.	Construction Industry Research and Information Association (CIRIA)	Dr B.W. Staynes
2.	Plastics Design and Engineering	Mr B.D. Gray
3.	Laing Technology Group Limited (LTG)	Mr G.P. Hammersley
4.	Deutsche Forschungsanstalt fur Luft und Raumfahrt (DLR) (Germany)	Dr A.F. Johnson
5.	National Physical Laboratory (NPL)	Dr G.D. Sims
6.	ERA Technology Ltd	Mr D.P. Bashford
7.	Armfibre Ltd	Mr P.E. Ball

(ii) Academic Organisations

1.	University of Bristol	Professor R.D. Adams
2.	Ecole Nationale Superieure des Mines de Paris	Dr R.A. Bunsell
3.	Latvian Academy of Sciences	Professor V. Tamuzs

DISCLAIMER

This EUROCOMP Design Code is the product of the work of the EUROCOMP Project Steering Committee, which represents the views of a wide body of designers, academic and manufacturing organisations as to what is considered to be current good practice.

The advice given has no legal standing and the EUROCOMP Group accepts no responsibility for the adequacy of the contents or any omissions.

Part 1 EUROCOMP Design Code

CONTENTS
EUROCOMP DESIGN CODE

CONTENTS

1 INTRODUCTION

1 INTRODUCTION

1.1 SCOPE OF EUROCOMP DESIGN CODE

P(1) The EUROCOMP Design Code applies to the structural design of buildings and civil engineering works in glass fibre reinforced polymeric composites.

P(2) This Code is only concerned with the requirements for resistance, serviceability and durability of structures. Other requirements, e.g. concerning thermal or sound insulation, are not considered.

P(3) Construction and workmanship is covered to the extent that is necessary to indicate the quality of the materials and products which should be used and the standard of workmanship needed during all stages of production and erection to comply with the assumptions of the design rules. Construction and workmanship are covered in Chapter 6, and are to be considered as minimum requirements which may have to be further developed for particular types of buildings or civil engineering works.

P(4) The EUROCOMP Design Code does not cover the special requirements of seismic design.

P(5) The EUROCOMP Design Code aims to be harmonious with Eurocodes. Numerical values of the actions on buildings and civil engineering works to be taken into account in the design are not given in the EUROCOMP Design Code. They are provided in the Eurocode 1 "Bases of Design and Actions on Structures"[1] applicable to the various types of construction.

(6) The principles of this EUROCOMP Design Code should be applicable to any fibre reinforced polymeric composite. However, the design methods and design data are specific to the use of glass fibres.

(7) The attached EUROCOMP Handbook is intended to provide additional information to supplement the EUROCOMP Design Code.

1.2 DISTINCTION BETWEEN PRINCIPLES AND APPLICATION RULES

P(1) Depending on the character of the individual clauses, distinction is made in the EUROCOMP Design Code between *Principles* and *Application Rules*.

[1] At present at the draft stage.

P(2) The *Principles* comprise:

- general statements and definitions for which there is no alternative, as well as
- requirements and analytical models for which no alternative is permitted unless specifically stated.

P(3) In this code the *Principles* are preceded by the letter P.

P(4) The *Application Rules* are generally recognised rules which follow the *Principles* and satisfy their requirements.

P(5) It is permissible to use alternative design rules different from the *Application Rules* given in the EUROCOMP Design Code, provided that it is shown that the alternative rules accord with the relevant *Principles* and are at least equivalent with regard to the resistance, serviceability and durability achieved for the structure with the EUROCOMP Design Code.

(6) Additional information is contained in the EUROCOMP Handbook which accompanies the EUROCOMP Design Code.

1.3 ASSUMPTIONS

P(1) The following assumptions apply:

- structures are designed by appropriately qualified and experienced personnel
- adequate supervision and quality control is provided in factories, in plants, and on site
- construction is carried out by personnel having the appropriate skill and experience
- the construction materials and products are used as specified in the EUROCOMP Design Code or in the relevant material or product specifications
- the structure will be adequately maintained
- the structure will be used in accordance with the design brief.

P(2) The design procedures are valid only when the requirements for construction and workmanship given in Chapter 6 are also complied with.

1.4 DEFINITIONS

1.4.1 Terms common to all Eurocodes

P(1) Unless otherwise stated in the following, the terminology used in International Standard ISO 8930 applies.

1.4.2 Special terms used in the EUROCOMP Design Code and the EUROCOMP Handbook

- **Accelerator** - A material which when mixed with a resin will speed up the chemical reaction between the resin and the curing agent.
- **Adherend** - A member of a bonded joint.
- **Adhesive** - A substance capable of holding two materials together by surface attachment.
- **Anisotropic** - Not isotropic; having mechanical and/or physical properties which vary with direction at a point in the material.
- **Aramid** - A manufactured fibre in which the fibre-forming substance consists of a long-chain synthetic aromatic polyamide in which at least 85% of the amide linkages are attached directly to two aromatic rings.
- **Areal Weight of Fibre Reinforcement** - The weight of fibre per unit area of mat or fabric; this is often expressed as grams per square metre (g/m^2).
- **Aspect Ratio** - In an essentially two-dimensional rectangular structure, the ratio of the long dimension to the short dimension. Also, in micro-mechanics, the ratio of length to diameter of the fibre.
- **Assembly** - A series of components connected together to form a part of the final structure. It may be assembled in its final location or assembled elsewhere and transported to its final location.
- **Autoclave** - A closed vessel for producing an environment of fluid pressure, with or without heat, to an enclosed object which is undergoing a chemical reaction or other operation.
- **Autoclave Moulding** - A process similar to the pressure bag technique. The lay-up is covered by a pressure bag, and the entire assembly is placed in an autoclave capable of providing heat and pressure for curing the part. The pressure bag is normally vented to the outside.
- **B-Stage** - An intermediate stage in the reaction of some thermosetting resins, in which the material softens when heated and swells when in contact with certain liquids but does not entirely fuse or dissolve. Materials are sometimes precured to this stage, called prepregs, to facilitate handling and processing prior to final cure.

- **Balanced Laminate** - A composite laminate in which the piles are stacked so that the balance of $+\theta$ layers and $-\theta$ layers is maintained.

- **Binder** - A bonding resin used to hold strands together in a mat or preform during manufacture of a moulded part.

- **Bond** - The adhesion of one surface to another, with the use of an adhesive or bonding agent.

- **Bonded Insert Joint** - A joint where loads from a structural element are transferred through an adhesive material and an insert to a mechanical attachment.

- **Bonded Joint** - A joint where the surfaces are held together by means of structural adhesive or matrix polymer.

- **Brittle Material Behaviour** - Material deformation with negligible plastic deformation before fracture.

- **Bundle** - A general term for a collection of essentially parallel filaments or fibres.

- **Carbon Fibres** - Fibres produced by the pyrolysis of organic precursor fibres such as rayon, polyacrylonitrile (PAN), or pitch in an inert atmosphere. The term is often used interchangeably with "graphite". However, carbon fibres and graphite fibres differ in the temperature at which the fibres are made and heat-treated, and the carbon content.

- **Cast-in Joint** - A joint where one load bearing member is partly embedded into the other load bearing member by means of bonding. This joint is also know as an intersecting wall lap joint.

- **Chopped Strand Mat (CSM)** - Non-woven mat in which the glass fibre strands are chopped into short lengths of approximately 50 mm and fairly evenly distributed and randomly orientated. The mat is held together by a binder.

- **Cold Press Moulding** - A low pressure, low temperature process in which fibres are impregnated with a cold cure resin and then pressed between matched dies.

- **Combined Joint** - A joint where the surfaces are held together by both adhesive and mechanical means.

- **Component** - Any structural element, made of one or more laminates, with or without a core made of another material. It may by itself form a structure, such as a simple post or plank, or be connected to other components to form a more complex structure.

- **Composite or Composite Material** - A combination of high modulus, high strength and high aspect ratio reinforcing material encapsulated by and acting in concert with a polymeric matrix.

- **Continuous Filament** - A yarn or strand in which the individual filaments are substantially the same length as the strand.

- **Continuous Filament Mat (CFM) -** A non-woven material similar to chopped strand mat except that the fibres are swirled at random and are continuous.

- **Coupling Agent** - Any chemical substance designed to react with both the reinforcement and matrix phases of a composite material to form or promote a stronger bond at the interface. Coupling agents are applied to the reinforcement phase from an aqueous or organic solution or from a gas phase, or added to the matrix as an integral blend.

- **Crazing** - Fine cracks at or near the surface of a plastic material.

- **Cure** - To change the properties of a thermosetting resin irreversibly by chemical reaction, i.e. condensation, ring-closure, or addition. Cure may be accomplished by addition of curing (cross-linking) agents, with or without catalyst, and with or without heat.

- **Cure Cycle** - The schedule of time periods at specified conditions to which a reacting thermosetting material is subjected, in order to reach a specified property level.

- **Cure Stress** - A residual internal stress produced during the curing cycle of composite structures. Normally, these stresses originate when different components have different thermal coefficients of expansion.

- **Delamination** - The separation of the layers of material in a laminate. This may be local or may cover a large area of the laminate. It may occur at any time in the cure or subsequent life of the laminate and may arise from a wide variety of causes.

- **Dough Moulding Compound (DMC) -** A mixture of chopped fibres, fillers and thermosetting resin, used in hot press moulding and injection moulding.

- **Drop Off** - Change in thickness of a laminate due to the ending of a reinforcement layer.

- **Dry Fibre Area** - Area of fibre not totally encapsulated in resin.

- **Ductile Material Behaviour -** Material deformation with significant plastic deformation before fracture.

- **End** - A single fibre, strand, roving or yarn being already incorporated into a product. An end may be an individual wrap yarn or cord in a woven fabric. In referring to aramid and glass fibres, an end is usually an untwisted bundle of continuous filaments.

- **Epoxy Resins** - Resins which may be of widely different structures but which are characterised by the reaction of the epoxy group to form a cross-linked hard resin.

- **Fabric, Non-woven** - A textile structure produced by bonding or interlocking of fibres, or both, accomplished by mechanical, chemical, thermal or solvent means, and combinations thereof.

- **Fabric, Woven** - A generic material construction consisting of interlaced yarns or fibres, usually a planar structure. In a fabric lamina, the warp direction is considered to be the longitudinal direction, analogous to the filament direction in a filamentary lamina.

- **Fibre** - A general term used to refer to filamentary materials. Often, "fibre" is used synonymously with "filament".

- **Fibre Content** - The amount of fibre present in a composite. This is usually expressed as a percentage volume or weight fraction of the composite.

- **Fibre Count** - The number of fibres per unit width of ply present in a specified section of composite.

- **Fibre Direction** - The orientation or alignment of the longitudinal axis of the fibre with respect to a stated reference axis.

- **Filament** - The smallest unit of a fibrous material. The basic units formed during spinning and which are gathered into strands of fibre. Filaments usually are of extreme length and of very small diameter.

- **Filament Winding** - A reinforced-plastics process that employs a series of continuous, resin-impregnated fibres applied to a mandrel in a predetermined geometrical relationship under controlled tension.

- **Filament Wound** - Pertaining to an object created by the filament winding method of fabrication.

- **Fill** - See Weft.

- **Filler** - A relatively inert substance added to a material to alter its physical, mechanical, thermal, electrical or other properties or to lower cost. Sometimes the term is used specifically to mean particulate additives.

- **Finish (or Size System)** - A material, with which filaments are treated, which contains a coupling agent to improve the bond between the filament surface and the resin matrix in a composite material. In addition, finishes often contain ingredients which provide lubricity to the filament surface, preventing abrasive damage during handling, and a binder which promotes strand integrity and facilitates packing of the filaments.

- **Flash** - Excess material which forms at the parting line of a mould or die, or which is extruded from a closed mould.

- **FRP** - Fibre Reinforced Plastic (or Polymer).

- **Gel** - The initial jelly-like solid phase that develops during formation of a resin from a liquid. Also a semi-solid system consisting of a network of solid aggregates in which liquid is held.

- **Gel Coat** - A quick-setting resin used in moulding processes to provide an improved surface for the composite; it is the first resin applied to the mould after the mould-release agent.

- **Gel Point** - The stage at which a liquid begins to exhibit pseudo-elastic properties.

- **Gel Time** - The period of time from a pre-determined starting point to the onset of gelation (gel point) as defined by a specific test method.

- **Glass** - An inorganic product of fusion which has cooled to a rigid condition without crystallising.

- **Glass Cloth** - Conventionally-woven glass fibre material.

- **Glass Fibre** - A fibre spun from an inorganic product of fusion which has cooled to a rigid condition without crystallising.

- **Glass Transition** - The reversible change in an amorphous polymer or in amorphous regions in partially crystalline polymer from (or to) a viscous or rubbery condition to (or from) a hard and relatively brittle one.

- **Glass Transition Temperature** - The approximate midpoint of the temperature range over which the glass transition takes place, below which the polymer fails in a brittle manner and above which it behaves as a rubbery solid.

- **Hand Lay-up** - A process in which resin and reinforcement are applied either to a mould or to a working surface, and successive plies are built up and worked by hand.
- **Heat Distortion Temperature** - The temperature at which a standard beam under controlled heating conditions reaches a prescribed deflection (ISO 75).
- **Homogeneous** - Material properties are independent of the position in the structure.
- **Honeycomb** - Manufactured product of resin-impregnated sheet material or sheet metal, formed into hexagonal-shaped cells or similar. Used as a core material in sandwich constructions.
- **Hot Press Moulding** - Heated matched dies are loaded with thermosetting compound and pressed together until cured.
- **Hybrid** - A composite laminate comprised of laminae of two or more composite material systems. Or, a combination of two or more different fibres into a structure.
- **Impregnate** - The process of combining the reinforcement with resin.
- **Inhibitor** - A substance which retards a chemical reaction; used in certain types of monomers and resins to prolong storage life or to control the curing process.
- **Insert** - An integral part of the FRP component consisting of metal or other material which may be moulded or bonded into position or pressed into the component after completion.
- **Interface** - The boundary between the individual, physically distinguishable constituents of a composite.
- **Interlaminar** - Descriptive term pertaining to some object (e.g. void), event (e.g. fracture) or potential field (e.g. shear stress) referenced as existing or occurring between two or more adjacent laminae.
- **Interlaminar Shear** - Shearing force tending to produce a relative displacement between two laminae in a laminate along the plane of their interface.
- **Isotropic** - Having the same properties irrespective of direction.
- **Lamina** - A single ply or layer in a laminate of layers.
- **Laminae** - Plural of lamina.

- **Laminate** - A product made by bonding together two or more layers or laminae of material or materials.
- **Laminated Joint** - A joint where loads are primarily transferred from one member to another through additional doublers, which are laminated on to cured members.
- **Lay-up** - A process of fabrication involving the assembly of successive layers of resin-impregnated material.
- **Mandrel** - A form fixture or male mould used for the base in the production of a part by lay-up, filament winding or pultrusion.
- **Mat** - A fibrous material consisting of randomly orientated chopped or swirled continuous filaments loosely held together with a binder.
- **Matrix** - The essentially homogeneous material in which the fibre system of a composite is embedded. Consists of a resin system with or without additives or fillers.
- **Mechanical Joint** - A joint where the surfaces are held together by bolts or similar or with other type of mechanical interlocking.
- **Milled Fibres** - Continuous glass strands hammer-milled into small modules of filamentised glass. Useful as anticrazing reinforcing fillers for adhesives.
- **Monomer** - A compound consisting of molecules each of which can provide one or more constitutional units.
- **Moulding** - The forming of a polymer or composite into a solid mass of prescribed shape and size within or on a mould, possibly accomplished under heat and pressure; sometimes used to denote the finished part.
- **Moulded joint** - A joint where loads are transferred from one member to another by using a moulded connection member.
- **Mould Release Agent** - A lubricant applied to mould surfaces to facilitate release of the moulded article.
- **Orthotropic** - Having three mutually perpendicular planes of elastic symmetry.
- **Peel Ply** - The outside layer of laminate which is removed or sacrificed to achieve improved bonding of additional plies.
- **Pick Count** - The number of filling yarns per unit length of woven roving.

- **Plastic** - A material that contains one or more organic polymers of large molecular weight, is a solid in its finished state and, at some state in its manufacture or processing into finished articles, can be shaped by flow.

- **Ply** - synonymous with lamina (both terms are used in the documents).

- **Plymol** - A variation of the pressure bag moulding process, used to produce cylinders.

- **Polyester Resin** - Thermosetting resins produced by dissolving unsaturated, generally linear, alkyd resins in a vinyl type monomer such as styrene and capable of being crosslinked by vinyl polymerisation using initiators and promoters.

- **Polymer** - An organic material composed of molecules characterised by the repetition of one or more types of monomeric units.

- **Polymerisation** - A chemical reaction in which the molecules of monomers are linked together to form polymers.

- **Porosity** - The ratio of the volume of air or other gases contained within the boundaries of a material to the total volume (solid material plus air or void), expressed as a percentage.

- **Postcure** - Additional elevated temperature cure, usually without pressure, to improve final properties or complete the cure or both.

- **Pot Life** - The period of time during which a reacting thermosetting composition remains suitable for its intended processing after mixing with a reaction initiating agent.

- **Precursor (for Carbon Fibre)** - Either the PAN or pitch fibres from which carbon fibres are derived.

- **Preform** - A preshaped fibrous reinforcement formed by distribution of chopped fibres by air, water flotation or vacuum over the surface of a perforated screen to the approximate contour and thickness of the final part. Also, a preshaped fibrous reinforcement of mat or cloth formed to desired shape on a mandrel or mock-up prior to being placed in a mould press.

- **Prepreg** - Ready to mould or cure material in sheet form which may be fibre, cloth or mat impregnated with resin and stored for use. The resin is partially cured to a B-stage and supplied to the fabricator for lay-up and cure.

- **Pressure Bag Moulding** - A similar process to vacuum bag moulding, with pressure being applied to a rubber bag to aid consolidation of a laminate.
- **Pullforming** - A combination of compression moulding and pultrusion used to produce products with varying cross-sections not possible with conventional pultrusion.
- **Pullwinding** - A combination of filament winding and pultrusion used to produce tubes.
- **Pultrusion** - Process for the manufacture of composite profiles by pulling layers of fibrous materials, impregnated with a synthetic resin, through a heated die, thus forming the ultimate shape of the profile. Used for the manufacture of rods, tubes and structural shapes of constant cross-section.
- **Quasi-isotropic Laminate** - A laminate approximating isotropy in the plane of the laminate by orientation of plies in several directions, typically 0°/ 90°/ ±45°.
- **Reinforced Plastic** - A plastic with fibres of relatively high stiffness or very high strength embedded in the composition. This improves the mechanical properties over those of the base resin.
- **Release Agent** - An additive which promotes release see also Mould Release Agent.
- **Resin** - The polymeric material used to bind together the reinforcing fibres in FRP.
- **Resin Content** - The amount of matrix present in a composite either by percent weight or by percent volume.
- **Resin System** - A mixture of resin, with ingredients such as catalyst, initiator (curing agent), diluents, etc. required for the intended processing and final product.
- **Resin Transfer Moulding (RTM)** - A closed mould process in which dry reinforcement in the form of mat or cloth is placed into a matched mould. Resin is then injected in to fill the cavity and flows through the fibres to fill the mould space.
- **Roving** - Strands of continuous fibres.
- **Sandwich Construction** - A structural panel concept consisting in its simplest form of two relatively thin, parallel sheets of structural material bonded to, and separated by, a relatively thick light-weight core.

- **Sheet Moulding Compound (SMC)** - A mixture of fibres, fillers and thermosetting resin in sheet form similar to DMC, but having longer fibres and higher fibre content. It is used in hot press moulding processes.

- **Size** - Any treatment which is applied to yarn or fibres at the time of formation to protect the surface and aid the process of handling and fabrication, or to control the fibre characteristics. The treatment contains ingredients which provide surface lubricity and binding action but, unlike a finish, contain no coupling agent. Before final fabrication into a composite, the size is usually removed and a finish applied.

- **Spray-up** - Fabrication technique where a mixture of chopped fibres and resin can be simultaneously deposited on a mould.

- **Stacking Sequence** - The configuration of a composite laminate with regard to the angles of lay-up, the number of lamina at each angle, and the exact sequence of the laminae lay-up.

- **Strand** - Normally an untwisted bundle or assembly of continuous filaments used as a unit, including tow, ends, yarn, etc.

- **Structural Adhesive** - A material employed to form high strength bonds in structural assemblies which perform load bearing function.

- **Surface or Surfacing Mat** - A thin mat of fine fibres used primarily to produce a smooth surface on an organic matrix composite.

- **Symmetric Laminate** - A composite laminate in which each lamina type, angles and areal mass is exactly mirrored about the laminate neutral axis.

- **Tex** - A unit for expressing linear density equal to the mass or weight in grams of 1000 metres of filament, fibre yarn or other textile strand.

- **Tg** - see Glass Transition Temperature.

- **Thermoplastic** - A plastic that repeatedly can be softened by heating and hardened by cooling through a temperature range characteristic of the plastic, and when in the softened stage can be shaped by flow into articles by moulding or extrusion.

- **Thermoset** - A plastic that is substantially infusible and insoluble after being cured by heat or other means, e.g. polyester, epoxy, phenolic resin.

- **Top Coat** - A surface layer of resin to give additional protection or for aesthetic purposes, also called flow coat.

- **Tow** - An untwisted bundle of continuous filaments. Commonly used in referring to man-made fibres, particularly carbon.
- **Unidirectional Laminate** - A reinforced laminate in which substantially all of the fibres are in the same direction.
- **Vacuum Bag Moulding** - A process in which the lay-up is cured under pressure generated by drawing a vacuum in the space between the lay-up and a flexible sheet placed over it and sealed at the edges.
- **Veil** - A thin layer of mat similar to a surface mat, often composed of organic fibres as well as glass fibres.
- **Vinylester Resin** - Thermosetting resins that consist of a polymer backbone with an acrylate or methacrylate termination.
- **Warp** - The longitudinally orientated yarn in a woven fabric; a group of yarns in long lengths and approximately parallel.
- **Weft or Fill** - The transverse threads or fibres in a woven fabric; those fibres running perpendicular to the warp.
- **Wet Lay-up** - A method of making a reinforced product by applying a liquid resin system while the reinforcement is put in place, typically layer by layer.
- **Wet Winding** - A method of filament winding in which the fibre reinforcement is impregnated with the resin system as a liquid just prior to wrapping on a mandrel.
- **Woven Fabrics** - See Fabric, Woven
- **Woven Roving (WR)** - Glass fibre material made by the weaving of roving.
- **Yarn** - A generic term for strands or bundles of continuous filaments or fibres, usually twisted and suitable for making textile fabric.

A number of abbreviations are commonly used by the composites industry to describe their products including:

- FRC Fibre Reinforced Composites
- FRP Fibre Reinforced Plastic
- GRP Glass Reinforced Plastic
- PMC Polymer Matrix Composites
- RFC Resin Fibre Composites

This last term has been used throughout the EUROCOMP Design Code and the EUROCOMP Handbook.

1.5 S.I. UNITS

P(1) S.I. Units shall be used in accordance with ISO 1000.

(2) For calculations, the following units are recommended:

- forces and loads : kN, kN/m, kN/m^2
- unit mass : kg/m^3
- unit weight : kN/m^3
- stresses, strengths and modulus : N/mm^2 (= MN/m^2 or MPa)
- moments (bending ...) : kNm

1.6 SYMBOLS COMMON TO ALL EUROCODES

1.6.1 Latin upper case letters

A	Accidental action; Area
C	Fixed value
E	Modulus of elasticity; Effect of action
F	Action; Force
G	Permanent action; Shear modulus
I	Second moment of area
M	Moment
N	Axial force
Q	Variable action
R	Resistance
S	Internal forces and moments
T	Torsional moment
V	Shear force
W	Section modulus
X	Value of a property of a material

1.6.2 Latin lower case letters

a	Distance; Geometrical data
Δa	Additive or reducing safety element for geometrical data
b	Width
d	Diameter; Depth
e	Eccentricity
f	Strength (of a material)
h	Height
i	Radius of gyration
k	Coefficient; Factor
l[2]	Length; Span

[2] l can be replaced by L or handwritten for certain lengths or to avoid confusion with 1 (numeral).

m	Mass
r	Radius
t	Thickness
u, v, w	Components of the displacement of a point
x, y, z	Co-ordinates

1.6.3 Greek lower case letters

α	Angle; Ratio
β	Angle; Ratio
γ	Partial safety factor
ε	Strain
λ	Slenderness ratio
μ	Coefficient of friction
υ	Poisson's ratio
ρ	Mass density
σ	Normal stress
τ	Shear stress
ψ	Factors defining representative values of variable actions
	ψ_0 for combination values
	ψ_1 for frequent values
	ψ_2 for quasi-permanent values

1.6.4 Subscripts

c	Compression
cr (or crit)	Critical
d	Design
dir	Direct
dst	Destabilising
eff	Effective
ext	External
f	Flange
F (or P)	Action
g (or G)	Permanent action
h	High; Higher
ind	Indirect
inf	Inferior; Lower
int	Internal
k	Characteristic
l	Low; Lower
m (or M)	Material; Matrix
max	Maximum
min	Minimum
nom	Nominal
q (or Q)	Variable action
R	Resistance
rep	Representative
S	Internal moments and forces
stb	Stabilising

sup	Superior; Upper
t (or ten)	Tension
t (or tor)	Torsion
u	Ultimate
v	Shear
w	Web
x, y, z	Co-ordinates

1.7 SPECIAL SYMBOLS USED IN THE EUROCOMP DESIGN CODE

In general, the symbols used in the EUROCOMP Design Code are based on the common symbols in 1.6 and on derivatives of these as, for example:

$G_{d,sup}$ Upper design value of a permanent action

Such derivations are defined in the text where they occur, at the start of the relevant chapter or section.

1.8 CO-ORDINATE SYSTEMS

The systems used for the co-ordinates for components, elements and individual plies in the EUROCOMP Design Code shown in Figure 1.1.

Co-ordinate systems

XYZ Component axes
x y z Axes of the separate elements
1 2 3 Axes of individual plies of an element.

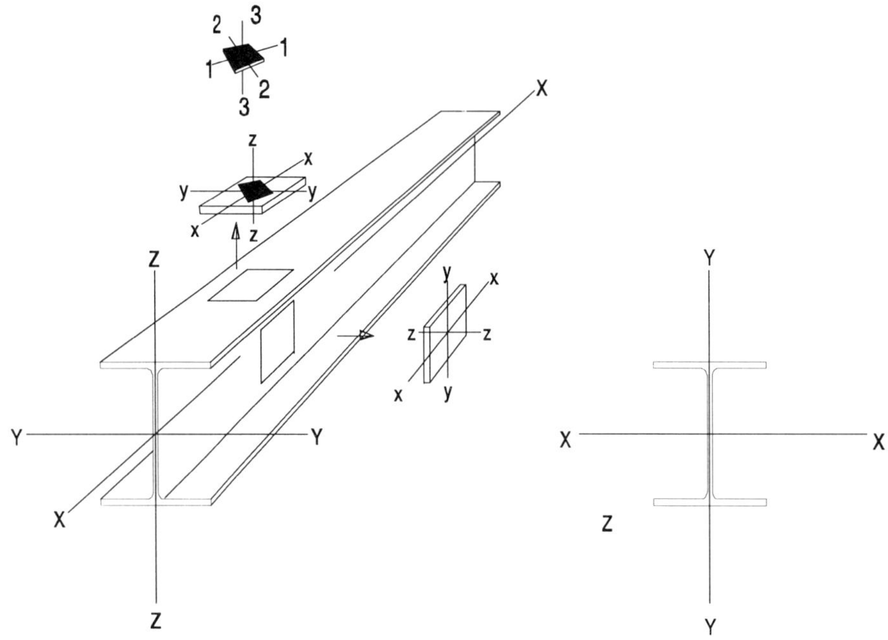

Figure 1.1 Co-ordinate systems

CONTENTS

2 BASIS OF DESIGN

2 BASIS OF DESIGN

2.0 NOTATION - SECTIONS 2.1 TO 2.4 (SEE ALSO SECTIONS 1.6 AND 1.7)

A	Accidental actions
A_d	Design value of accidental action
A_k	Characteristic value of accidental action
C_d	Nominal value of, or a function of certain design properties of, materials
C_m	Coefficient of model uncertainty
D_d	Design value of the damage indicator (fatigue)
$E_{d,dst}$	Design effects of destabilising actions
$E_{d,stb}$	Design effects of stabilising actions
F_d	Design values of an action
F_k	Characteristic value of an action
$G_{d,inf}$	Lower design value of a permanent action
$G_{s,sup}$	Upper design value of a permanent action
G_{ind}	Indirect permanent action
G_k	Characteristic value of a permanent action
$G_{k,inf}$	Lower characteristic value of a permanent action
$G_{k,sup}$	Upper characteristic value of a permanent action
$G_{k,j}$	Characteristic values of permanent actions
Q_{ind}	Indirect variable action
Q_k	Characteristic value of a variable action
$Q_{k,1}$	Characteristic value of one of the variable actions
$Q_{k,i}$	Characteristic value of the other variable actions
a_d	Design values of geometrical data
a_{nom}	Nominal value of geometrical data
Δa	Change made to nominal geometrical data for particular design purposes (e.g. assessment of effects of imperfections)
γ_f	Partial safety factor for actions
$\gamma_{G,inf}$	Partial safety factor for permanent actions, in calculating the lower design values
$\gamma_{G,sup}$	Partial safety factor for permanent actions, in calculating the upper design values
$\gamma_{GA}, \gamma_{GA,j}$	Partial safety factors for permanent actions, for accidental design situations
γ_m	Partial safety factor for material properties
$\gamma_{m,i}$	Partial safety coefficient for condition i
$\gamma_{G,j}$	Partial safety factor for any permanent action j
$\gamma_{Q,i}$	Partial safety factor for any variable action i
$\gamma_{Q,1}$	Partial safety factor for the basic most unfavourable variable action

2.1 FUNDAMENTAL REQUIREMENTS AND WARNING OF FAILURE

2.1.1 Fundamental requirements

P(1) A structure shall be designed and constructed in such a way that:

- with acceptable probability, it will remain fit for the use for which it is required, having due regard to its intended life and its cost
- with appropriate degrees of reliability, it will sustain all actions and influences likely to occur during execution and use and have adequate durability in relation to maintenance costs.

P(2) A structure shall also be designed in such a way that it will not be damaged by events like explosions, impact or consequences of human errors, to an extent disproportionate to the original cause.

(3) The potential damage should be limited or avoided by appropriate choice of one or more of the following:

- avoiding, eliminating or reducing the hazards which the structure is to sustain
- selecting a structural form which has low sensitivity to the hazards considered
- selecting a structural form and design that can survive adequately the accidental removal of an individual element
- tying the structure together.

P(4) The above requirements shall be met by the choice of suitable materials, by appropriate design and detailing and by specifying control procedures for production, construction and use as relevant to the particular project.

2.1.2 Warning of failure

P(1) Structures made with FRP composites shall be so designed as to give reasonable and adequate warning of failure prior to reaching an ultimate limit state.

(2) In general, FRP composites exhibit little or no ductile behaviour beyond a point of linear stress - strain behaviour of the material: failure of the material may occur locally soon after this point has been reached. The design should take account of this behaviour by ensuring that a serviceability limit state is reached prior to its ultimate limit state for the mode of failure being considered.

(3) Such a serviceability limit state might be one of the following:

- excessive deflection/deformation
- buckling or wrinkling
- local damage
- environmental damage.

2.2 DEFINITIONS AND CLASSIFICATIONS

2.2.1 Limit states and design situations

2.2.1.1 Limit states

P(1) Limit states are states beyond which the structure no longer satisfies the design performance requirements.

Limit states are classified into:

- ultimate limit states
- serviceability limit states.

P(2) Ultimate limit states are those associated with collapse, or with other forms of structural failure which may endanger the safety of people or the continued performance of associated structures or components.

P(3) States prior to structural collapse which, for simplicity, are considered in place of the collapse itself are also treated as ultimate limit states.

(4) Ultimate limit states which may require consideration include:

- loss of equilibrium of the structure or any part of it, considered as a rigid body
- failure by excessive deformation, rupture or loss of stability of the structure or any part of it, including supports and foundations.

P(5) Serviceability limit states correspond to states beyond which specified service requirements are no longer met.

(6) Serviceability limit states which may require consideration include:

- deformations or deflections which affect the appearance or effective use of the structure (including the malfunction of machines or services) or cause damage to finishes or non-structural elements

- vibration which causes discomfort to people, damage to the building or its contents, or which limits its functional effectiveness
- cracking or delamination of the FRP composite which is likely to affect appearance, durability or water-tightness
- local damage of FRP composite (due, for example, to impact or local bearing failure) in the presence of excessive stress which is likely to lead to loss of durability.

2.2.1.2 Design situations

P(1) Design situations are classified as:

- persistent situations corresponding to normal conditions of use of the structure
- transient situations; for example, during construction or repair
- accidental situations.

2.2.2 Actions

2.2.2.1 Definitions and principal classifications

(Fuller definitions of the classifications of actions will be found in the Eurocode for Actions.)

P(1) An action (F) is:

- a force (load) applied to the structure (direct action), or
- an imposed deformation (indirect action); for example, temperature effects or settlement.

P(2) Actions are classified:

(i) by their variation in time

- permanent actions (G), e.g. self-weight of structures, fittings, ancillaries and fixed equipment
- variable actions (Q), e.g. imposed loads, wind loads or snow loads
- accidental actions (A), e.g. blast, explosions or impact from projectiles or vehicles.

(ii) by their spatial variation

- fixed actions, e.g. self-weight (but see 2.3.2.3(3) for structures very sensitive to variation in self-weight)

- free actions, which result in different arrangements of actions, e.g. movable imposed loads, wind loads, snow loads.

(3) Indirect actions may be either permanent G_{ind} (e.g. settlement of support) or variable Q_{ind} (e.g. temperature) and are treated accordingly.

(4) Supplementary classifications relating to the response of the structure are given in the relevant clauses.

2.2.2.2 Characteristic values of actions

P(1) Characteristic values F_k are specified:

- in the Eurocode for Actions or other relevant loading codes, or
- by the client, or the designer in consultation with the client, provided that minimum provisions, specified in the relevant codes or by the competent authority, are observed.

P(2) For permanent actions where the coefficient of variation is large or where the actions are likely to vary during the life of the structure (e.g. for some superimposed permanent loads), two characteristic values are distinguished, an upper ($G_{k,sup}$) and a lower ($G_{k,inf}$). Elsewhere a single characteristic value (G_k) is sufficient.

(3) The self-weight of the structure may, in most cases, be calculated on the basis of the nominal dimensions and mean density or unit masses.

P(4) For variable actions the characteristic value (Q_k) corresponds to either:

- the upper value with an intended probability of not being exceeded, or the lower value with an intended probability of not being reached, during some reference period, having regard to the intended life of the structure or the assumed duration of the design situation, or
- the specified value.

(5) For accidental actions the characteristic value A_k (when relevant) generally corresponds to a specified value.

2.2.2.3 Representative values of variable actions

(Fuller definitions of the classifications of actions will be found in the Eurocode for Actions.)

P(1) The main representative value is the characteristic value Q_k.

P(2) Other representative values are expressed in terms of the characteristic value Q_k by means of a factor ψ_i. These values are defined as:

- combination value : $\psi_0 Q_k$
- frequent value : $\psi_1 Q_k$
- quasi-permanent value : $\psi_2 Q_k$

P(3) Supplementary representative values are used for fatigue verification and dynamic analysis.

P(4) The factors ψ_i are specified:

- in the Eurocode for Actions or other relevant loading codes, or
- by the client or the designer in conjunction with the client, provided that minimum provisions, specified in the relevant codes or by the competent public authority, are observed.

(5) Subject to P(4) above, for building structures the simplified rules given in 2.3.3.1 may be used.

2.2.2.4 Design values of actions

P(1) The design value F_d of an action is expressed in general terms as:

$$F_d = \gamma_f F_k$$

P(2) Specific examples are:

$$G_d = \gamma_g G_k$$
$$Q_d = \gamma_Q Q_k \text{ or } \gamma_Q \psi_i Q_k \qquad (2.1)$$
$$A_d = \gamma_A A_k \text{ (if } A_d \text{ is not directly specified)}$$

where γ_f, γ_G, γ_Q and γ_A are the partial safety factors for the action considered, taking account of, for example, the possibility of unfavourable deviations of the actions, the possibility of inaccurate modelling of the actions, uncertainties in the assessment of the effects of the actions, and uncertainties in the assessment of the limit state considered (see 2.3.3.1).

P(3) The upper and lower design values of permanent actions are expressed as follows (see 2.2.2.2):

- where only a single characteristic value G_k is used, then:
  $$G_{d,sup} = \gamma_{G,sup} G_k$$
  $$G_{d,inf} = \gamma_{G,inf} G_k$$
- where upper and lower characteristic values of permanent actions are used, then:
  $$G_{d,sup} = \gamma_{G,sup} G_{k,sup}$$
  $$G_{d,inf} = \gamma_{G,inf} G_{k,inf}$$

where $G_{k,sup}$ and $G_{k,inf}$ are the upper and lower characteristic values of permanent actions
and $\gamma_{G,sup}$ and $\gamma_{G,inf}$ are the upper and lower values of the partial safety factor for the permanent actions.

2.2.2.5 Design values of the effects of actions

P(1) The effects of actions (E) are responses (for example internal forces and moments, stresses, strains) of the structure to the actions. Design values of the effects of actions (E_d) are determined from the design values of the actions, geometrical data and material properties where relevant:

$$E_d = E_d\,(F_d,\ a_d,\ ....) \qquad (2.2)$$

where a_d is defined in 2.2.4.

(2) In some cases, in particular for non-linear analysis or the design of connections, the effect of the randomness of the intensity of the actions and the uncertainty associated with the analytical procedures, e.g. the models used in the calculations, should be considered separately. This may be achieved by the application of a coefficient of model uncertainty, C_m, applied either to the actions or to the internal forces and moments, or incorporated in the design expression or condition to be satisfied, provided that the purpose and value of such a factor are defined where used.

2.2.3 Material properties

2.2.3.1 Characteristic values

P(1) A material property shall wherever possible, be represented by a characteristic value X_k corresponding to a fractile or other limiting condition in the assumed statistical distribution of the particular property of the material, specified by relevant standards and tested under specified conditions.

(2) In certain cases, for example where insufficient tests have been carried out to obtain a statistically significant number of results, a nominal value may be used as the characteristic value, provided that an appropriate value of γ_m is used (see 2.3.3.2).

(3) A material strength, or other property may have two characteristic values, an upper and a lower. In most cases only the lower value will need to be considered. In some cases, higher values may be adopted, depending on the type of problem considered, e.g. the modulus of elasticity when considering the effects of indirect actions and strength where overstrength effects may lead to a reduction in safety.

P(4) The approach in P(1) above does not apply to fatigue (see 4.13).

2.2.3.2 Design values

P(1) The design value X_d of a material property is generally defined as:

$$X_d = X_k/\gamma_m \tag{2.3}$$

where γ_m is the partial safety factor for the material property, given in 2.3.3.2 and 2.3.4.

Other definitions are given in the appropriate sections.

P(2) Design values for the material properties, geometrical data and effects of actions, where relevant, shall be used to determine the design resistance R_d from:

$$R_d = R(X_d, a_d, \ldots) \tag{2.4}$$

(3) The design value R_d may be determined from tests. Guidance is given in Chapters 4 and 7 (see also 2.5.1.5).

2.2.4 Geometrical properties

P(1) Geometrical data describing the structure are generally represented by their nominal values:

$$a_d = a_{nom} \tag{2.5a}$$

P(2) In some cases the geometrical design values are defined by:

$$a_d = a_{nom} + \Delta a \tag{2.5b}$$

The values of Δa are given in the appropriate clauses.

P(3) For imperfections to be adopted in the global analysis of the structure, see 2.5.1.3 and 2.5.2.2.

2.2.5 Load arrangements and load cases

(Detailed rules on load arrangements and load cases are given in the Eurocode for Actions.)

P(1) A load arrangement identifies the position, magnitude and direction of a free action (see 2.5.1.2).

P(2) A load case identifies compatible load arrangements, sets of deformations and imperfections considered for a particular verification.

2.3 DESIGN REQUIREMENTS

2.3.1 General

P(1) It shall be verified that no relevant limit state is exceeded.

P(2) All relevant design situations and load cases shall be considered.

P(3) Possible deviations from the assumed directions or positions of actions shall be considered.

P(4) Calculations shall be performed using appropriate design models (supplemented, if necessary, by tests) involving all relevant variables. The models shall be sufficiently precise to predict the structural behaviour, commensurate with the standard of workmanship likely to be achieved, and with the reliability of the information on which the design is based.

(5) The design process should follow the flow diagram shown in Figure 2.1.

2.3.2 Ultimate limit states

2.3.2.1 Verification conditions

P(1) When considering a limit state of static equilibrium or of gross displacements or deformations of the structure, it shall be verified that:

$$E_{d,dst} < E_{d,stb} \quad (2.6a)$$

where $E_{d,dst}$ and $E_{d,stb}$ are the design effects of destabilising and stabilising actions, respectively.

P(2) When considering a limit state of rupture or excessive deformation of a section, member or connection (fatigue excluded) it shall be verified that:

$$S_d \leq R_d \quad (2.6b)$$

where S_d is the design value of an internal force or moment or strain or deformation (or of a respective vector of some of all these) and R_d is the corresponding design resistance or limiting value of strain or deformation, associating all structural properties with the respective design values (see 2.5.3).

P(3) When considering a limit state of transformation of the structure into a mechanism, it shall be verified that a mechanism does not occur unless actions exceed their design values, associating all structural properties with the respective design values.

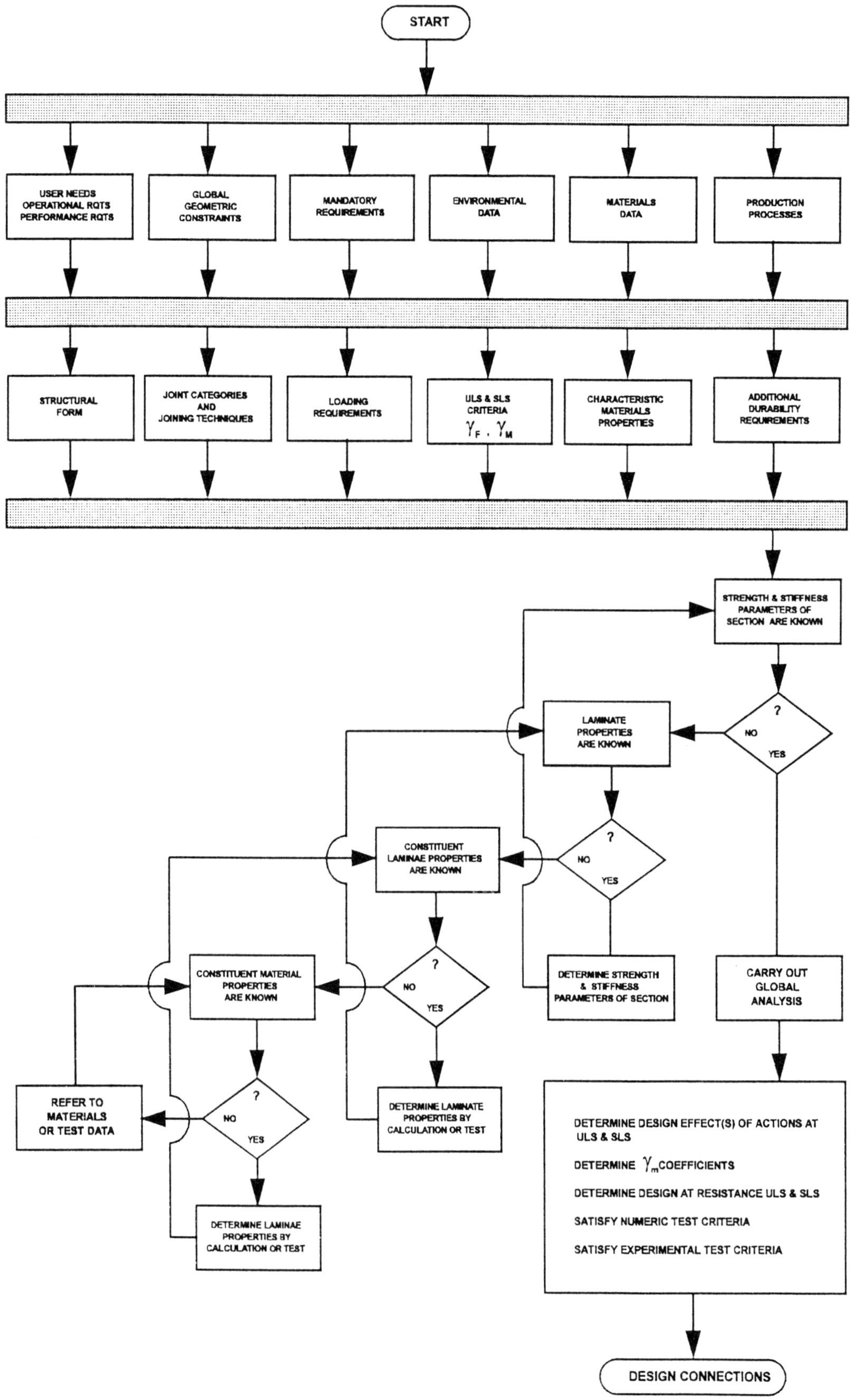

Figure 2.1 Flow diagram for section/member design in FRP composites.

P(4) When considering a limit state of stability induced by second-order effects it shall be verified that instability does not occur unless actions exceed their design values, associating all structural properties with the respective design values. In addition, sections shall be verified according to P(2) above.

P(5) When considering a limit state of rupture induced by fatigue it shall be verified that:

$$D_d \leq 1 \tag{2.6c}$$

where D_d is the design value of the damage indicator (see 4.13).

2.3.2.2 Combinations of actions

P(1) For each load case, design values E_d for the effects of actions shall be determined from combination rules involving design values of actions as identified by Table 2.1.

Table 2.1 Design values for actions for use in the combination of actions.

Design situation	*Permanent actions G_d*	*Variable Actions*		*Accidental actions A_d*
		One with its characteristic value	*Others with their combination value*	
Persistent and Transient	$\gamma_G G_k$	$\gamma_Q Q_k$	$\psi_0 \gamma_Q Q_k$	-
Accidental *	$\gamma_{GA} G_k$	$\psi_1 Q_k$	$\psi_2 Q_k$	$\gamma_a A_k$ (if A_d is not specified directly)

* if not specified differently elsewhere.

P(2) The design values of Table 2.1 shall be combined using the following expressions (given in symbolic form).
(Fuller definition of the classifications of actions will be found in the Eurocode for Actions.)

- Persistent and transient design situations for verifications other than those relating to fatigue (fundamental combinations):

$$\Sigma\gamma_{G,j}\, G_{k,j} + \gamma_{Q,1}\, Q_{k,1} + \Sigma\gamma_{Q,i}\, \psi_{0,i}\, Q_{k,i}\ (i > 1) \tag{2.7a}$$

- Accidental design situations (if not specified differently elsewhere):

$$\Sigma\gamma_{GA,j}\, G_{k,j} + A_d + \psi_{1,1}\, Q_{k,1} + \Sigma\psi_{2,i}\, Q_{k,i}\ (i > 1) \qquad (2.7b)$$

where:
$G_{k,j}$ = characteristic values of permanent actions
$Q_{k,1}$ = characteristic value of one of the variable actions
$Q_{k,i}$ = characteristic values of the other variable actions
A_d = design value (specified value) of the accidental action
$\gamma_{G,j}$ = partial safety factors for permanent action j
$\gamma_{GA,j}$ = as $\gamma_{G,j}$, but for accidental design situations
$\gamma_{Q,i}$ = partial safety factors for variable action i
ψ_0, ψ_1, ψ_2 = coefficients defined in 2.2.2.3

Imposed deformations should be considered where relevant.

P(3) Combinations for accidental design situations either involve an explicit accidental action A (e.g. shock) or refer to a situation after an accidental event (A = 0). Unless specified otherwise γ_{GA} = 1 may be used.

P(4) For fatigue, see 4.13.

(5) Simplified equations for building structures are given in 2.3.3.1.

P(6) For fire design, see 4.16.

2.3.2.3 Design values of permanent actions

P(1) In the various combinations defined above, those permanent actions that increase the effect of the variable actions (i.e. produce unfavourable effects) shall be represented by their upper design values, those that decrease the effect of the variable actions (i.e. produce favourable effects) by their lower design values (see 2.2.2.4).

P(2) Except for the case in P(3) below, either the lower or the upper design value (whichever gives the more unfavourable effect) shall be applied throughout the structure.

P(3) Where the results of a verification may be very sensitive to variations of the magnitude of a permanent action from place to place in the structure, the unfavourable and favourable parts of this action shall be considered as individual actions. This applies in particular to the verification of static equilibrium. In the aforementioned cases specific γ_G values need to be considered (see 2.3.3.1 for building structures).

(4) For continuous beams without cantilevers the same design value of the self-weight (evaluated as in 2.2.2.2) may be applied to all spans.

2.3.3 Partial safety factors for ultimate limit states

2.3.3.1 Partial safety factors for actions on building structures

P(1) Partial safety factors for persistent and transient design situations are given in Table 2.2.

P(2) For accidental design situations to which expression (2.7b) applies, the partial safety factors for variable actions may be put equal to unity.

(3) Where, according to 2.3.2.3 P(3), favourable and unfavourable parts of a permanent action need to be considered as individual actions, the favourable part should be associated with $\gamma_{G,inf}$ = 1.1 and the unfavourable part with $\gamma_{G,sup}$ = 1.35.

(4) Imposed deformations. Where linear behaviour may be assumed, the factor for unfavourable effects may be reduced by 20% (i.e. γ_Q = 1.35).

(5) Vectorial effects. Where components of a vectorial force act independently, the factor applied to any favourable component shall be reduced by 20%.

Table 2.2 Partial safety factors for actions on building structures for persistent and transient design situations.

	Permanent actions γ_G *	*Variable actions* γ_Q	
		One with its characteristic value	*Others with their combination value*
Favourable effect	1.0	**	**
Unfavourable effect	1.35	1.5	1.5

* See also (3) below.
** See Eurocode for Actions; in normal cases for building structures $\gamma_{Q,inf}$ = 0.

(6) By adopting the γ values given in Table 2.2, the expression (2.7a) may be replaced by:

- design situations with only one variable action $Q_{k,1}$

$$\Sigma\gamma_{G,j}\, G_{k,j} + 1.5\, Q_{k,1} \qquad (2.8a)$$

- design situation with two or more variable actions $Q_{k,i}$

$$\Sigma\gamma_{G,j}\, G_{k,j} + 1.35\, Q_{k,i} \text{ (i = 1, 2, ..., n and n > 1)} \qquad (2.8b)$$

whichever gives the larger value.

2.3.3.2 Partial safety factors for materials

P(1) Partial safety factors for material properties are given in Table 2.3.

Table 2.3 Partial safety factors for materials.

Partial safety coefficient	*Description (condition i)*	*Value of $\gamma_{m,i}$*		*To be read with notes*
		Max	*Min*	
$\gamma_{m,1}$	Derivation of material properties from test values (level of uncertainty)	2.25	1.0	(a), (b) and (c)
$\gamma_{m,2}$	Material and production process	2.7	1.1	(a), (b) and (d)
$\gamma_{m,3}$	Environmental effects and duration of loading	3	1	(a), (b) and (e)

NOTES (to be read with Table 2.3)

(a) The partial safety factor (γ_m) for material properties, at the ultimate limit state, is given by the expression:

$$\gamma_m = \gamma_{m,1}\ \gamma_{m,2}\ \gamma_{m,3}$$

where $\gamma_{m,1}$ ... $\gamma_{m,3}$ are the partial safety coefficients obtained from Table 2.3 and from these notes.

For building structures γ_m shall not be taken as less than 1.5 and need not be taken as greater than 10.

(b) The design situations for which the coefficients given in Table 2.3 are applicable are those where the Actions are persistent or transient (see 2.3.2.2).

Where an accidental loading is being considered, the value of γ_m, calculated as in (a) above, may be reduced by an amount not exceeding one third, unless more detailed information or test data justifies a greater reduction, for the circumstances being considered, provided that the limits for γ_m given in note (a) are not exceeded.

The assumption of a higher value of γ_m, calculated as in (a) above, may create an undesirable increase in the calculated resistance of the fibre or matrix, lamina or laminate or, in the case of a pultrusion or whole panel, of the element as a whole. This may be at the section being considered or elsewhere in the member. In such cases, the value of γ_m so calculated shall be reduced by not less than one third, unless more detailed information or test data justifies a lesser reduction, for the circumstances being considered.

(c) The partial safety coefficient ($\gamma_{m,1}$) relating to the level of uncertainty in deriving the laminate, panel or pultruded section characteristics shall be as in Table 2.4, unless more detailed information or test data justifies a lesser value, for the circumstances being considered.

Table 2.4 Values for $\gamma_{m,1}$.

Derivation of properties	$\gamma_{m,1}$
Properties of constituent materials (i.e. fibre and matrix) are derived from test specimen data.	2.25
Properties of individual laminae are derived from theory.	2.25
Properties of the laminate, panel or pultrusion are derived from theory.	2.25
Properties of individual plies are derived from test specimen data.	1.5
Properties of the laminate, panel or pultrusion are derived from theory.	1.5
Properties of the laminate, panel or pultrusion are derived from test specimen data.	1.15

(d) The partial safety coefficient ($\gamma_{m,2}$) relating to the materials and the production process shall be as in Table 2.5, unless more detailed information or test data justifies a lesser value, for the circumstances being considered.

Table 2.5 Values for $\gamma_{m,2}$.

Method of manufacture	$\gamma_{m,2}$ Fully post-cured at works	Not fully post-cured at works
Hand-held spray application	2.2	3.2
Machine-controlled spray application	1.4	2.0
Hand lay-up	1.4	2.0
Resin transfer moulding	1.2	1.7
Pre-preg lay-up	1.1	1.7
Machine-controlled filament winding	1.1	1.7
Pultrusion	1.1	1.7

(e) The partial safety coefficient ($\gamma_{m,3}$) relating to environmental effects and duration of loading shall be as in Table 2.6, unless more detailed information or test data justifies a lesser value, for the circumstances being considered.

Table 2.6 Values for $\gamma_{m,3}$.

Operating design temperature** (°C)	HDT* (°C)	$\gamma_{m,3}$ Short-term loading	Long-term loading
25-50	55-80	1.2	3.0
	80-90	1.1	2.8
	>90	1.0	2.5
0-25	55-70	1.1	2.7
	70-80	1.0	2.6
	>80	1.0	2.5

* Heat distortion temperature.

** For operating temperatures outside this range seek specialist advice.

(2) These values are assumed to take account of differences between the strength of test specimens of the structural materials and their strength *in situ*.

(3) The values given above are valid when the quality control procedures given in Chapter 8 are followed. They are applicable to characteristic values defined here in Chapter 2 and for design data as described in Chapter 4.

(4) Higher or lower values of γ_m may be used if these are justified by adequate control procedures.

P(5) These values do not apply for fatigue verification, where structural properties are determined by testing, see 4.13.

2.3.4 Serviceability limit states

P(1) It shall be verified that:

$$E_d \leq C_d \text{ or } E_d \leq R_d$$

where:

C_d = a nominal value or a function of certain design properties of materials (other than strength or resistance) related to the design effects of actions considered, and

E_d = the design effect of actions, determined on the basis of one of the combinations defined below.

The required combination is identified in the particular clause for serviceability verification (see Chapter 4).

P(2) Three combinations of actions for serviceability limit states are defined by the following expressions:

Rare combination

$$\Sigma G_{k,j} + Q_{k,1} + \Sigma \psi_{0,i} Q_{k,i} \; (i > 1) \tag{2.9a}$$

Frequent combination

$$\Sigma G_{k,j} + \psi_{1,1} Q_{k,1} + \Sigma \psi_{2,i} Q_{k,i} \; (i > 1) \tag{2.9b}$$

Quasi-permanent combination

$$\Sigma G_{k,j} + \Sigma \psi_{2,i} Q_{k,i} \; (i \geq 1) \tag{2.9c}$$

where the notation is defined in 2.3.2.2(2).

Imposed deformations shall be considered when relevant.

(3) Upper limits of tensile, compressive and shear stresses in the composite, in the presence of rare and quasi-permanent combinations, may be fixed to avoid damage to the composite and excessive creep deformations (see 2.5.5 and 4.7).

(4) Where simplified compliance rules are given in the relevant clauses dealing with serviceability limit states, detailed calculations using combinations of actions are not required.

(5) Where the design considers the compliance of serviceability limit states by detailed calculations, simplified expressions may be used for building structures.

(6) For building structures the rare combination may be simplified to the following expressions, which may also be used as a substitute for the frequent combination:

- design situations with only one variable action, $Q_{k.1}$

$$\Sigma G_{k,j} + Q_{k,1} \tag{2.9d}$$

- design situations with two or more variable actions, $Q_{k.i}$

$$\Sigma G_{k,j} + 0.9\Sigma Q_{k,i} \; (i = 1,2, ..., n \text{ and } n > 1) \tag{2.9e}$$

whichever gives the larger value.

P(7) Values of γ_m shall be taken as not less than 1.3, except where stated otherwise in particular clauses.

2.4 DURABILITY

P(1) To ensure an adequately durable structure, the following interrelated factors shall be considered:

- the use of the structure
- the required performance criteria
- the expected environmental conditions
- the intended life of the structure
- the composition, properties and performance of the materials
- the shape of members and the structural detailing
- the quality of workmanship, and level of control
- the particular protective measures
- the likely maintenance during the intended life.

P(2) The environmental conditions shall be estimated at the design stage to assess their significance in relation to durability and to enable adequate provisions to be made for protection of the materials.

2.5 ANALYSIS

2.5.1 General provisions

2.5.1.0 Notation (see also 1.6 and 1.7)

H_{fd} Additional horizontal force to be considered in the design of horizontal structural elements, when taking account of imperfections.

ΔH_j Increase in the horizontal force acting on the floor of a frame structure, due to imperfections.

L Total height of a structure in metres.

N_{ba}, N_{bc} Design axial forces on columns or walls adjacent to a horizontal load transferring element, when considering imperfections.

n Number of vertical continuous members acting together.

α Angle of inclination of a structure, assumed in assessing the effects of imperfections.

α_n Reduction coefficient in calculating α (Equation 2.11).

2.5.1.1 General

P(1) The purpose of analysis is the establishment of the distribution of either internal forces and moments, or stresses, strains and displacements, over the whole or part of a structure. Additional local analysis shall be carried out where necessary.

(2) In most normal cases analysis will be used to establish the distribution of internal forces and moments; however, for certain complex elements, the methods of analysis used (e.g. finite element analysis) give stresses, strains and displacements rather than internal forces and moments. Special methods may be required to use these results to obtain the internal forces and moments.

P(3) Analyses are carried out using idealisations of both the geometry and the behaviour of the structure. The idealisations selected shall be appropriate to the problem being considered.

(4) The geometry is commonly idealised by considering the structure to be made up of linear elements, two dimensional elements and, occasionally, shells. Geometrical idealisations are considered in 2.5.2.

(5) Common behavioural idealisations used for analysis are:-

- elastic behaviour (2.5.3.2 - 2.5.3.3)
- elastic behaviour considering second-order effects
- non-linear behaviour (2.5.5).

(6) Additional local analyses may be necessary where the assumption of linear strain distribution does not apply, e.g.

- when conditions of isotropic behaviour can not be assumed
- when shear deformations are appreciable
- when specially formulated connections are used.

(7) Depending on the type of structure, its function or the method of construction, design may be carried out primarily for the serviceability limit state, i.e. maximum strain. In many cases, provided that checks for this limit state have been carried out, checks for the ultimate limit state may be dispensed with as compliance can be seen by inspection.

P(8) Design of connections shall always be checked at the ultimate limit state.

(9) For linear elements and plates, the effects of shear and longitudinal forces on the deformations may be ignored where these are likely to be less than 10% of those due to bending alone.

2.5.1.2 Load cases and combinations

P(1) For the relevant combinations of actions, sufficient load cases shall be considered to enable the critical design conditions to be established at all sections within the structure or part of the structure considered.

(2) Simplified combinations of actions and load cases may be used, if based on a reasonable interpretation of the structural response.

(3) For continuous beams and slabs in buildings without cantilevers, subjected to predominantly uniformly distributed loads, it will generally be sufficient to consider only the following load cases (see 2.3.2.2).

(a) alternate spans carrying the design variable and permanent load ($\gamma_Q Q_k + \gamma_G G_k$), all other spans carrying only the design permanent load $\gamma_G G_k$.

(b) any two adjacent spans carrying the design variable and permanent loads ($\gamma_Q Q_k + \gamma_G G_k$), all other spans carrying only the design permanent load $\gamma_G G_k$.

2.5.1.3 Imperfections

P(1) In the ultimate limit state, consideration shall be given to the effects of possible imperfections in the geometry of the unloaded structure. Where significant, any possible unfavourable effects of such imperfections shall be taken into account.

P(2) Individual sections shall be designed for the internal forces and moments arising from global analysis, combining effects of actions and imperfections of the structure as a whole.

(3) In the absence of other provisions, the influence of structural imperfections may be assessed by representing them as an effective geometrical imperfection using a procedure such as that given in (4) to (8) below.

(4) Where a structure is being analysed as a whole, the possible effects of imperfections may be assessed by assuming that the structure is inclined at an angle α to the vertical, where:

$$\alpha = 1/(100 \times L^{0.5}) \text{ (radians)} \tag{2.10}$$

where L is the total height of the structure in metres. α should not be taken as less than 1/250 for cases where second order effects are insignificant or 1/100 where they have to be taken into account.

(5) For cases where n vertically continuous members act together, α, given by (4) above, may be reduced by the factor α_n given by Equation (2.11):

$$\alpha_n = ((1 + 1/n)/2)^{0.5} \tag{2.11}$$

(6) If more convenient, the deviations from the vertical given by Equation (2.10) may be replaced by equivalent horizontal forces which should be taken into account in the design of the overall structure, bracing elements, supports and ties.

(7) Structural elements which are assumed to transfer stabilising forces from the elements of a structure to be braced to the bracing elements should be designed to carry an additional horizontal force H_{fd} such that:

$$H_{fd} = (N_{bc} + N_{ba}) \times \alpha/2 \tag{2.12}$$

where N_{bc} and N_{ba} denote the design axial forces on the adjacent columns or walls, acting on the load-transferring element being considered. H_{fd} should not be taken into account in the design of the bracing elements.

(8) Where the effects of imperfections are smaller than the effects of design horizontal actions, their influence may be ignored. Imperfections need not be considered in accidental combinations of actions.

2.5.1.4 Second order effects (see also 2.5.3.2.2)

P(1) Second order effects shall be taken into account where they may significantly affect the overall stability of a structure or the attainment of the ultimate limit state at critical sections.

(2) For normal structures, second order effects may be neglected where they increase the moments, calculated ignoring displacements, by not more than 10%.

(3) For individual members, second order effects may be neglected where the deflection of the member or part of the member transverse to the direction of force in the member or joint of the member is less than half the thickness of the member or outstanding part of the member measured in the direction of the deflection. Such deflection shall be measured relative to a straight line drawn between points of adequate transverse support.

2.5.1.5 Design by testing

P(1) The design of structures or structural elements may be based on testing, as detailed in Chapter 7.

2.5.2 Idealisation of the structure

2.5.2.1 Structural models for overall analysis

(1) The elements of a structure are normally classified, by consideration of their nature and function, as beams, columns, slabs, walls, plates, arches, shells, etc. Rules are provided for the analysis of the more common of these elements and of structures consisting of combinations of these elements.

(2) A beam whose span is less than 25 times its depth is considered as a deep beam for which, in general, shear effects will need to be taken into account.

(3) A plate subjected to dominantly uniformly distributed loads may be considered to be one-way spanning if either:

- it possesses two free (unsupported) and sensibly parallel edges, or
- it is the central part of a sensibly rectangular plate supported on four edges with a ratio of the longer to shorter span greater than 2 and is so reinforced that the unit stiffness of the plate in the direction of the assumed one-way spanning is greater than or equal to that in the transverse direction.

2.5.2.2 Geometrical data

(1) For the purposes of global analysis at the ultimate limit state and the serviceability limit state, to determine the forces in the elements forming the structure the section properties may be those corresponding to the nominal dimension of the said elements.

P(2) For the purposes of design of the said elements at the ultimate limit state, the section properties shall be those corresponding to the nominal dimensions less the manufacturing tolerances given in Table 6.1.

(3) For the purposes of design of the said elements at the serviceability limit state, the section properties may be those corresponding to the nominal dimensions of the said elements.

2.5.3 Calculation methods

2.5.3.1 Basic considerations

P(1) All methods of analysis shall satisfy equilibrium.

P(2) Compatibility conditions shall be checked directly for the limit states considered.

(3) Normally, equilibrium will be checked on the basis of the undeformed structure (first order theory). However, in cases where deformations lead to a significant increase in the internal forces and moments, equilibrium shall be checked considering the deformed structure (second order theory) (see 2.5.3.2.2).

(4) Global analysis for imposed deformations, such as temperature and shrinkage effects, may be omitted where structures are divided by joints into sections chosen to accommodate the deformations.

(5) In normal cases, the maximum dimension of such sections should not exceed 10 m.

2.5.3.2 Types of structural analysis

(a) Ultimate Limit States

P(1) Depending on the specific nature of the structure, the limit state being considered and the specific conditions of design or execution, for the ultimate limit states a linear elastic or a second-order analysis method shall be used.

P(2) The method used should be formulated so that, within its defined field of validity, the level of reliability generally required by this code is achieved, taking account of the particular uncertainties associated with the method.

P(3) In this section, the term "non-linear analysis" relates to analyses which take account of non-linearity of the material behaviour or of specially formulated connections. Analyses which take account of non-linear behaviour resulting from the deflection of elements are termed "second order analyses" (thus a "non-linear second order analysis" takes account of both effects).

(4) Wherever possible, joints should be located away from critical sections. If this is not possible, the deformation or rotation capacity of the joint should be assessed.

P(5) Full account shall be taken of the elastic stability of structural members, e.g.:

- axial load (Euler buckling and torsional buckling)
- lateral torsional stability
- web buckling (shear and diagonal compression)
- local flange stability
- the effect of axial load on the flexural stiffness of members.

(b) Serviceability analysis

(1) Analyses carried out in connection with serviceability limit states should normally be based on linear elastic theory.

(2) Stiffness of members and elastic modulus may be based on short term properties. Allowance for time-dependent effects should be made if these are likely to be significant (see 2.5.5 and 4.7).

P(3) Where shear of the composite material has a significant effect on the performance of the structure or member considered, it shall be taken into account in the analysis (see 2.5.2.1 (2) and 4.5.2 P(6)).

2.5.3.3 Simplification

P(1) Simplified methods or design aids based on appropriate simplifications may be used for analysis provided they have been formulated to give the level of reliability implicit in the methods given in this code over their stated field of validity.

P(2) No redistribution of forces or the effects of actions within the structure shall be permitted.

2.5.4 Effects of anisotropy

2.5.4.0 Notation (see also 1.6)

θ The angle of orientation of the fibres in the plane of the laminate (the x-y plane) relative to the x-axis.

2.5.4.1 Definitions (see also 1.4)

P(1) Anisotropic, isotropic and orthotropic properties of laminates shall be as defined in 1.4.

P(2) Symmetrical laminates are those in which the plies are stacked symmetrically with respect to the mid-plane of the laminate.

P(3) Laminates that are subject to bending-stretching-torsion coupling are defined as generally orthotropic. Typically, generally orthotropic laminates have a symmetrical and balanced stacking sequence of plies, with an appreciable amount (more than about 20%) of reinforcement orientated at other than 0° and 90° (e.g. +45° or −45°).

P(4) Specially orthotropic laminates are those with uncoupled in-plane and flexural behaviour and that do not exhibit bending-stretching-torsion coupling. Normally this is reached with a laminate structure that consists of 0° and/or 90° plies only.

P(5) Isotropic laminates have plies that are in-plane isotropic and are reinforced with either CSM or CFM only.

P(6) Pultruded sections consist of a sequence, usually symmetrical and balanced, of specially orthotropic and isotropic layers. (The term "layer" means here that the different reinforcement types do not necessarily appear as plies of constant thickness).

2.5.4.2 *General* **(see also 4.4 and 4.5)**

P(1) The effects of type, volume fraction and orientation of the reinforcement in each lamina (ply) or layer, and the thickness, stacking sequence and number of laminae or layers on the structural behaviour of the laminate or pultruded section shall be assessed.

P(2) The design of certain members (e.g. in which appreciable bending-stretching-torsion coupling, or some combination of these, occurs) is outside the scope of the EUROCOMP Design Code. Reference should be made to the specialist literature or design programmes (see Chapter 4) for information and recommendations on the applications, analysis and design of such members.

P(3) Bending-stretching-torsion coupling need not be considered specifically in members that are symmetrical, balanced, and otherwise comply with the guidelines for minimising the effects of such coupling given later in this section. For such members, the closed-form expressions given in Chapter 4 may be used.

(4) The following approach to the selection of the stacking sequence may be followed in order to minimise the effects of anisotropy on laminates comprising two or more plies. The recommendations are not applicable to pultruded sections.

(a) Plies of a laminate should be stacked symmetrically, and overall balance must be maintained to avoid bending-stretching-torsion coupling, comprising in-plane/out-of-plane coupling or stretching/shear coupling.

(b) Except when using woven fabrics, adjacent plies should be oriented (when possible) with no more than 60° between them.

(c) If possible, avoid grouping 0° and 90° plies together; separate them by ±45° plies to minimise interlaminar shear and through- thickness stresses when subject to in-plane loads.

(d) Wherever possible avoid excessive grouping of similar plies.

(e) Exterior surface plies should be continuous and orientated at 45°, if possible, so as to minimise interlaminar shear effects.

(f) If possible, curtailment of plies should be symmetric about the laminate mid-plane.

(5) In addition to influencing the anisotropic response of the laminate, the stacking sequence can cause interlaminar through-thickness stress to occur at the free edge of the laminate. Interlaminar tension stresses can cause delamination under static, cyclic and impact loading.

(6) Testing or proven details should be used to ensure that at the free edge of a laminate, sides of a laminate or holes, the interlaminar shearing stress does not cause delamination of the regions.

(7) Particular attention should be paid to interlaminar through-thickness and shear stresses because strengths in these modes are low for FRP and depend principally on the resin matrix and not on the fibres.

2.5.5 Determination of the effects of time dependent deformation of the composite

2.5.5.0 Notation (see also 1.6)

E(t) Modulus of elasticity of the composite at time t.

2.5.5.1 General (see also 4.12)

P(1) The accuracy of the procedures for the calculation of the effects of creep and shrinkage of the composite shall be consistent with the reliability of the data available for the description of these phenomena and the importance of their effects on the limit state considered.

(2) At the ultimate limit state the effects of creep (e.g. creep rupture) shall be taken into account by the choice of the appropriate value of $\gamma_{m,3}$ (see 2.3.3.2).

(3) Otherwise, provided stresses are kept within the limits corresponding to normal service conditions, the effects of creep and shrinkage need be taken into account only for the serviceability limit states, except where their influence on second order effects at the ultimate limit state is likely to be significant (see 2.5.1.4).

P(4) Special investigations shall be considered when the composite is subjected to extremes of temperature.

(5) The following assumptions may be adopted to give an acceptable estimate of the behaviour of a composite material if the stresses are kept within the limits corresponding to the normal service conditions:

- creep and shrinkage are independent
- a linear relationship is assumed between creep and the stress causing the creep
- non-uniform temperature and moisture effects are neglected
- the principle of superposition is assumed to apply for actions occurring at different ages
- the above assumptions apply to tensile, compressive and shear stresses.

(6) If the stresses in the composite vary only slightly, the deformations may be calculated using an effective modulus of elasticity, E(t), which may, for example, be determined from Figure 4.13.

(7) For the method of analysis of the effects of time dependent deformation of FRP composites, see 4.12.

CONTENTS

3 MATERIALS

3 MATERIALS

3.0 GENERAL

P(1) As the components of a composite material are fundamental to the behaviour of the composite in a structure and as the specification of materials may change, the designer should always seek specialist advice from polymer, reinforcement and manufacturing suppliers.

3.1 REINFORCEMENT

3.1.1 Fibres

(1) Fibres are used to convey structural stiffness and strength to composite materials. Selection of fibres, specification of the form of the reinforcement and choice of the process by which the reinforcement is incorporated into the composite is set by the properties required in the composite material.

(2) Strength, stiffness and stress-strain properties of composites are a function of the volume fraction of fibres in the section of the composite, the matrix resin used and the directionality of the fibres with respect to the external loads. The volume fraction of fibres attained in the composite is a function of the form of the reinforcement and manufacturing process. Strength properties of glass fibres can be reduced by weaving and handling although processing finishes are also used to limit this effect.

P(3) The reinforcement material shall be made from a suitable grade of glass fibre having a surface finish compatible with the resin used to promote wetting, to control adhesion and to obtain interface stability. These shall either comply with the relevant ISO or National Standard as appropriate, or be the subject of agreement between the purchaser and the manufacturer.

P(4) For compliance with this EUROCOMP Design Code only the following reinforcements may be used:

E-glass
C-glass
ECR-glass

(5) Typical properties of fibres before processing are given in Table 3.1

3.1.1.1 E-glass

P(1) E-glass fibre shall be selected when the highest strength and electrically resistant composites are required. It is the most widely used fibre type in composites.

P(2) The relevant standard is ISO 3598: Textile glass - Yarn - Basis for a specification.

Table 3.1 Typical properties of fibres before processing (references 3.1, 3.2).

	E-glass	*C-glass*	*ECR-glass*
Specific gravity	2.54	2.50	2.71
Tensile strength N/mm^2 (22°C)	3400	3000	3300
Tensile modulus kN/mm^2 (22°C)	72	69	72
Elongation %	4.8	4.8	4.8
Coeff. of thermal expansion 10^{-6}/°C	5.0	7.2	5.9

3.1.1.2 C-glass

(1) C-glass fibres may be selected where their better chemical resistance is required and fibre properties are adequate for the design. They have very good chemical resistance but slightly lower strength compared to E-glass. They are usually used in the form of surface tissue.

3.1.1.3 ECR-glass

(1) ECR-glass fibres may be selected as an alternative to E-glass. They are boron free, very resistant to chemical attack and have similar properties to E-glass.

3.1.2 Rovings

(1) Rovings are continuous strands of multifibres supplied on creels and used directly and extensively in automated composite processes such as pultrusion and filament winding. Use of rovings is restricted to processes where tension can be applied to control orientation and consolidation. Fibre strength properties are highest in rovings but are reduced when these materials are used as woven reinforcements.

P(2) The relevant standard is ISO 2797: Glass fibre rovings for the reinforcement of polyester and epoxide resin systems.

3.1.3 Mats

3.1.3.1 Chopped strand mat (CSM)

(1) This is a non-woven planar material in which the glass fibre strands are chopped into short lengths and fairly evenly distributed and randomly orientated. The mat is held together by a binder which may dissociate during the impregnation stage. The non-aligned nature of these materials with the random crossing of fibres does not allow fibre volume fractions to exceed about 25% and the method of composite manufacture will significantly determine the level.

(2) The effective fibre fraction that determines the structural reinforcing effect in any in-plane direction is usually not more than 10% by volume and the use of discontinuous fibres leads to a greater dependency on the resin for load transfer. Lower long term strength properties compared with those of continuous fibres may result.

P(3) Composite materials manufactured from CSM reinforcements shall be limited to low stressed and low stiffness applications.

P(4) The nature of the binder will determine the integrity of these materials where there is exposure to water and selection of material should ensure compatibility with the resin to avoid osmosis and potential blistering.

P(5) The relevant standard is ISO 2559: E-glass chopped strand mat for the reinforcement of polyester resin systems.

3.1.3.2 Continuous filament mat (CFM)

P(1) CFM may be used as an alternative to CSM. Properties and application potential are similar to CSM. CFM is a non-woven material similar to chopped strand mat except that the fibres are swirled at random and are continuous. The swirled strands are interlocked with a binder. The user should check with the supplier on the compatibility of the binder for a specific resin system.

3.1.4 Woven rovings (WR)

(1) Woven rovings are bidirectional reinforcements constructed from untwisted fibres in parallel tows having a warp array (roll direction) interspersed with a weft array (width of roll). Woven materials allow ease of handling for the construction of large areas of composite. Woven materials produce higher strength and stiffness composites than random fibre reinforcements but the presence of crimp in heavyweight reinforcements may restrict fibre volume fractions to lower than 40% unless process compaction is high. Thus in a balanced woven roving the effective fibre volume fraction will be 20% in each of the warp and weft directions.

P(2) Woven materials shall be considered where the advantages of continuous fibres and the aligned bidirectional character of the reinforcement are suitable for the loading cases of the application.

P(3) The relevant standard is ISO 2113: Woven glass fibre rovings fabrics of E-glass fibre for the reinforcement of polyester resin systems.

3.1.5 Fabrics

(1) Woven fabrics are constructed by interlacing warp and weft yarns, fibres or filaments in a variety of patterns to form fabric styles such as plain, twill, satin, unidirectional and others. Woven fabrics are usually lighter than woven rovings, have less crimp and may achieve volume fractions of over 50% depending on the method of composite compaction.

P(2) Fabrics should be considered where the weaving pattern style or drape characteristics confers advantages to the performance or manufacture of the composite.

P(3) The relevant standard is ISO 2113: Woven glass fibre fabrics for plastics reinforcement.

3.1.6 Non-crimp fabrics

(1) Materials made from unidirectional fibre tows laid in parallel to each other or held at precise predetermined orientations should allow volume fractions of fibre to exceed 50% within plies and highly aligned properties to be generated. Materials with no crimp are beneficial to creep behaviour.

P(2) Non-crimp fabrics should be considered when the highest materials properties are required.

3.1.7 Prepregs

(1) Prepregs are fibre reinforcements with resins already infiltrated but not fully cured. Manufacture of composites from prepregs requires temperatures in the range of 70°C to 150°C depending on the resin system and the means of applying a consolidating pressure. Fibre volume fractions well in excess of 50% are normally achievable.

P(2) Prepregs should be considered where the design and manufacturing process requires a high fibre volume fraction, low void content and composite premium performance. Elevated temperatures are essential to cure these materials and achieve the highest mechanical properties and environmental resistance.

3.2 RESINS

3.2.1 General

P(1) The selection of polymer resins for use in structural composites will be determined by a number of factors and should not be made without full consultation with materials suppliers and fabricators. Properties required are usually dominated by strength, stiffness, toughness and durability. Account should be taken of the application, service temperature and environment, method of fabrication, cure conditions and level of properties required. A knowledge of service temperature is required to select an appropriate stable resin system as with all polymers loss of stiffness and significant creep will occur if the service temperature is close to the resin second order glass transition temperature. The latter is related to the heat distortion temperature (HDT).

P(2) The performance of structural composites is a function of polymer properties and the quality of the manufactured materials. Low viscosity polymers must be used to ensure fibre wetting and complete impregnation to obtain low void composites.

P(3) Control of the curing process and attainment of full cure of the polymer is essential for attaining optimum mechanical properties, preventing heat softening, limiting creep, reducing moisture diffusion and minimising plasticisation effects. Full cure is only achieved by thorough and accurate blending of the components prior to forming the composite. Room temperature and elevated temperature curing resin systems are available to suit a particular composite processing method.

P(4) The curing reaction is exothermic, providing heat which aids the curing process but which may, if thickness is excessive, cause a temperature build-up sufficient to damage the composite. Production of thick sections should be accomplished through the build-up of thinner sections where this effect is expected. The user should consult with materials suppliers for guidance on the maximum thickness that can be applied at each step and the maximum time the laminate is allowed to cure between lay-up steps to obtain good secondary bonding without the need for special treatments. This is required to ensure a high level of adhesion between the plies of the laminate.

3.2.2 Polyester resins

P(1) For less demanding end-use areas the so-called general purpose orthophthalic resins may be considered. These are the least expensive polyester systems and are widely used. The orthophthalic acid based resin has a reliable combination of good mechanical properties, moderate service temperature capability and chemical resistance.

There is no ISO Standard.

P(2) Isophthalic acid (IPA) based resins, where the isophthalic acid replaces the phthalic acid anhydride, should be considered when better water resistance, chemical resistance or heat resistance is required. They have superior mechanical properties with the possibility of a slightly higher elongation at break despite their more rigid structure but are more expensive than orthophthalic resins.

P(3) Bisphenol-A (BPA) polyester resins, where the phthalic acid or anhydride is partly or completely replaced by bisphenol A, should be considered where the most durable polyester composites are required. They are used almost exclusively in high performance applications where their substantially higher cost can be justified.

(4) Chlorendic polyester resins made from HET acid and maleic anhydride may also be considered. They have increased fire resistance due to the presence of chlorine. Strength and toughness properties are lower than in isophthalic resins.

(5) Typical range of properties of polyester resins are listed in Table 3.2.

3.2.3 Vinyl ester resins

(1) Vinyl ester resins may be derived from backbone components of polyester or urethane resins but those based on epoxide resins are of particular commercial significance. They resemble polyesters in their processing with use of styrene as a reactive diluent, allowing cold curing by a free radical mechanism with initiation through a peroxide catalyst and cobalt salt accelerator. Although the curing mechanism is similar in principle, the curing process is less forgiving in that the window of initiator content and curing temperature is narrower than for polyesters and this can sometime lead to manufacturing difficulties.

(2) Vinyl ester resins derived from bisphenol-A epoxies are suitable for chemical resistant FRP structures, filament winding and pultrusion with failure elongations of about 6%.

(3) Vinyl ester resins derived from phenolic novolac epoxies are suitable for solvent resistant applications and use at higher service temperatures but have lower failure elongations of about 3%.

P(4) Vinyl esters should also be considered for higher performance applications than isophthalic polyesters, as they offer superior resistance to water and chemical attack, better retention of strength and stiffness at elevated temperatures and greater toughness. Failure strains are up to twice that of standard ortho- and isophthalic polyesters.

(5) Typical properties of vinyl ester resins are given in table 3.3.

Table 3.2 Typical range of properties of polyester resins (supplied by Neste, reference 3.3) determined from resin castings.

PROPERTY	*Ortho-phthalic polyester G 105E*	*High-grade ortho-phthalic polyester G 300*	*Chemical resistant, iso-phthalic polyester K 530*	*Elevated temperature iso-phthalic polyester S 599*	*Fire retardant HET-acid based polyester F 892*	*Fire retardant brominated polyester F 820*	
Physical							
Density, g/cm^3	1.1	1.1	1.1	1.1	1.3	1.3	
Water absorption mg (24h)	19	24	18	28	-	23	ISO 62
Heat deflection temperature (HDT) °C	66	95	93	125	62	60	ISO 75 Method A
Viscosity* 23°C Ns/mm^2	180	250	800	1100	500	400	Brookfield
Mechanical							
Tensile strength N/mm^2	55	70	65	55	50	45	ISO 527
Tensile modulus kN/mm^2	3.6	3.3	4.1	3.7	3.4	3.6	ISO 527
Flexural strength N/mm^2	90	110	125	100	75	80	ISO 178
Flexural modulus kN/mm^2	4.1	3.8	3.7	3.7	3.2	3.8	ISO 178
Elongation at break %	2.0	3.5	2.5	1.5	1.7	1.4	ISO 527

Note that relative resin ratings do not necessarily carry over directly to composite properties.

* Liquid resin.

3.2.4 Modified acrylic resins (Modar)

(1) These thermosetting resins contain a urethane prepolymer in methyl methacrylate monomer. Although they are cured in a similar fashion to polyesters, their initial viscosity is much lower and the cure reaction is more rapid. The speed of cure along with the odour of the monomer preclude the use of open mould processes, but these resins are especially suitable for fast processing by methods including pultrusion and resin

Table 3.3 Typical properties of vinyl ester cast resins (reference 3.4).

	Bisphenol-A vinyl ester	*Novolac vinyl ester*
Specific gravity	1.12	1.16
Tensile strength N/mm^2	82	68
Tensile modulus kN/mm^2	3.5	3.5
Elongation at break %	6	3-4
Flexural strength N/mm^2	131	125
Heat distortion temperature °C	102	150

Note that relative resin ratings do not necessarily carry over directly to composite properties.

transfer moulding. The low viscosity of the resin allows high concentrations of the fire retardant filler alumina hydrate to be incorporated, leading to composites with good flammability characteristics second to those based on phenolic resins.

P(2) Methacrylate resins may be considered but as experience of use is much more limited than with polyesters the user should consult with materials suppliers for information on their application, processing and properties.

3.2.5 Phenolic resins

P(1) Phenolic resins should be considered when the overriding emphasis is on performance under heat, retention of properties under fire conditions or low emission of toxic fumes (about 10% that of an isophthalic polyester based glass FRP). The attractive characteristics of composites made from phenolic resins are low flammability, low spread of flame and little smoke. The resole phenolics allow similar curing to polyesters with the addition of an acid based catalyst or heat. Though cure at room temperature is possible, an elevated temperature post cure above 80°C is required to obtain dimensionally stable materials. The use of acid based catalysts may lead to moulding tool corrosion.

(2) The mechanical properties of the resole phenolics are comparable with those of orthophthalic polyesters. Typical values are given in Table 3.4. However the high heat distortion temperature and fire resistance are leading to greater use of these materials. Low shrinkage compared with polyesters is a characteristic of this resin.

(3) All phenolics cure by a condensation mechanism which results in the evolution of water as a by-product of the cure reaction and extensive microvoiding within the matrix. Experience indicates that the microvoids have little effect on the composite properties except that significantly higher

water uptake is observed. High water content can cause structures to delaminate when exposed to heat.

Table 3.4 Typical properties of phenolic cast resins (reference 3.5).

	Resole - acid cured
Specific gravity	1.24
Tensile strength N/mm^2	24-40
Tensile modulus kN/mm^2	1.5-2.5
Elongation at break %	1.8
Flexural strength N/mm^2	60-80
Heat distortion temperature °C	250

3.2.6 Epoxy resins

(1) Epoxies offer mechanical properties and water resistance superior to those of polyesters, with less shrinkage during cure, but at higher cost by a factor of 2-3 in the case of standard epoxy laminating resins and as much as 10-15 times in the case of higher performance aerospace prepreg materials. They are generally more associated with automated methods of manufacturing fibre composites, being readily incorporated into pultrusion, filament winding, resin transfer moulding and compression moulding processes.

P(2) Epoxy resins should be considered where higher shear strength than is available with polyester resins is required and the application requires good mechanical properties at elevated temperatures or durability. Epoxies occur in many varieties, feature high strength properties (as seen in some of the most effective engineering adhesives) and have been developed for water resistant coatings.

P(3) Epoxy resins should only be considered where the manufacturing method and size of component allows elevated temperature curing to be used. Optimum properties are not developed until the composite has been post cured in the range 60 to 150°C.

P(4) Some epoxy resins have low ultra-violet resistance and may need special surface protection.

(5) Typical properties of epoxy resins are given in Table 3.5.

Table 3.5 Typical properties of epoxy cast resins (reference 3.3).

	DGEBA-APTA amine cured 20°C Cure	*DGEBA-MDA amine cured 120°C*	*DGEBF-rubber modified epoxy MDA/MPDA 120°C*
Specific gravity	1.2	1.2	1.2
Tensile strength N/mm^2	62	90	125
Tensile modulus kN/mm^2	3.2	3.0	4.1
Elongation at break %	2	8	5
Shear strength N/mm^2	61	52	84
Heat distortion temperature °C	62	121	110

Note that relative resin ratings do not necessarily carry over directly to composite properties.

3.3 CORES

3.3.1 General

P(1) Core materials may be load bearing or simply used as formers for shaping FRP sections. Structural cores should be used for efficient sandwich construction design. These may be foam, honeycomb or solid materials.

P(2) Properties required in a structural core material are low density, high shear modulus and shear strength, good strength in the through-thickness direction, thermal and dimensional stability, fatigue resistance, impact strength, resistance to moisture, ease of shaping and good surface adhesive bonding strength.

(3) The user should consult with materials suppliers and fabricators on the selection and suitability of particular core systems. Indicative information on core properties may be found in the EUROCOMP Handbook.

3.3.2 Foam

3.3.2.1 General

(1) Foams consist of open or closed cell materials made from many plastics and may be used in sandwich panel manufacture from slab stock or by foaming in situ. The latter presents more problems of attaining uniformity of properties. Commonly used core materials include polyurethanes and rigid polyvinyls and polymethacrylimides (PMI) available in lightweight and industrial grades.

(2) There is no ISO Standard.

(3) Densities range from 5 kg/m^3 to the density of the solid plastic. Mechanical properties are dependent on density and are nonlinear.

3.3.2.2 Reinforced foams

(1) Plastic foams may be reinforced with glass or other short fibres. Improvements in mechanical and physical properties are dependent on fibre type, amount of fibre added and fibre distribution. Consistency of properties is more difficult to control.

3.3.2.3 Structural foams

(1) Structural foams with graded densities may be made with solid surfaces but cellular cores. The inherent sandwich construction produces more efficient structural properties. They may be finished with FRP skins.

3.3.2.4 Others

(1) Syntactic foams may be appropriate and can offer higher compression strength than other foams. The foam nature is produced by mixing hollow microspheres of glass, ceramic, or polymers with other liquid polymers either to cast or to form a moulding compound. Such materials usually are lightweight and strong.

(2) They can usually be shaped or pressed into cavities and moulds where use of rigid foams is not practical.

3.3.3 Honeycombs

3.3.3.1 General

P(1) Honeycomb cores should be considered where weight critical efficient structures are required. Various grades and types of honeycombs are available but users should consult with suppliers for properties, specific application information and compliance with appropriate standards.

3.3.3.2 *Alumium*

P(1) If aluminium honeycomb is used with electrically conducting composites (e.g. carbon fibre composites), construction shall ensure that the honeycomb is electrically insulated from the core by using a glass fibre interlayer to prevent potential internal galvanic corrosion problems occurring in wet environments.

(2) The most commonly used grade is 5052 H39 aluminium alloy with a corrosion resistant coating applied for general purpose applications. Properties are dependent on foil thickness, cell geometry and size.

(3) Grade 5056 H39 aluminium alloy with a corrosion resistant coating offers slightly higher mechanical properties than the 5052 alloy honeycomb.

3.3.3.3 *Aramid*

(1) Aramid fibre dipped in liquid phenolic, polyester or polyamide resin to form a paper can be used to form a honeycomb (NOMEX). Properties are dependent on paper thickness, cell geometry and size. Advantages of non-metallic honeycomb cores are absence of internal corrosion and impact resilience.

3.3.3.4 *Others*

(1) Other non-metallic materials may be considered for forming honeycombs for specific applications. Glass FRP, aramid FRP and Kraft paper honeycombs have been used.

3.3.4 Solids

(1) Solid materials may be used as cores but in sandwich construction will only be efficient if lightweight strong materials are used. End grain balsa-wood has been widely used in sandwich constructions and can offer advantages to foams in compression and shear critical loadings.

3.4 GEL COATS

P(1) Gel coats and top finishing flow coats shall be considered and incorporated into the total composite system specification where there is advantage to increasing durability or the need to obtain a particular surface finish. These coats give additional thickness to the structure material. They should be applied sequentially as part of the composite build process.

P(2) Selection and specification of gel coats and thickness will be determined by its prime functions and advice must be sought from suppliers of materials. Users of gel coats and top coats should ensure compatibility with the underlying structural FRP, with control during composite build to obtain good adhesion of gel coat. Gel coats should not be allowed to cure past their "green" state before build up of the FRP. The quality of the gel coat is critical for controlling moisture ingress and preventing blister formation from osmosis. Some combinations of chopped strand mat reinforcements and resins may give rise to this phenomenon when continuously immersed in water.

P(3) Gel coats and top coats shall be regarded as nonstructural.

(4) Gel coats and top coats may be added to the surface of a composite structure for a variety of reasons: to filter out ultraviolet radiation and improve weathering, to add flame retardancy or provide an increased thermal barrier, to improve chemical resistance, to improve erosion, to provide electrical insulation (relevant to carbon fibre composites), to provide an increased barrier to moisture, or to provide a colour scheme and improve general finish.

(5) Top coats are more difficult to apply and it is more difficult to obtain a uniform surface than with a gel coat. The quality is dependent on the nature of the underlying surface and final coating thickness may be variable. Thus protection and functional behaviour may differ from gel coats. Top coats should be applied in conjunction with build of the composite structure.

3.5 SURFACE VEILS

P(1) Surfacing veils should be considered where their inclusion is beneficial to improving the appearance and properties of surface layers and coatings. They should be used where high resistance to chemical environments is required. C-glass materials are commonly used.

P(2) Surface veils should not be used on surfaces to be joined by bonding. If they are present, the whole layer thickness shall be removed by grinding over the area required for bonding.

(3) Veils can be used to control the thickness of a resin rich surface layer. They may be used in conjunction with gel coats and top coats to provide reinforcement support and increased retention of resin.

3.6 ADDITIVES

3.6.1 Fillers

P(1) Particulate fillers are non-structural and their use should be limited in the body of composites as they usually reduce long term structural properties and durability. The user should consult with materials suppliers on application of fillers.

(2) Fillers are inorganic particulates that can be added to the resin to reduce shrinkage, to reduce the peak exotherm during cure, to increase viscosity, to increase local hardness, to reduce flammability and to reduce cost. They can increase modulus and compressive strength and may be included in

surface coatings for improving specific properties.

3.6.2 Pigments

P(1) Pigments can be used as part of the basic resin or gel coat system to obtain a desired colour scheme. Users should consult with materials suppliers to ensure compatibility in processing and suitability for the intended application, including colour fastness to weathering, and whether their use will affect the long term structural stability. Some metallic salt based pigments may seriously lower the adhesion between fibre and matrix.

(2) Pigments may be used to impart ultra-violet blocking.

3.6.3 Flame retardants

P(1) The user shall ensure that full account is taken of the effect of fire retardant additives on structural properties in consultation with materials suppliers. The incorporation of such materials as additives to the basic resin can produce lower than expected mechanical properties and reduced weathering resistance.

There is no ISO Standard.

(2) Flame retardants may be incorporated as additives in particulate or coating form. They may be halogenated or non-halogenated. Their use must be considered in conjunction with their effect on mechanical properties, especially for durability. There is a need to match all the flammability characteristics, ignitability, heat development, smoke and toxicity and to ensure continued structural integrity. Some fire retardants may improve one of these fire performance characteristics at the expense of others.

REFERENCES

3.1 *Handbook of Composites,*
Edited by G Lubin, Von Nostrand Reinhold, New York, 1982.

3.2 *E and ECR-Glass Technical Status*, Owens Corning Data Sheet, 1991.

3.3 *Polyester Technical Information Data Sheets*, NESTE Chemicals, Helsinki:

G105E resins - UP 27 1991
G300 resins - G300 1989
K530 resins - UP 26 1991
S599 resins - UP 29 1993
F892 resins - UP 20 1992
F820 resins - UP 19 1992

3.4 *Derakane Vinyl Ester Resins - Product and Usage Guide,*
Dow Chemicals Publication No: CH-224-078-E-892, 1982.

3.5 *T M Mayer and N L Hancox - Design Data for Reinforced Plastics - A Guide for Engineers and Designers*,
Chapman and Hall, London, 1994.

CONTENTS

4 SECTION AND MEMBER DESIGN

4 SECTION AND MEMBER DESIGN

4.0 NOTATION

C_d	Design limiting condition other than strength (usually serviceability limit state, SLS) (see 2.3.4)
D_x	Plate flexural rigidity x - x direction (see Table 4.1)
D_y	Plate flexural rigidity y - y direction (see Table 4.1)
E_d	Effect of actions
F_d	Design value of the action
R_d	Design resistance (usually ultimate limit state, ULS)

4.1 ULTIMATE LIMIT STATE

4.1.1 Basic conditions to be satisfied

P(1) For the states of stress to be considered, the following condition shall be satisfied:

$E_d \leq R_d$

where:

E_d is as defined in 4.0

$= E_d(F_d, a_{d,} ...)$ (see 2.2.2.5)

in which:

F_d is the design value of the action (see 2.2.2.4)

$= \gamma_f F_k$

$= \gamma_G G_k + \psi_i \gamma_Q Q_k$

where γ_f, γ_G, γ_Q and ψ_i are appropriate to the ULS (see 2.2.2.4).

a_d represents the geometry of the section being designed (limits on these for validity of design criterion to be stated)

R_d is the design resistance (see 2.2.3.2)

$= R(X_d, a_d, K_d, ...)$

in which:

X_d	=	the design material property (see 2.2.3.2) $=X_k/\gamma_m$
C_m	=	an analytical uncertainty factor.

P(2) The ultimate limit states to be considered for a particular element shall be as set down for the type of element (see 4.3 and following sections).

Table 4.1 Plate stiffness equations.

Isotropic plates

$$D = \frac{Et^3}{12(1 - \nu^2)}$$

Specially orthotropic plates

$$D_x = \frac{E_x t^3}{12\,(1 - \nu_{xy}\nu_{yx})}$$

$$D_{xy} = D_{yx} = \frac{\nu_{yx} E_x t^3}{12\,(1 - \nu_{xy}\nu_{yx})}$$

$$D'_{xy} = \frac{G_{xy} t^3}{12}$$

$$D_y = \frac{E_y t^3}{12(\,1 - \nu_{xy}\nu_{yx})}$$

where:

t	=	thickness of laminate
D_{xy}	=	shear stiffness
E	=	modulus of elasticity in bending
E_x	=	modulus of elasticity in bending (x direction)
E_y	=	modulus of elasticity in bending (y direction)
G_{xy}	=	in-plane shear modulus
ν_{xy}	=	major Poisson's ratio
ν_{yx}	=	minor Poisson's ratio

4.1.2 Determination of the effect of actions

P(1) The effect of actions shall be determined in accordance with the principles set down in Chapter 2.

4.1.3 Determination of design resistance

P(1) The design resistance R_d of a member or section shall, wherever possible, be based on a design value X_d of a material property of the FRP composite derived from a characteristic value X_k of that material property (see 2.2.3.1 and 2.2.3.2).

P(2) The characteristic value of the material property may be determined from one or more of the following:

- the corresponding property of the constituent materials (i.e. the fibre and matrix), with the properties of the individual plies or laminae and those of the laminate derived from theory
- coupons made as or cut from each of the plies or laminae forming the laminate, due regard being given to the orientation of the main and transverse fibre reinforcement, with the properties of the laminate derived from theory
- the complete laminate forming all or part of the member section
- prototypes of the section/member or complete structure.

(See 2.3.3.2 : Table 2.3)

P(3) The value of γ_m applied to the characteristic material property X_k to derive the design value X_d shall be appropriate to the level of uncertainty involved in determining the design resistance (see 2.3.3.2).

4.2 SERVICEABILITY LIMIT STATE

4.2.1 Basic conditions to be satisfied

P(1) Serviceability limit states are:

- deformations or deflections which adversely affect the appearance or effective use of the structure (including the proper functioning of machines or sevices)
- vibration, oscillation or sway which causes discomfort to the occupants of a building or damage to its contents
- deformations, deflections, vibrations, oscillation or sway which causes damage to finishes or non-structural elements.

P(2) To avoid exceeding these limits, it is necessary to limit deformations, deflections and vibrations.

(3) Except when specific limiting values are agreed between the client, the designer and the competent authority, the limiting values given in this chapter should be applied.

4.2.2 Deflections

4.2.2.1 Requirements

P(1) Structures and components shall be so proportioned that deflections are within the limits agreed between the client, the designer and the competent authority as being appropriate to the intended use and occupancy of the building and the nature of the materials supported.

(2) Recommended limits for deflections are given in 4.5.2 In some cases more stringent limits (or exceptionally less stringent limits) will be appropriate to suit the use of the building or the characteristics of the cladding materials or to ensure the proper operation of installed equipment.

(3) The deflections should be calculated making due allowance for any second order effects and the rotational stiffness of any semi-rigid joints.

4.2.2.2 Limiting values

(1) The limiting values for the vertical deflections below are illustrated by reference to the simply supported beam in Figure 4.1 in which:

δ_{max} = the sagging in the final state relative to the straight line joining the supports

δ_1 = the variation of the deflection of the beam due to the permanent loads immediately after loading

δ_2 = the variation of the deflection of the beam due to the variable loading plus any time dependent deformations due to the permanent load

δ_0 = the pre-camber (hogging) of the beam in the unloaded state.

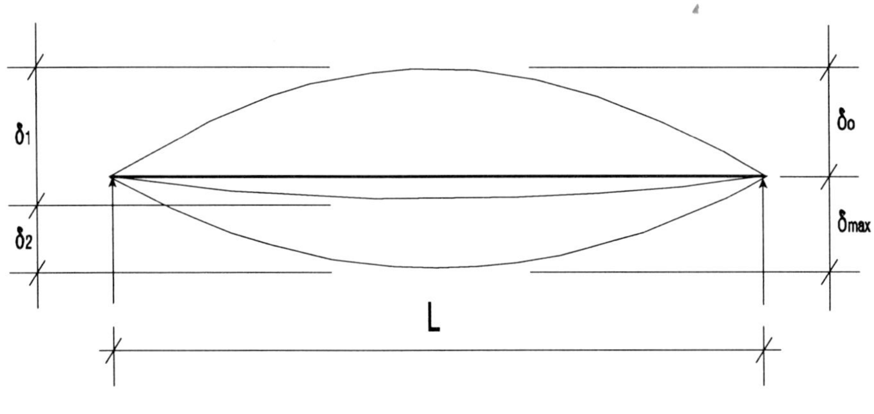

Figure 4.1 Vertical deflections.

4.2.3 Stresses and strains

P(1) For the states of stress and strain to be considered, the following condition shall be satisfied:

$E_d \leq C_d$

where:

C_d and E_d are as defined in 4.0, i.e.

C_d = design limiting condition for strain at SLS and
E_d = $E_d(F_d, a_d, E \ldots)$ (see 2.2.2.5)

in which:

F_d = the design value of the action (see 2.2.2.4)

$= \gamma_f F_k$

$= \gamma_G G_k + \psi_i \gamma_Q Q_k$

where γ_f, γ_G, γ_Q and ψ_i are appropriate to the SLS (see 2.2.2.4).

a_d represents the geometry of the section being designed.

4.3 MEMBERS IN TENSION

4.3.0 Notation

A Area
A_{net} Net cross-sectional area
$E_{x,t,k}$ Characteristic tensile modulus of the member in the axial direction
$N_{t,Sd}$ Design tensile force
$N_{t,Rd}$ Design tension resistance of the cross-section
γ_m Partial safety factor for material resistance
$\sigma_{x,t,k}$ Gross cross-sectional area tensile strength of laminate
ε_{mf} Strain to failure of the matrix
ε Axial strain, generally

4.3.1 Scope and definitions

P(1) This section refers to elements which are subject only to tension forces. Such elements are aligned with their longitudinal axis parallel to the x-coordinate of the composite material.

4.3.2 Design procedure

P(1) The design value of the tensile force $N_{t,Sd}$ at each cross-section shall satisfy:

$$N_{t,\,Sd} \leq N_{t,\,Rd} \tag{4.1}$$

where $N_{t,\,Rd}$ is the design tension resistance of the cross-section, taken as the smaller of:

- the design tensile resistance of the minimum gross cross-section

$$N_{t,\,Rd} = A\sigma_{x,\,t,\,k}/\gamma_m \tag{4.2}$$

- the design tensile resistance of the net cross-section at holes

$$N_{t,\,Rd} = 0.9\ A_{net}\ \sigma_{x,\,t,\,k}/\gamma_m \tag{4.3}$$

P(2) Angles connected through one leg and other types of sections connected through outstands such as T-sections and channels shall comply with the requirements of Chapter 5.

P(3) End connections and splices shall be suitably designed for tensile forces transferred across the joint (see Chapter 5).

P(4) If the component requires resistance to chemical attack in service, then the axial strain (ε) due to the tensile force $N_{t,Sd}$ shall satisfy:

$\varepsilon \leq 0.2\%$ or

$\varepsilon \leq 0.1\ \varepsilon_{mf}$

where $$\varepsilon = N_{t,Sd}\ L/A_{net}\ E_{x,t,k} \tag{4.4}$$

4.4 MEMBERS IN COMPRESSION

4.4.0 Notation

A	Area
A_{eff}	Effective area
$N_{c,Sd}$	Design compressive force
$N_{c,Rd}$	Design compression resistance of the cross-section
$N_{c,cr}$	Buckling resistance of member
γ_m	Partial safety factor for material resistance
$\sigma_{c,k}$	Characteristic compressive strength of laminate, i.e. $\sigma_{x,c,k}$ or $\sigma_{y,c,k}$
$\sigma_{c,cr}$	Buckling strength of section

4.4.1 Scope and definitions

P(1) This section applies to members subjected to axial compression, with or without bending, for which the effects of torsion can be neglected.

4.4.2 Design procedure

P(1) The influence of second order effects, for example increased moments due to deflections, see 2.5.3.2.2, shall be considered if the increase above the first order effects exceeds 10%.

P(2) Design for structural stability taking account of second order effects shall ensure that, for the most unfavourable combinations of actions at the ultimate limit state, loss of static equilibrium (locally or for the structure as a whole) does not occur or the resistance of individual cross-sections subjected to bending and longitudinal forces are not exceeded.

P(3) The structural behaviour shall be considered in any direction in which failure due to second order effects may occur.

P(4) In the case of unsymmetrical sections due allowance shall be made for the additional moment due to the eccentricity of the centroidal axis of the effective section.

P(5) End connections and joints shall be suitably designed for the compressive forces to be transferred across them (see Chapter 5).

P(6) Possible uncertainties in the restraints at connections shall be considered.

P(7) For members in axial compression, the design value of the compressive force $N_{c,Sd}$ at each cross-section shall satisfy:

$$N_{c,Sd} \le N_{c,Rd} \qquad (4.5)$$

where $N_{c,Rd}$ is the design compression resistance of the cross-section, taken as the smallest of:

- the design ultimate resistance of the cross-section

$$N_{c,Rd} = A\sigma_{c,R}/\gamma_m \qquad (4.6)$$

- the member buckling resistance

$$N_{c,Rd} = k\pi^2 E_{x,d} I_{ZZ} \,/\, L^2\gamma_m \qquad (4.7)$$

where:

I_{ZZ} = 2nd moment of area (about weak Z - Z axis)

k = 1 for columns with pinned ends

k = 4 for columns with built-in ends

- the design local buckling resistance of the cross-section

$$N_{c,Rd} = A_{eff}\sigma_{c,cr}/\gamma_m \tag{4.8}$$

where:

A_{eff} is the effective area of the cross-section

$\sigma_{c,cr}$ is the least critical buckling resistance of the individual elements of the section determined from the following:

for an element of the section which can be defined as a long rectangular plate with the two longer edges simply supported:

$$\sigma_{c,cr,y} = 2\pi^2\{(D_xD_y)^{1/2} + H_0\}/tb^2 \tag{4.9}$$

where:

$H_0 = ½\ (\nu_{xy}D_y + \nu_{yx}D_x) + 2(G_{xy}t^3/12)$

b = effective width of the plate element

t = thickness of the plate element

D_x and D_y are plate stiffnesses and are given in Table 4.1

for an element of the section which can be defined as a long rectangular plate with one of the longer edges pinned and the other free:

$$\sigma_{c,cr,y} = \pi^2\{(D_x(b/a)^2)+(12D'_{xy}/\pi^2)\}/tb^2 \tag{4.10}$$

where:

a = the half wavelength of the buckle and is taken to be the length of the plate

b = effective width of the plate element

t = thickness of the plate element

D_x, D_y and D'_{xy} are plate stiffnesses and are given in Table 4.1

for tubular sections with isotropic construction and r/t > 10

$$\sigma_{c,cr,y} = 0.25E_{x,c,k}tr \tag{4.11}$$

where:

$E_{x,c,k}$ = characteristic axial compressive modulus

t = thickness of the tube

r = mean radius of the tube.

4.5 MEMBERS IN FLEXURE

4.5.0 Notation

M_{Sd}	Design internal bending moment
M_{Rd}	Design internal bending moment resistance of the cross-section
$M_{c,cr}$	Local buckling resistance of the cross-section
$M_{b,cr}$	Lateral torsional buckling resistance moment of member
W_t, W_c	Section modulus for tension, compression
γ_m	Partial safety factor for material resistance
$\sigma_{t,k}$, $\sigma_{c,k}$	Characteristic tensile, compressive strength of laminate or panel

4.5.1 General

P(1) The design resistance shall be equal to or exceed the following ultimate limit states:

- flexural strength
- axial strength
- shear strength
- bearing stress
- web buckling due to flexure
- web buckling due to shear
- web buckling due to flexure and shear
- web crippling
- compression flange buckling
- lateral torsional buckling.

For the last 6 items see 4.7, Stability

P(2) The design resistance shall be equal to or exceed that for the serviceability limit state of deformation due to bending plus deformation due to shear.

P(3) The second moment of area (I) and the shear area shall be determined such that maximum allowable deformation due to bending and shear shall not be exceeded.

P(4) The section modulus (W) shall be determined such that the maximum allowable tensile and compressive stresses in the section are not exceeded.

P(5) The minimum ratio of flange width to flange thickness shall be determined to sustain the required compressive stress in the flange without local buckling.

P(6) The minimum ratio of web width to web thickness shall be determined to sustain the required shear stress in the web without local buckling.

P(7) The resistance of the member shall be adequate to prevent lateral torsional buckling.

(8) Load tables which comply with this EUROCOMP Design Code may be used in order to simplify some parts of the design procedure.

(9) Load tables shall be appropriate to the end conditions of the beam and the applied loading case.

4.5.2 Design procedure - serviceability limit state - deformation

P(1) The deformation of a member shall be such that it does not adversely affect its proper function or appearance.

P(2) The sum of all relevant deformations due to short and long term loading actions shall not exceed the maximum allowable deformation.

(3) Deformations should not exceed those which can be accommodated by other connected elements such as partitions, glazing, cladding, services or finishes. In some cases limitation may be required to ensure the proper functioning of machinery or apparatus supported by the structure or to avoid ponding on flat roofs. Vibration may also require limitation as it can cause discomfort or alarm to users of a building and, in extreme cases, structural damage.

P(4) Appropriate limiting values of deflection taking into account the nature of the structure, finishes, partitions and fixings, and the function of the structure shall be agreed with the client or taken from Table 4.2.

(5) The conventional engineering equations for bending of isotropic, homogeneous beams may be used for composite materials. For simple beams they take the form:

$$\text{Deflection (bending)} = k_1 F_v L^3/(EI) \tag{4.12}$$

where :

EI = appropriate flexural rigidity of the full section

F_v = total vertical load on the beam

k_1 = a factor depending on the type of loading and the end conditions. A set of factors is given in Table 4.3.

Table 4.2 Recommended limiting values for deflection.

Typical conditions	*Limits (see Figure 4.1)*	
	δ_{max}	δ_2
Walkways for occasional non-public access	L/150	L/175
General non-specific applications	L/175	L/200
General public access flooring	L/250	L/300
Floors and roofs supporting plaster or other brittle finish or non-flexible partitions	L/250	L/350
Floors supporting columns (unless the deflection has been included in the global analysis for the ultimate limit state)	L/400	L/500
Where δ_{max} can impair the appearance of the structure	L/250	-

Table 4.3 Selected values for k_1 and k_2.

End conditions	*Loading type*	k_1	k_2
Cantilever	Point load at end	1/3	1
Cantilever	Uniformly distributed	1/8	1/2
Supported at ends	Point load at centre	1/48	1/4
Supported at ends	Uniformly distributed	5/384	1/8
Fixed at ends	Uniformly distributed	1/384	1/24

P(6) In beams with a span/depth ratio of less than 25, deflection due to shear shall be allowed for.

(7) Equations for deflection due to shear take the form:

$$\text{Deflection (shear)} = k_2 F_v L / A_v G_{xy} \qquad (4.13)$$

where:

k_2 = a factor depending on the type of loading and the end conditions. A set of factors is given in Table 4.3
F_v = total vertical load
A_v = shear area of the web
G_{xy} = in-plane shear modulus of the web.

4.5.3 Strength in bending

P(1) The design value of the internal bending moment M_{Sd} at each cross-section shall satisfy:

$$M_{Sd} \leq M_{Rd} \qquad (4.14)$$

where M_{Rd} is the design moment resistance of the cross-section, taken as the smallest of:

- the design ultimate resistance moment of the gross section

$$M_{Rd} = W_t\sigma_{t,k}/\gamma_m \qquad (4.15)$$

- the design local buckling resistance of the gross section

$$M_{Rd} = M_{c,cr}/\gamma_m \text{ or } W_c\sigma_{c,k}/\gamma_m \qquad (4.16)$$

where W_c is the section modulus in compression

- the lateral torsional buckling resistance moment of the member

$$M_{Rd} = M_{b,cr}/\gamma_m \qquad (4.17)$$

P(2) End connections and splices shall be suitably designed for moments transferred across the joint (see Chapter 5).

P(3) Members shall be analysed at a sufficient number of cross-sections to ensure that the requirements of the EUROCOMP Design Code are satisfied at all cross-sections along the member.

4.6 MEMBERS IN SHEAR

4.6.0 Notation

A_v	Shear area
A	Cross-section area
V_{Sd}	Design internal shear force of the section
V_{Rd}	Design shear resistance of the section
b	Overall breadth
d_w	Depth of web
t_w	Web thickness
$\tau_{xy,k}$	Characteristic in-plane shear strength of laminate

4.6.1 General

P(1) The design value of the shear force V_{Sd} at each cross-section shall satisfy:

$$V_{Sd} \leq V_{Rd} \tag{4.18}$$

where V_{Rd} is the design shear resistance given by:

$$V_{Rd} = A_v \tau_{xy,k} / \gamma_m \tag{4.19}$$

(2) The shear area A_v may be taken as follows:

- For fabricated I, H and hollow box sections, loaded parallel to the web:

$$A_v = \Sigma(d_w t_w) \tag{4.20}$$

- For I, H and channel sections, loaded parallel to flanges

$$A_v = A - \Sigma(d_w t_w) \tag{4.21}$$

- For circular hollow sections and tubes of uniform thickness

$$A_v = 2A/\pi \tag{4.22}$$

- For plates and solid bars

$$A_v = A \tag{4.23}$$

(3) For other cases, A_v should be determined in a manner similar to the above.

(4) In appropriate cases, the formulae in (2) may be applied to components of a built-up section.

P(5) If the web thickness is not constant, t_w shall be taken as the minimum thickness.

P(6) The shear buckling resistance shall be verified as specified in 4.7.2.

P(7) End connection and splices shall be suitably designed for shear forces transferred across the joint (see Chapter 5).

4.7 STABILITY

4.7.0 Notation

D_x	Flexural rigidity in longitudinal direction (see Table 4.1)
D_y	Flexural rigidity in transverse direction (see Table 4.1)
$E_{y,d}$	Modulus of elasticity (design) in transverse direction
a	Length of plate, longitudinal direction
b	Width of plate, transverse direction
b_s	Longitudinal spacing of the stiffeners
d_w	Depth of web = d $-2t_f$
t_w	Thickness of the web
ν_{xy}	Poisson's ratio
$\sigma_{x,b}$	In-plane stress due to bending
$\sigma_{x,cr,c}$	Critical buckling stress
$\sigma_{x,cr,b}$	Critical buckling in-plane bending stress without shear stress
τ_{xy}	Shear stress in xy plane
τ_{cr}	Critical buckling shear stress without normal stress

4.7.1 Critical flexural stress in the web

P(1) The in-plane bending stress shall satisfy the relationship:

$$\sigma_{x,b} \leq \sigma_{x,cr,b} \tag{4.24}$$

(2) The critical buckling stress of rectangular panels subject to in-plane flexure may be taken as:

- for isotropic materials

$$\sigma_{x,cr,b} = \frac{k\pi^2 E(t_w/d_w)^2}{12(1 - \nu_{xy}^2)} \tag{4.25}$$

where k = 23.9

- for specially orthotropic materials

$$\sigma_{x,cr,b} = \frac{k\pi^2 D_x}{d^2_w t_w} \tag{4.26}$$

where if the web may be assumed to be clamped to the flange

then k = 50 if $D_y/D_x = 1$
or k = 20 if $D_y/D_x = 0.5$

4.7.2 Critical shear stress in the web

P(1) The shear stress in the web shall satisfy the relationship:

$$\tau_{xy} \leq \tau_{xy,cr,b} \quad (4.27)$$

The critical buckling stress of rectangular plates subjected to in-plane shear stress may be taken as:

- for isotropic materials:

  $$\tau_{xy,cr,b} = k\pi^2 E(t_w/d_w)^2/12(1 - v^2) \quad (4.28)$$

  where, for plates with simply supported edges and a/b > 10:

  k = 5.35

  or, for plates with built-in edges and a/b > 10:

  k = 8.98

- and for specially orthotropic materials:

  $$\tau_{cr} = 4k(D_x D^3_y)^{0.25}/d^2_w\ t_w \quad (4.29)$$

  where:

  k = 8

  D_x and D_y are as given in Table 4.1.

4.7.3 Combined shear and in-plane bending in the web

P(1) Unstiffened and stiffened webs of members subject to combined bending and shear shall be so dimensioned that:

$$(\tau_{xy}/\tau_{xy,cr,b})^2 + (\sigma_{b,x}/\sigma_{x,cr,b})^2 \leq 1 \quad (4.30)$$

(2) The buckling stress may be increased by the incorporation of suitable web stiffeners provided the in-plane stress is not greater than the in-plane strength.

P(3) Stiffeners shall be properly spaced and must not deform excessively.

(4) Isotropic web-stiffeners, with the same elastic properties as the member to which they are attached, may be dimensioned as follows:

the minimum second moment of inertia of the stiffener about the plane of the web I_s satisfies the relationship:

$$I_s = 0.34d^4_w(t_w/b_s)^3 \tag{4.31}$$

4.7.4 Resistance of web to transverse forces

4.7.4.0 Notation

$E_{y,d}$	Modulus of elasticity
S_s	Length of the stiff bearing from Figure 4.2
$V_{y,Rd}$	Design crushing resistance of web of I, H or U section
b_f	Flange breadth
d	Depth
h	Overall depth of the beam
t_f	Flange thickness
t_w	Web thickness
γ_m	Partial safety factor for material property
$\sigma_{y,c,k,f}$	Characteristic transverse compressive strength of flange material
$\sigma_{y,c,k,w}$	Characteristic transverse compressive strength of web material

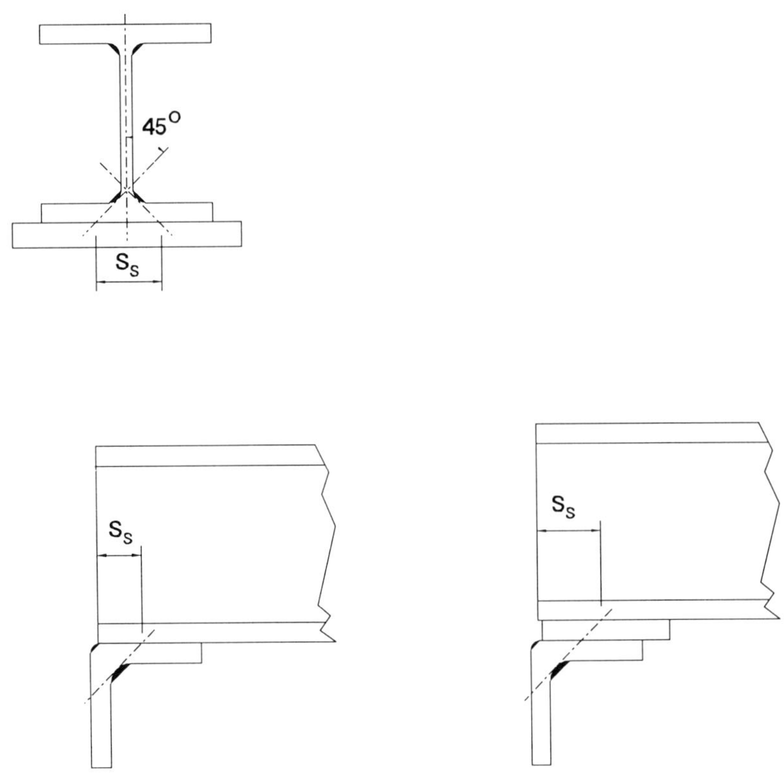

Figure 4.2 Length of bearing.

4.7.4.1 Basis

P(1) The resistance of an unstiffened web to transverse forces applied through a flange, shall be checked for the following modes of failure:

- crushing of the web
- buckling of the web over most of the depth of the member.

P(2) Distinction shall be made between two types of load application, as follows:

- forces applied through one flange and resisted by shear forces in the web
- forces applied to one flange and transferred through the web directly to the other flange.

P(3) In addition, the effect of the transverse force on the moment resistance of the member shall be considered.

4.7.4.2 Length of the stiff bearing

P(1) The length of the stiff bearing, S_s, over which the applied force is effectively distributed, should be determined by dispersion at a slope of 1:1 of the load through solid composite material which is properly fixed in place (see Figure 4.2).

P(2) No dispersion shall be taken through loose packs.

P(3) The minimum length of the stiff bearing shall be determined from:

$$S_s = R_u/(t_w \sigma_{y,c,d,w}) \tag{4.32}$$

where:

R_u = reaction at the support.

4.7.4.3 Crushing resistance

(1) The design crushing resistance $V_{y,Rd}$ of the web of an I, H or U section may be obtained from:

$$V_{y,Rd} = (S_s + S_n)t_w\,\sigma_{y,c,k,w}/\gamma_m \tag{4.33}$$

where:
S_s = length of the stiff bearing from Figure 4.2
S_n = length obtained by dispersion at 45° through half the depth of the section.

4.7.4.4 Buckling resistance

(1) The design buckling resistance $R_{b,Rb}$ of the web of an I, H or U section may be obtained by considering the web as a virtual compression member with an effective breadth b_{eff} obtained from:

$$b_{eff} = (h^2 + S_s^2)^{0.5} \tag{4.34}$$

P(2) Near the ends of a member (or at openings in the web) the effective breadth b_{eff} shall not be taken as greater then the breadth actually available, measured at mid-depth.

(3) The critical buckling stress (σ_{cr}) may be determined from:

$$\sigma_{cr} = \frac{k\pi^2\sqrt{D_x D_y}}{b_{eff}^2 t_w} \tag{4.35}$$

where:

$$k = 2\left(1 + \frac{H}{\sqrt{D_x D_y}}\right)$$

and

$$H_o = 1/2(\nu_{xy} D_y + \nu_{yx} D_x) + \frac{(G t_w^3)}{6}$$

Note: for the isotropic case $D_x = D_y = H_0$ and $k = 4$.

P(4) The flange through which the load is applied shall wherever possible be restrained in position at the point of load application. Where this is not practicable, a special buckling investigation shall be carried out.

4.7.4.5 Buckling of compression flange (one edge pinned, one edge free)

P(1) The longitudinal compression flange shall satisfy the following relationship at all points in the member:

$$\sigma_{c,x} \leq \sigma_{x,cr,c}/\gamma_m \tag{4.36}$$

where $\sigma_{x,cr,c}$ is the critical buckling stress.

(2) For the isotropic case in which the flange is considerably longer than its breadth, in the absence of a more rigorous analysis, the critical buckling stress $\sigma_{x,cr,c}$ in the flange may be taken as:

$$\sigma_{x,cr,c} = G_{xy,d}\,(2t_f/b_f)^2 \tag{4.37}$$

where:

G_{xy} = shear modulus in the plane of the flange.

(3) For orthotropic cases the local buckling stress for a long plate with one "free" and one "pinned" longitudinal edge is:

$$\sigma_{x,cr,b} = \pi^2\{(D_x(b_f/a)^2)+(12D'_{xy}/\pi^2)\}/t_f b_f^2 \quad (4.38)$$

where:

a = half wavelength of the buckle = length of the flange.

4.7.4.6 Lateral torsional buckling

(1) The critical buckling moment may be obtained from:

$$M_b C_1 = P_{ey}\left[K\frac{I_w}{I_{zz}} + \frac{GJ}{P_{ey}}\right]^{0.25} \quad (4.39)$$

P_{ey} is the Euler column buckling load about the weak axis:

$$P_{ey} = \pi^2\, E_{z,b,d}\, I_{zz}/(kL)^2 \quad (4.40)$$

where:

L = length of beam between points that have lateral restraint
I_{zz} = second moment of area about minor axis
J = torsion constant
I_w = warping constant (Equation 4.41)
$E_{z,b,k}$ = modulus of elasticity, bending about minor axis
G_{xy} = shear modulus of flange material
K = effective length factor referring to end rotation about minor axis. It is 0.5 for full fixity and 1.0 for no fixity
C_1 = a factor depending on the loading and end restraint conditions.

For a beam simply supported about the major axis:

Uniformly distributed loading and k is 1.0, C_1 = 1.132
Uniformly distributed loading and k is 0.5, C_1 = 0.972
Central concentrated point load and k is 1.0, C_1 = 1.365
Central concentrated point load and k is 0.5, C_1 = 1.070

Equation (4.39) is valid only for a doubly symmetrical cross-section with the loading through the shear centre and end warping taken to have no fixity.

I_w may be determined from:

$$I_w = I_f \frac{(D - t)^2}{2}$$

where

$$I_f = \frac{b_t^t}{12}$$

4.8 COMBINATION MEMBERS

4.8.0 Notation

(1) For notation see 4.3, 4.4 and 4.5.

4.8.1 Scope and definitions

P(1) This section applies to members subjected to axial compression or tension parallel to the longitudinal axis of the member, and to bending about either or both the transverse axes.

P(2) For the purpose of this section, the x-axis is the longitudinal axis, the y-axis is the principal transverse axis for bending and the z-axis is the secondary transverse axis for bending (see 1.8).

P(3) Other definitions are as the appropriate sections on tension, compression and flexure.

4.8.2 Design procedure

P(1) For tension and bending at the ultimate limit state, the following conditions shall be satisfied at each cross-section:

$$\frac{N_{t,Sd,x}}{N_{t,Rd,x}} + \frac{M_{Sd,y}}{M_{Rd,y}} + \frac{M_{Sd,z}}{M_{Rd,z}} \leq 1 \qquad (4.42a)$$

$$\frac{N_{t,Sd,x}}{N_{t,Rd,x}} + \frac{M_{Sd,y}}{M_{b,cr,y}} + \frac{M_{Sd,z}}{M_{b,cr,z}} \leq 1 \qquad (4.42b)$$

P(2) For tension and bending at the serviceability limit state, the maximum combined tensile strain at any cross-section of the member shall not exceed the appropriate values for a tension member under similar environmental and service conditions.

P(3) For compression and bending at the ultimate limit state, the following conditions shall be satisfied at each cross-section:

$$\frac{N_{c,Sd,x}}{N_{c,Rd,x}} + \frac{M_{Sd,y}}{M_{Rd,y}} + \frac{M_{Sd,z}}{M_{Rd,z}} \leq 1 \qquad (4.43a)$$

$$\frac{N_{c,Sd,x}}{N_{c,Rd,x}} + \frac{M_{Sd,y}}{M_{c,cr,y}} + \frac{M_{Sd,z}}{M_{c,cr,z}} \leq 1 \qquad (4.43b)$$

$$X_x + X_y + X_z \leq 1 \qquad (4.43c)$$

where

$$X_x = MAX \begin{pmatrix} \dfrac{N_{c,Sd,x}}{0.7N_{cr,y}} \\ \dfrac{N_{c,Sd,x}}{0.7N_{cr,z}} \end{pmatrix} \qquad (4.44)$$

in which:
$N_{cr,y}$ = critical buckling load with bending about the (major) y-axis
$N_{cr,z}$ = critical buckling load with bending about the (minor) z-axis.

$$X_y = MAX \begin{pmatrix} M_{Sd,y} / M_{Rd,y} \left(1 - \dfrac{N_{c,Sd,x}}{N_{cr,y}}\right) \\ M_{Sd,y} / M_{b,cr,y} \left(1 - \dfrac{N_{c,Sd,x}}{N_{cr,y}}\right) \end{pmatrix} \qquad (4.45a)$$

and

$$X_z = MAX \begin{pmatrix} M_{Sd,z} / M_{Rd,z} \left(1 - \dfrac{N_{c,Sd,x}}{N_{cr,z}}\right) \\ M_{Sd,z} / M_{b,cr,z} \left(1 - \dfrac{N_{c,Sd,x}}{N_{cr,z}}\right) \end{pmatrix} \qquad (4.45b)$$

P(4) The effects of creep (see 4.12) on the values of the critical buckling loads used in the above expressions shall be taken into account.

(5) For compression and bending at the serviceability limit state, provided that all the conditions in P(3) and P(1) above are satisfied, only the limit state of deflection, due account being taken of additional deflections resulting from creep, need be considered. The limit state of local damage need not in general be considered for sections in compression and bending.

4.9 PLATES

4.9.0 Notation

D_{ij} Bending terms of plate stiffness matrix
M_x, M_y Moment stress resultants
a, b Plate dimensions (x-direction, y-direction)
q_o Load intensity
w Deflection
α Stiffness parameter
β_x, β_y Moment stress resultant parameters

4.9.1 General

P(1) A plate shall be defined as a planar load carrying component spanning two directions whose thickness is significantly less than its side lengths.

(2) In the EUROCOMP Design Code plates are classified as Type I, Type II or Type III. (Detailed descriptions are included in the EUROCOMP Handbook). Analysis and design of plates may be carried out using the Tables given in the EUROCOMP Handbook or alternatively by advanced numerical techniques (see also 4.10.6).

(3) If the plate, Figure 4.3, is supported at the edges and loaded by a transverse load q_o, the deflection w at the centre of the plate is given by the following design formula:

$$w = \alpha q_o b a^3 / D_{22} \tag{4.46}$$

and the resultant bending moments at the centre of the plate are:

$$M_x = \beta_x q_o a^2 \tag{4.47}$$

and

$$M_y = \beta_y q_o a^2 \tag{4.48}$$

The plate stiffness parameter α and stress concentration parameters β_x and β_y depend on the plate geometry a/b, edge and loading conditions, and material elastic constants E_x, E_y, G_{xy} and ν_{xy}.

The coefficients a, β_x and β_y in Equations (4.46) - (4.48) are based on small deflection analysis and are valid for a maximum deflection less than or equal to half the plate thickness. For deflections greater than half the plate thickness a non-linear analysis which takes account of the plate membrane action is required.

(4) The coefficients α, β_x and β_y for various D_{11}/D_{22} and a/b ratios are given in Tables 4.1 to 4.38 in the EUROCOMP Handbook.

The following steps can be used to design plates in bending:

Step 1

If the elastic and strength properties of the plies for the laminated plate are known from experimental values or manufacturers' data, go to Step 2.

From pre-defined reinforcement and resin types determine the elastic and strength properties of the plies of the laminated plate from 4.10.2 and 4.10.4 respectively.

Step 2

If the elastic and strength properties of the laminated plate are known from experimental values or manufacturers' data, go to Step 3.

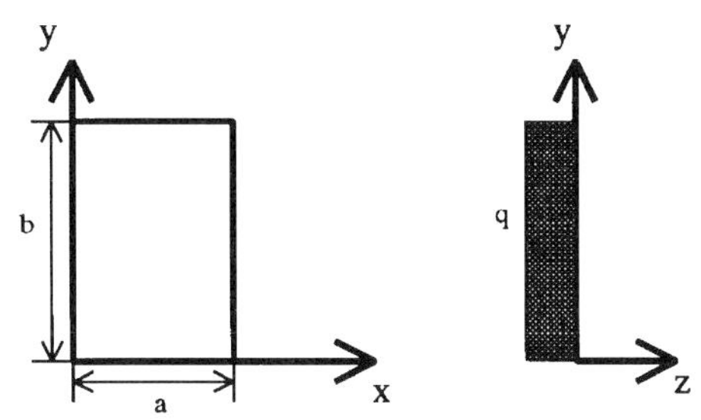

Uniformly distributed load (over entire plate)

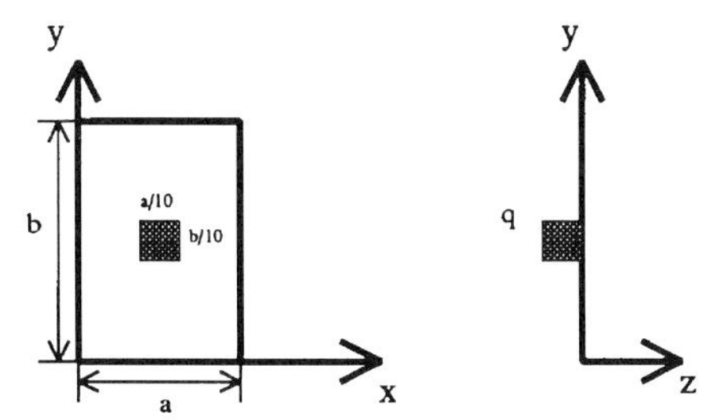

Uniformly distributed load (over central rectangular area)

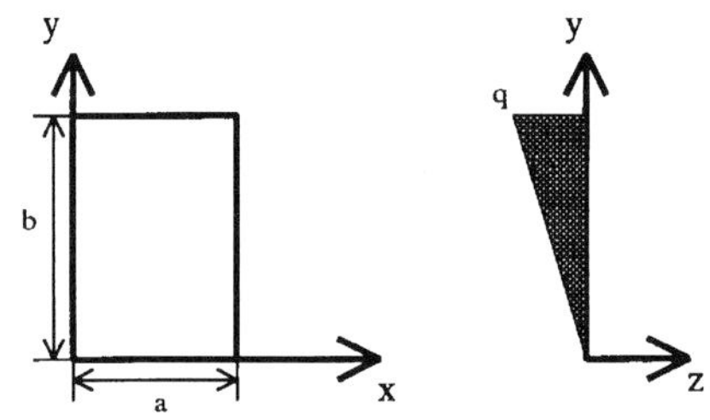

Triangularly distributed load (increasing parallel to side 'b')
Figure 4.3 Plate geometry and loading descriptions.

From the pre-determined elastic and strength properties of the plies determine the laminate plate stiffness and strength properties from 4.10.3 and 4.10.5 respectively.

Step 3

Determine maximum values of the deflection and bending moments of the plate.

The plate equations (4.46) to (4.48) can be used in conjunction with the values of the coefficients given in Tables 4.1 to 4.38 of the EUROCOMP Handbook to determine the maximum deflection and bending moments of the plate.

Step 4

Check requirements of serviceability and ultimate limit state. If not satisfied change the design of the laminated plate and go to Step 1.

4.9.2 Plates Type I (isotropic plates)

P(1) Plates Type I are isotropic plates. Their materials properties are such that $D_{16} = D_{26} = 0.0$ and $D_{11} = D_{22}$.

(2) Determination of deflections, stresses and strains of this category of plates under any loading regime can be carried out using numerical techniques (see also 4.10.6).

(3) Theoretical equations, closed formed solutions and design tables are available in standard text and design books for standard support conditions and limited types of loading.

(4) Deflection and moment coefficients for different loading and edge conditions are given in Tables 4.1 and 4.2 in the EUROCOMP Handbook. The requisite tables, for a given loading and material description, are identified using Table 4.4.

Table 4.4 Tables in EUROCOMP Handbook giving deflection and moment coefficients for plates.

Plate	*Simply supported around four edges*	*Clamped around four edges*
Type I (isotropic)	Table 4.1	Table 4.2
Type II	Tables 4.3-4.6 UDL Tables 4.7-4.10 TL Tables 4.11-4.14 PL	Tables 4.15-4.18 UDL Tables 4.19-4.22 TL Tables 4.23-4.26 PL
Type III	Tables 4.27-4.38 UDL	-

Notes:

(i) Plate dimensions:- Coefficients are given for ratios a/b = 0.25, 0.5, 0.75 and 1.0, where a, b are the plate dimensions.

(ii) Loading:-
UDL, Uniformly distributed load, intensity 'q_0'
TL, Linearly varying triangular load, maximum intensity 'q_0'
PL, Patch load, 'q_0' of dimensions a/10 by b/10 applied at the centre of the plate.

(iii) Plate properties:- For Plates Type I, moment and deflection coefficients are given for values of D_{12}/D_{11} = 0.15, 0.3.

For Plates Type II, moment and deflection coefficients are given for the following combinations:
D_{11}/D_{22} = 0.5, 2.0, 4.0, 10.0 with
D_{12}/D_{11} = 0.15, 0.3 and D_{66}/D_{22} = 0.1, 0.3, 0.6, 1.0

For Plates Type III, sets of tables are given for D_{16}/D_{11} = D_{26}/D_{11} = 0.35, 0.65 and 1.0.
Moment and deflection coefficients are given for the following combinations:
D_{11}/D_{22} = 0.5, 2.0, 4.0, 10.0 with
D_{12}/D_{11} = 0.15, 0.3 and D_{66}/D_{22} = 0.1, 0.3, 0.6, 1.0

4.9.3 Plates Type II

P(1) Plates Type II, a particular type of specially orthotropic plates, are defined as plates whose material properties are such that $D_{16} = D_{26} = 0.0$ and $D_{11} \neq D_{22}$.

(2) Plates Type II will usually be constructed from cross-plied or unidirectional laminates.

(3) Deflection and moment coefficients for different loading and edge conditions are given in Tables 4.3 to 4.26 in the EUROCOMP Handbook. The requisite tables, for a given loading and material description, are identified using Table 4.4.

4.10 LAMINATE DESIGN

4.10.0 Notation

A_{ij}	Extensional or membrane terms of laminate stiffness matrix
B_{ij}	Coupling terms of laminate stiffness matrix
D_{ij}	Bending terms of laminate stiffness matrix
E_1	Longitudinal Young's modulus of the lamina
E_2	Transverse Young's modulus of the lamina
G_{12}	In-plane shear modulus of the lamina
G_{13}	Out-of-plane shear modulus of lamina (in the 1 - 3 plane)
G_{23}	Out-of-plane shear modulus of lamina (in 2 - 3 plane)
M_x, M_y, M_{xy}	Moment stress resultants per unit width
N_x, N_y, N_{xy}	Force stress resultants per unit width
Q_{ij}	Laminate reduced stiffness terms
$\bar{Q}_{ij}$	Transformed reduced stiffness terms
a_{ij}	Extensional or membrane terms of laminate compliance matrix
b_{ij}	Coupling terms of laminate compliance matrix
d_{ij}	Bending terms of laminate compliance matrix
h_k	Distance from middle surface to upper surface of kth layer
h	Total laminate thickness
n	Number of laminae in a laminate
V_f	Volume fraction of fibres
V_m	Volume fraction of matrix
ε_{1t}	Longitudinal failure strain of lamina in tension
ε_{1c}	Longitudinal failure strain of lamina in compression
ε_{2t}	Transverse failure strain of lamina in tension
ε_{2c}	Transverse failure strain of lamina in compression
ε_{ft}	Failure strain of fibres in tension
ν_{12}	Major Poisson's ratio of the lamina
ν_{21}	Minor Poisson's ratio of the lamina
ν_m	Poisson's ratio of the matrix
σ_{1t}	Longitudinal tensile strength of the lamina
σ_{1c}	Longitudinal compressive strength of the lamina
σ_{2t}	Transverse tensile strength of the lamina
σ_{2c}	Transverse compressive strength of the lamina

σ_{12s}	Shear strength of the lamina
σ_{ft}	Tensile strength of fibres
σ_m	Tensile/compressive strength of matrix
σ_{12m}	Shear strength of matrix

4.10.1 Basic conditions to be satisfied

P(1) The design of the laminate shall satisfy the conditions of 4.1 at the ultimate limit state and of 4.2 at the serviceability limit state.

P(2) The determination of the stresses and strains in the laminate shall be carried out using an appropriate means of analysis.

P(3) The analysis shall be carried out using the stiffnesses of the individual lamina appropriate to the limit state being considered.

(4) A method for determining the laminate stiffness is given in 4.10.3.

(5) A method for determining the laminate strength is given in 4.10.5.

P(6) The stiffness of the individual laminae shall be determined in accordance with 4.10.2.

P(7) The strength of the individual laminae shall be determined in accordance with 4.10.4.

P(8) At no point in the laminate shall the stress exceed the design strength at the particular limit state being considered.

P(9) Similarly, at no point shall the strain exceed the limiting design value.

4.10.2 Lamina stiffness

4.10.2.1 Unidirectional reinforcement

P(1) A unidirectional continuous lamina is by definition one that:

- has higher stiffness and strength in the direction of the fibres
- has stiffness and strength dominated by the matrix properties in the transverse direction
- is idealised as orthotropic (as a result of the dissimilarity of the fibre and matrix mechanical properties).

P(2) The stiffness parameters of a lamina shall be determined by experimental testing or taken from manufacturers' data.

(3) In the absence of experimental testing and manufacturers' data, lamina stiffness may be calculated using constituent properties and the Halpin-Tsai method.

(4) The Halpin-Tsai method determines the effective property of a composite in terms of the properties of the fibres and matrix as follows:

$$P = \frac{P_m [P_f + \xi P_m + \xi v_f (P_f - P_m)]}{[P_f + \xi P_m - v_f (P_f - P_m)]} \qquad (4.49)$$

where:

P = effective property of the composite (E_1, E_2, G_{12}, G_{23}, ...)
P_f = corresponding properties of the fibres
P_m = as in P_f but of matrix
ξ = reinforcing efficiency parameter of the composite material, indicating the extent to which the applied force is transmitted to the reinforcing phase.

Values of ξ for various directional properties are given in Table 4.5.

Special cases exist when ξ has limiting values. When $\xi \rightarrow \infty$ this method models continuous systems of fibre and matrix materials stacked parallel to the load direction. This idealisation, which is known to be accurate, can be used to determine E_1 and ν_{12}.

$$E_1 = v_f E_f + (1 - v_f)E_m \qquad (4.50)$$

When $\xi \rightarrow 0$ the method models continuous systems of fibre and matrix materials stacked perpendicular to the load direction.

$$E_2 = E_m E_f/(E_f (1 - v_f) + E_m v_f) \qquad (4.51)$$

The value of E_2 is always underestimated using Equation (4.51). Equations (4.50) and (4.51) are the upper and lower bounds for the modulus values of a two phase solid - solid composite system.

Table 4.5 Values of 'ξ' for use with Halpin-Tsai equations.

Material property	*Value of 'ξ' for use with Halpin-Tsai equations*
E_1	∞
ν_{12}	∞
E_2, $v_f < 0.65$	2.0
E_2, $v_f \geq 0.65$	$2.0 + 40\ v_f^{10}$
G_{12}, $v_f < 0.65$	1.0
G_{12}, $v_f \geq 0.65$	$1.0 + 40\ v_f^{10}$
G_{23}	$(4 - 3\ \nu_m)^{-1}$

Figures 4.4 to 4.10 give design charts, derived from the Halpin-Tsai equations, for typical E-glass/polyester resin composite laminae, using the material properties given in Table 4.6. For the purposes of generating the graphs it has been assumed that the fibres and matrix are isotropic.

Table 4.6 Assumed material properties used in the derivation of Figures 4.4 to 4.10.

Property	*Units*	*Fibres*	*Matrix*
Density	kg/m^3	2550	1230
Modulus of elasticity	kN/mm^2	72.4	3.7
Poisson's ratio	-	0.32	0.42
Shear modulus	kN/mm^2	27.4	1.3
Thermal expansion	$10^{-6}/^{\circ}C$	2.8	100

(5) Alternative methods of calculating lamina stiffness, namely the *Simple Rule of Mixtures* and the *Self-Consistent Doubly Embedded* method are given in the EUROCOMP Handbook. These methods may alternatively be used for preliminary design calculations.

4.10.2.2 Other typical reinforcements

P(1) Elastic properties of reinforcements should be determined by experimental testing or taken from manufacturers' data.

(2) In the absence of test results or manufacturers' data, representative properties for woven rovings may be calculated using the linear equations given in Table 4.7. The set of equations are valid for a glass content by weight in the range of 40 - 55%.

(3) Elastic properties of woven rovings vary considerably, owing to basic scatter in the fibre and matrix properties.

(4) Elastic properties of chopped strand mat laminates vary considerably, owing to basic scatter in the fibre and matrix properties.

P(5) In the absence of test results or manufacturers' data, representative properties of chopped strand mat can be calculated using the linear equations given in Table 4.8, which are valid for a glass content by weight in the range of 20 - 30%.

(6) Chopped strand mat may be considered isotropic for design.

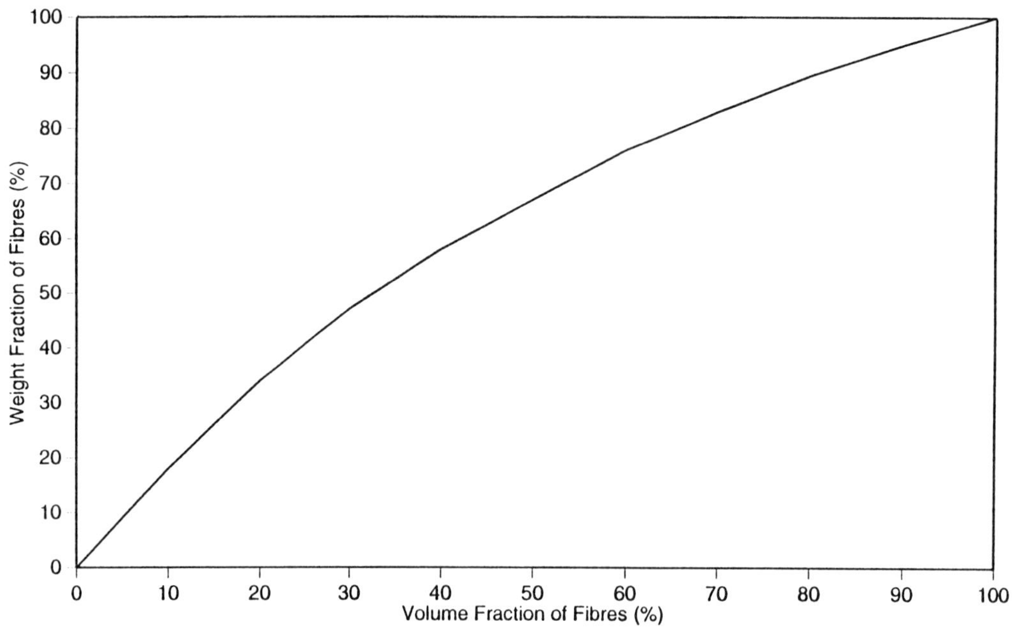

Figure 4.4 Design of unidirectional composite laminae: Halpin-Tsai rule of mixtures. Relationship of weight to volume fraction.

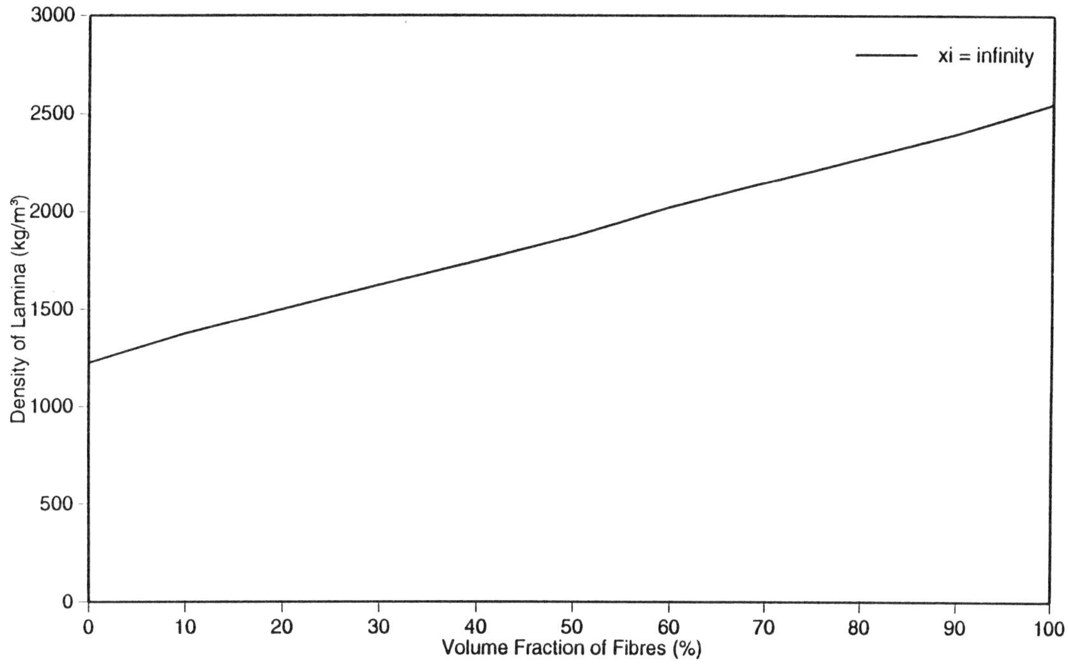

Figure 4.5 Design of unidirectional composite laminae: Halpin-Tsai rule of mixtures. Relationship of density to volume fraction.

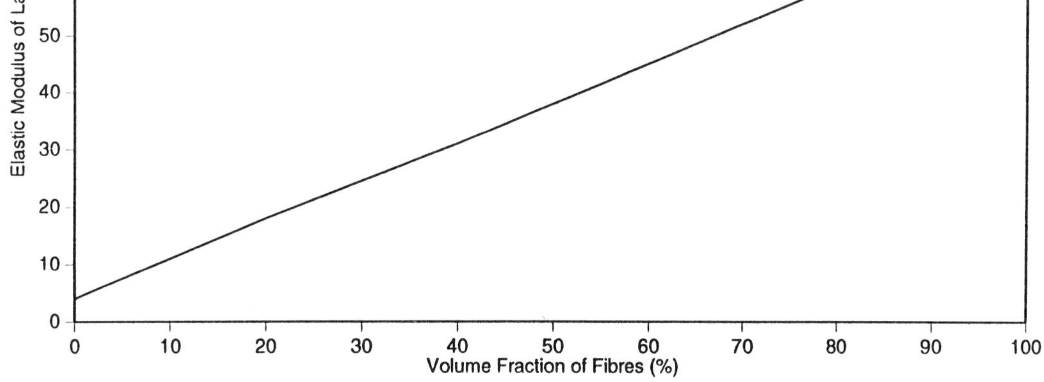

Figure 4.6 Design of unidirectional composite laminae: Halpin-Tsai rule of mixtures. Relationship of elastic modulus to volume fraction.

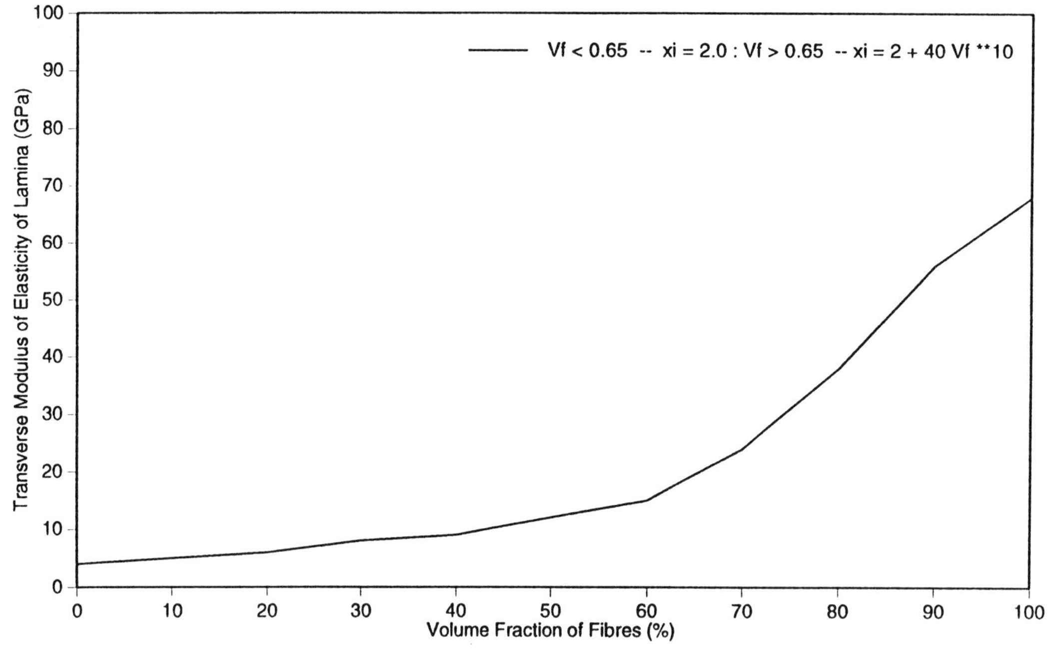

Figure 4.7 Design of unidirectional composite laminae: Halpin-Tsai rule of mixtures. Relationship of transverse modulus to volume fraction.

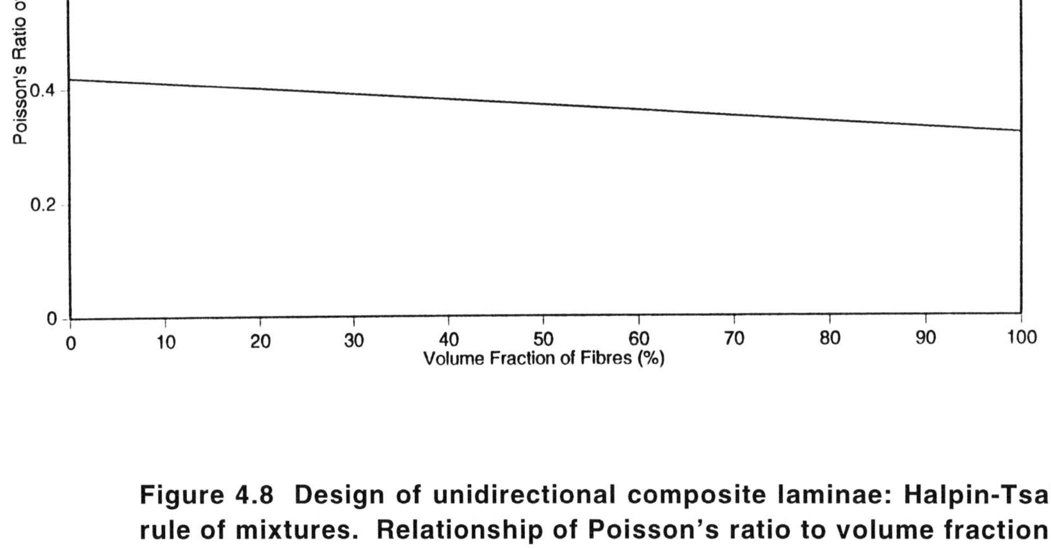

Figure 4.8 Design of unidirectional composite laminae: Halpin-Tsai rule of mixtures. Relationship of Poisson's ratio to volume fraction.

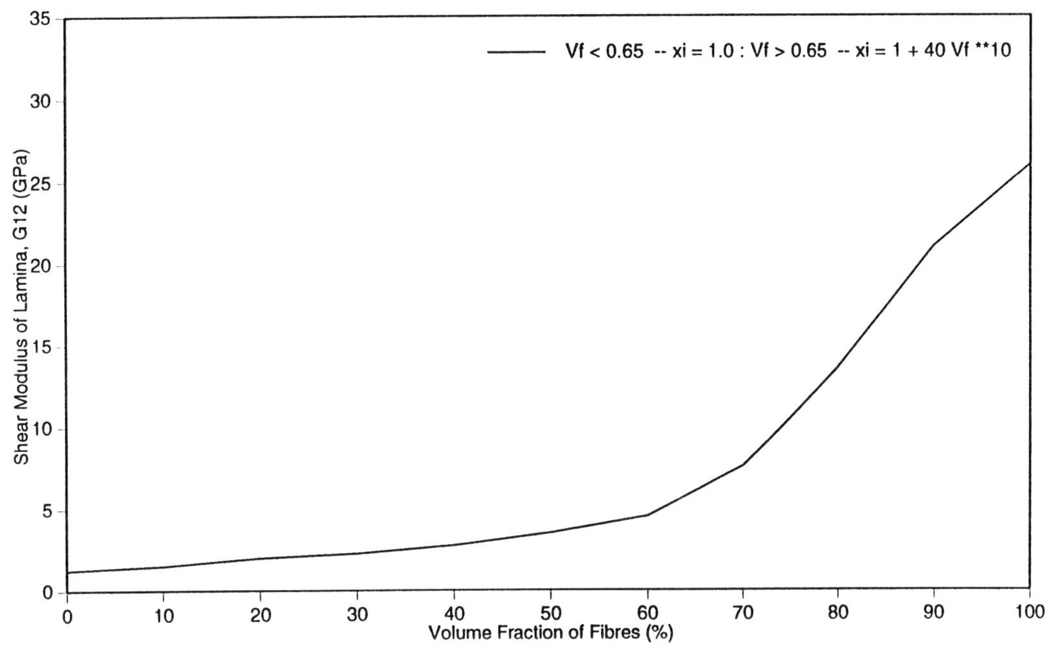

Figure 4.9 Design of unidirectional composite laminae: Halpin-Tsai rule of mixtures. Relationship of shear modulus to volume fraction.

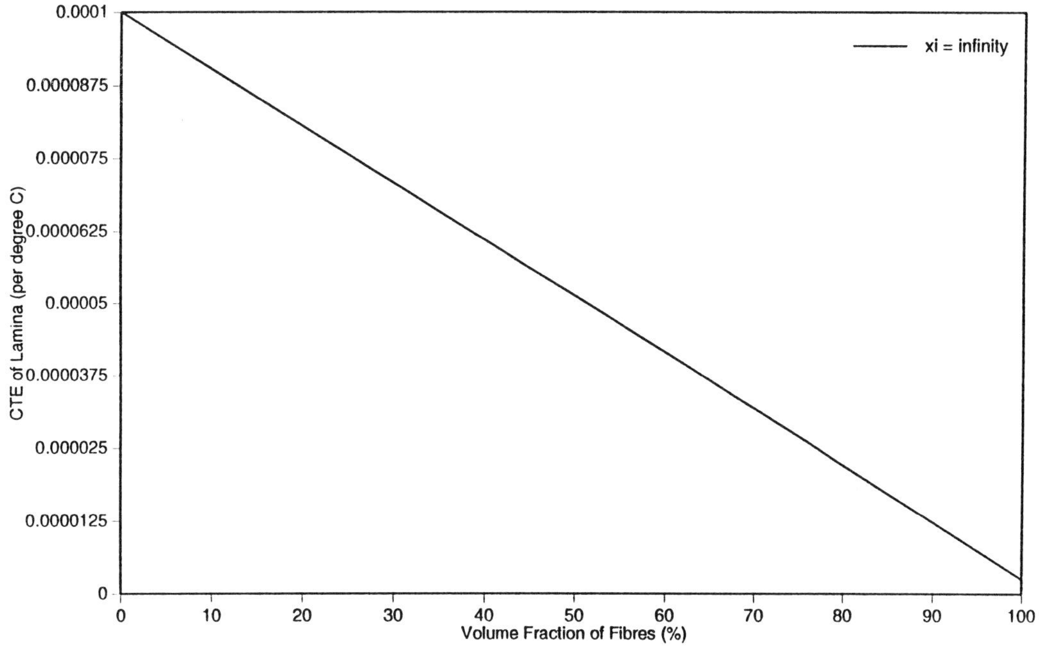

Figure 4.10 Design of unidirectional composite laminae: Halpin-Tsai rule of mixtures. Relationship of coefficient of thermal expansion to volume fraction.

Table 4.7 Elastic properties of typical E-glass WR/polyester hand lay- up systems (fibre applicability range 40-55% by weight).

Elastic property	*Weight fraction of gass (w_f)*
Tensile modulus, E_t (kN/mm²)	
lower bound	$30w_f - 4$
upper bound	$36w_f - 2$
Flexural modulus, E_f (kN/mm²)	$w_f + 13$
In plane shear modulus, G_{12} (kN/mm²)	3.6
Poisson's ratio, ν_{12}	
compressive	0.25
tensile	0.13

Example: for 50% weight content, $E_{t(lower)} = 30 \times 50/(100) - 4 = 11$ kN/mm^2.

(The volume fraction can be related to the weight fraction using the graph in Figure 4.4).

Table 4.8 Elastic properties of typical E-glass CSM/ polyester hand lay-up systems (fibre applicability range 20-30% by weight).

Elastic property	*Weight fraction of glass (w_f)*
Tensile modulus, E_t (kN/mm²)	
lower bound	$21\ w_f + 1$
upper bound	$21\ w_f + 3$
Flexural modulus, E_f (kN/mm²)	$11\ w_f + 3$
In plane shear modulus, G_{12} (kN/mm²)	2.76
Poisson's ratio, ν_{12}	
compressive	0.42
tensile	0.32

Example: for 25% weight content, $E_{t(lower)} = 21 \times 25/(100) + 1 = 6.25$ kN/mm^2.

(The volume fraction can be related to the weight fraction using the graph in Figure 4.4).

4.10.3 Laminate stiffness

P(1) Laminate stiffnesses should be determined by experimental testing or taken from manufacturers' data.

P(2) In the absence of experimental or manufacturers' data the material properties may be calculated using classical lamination theory (CLT) and the respective directional properties of the constituent laminae.

(3) The following steps may be used to determine the elastic properties of a laminate. A more detailed description is given in the Eurocomp Handbook.

Step 1

From the pre-determined elastic properties E_1, E_2, G_{12}, and ν_{12} of each lamina, determine the reduced stiffness terms Q_{11}, Q_{22}, Q_{66} and Q_{12} from Equations (4.52):

$$\begin{aligned} Q_{11} &= E_1/(1 - \nu_{12}\nu_{21}) \\ Q_{22} &= E_2/(1 - \nu_{12}\nu_{21}) \\ Q_{66} &= G_{12} \\ Q_{12} &= \nu_{21}E_1/(1 - \nu_{12}\nu_{21}) \end{aligned} \qquad (4.50)$$

where $\nu_{21} = (E_2/ E_1)\ \nu_{12}$

Step 2

Having obtained the lamina reduced stiffness terms Q_{ij} for each lamina, calculate the transformed lamina reduced stiffness terms $\bar{Q}_{ij}$ for a given angle of orientation

	Q_{11}	Q_{22}	Q_{12}	Q_{66}
$\bar{Q}_{11}$	n^4	n^4	$2m^2n^2$	$4m^2n^2$
$\bar{Q}_{22}$	n^4	m^4	$2m^2n^2$	$4m^2n^2$
$\bar{Q}_{12}$	m^2n^2	m^2n^2	m^4+n^4	$-4m^2n^2$
$\bar{Q}_{66}$	m^2n^2	m^2n^2	$-2m^2n^2$	$(m^2-n^2)^2$
$\bar{Q}_{16}$	$-m^3n$	mn^3	$-mn(m^2-n^2)$	$-2mn(m^2-n^2)$
$\bar{Q}_{26}$	$-mn^3$	m^3n^2	$mn(m^2-n^2)$	$2mn(m^2-n^2)$

(e.g.: $\bar{Q}_{11} = m^4Q_{11} + n^4Q_{22} + 2m^2n^2Q_{12} + 4m^2n^2Q_{66}$)

where:

$m = \cos ø$

$n = \sin ø$

θ = angle of orientation of fibres

Step 3
Determine h_{k-1}, the distance from the mid-plane of the laminate to the lower surface of the kth layer, and h_k, the distance to the upper surface of the kth layer for each lamina in a laminate configuration.

Step 4
Determine the A_{ij}, B_{ij} and D_{ij} terms for the laminate using Equations (4.53):

$$A_{ij} = \sum_{k=1}^{n} (\overline{Q}_{ij})_k (h_k - h_{k-1}) \qquad i,j = 1,2,6$$

$$B_{ij} = 1/2 \sum_{k=1}^{n} (\overline{Q}_{ij})_k (h_k^2 - h_{k-1}^2) \qquad i,j = 1,2,6$$

(4.51)

$$D_{ij} = 1/3 \sum_{k=1}^{n} (\overline{Q}_{ij})_k (h_k^3 - h_{k-1}^3) \qquad i,j = 1,2,6$$

where:

A_{ij} = extensional or membrane stiffness terms of a laminate
B_{ij} = coupling stiffness terms of a laminate
D_{ij} = bending stiffness terms of a laminate
n = the number of laminae in a laminate.

The A_{ij}, B_{ij} and D_{ij} terms relate the stress and moment resultants to strains and curvatures in the following matrix:

	ε_x	ε_y	ε_{xy}	χ_x	χ_y	χ_{xy}
N_x	A_{11}	A_{12}	A_{16}	B_{11}	B_{12}	B_{16}
N_y	A_{21}	A_{22}	A_{26}	B_{12}	B_{22}	B_{26}
N_{xy}	A_{16}	A_{26}	A_{66}	B_{16}	B_{26}	B_{66}
M_x	B_{11}	B_{12}	B_{16}	D_{11}	D_{12}	D_{16}
M_y	B_{12}	B_{22}	B_{26}	D_{12}	D_{22}	D_{26}
M_{xy}	B_{16}	B_{26}	B_{66}	D_{16}	D_{26}	D_{66}

where χ denotes curvature.

Step 5
Having obtained the laminate stiffness terms A_{ij}, B_{ij} and D_{ij}, calculate the corresponding compliance terms a_{ij}, b_{ij} and d_{ij} by inversing the laminate stiffness matrices.

Step 6
Having obtained the compliance terms a_{ij}, b_{ij} and d_{ij}, calculate the laminate membrane and bending equivalent elastic constants.

The equivalent membrane elastic constants are:

$$\begin{aligned} E_x &= 1/ha_{11} \\ E_y &= 1/ha_{22} \\ G_{xy} &= 1/ha_{66} \\ \nu_{xy} &= -a_{12}/a_{11} \\ \nu_{yx} &= -a_{12}/a_{22} \end{aligned} \tag{4.54}$$

The equivalent bending elastic constants are:

$$\begin{aligned} E_x &= 12/h^3d_{11} \\ E_y &= 12/h^3d_{22} \\ G_{xy} &= 12/h^3d_{66} \\ \nu_{xy} &= -d_{12}/d_{11} \\ \nu_{yx} &= -d_{12}/d_{22} \end{aligned} \tag{4.55}$$

The above steps for calculating laminate stiffness and equivalent elastic properties are summarised in the following flow chart, Figure 4.11.

4.10.4 Lamina strength

4.10.4.1 Unidirectional

P(1) The Hart-Smith failure criterion should be used for design. Design strengths shall be determined by experimental testing or taken from manufacturers' data, and divided by the appropriate partial safety factor.

(2) In the absence of experimental or manufacturers' data the Tsai-Wu failure criterion may be used.

(a) The Hart-Smith strain failure criterion

P(1) The Hart-Smith failure envelope for a single ply shall be constructed as shown in Figures 4.13 (a)-(d) in the EUROCOMP Handbook.

P(2) The respective failure strains shall be obtained by experimental testing or from manufacturers' data.

P(3) The lamina shall be deemed to have failed for any combination of strains lying outside the failure envelope.

Obtain lamina properties E_1, E_2, G_{12}, ν_{12}, h, angle of orientation θ

⇓

Determine lamina reduced stiffness values Q_{ij} from Eqn 4.52

⇓

Calculate lamina transformed reduced stiffness Q_{ij}

⇓

Determine coordinates h_k for each lamina from the bottom surface

⇓

For a given angle of orientation per lamina calculate
A_{ij}, B_{ij} and D_{ij} from Eqn (4.53)

⇓

Invert the laminate stiffness matrix and calculate laminate compliances
a_{ij}, b_{ij} and d_{ij}

⇓

Calculate equivalent laminate elastic properties E_x, E_y, G_{xy},
ν_{xy}, ν_{yx} from Eqns (4.54) & (4.55)

Figure 4.11 Flow chart for calculation of laminate stiffness and equivalent elastic properties.

(b) The Tsai-Wu failure criterion

P(1) The Tsai-Wu failure criterion should be used for design in the absence of experimental or manufacturers' data. The design strengths, which shall be divided by the appropriate partial safety factors, may be estimated using the characteristic strengths of the fibres and the matrix. The respective lamina strengths calculated in this manner are given by:

longitudinal tensile strength,	$\sigma_{1t}v_f$	=	$\sigma_{ft} + \varepsilon_{ft}E_m\ (1 - v_f)$
longitudinal compressive strength,	σ_{1c}	=	σ_{1t} (as a first approximation)
transverse tensile and compressive strength,	σ_{2t}, σ_{2c}	=	σ_m
shear strength,	σ_{12}	=	σ_{12m}

The subscripts 'f' and 'm' in the above formulae relate to the fibres and matrix respectively; i.e. σ_{ft} is the strength of the fibres in tension. The characteristic lamina strengths, namely σ_{1t}, σ_{1c}, σ_{2t}, σ_{2c} and σ_{12} , which have been calculated from raw material data, can then be used with Equation (4.56) to check for possible failure of the lamina.

P(3) The lamina shall be deemed to have failed when the inequality in Equation (4.56) is no longer satisfied:-

$$\left(\frac{1}{\sigma_{1t}} + \frac{1}{\sigma_{1c}}\right)\sigma_1 + \left(\frac{1}{\sigma_{2t}} + \frac{1}{\sigma_{2c}}\right)\sigma_2 - \left(\frac{1}{\sigma_{1t}\,\sigma_{1c}}\right)\sigma_1^2 - \left(\frac{1}{\sigma_{2t}\,\sigma_{2c}}\right)\sigma_2^2 + \left(\frac{1}{\sigma_{12s}^2}\right)\sigma_{12}^2 - \left(\frac{0.5}{\sqrt{\sigma_{1t}\,\sigma_{1c}\,\sigma_{2t}\,\sigma_{2c}}}\right)\sigma_1\,\sigma_2 \leq 1.0 \qquad (4.56)$$

where the notation is as defined in 4.5.0 and σ_1 and σ_2 are the stresses parallel and perpendicular to the fibres respectively, and σ_{12} is the in-plane shear stress.

4.10.4.2 Other typical reinforcement

P(1) Strength of reinforcements shall be determined by experimental testing or taken from manufacturers' data.

(2) In the absence of experimental data the linear equations of Table 4.9 may be used to estimate strengths of typical E-glass WR/polyester hand lay-up systems. The set of equations are valid for glass content by weight ratio within the range of 40 to 55%.

(3) In the absence of test results or manufacturers' data the linear equations of Table 4.10 may be used to estimate strengths of typical E-glass chopped strand mat/polyester hand lay-up systems. The set of equations are valid for glass content by weight ratio within the range of 20 to 30%.

(4) The strengths shall be divided by the appropriate partial safety factor.

Table 4.9 Strengths of typical E-glass WR/polyester hand lay-up systems (fibre applicability range 40-55% by weight).

Strength property	*Weight fraction of glass* (w_f)
Tensile, σ_{1t}, σ_{2t} (N/mm²)	
lower bound	$784\ w_f - 212$
upper bound	$893\ w_f - 192$
Compressive, σ_{1c}, σ_{2c} (N/mm²)	
lower bound	$182\ w_f + 66$
upper bound	$190\ w_f + 109$
Flexural, σ_{1f}, σ_{2f} (N/mm²)	
lower bound	$533\ w_f - 16$
upper bound	$604\ w_f - 17$
In plane shear, σ_{12} (N/mm²)	$320\ w_f - 66$
Interlaminar shear, σ_{INT} (N/mm²)	$113\ w_f - 40$

Example: for 55% weight content, $\sigma_{t(upper)} = 893 \times 55/(100) - 192 = 299$ N/mm^2.

(The volume fraction can be related to the weight fraction using the graph in Figure 4.4).

4.10.5 Laminate strength

P(1) Laminate strength should be determined by experimental testing or taken from manufacturers' data. The Hart-Smith failure criterion shall be used for design.

P(2) The Hart-Smith failure envelope for a laminate shall be assembled in the same manner as for a single lamina.

P(3) In the absence of strength test or manufacturers' data for a laminate, the failure envelope shall be assembled using the failure envelopes of the constitutive laminae as illustrated in Figure 4.14 in the EUROCOMP Handbook.

(4) In the absence of strength data for the laminae and the laminate, failure of the laminate should be examined using the Tsai-Wu criterion, see 4.10.4.1(b).

Table 4.10 Strengths of typical E-glass CSM/polyester hand lay-up systems (fibre applicability range 20-30% by weight).

Strength property	*Proportion of glass content by weight* (w_f)
Tensile, σ_t (N/mm²)	$290\ w_f + 10$
Compressive, σ_c (N/mm²)	$275\ w_f + 60$
Flexural, σ_f (N/mm²)	
lower bound	$246\ w_f + 95$
upper bound	$410\ w_f + 98$
In plane shear, σ_{12} (N/mm²)	$157\ w_f + 34$
Interlaminar shear, σ_{INT} (N/mm²)	25

Example: for 25% weight content, $\sigma_{f(upper)} = 410 \times 25/100 + 98 = 200.5$ N/mm^2.

(The volume fraction can be related to the weight fraction using the graph in Figure 4.4).

(5) The following steps may be used to determine the strength of a laminate:

Step 1
Obtain the laminate membrane, bending and coupling stiffnesses A_{ij}, B_{ij}, D_{ij} (see Step 4, 4.10.3).

Step 2
Calculate corresponding compliance terms a_{ij}, b_{ij}, d_{ij} from A_{ij}, B_{ij}, D_{ij} (see Step 5, 4.10.3).

Step 3
Calculate laminate midplane deformations ε_x, ε_y, ε_{xy} and curvatures χ_x, χ_y, χ_{xy} in the laminate reference axes x - y from the following matrix:

	N_x	N_y	N_{xy}	M_x	M_y	M_{xy}
ε_x	a_{11}	a_{12}	a_{16}	b_{11}	b_{12}	b_{16}
ε_y	a_{21}	a_{22}	a_{26}	b_{12}	b_{22}	b_{26}
ε_{xy}	a_{16}	a_{26}	a_{66}	b_{16}	b_{26}	b_{66}
χ_x	b_{11}	b_{12}	b_{16}	d_{11}	d_{12}	d_{16}
χ_y	b_{12}	b_{22}	b_{26}	d_{12}	d_{22}	d_{26}
χ_{xy}	b_{16}	b_{26}	b_{66}	d_{16}	d_{26}	d_{66}

Step 4
For each lamina calculate the total strain arising from the membrane and bending strain contributions

$$\begin{Bmatrix} \overline{\varepsilon_x} \\ \overline{\varepsilon_y} \\ \overline{\varepsilon_{xy}} \end{Bmatrix} = \begin{Bmatrix} \varepsilon_x - z\chi_x \\ \varepsilon_y - z\chi_y \\ \varepsilon_{xy} - z\chi_{xy} \end{Bmatrix} \tag{4.57}$$

Step 5
Transform the lamina strains from laminate axis x - y to lamina axis 1 -

	$\bar{\varepsilon}_x$	$\bar{\varepsilon}_y$	$\bar{\varepsilon}_{xy}$
ε_1	m^2	n^2	mn
ε_2	n^2	m^2	−mn
ε_{12}	−2mn	2mn	$m^2 - n^2$

Step 6
Calculate lamina stresses in lamina axis 1 - 2 from total strains from the following matrix:

	ε_1	ε_2	ε_{12}
σ_1	Q_{11}	Q_{12}	0
σ_2	Q_{21}	Q_{22}	0
σ_{12}	0	0	Q_{66}

Step 7
Calculate lamina strength in accordance with 4.10.4.1(b). This will demonstrate whether any lamina has failed within the laminate, first ply failure.

The above steps for calculating laminate strength are summarised in the following flow chart, Figure 4.12.

4.10.6 Computer programs for analysis and design

4.10.6.1 General

(1) Computer programs have been developed specifically for the analysis and design of composite laminates and laminated structural elements. A program appropriate for the purpose should be available for designers, owing to the complex mechanical behaviour of composites.

4.10.6.2 Computing needs

(1) Computing needs are described in the EUROCOMP Handbook. Some identified analysis and design capabilities are useful but not necessary in the design of conventional composite structures.

From the given laminate configurations and initial loads determine laminate stiffnesses A_{ij}, B_{ij}, D_{ij}

⇓

Calculate laminate compliances a_{ij}, b_{ij}, d_{ij}

⇓

Determine total strains in laminate reference axis x - y

⇓

Transform strains into lamina axis 1 - 2

⇓

Convert lamina strains to lamina stresses in lamina axis 1 - 2

⇓

Use failure criterion to determine first ply failure load

Figure 4.12 Flow chart for calculation of laminate strength.

4.10.6.3 General requirements for a computer program

(1) General requirements for an effective and user-friendly analysis/design program are given in the EUROCOMP Handbook.

4.10.6.4 Programs available

(1) Specific programs developed for the analysis and design of composite laminates and laminated structural elements are identified in the EUROCOMP Handbook.

4.10.6.5 Program selection

(1) The selection of a program should be based on a specification that clearly defines the requirements set for the program.

(2) Both analysis/design capabilities and features affecting the use of the

program should be considered when the specification is prepared.

4.11 DESIGN DATA

The data in the following tables are characteristic values where:

Characteristic value = Mean value less 1.64 standard deviations.

The design material property is arrived at by dividing the characterstic value by the relevant partial coefficient.

The partial safety factor should be obtained from 2.3.3.2. As the quoted values are derived from test specimens, and are specific to particular manufacturing routes, γ_{m3} = 2 and γ_{m2} = 1.

Table 4.11 Characteristic material properties: Data sheet No. 1.

Property	*Units*	*Symbol*	*Characteristic value*
Tensile strength (long.)	N/mm^2	$\sigma_{x,t,k}$	410
Tensile strength (trans.)	N/mm^2	$\sigma_{y,t,k}$	44
Tensile modulus (long.)	kN/mm^2	$E_{x,t,k}$	27
Tensile modulus (trans.)	kN/mm^2	$E_{y,t,k}$	3.5
Compressive strength (long.)	N/mm^2	$\sigma_{x,c,k}$	270
Compressive strength (trans.)	N/mm^2	$\sigma_{y,c,k}$	
Compressive modulus (long.)	kN/mm^2	$E_{x,c,k}$	24
Compressive modulus (trans.)	kN/mm^2	$E_{y,c,k}$	4.5
Shear strength (in-plane)	N/mm^2	$\tau_{xy,k}$	15
Shear modulus (in-plane)	kN/mm^2	$G_{xy,k}$	4.2
Flexural strength (long.)	N/mm^2	$\sigma_{x,b,k}$	400
Flexural strength (trans.)	N/mm^2	$\sigma_{y,b,k}$	115
Flexural modulus (long.)	kN/mm^2	$E_{x,b,k}$	14
Flexural modulus (trans.)	kN/mm^2	$E_{y,b,k}$	8
Failure strain (long.)	%		1.8
Failure strain (trans.)	%		1.65
Poisson's ratio (long.)	-		0.2
Poisson's ratio (trans.)	-		0.1

Process	Pultrusion
Reinforcement	E-glass random mat plus roving in the ratio 1:4
Resin system	Polyester, vinylester, modar
Volume fraction of fibre	48% (typical)

Table 4.12 Characteristic material properties: Data sheet No. 2.

Property	*Units*	*Symbol*	*Characteristic value*
Tensile strength (long.)	N/mm^2	$\sigma_{x,t,k}$	207
Tensile strength (trans.)	N/mm^2	$\sigma_{y,t,k}$	48
Tensile modulus (long.)	kN/mm^2	$E_{x,t,k}$	17.2
Tensile modulus (trans.)	kN/mm^2	$E_{y,t,k}$	5.5
Compressive strength (long.)	N/mm^2	$\sigma_{x,c,k}$	207
Compressive strength (trans.)	N/mm^2	$\sigma_{y,c,k}$	103
Compressive modulus (long.)	kN/mm^2	$E_{x,c,k}$	17.2
Compressive modulus (trans.)	kN/mm^2	$E_{y,c,k}$	6.9
Shear strength (in-plane)	N/mm^2	$\tau_{xy,k}$	31
Shear modulus (in-plane)	kN/mm^2	$G_{xy,k}$	2.9
Flexural strength (long.)	N/mm^2	$\sigma_{x,b,k}$	207
Flexural strength (trans.)	N/mm^2	$\sigma_{y,b,k}$	69
Flexural modulus (long)	kN/mm^2	$E_{x,b,k}$	13.8
Flexural modulus (trans.)	kN/mm^2	$E_{y,b,k}$	5.5
Failure strain (long.)	%		
Failure strain (trans.)	%		
Poisson's ratio (long.)	-		0.33
Poisson's ratio (trans.)	-		0.11

Process	Pultrusion
Reinforcement	E-glass random mat plus roving in the ratio 1:1
Resin system	Polyester
Volume fraction of fibre	36% (typical)

Table 4.13 Characteristic material properties: Data sheet No. 3.

Property	*Units*	*Symbol*	*Characteristic value*
Tensile strength (long.)	N/mm^2	$\sigma_{x,t,k}$	690
Tensile strength (trans.)	N/mm^2	$\sigma_{y,t,k}$	
Tensile modulus (long.)	kN/mm^2	$E_{x,t,k}$	41
Tensile modulus (trans.)	kN/mm^2	$E_{y,t,k}$	
Compressive strength (long.)	N/mm^2	$\sigma_{x,c,k}$	414
Compressive strength (trans.)	N/mm^2	$\sigma_{y,c,k}$	
Compressive modulus (long.)	kN/mm^2	$E_{x,c,k}$	
Compressive modulus (trans.)	kN/mm^2	$E_{y,c,k}$	
Shear strength (in-plane)	N/mm^2	$\tau_{xy,k}$	38
Shear modulus (in-plane)	kN/mm^2	$G_{xy,k}$	
Flexural strength (long.)	N/mm^2	$\sigma_{x,b,k}$	690
Flexural strength (trans.)	N/mm^2	$\sigma_{y,b,k}$	
Flexural modulus (long.)	kN/mm^2	$E_{x,b,k}$	41
Flexural modulus (trans.)	kN/mm^2	$E_{y,b,k}$	
Failure strain (long.)	%		2.0
Failure strain (trans.)	%		
Poisson's ratio (long.)	-		
Poisson's ratio (trans.)	-		

Process	Pultrusion
Reinforcement	E-glass roving
Resin system	Polyester
Volume fraction of fibre	65% (typical)

Table 4.14 Characteristic material properties: Data sheet No. 4.

Property	*Units*	*Symbol*	*Character-istic value*
Tensile strength (long.)	N/mm^2	$\sigma_{x,t,k}$	80
Tensile strength (trans.)	N/mm^2	$\sigma_{y,t,k}$	80
Tensile modulus (long.)	N/mm^2	$E_{x,t,k}$	5600
Tensile modulus (trans.)	N/mm^2	$E_{y,t,k}$	5600
Compressive strength (long.)	N/mm^2	$\sigma_{x,c,k}$	80
Compressive strength (trans.)	N/mm^2	$\sigma_{y,c,k}$	80
Compressive modulus (long.)	N/mm^2	$E_{x,c,k}$	5600
Compressive modulus (trans.)	N/mm^2	$E_{y,c,k}$	5600
Shear strength (in-plane)	N/mm^2	$\tau_{xy,k}$	
Shear modulus (in-plane)	N/mm^2	$G_{xy,k}$	
Flexural strength (long.)	N/mm^2	$\sigma_{x,b,k}$	
Flexural strength (trans.)	N/mm^2	$\sigma_{y,b,k}$	
Flexural modulus (long.)	N/mm^2	$E_{x,b,k}$	
Flexural modulus (trans.)	N/mm^2	$E_{y,b,k}$	
Failure strain (long.)	%		
Failure strain (trans.)	%		
Poisson's ratio (long.)	-		0.33
Poisson's ratio (trans.)	-		0.33

Process	Hand lay-up
Reinforcement	E-glass chopped strand mat
Resin system	Polyester
Volume fraction of fibre	15.9% (typical) (2.5:1 R:G)

Table 4.15 Characteristic material properties: Data sheet No. 5.

Property	*Units*	*Symbol*	*Characteristic value*
Tensile strength (long.)	N/mm^2	$\sigma_{x,t,k}$	97
Tensile strength (trans.)	N/mm^2	$\sigma_{y,t,k}$	97
Tensile modulus (long.)	kN/mm^2	$E_{x,t,k}$	6.8
Tensile modulus (trans.)	kN/mm^2	$E_{y,t,k}$	6.8
Compressive strength (long.)	N/mm^2	$\sigma_{x,c,k}$	97
Compressive strength (trans.)	N/mm^2	$\sigma_{y,c,k}$	97
Compressive modulus (long.)	kN/mm^2	$E_{x,c,k}$	6.8
Compressive modulus (trans.)	kN/mm^2	$E_{y,c,k}$	6.8
Shear strength (in-plane)	N/mm^2	$\tau_{xy,k}$	
Shear modulus (in-plane)	kN/mm^2	$G_{xy,k}$	
Flexural strength (long.)	N/mm^2	$\sigma_{x,b,k}$	
Flexural strength (trans.)	N/mm^2	$\sigma_{y,b,k}$	
Flexural modulus (long.)	kN/mm^2	$E_{x,b,k}$	
Flexural modulus (trans.)	kN/mm^2	$E_{y,b,k}$	
Failure strain (long.)	%		
Failure strain (trans.)	%		
Poisson's ratio (long.)	-		0.33
Poisson's ratio (trans.)	-		0.33

Process	Hand lay-up
Reinforcement	E-glass chopped strand mat
Resin system	Polyester
Volume fraction of fibre	19% (typical) (2:1 R:G)

Table 4.16 Characteristic material properties: Data sheet No. 6.

Property	*Units*	*Symbol*	*Characteristic value*
Tensile strength (long.)	N/mm^2	$\sigma_{x,t,k}$	228
Tensile strength (trans.)	N/mm^2	$\sigma_{y,t,k}$	228
Tensile modulus (long.)	kN/mm^2	$E_{x,t,k}$	15.2
Tensile modulus (trans.)	kN/mm^2	$E_{y,t,k}$	15.2
Compressive strength (long.)	N/mm^2	$\sigma_{x,c,k}$	129
Compressive strength (trans.)	N/mm^2	$\sigma_{y,c,k}$	129
Compressive modulus (long.)	kN/mm^2	$E_{x,c,k}$	10.9
Compressive modulus (trans.)	kN/mm^2	$E_{y,c,k}$	10.9
Shear strength (in-plane)	N/mm^2	$\tau_{xy,k}$	65.6
Shear modulus (in-plane)	kN/mm^2	$G_{xy,k}$	65.6
Flexural strength (long.)	N/mm^2	$\sigma_{x,b,k}$	395
Flexural strength (trans.)	N/mm^2	$\sigma_{y,b,k}$	395
Flexural modulus (long.)	kN/mm^2	$E_{x,b,k}$	13.4
Flexural modulus (trans.)	kN/mm^2	$E_{y,b,k}$	13.4
Failure strain (long.)	%		2.8
Failure strain (trans.)	%		2.8
Poisson's ratio (long.)	-		
Poisson's ratio (trans.)	-		

Process	Hand lay-up
Reinforcement	E-glass woven roving
Resin system	Polyester
Volume fraction of fibre	35.6% (0.85:1 R:G)

Table 4.17 Characteristic material properties: Data sheet No 7.

Property	*Units*	*Symbol*	*Characteristic value*
Tensile strength (long.)	N/mm^2	$\sigma_{x,t,k}$	270
Tensile strength (trans.)	N/mm^2	$\sigma_{y,t,k}$	270
Tensile modulus (long.)	kN/mm^2	$E_{x,t,k}$	14.5
Tensile modulus (trans.)	kN/mm^2	$E_{y,t,k}$	14.5
Compressive strength (long.)	N/mm^2	$\sigma_{x,c,k}$	
Compressive strength (trans.)	N/mm^2	$\sigma_{y,c,k}$	
Compressive modulus (long.)	kN/mm^2	$E_{x,c,k}$	
Compressive modulus (trans.)	kN/mm^2	$E_{y,c,k}$	
Shear strength (in-plane)	N/mm^2	$\tau_{xy,k}$	14.6
Shear modulus (in-plane)	kN/mm^2	$G_{xy,k}$	
Flexural strength (long.)	N/mm^2	$\sigma_{x,b,k}$	350
Flexural strength (trans.)	N/mm^2	$\sigma_{y,b,k}$	350
Flexural modulus (long.)	kN/mm^2	$E_{x,b,k}$	12.5
Flexural modulus (trans.)	kN/mm^2	$E_{y,b,k}$	12.5
Failure strain (long.)	%		
Failure strain (trans.)	%		
Poisson's ratio (long.)	-		
Poisson's ratio (trans.)	-		

Process	Resin transfer moulding
Reinforcement	E-glass 0°/90° stitched roving
Resin system	Polyester
Volume fraction of fibre	38% (typical)

Table 4.18 Characteristic material properties: Data sheet No. 8.

Property	*Units*	*Symbol*	*Characteristic value*
Tensile strength (long.)	N/mm^2	$\sigma_{x,t,k}$	260
Tensile strength (trans.)	N/mm^2	$\sigma_{y,t,k}$	260
Tensile modulus (long.)	kN/mm^2	$E_{x,t,k}$	18.9
Tensile modulus (trans.)	kN/mm^2	$E_{y,t,k}$	18.9
Compressive strength (long.)	N/mm^2	$\sigma_{x,c,k}$	
Compressive strength (trans.)	N/mm^2	$\sigma_{y,c,k}$	
Compressive modulus (long.)	kN/mm^2	$E_{x,c,k}$	
Compressive modulus (trans.)	kN/mm^2	$E_{y,c,k}$	
Shear strength (in-plane)	N/mm^2	$\tau_{xy,k}$	16.1
Shear modulus (in-plane)	kN/mm^2	$G_{xy,k}$	
Flexural strength (long.)	N/mm^2	$\sigma_{x,b,k}$	320
Flexural strength (trans.)	N/mm^2	$\sigma_{y,b,k}$	320
Flexural modulus (long.)	kN/mm^2	$E_{x,b,k}$	12.5
Flexural modulus (trans.)	kN/mm^2	$E_{y,b,k}$	12.5
Failure strain (long.)	%		
Failure strain (trans.)	%		
Poisson's ratio (long.)	-		
Poisson's ratio (trans.)	-		

Process	Resin transfer moulding
Reinforcement	E-glass 0°/90° ± 45° stitched roving
Resin system	Polyester
Volume fraction of fibre	59% (typical)

4.12 CREEP

4.12.1 Considerations

4.12.1.1 Quantification of creep

P(1) Creep is the time dependant deformation of a material (or structure) under a constant load.

P(2) The creep modulus can be defined from the additional strain exhibited by a material after a given elapsed time. The creep modulus is the apparent stiffness as determined by the total deformation to the time defined.

(3) For the purposes of design it is convenient to describe the process of creep in a material by an effective reduction in modulus with time. It is not the instantaneous modulus of the material that may be exhibited if the material was loaded after unloading following extensive creep.

4.12.1.2 Factors affecting magnitude of creep

P(1) Creep is essentially a matrix dominated phenomenon for which the following factors will determine the extent of the creep modulus at a given time:

(a) Materials
Resin type and degree of cure
Interfacial bonding between fibre and matrix
Volume fraction of fibres
Form of the reinforcement (woven, non-woven, mats)
Orientation of the fibres with respect to the applied load
Processing method

(b) Environment:
Temperature
Loading regime (particularly compression, flexure and all shear modes)
Aggressive chemicals
Moisture content

In general terms, the greater the degree of cross-linking in the resin, the higher the volume fraction of reinforcement, the stronger the bond between fibre and matrix and the greater the margin between the working environment and the onset of the resin temperature susceptibility, the smaller the creep and the higher the creep modulus after a given time.

P(2) The applied load will influence creep in that the higher the stress the greater the creep rate (increase in strain with time.)

(3) Resins with a high heat distortion temperature (HDT) or glass transition temperature (Tg) usually exhibit a greater creep modulus than that of resins with a low HDT or Tg.

(4) Moisture and/or other solvents are likely to increase the creep rate by softening the matrix and by degrading the interfacial bond strength.

4.12.1.3 Design method

P(1) The design process should involve substituting in calculations for stiffness a value for the creep modulus that is suitably reduced to allow for long term deformation.

P(2) Ideally the creep modulus used should be obtained by direct experiments on the composite system in question.

(3) Where experimental data are available for equivalent materials it is important to check that the material and environmental parameters listed above do not differ significantly between the test data and the service conditions.

(4) Creep in reinforced composites is primarily a feature of the matrix deformation.

(5) In aligned fibre composites loading along the fibre direction is unlikely to lead to significant creep problems, while loading off axis to fibre direction may result in excessive deformation.

P(6) If data are not available then the guideline data presented in Figure 4.13 may be used. In order to accommodate composites with varying volume fraction and fibre orientations the figure shows normalised creep modulus against time. It is assumed that initial modulus and therefore creep modulus are linearly proportional to fibre volume fraction. This is a reasonable approximation in most circumstances.

4.12.1.4 Fibre orientation

P(1) The data in Figure 4.13 are indicative of changes to elastic modulus for three basic composite forms: unidirectional composites loaded along the fibres (in tension) and in shear and in-plane random mat and orthotropic composites (e.g. woven laminates).

P(2) No information is available on creep behaviour in compression. If compressive creep data cannot be generated, then the shear creep curve should be used as a guideline. Compressive deformation in composites involves shear deformation.

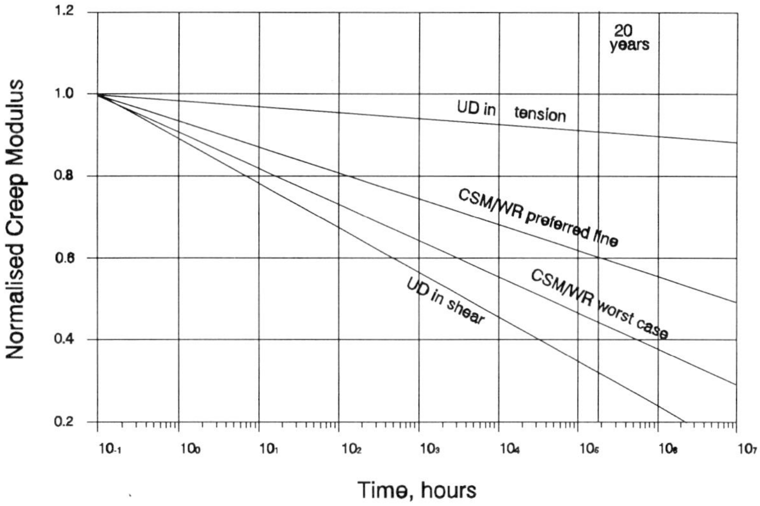

Figure 4.13 Indicative creep modulus vs time.
Note: These lines are derived from experimental data given in References 4.58 and 4.59 in the EUROCOMP Handbook. The normalised curves cover a stress range to 50% of ultimate and for times up to 10000 hours. Projections beyond 10000 hours assume no change in the mechanisms governing creep behaviour.

4.12.2 Accelerated testing

(1) Data may be generated to predict long term performance by testing at both elevated temperatures and high load levels compared to those normally allowed in the EUROCOMP Design Code. These methods are based upon the time - temperature superposition principle and applied stress - time superposition principle respectively. They are based upon the non-linear response of the creep compliance.

(2) The creep and relaxation test data can also be represented in the form of isochronous plots which are based upon a number of isochrones being projected from the time axes through a family of creep or stress relaxation curves. The resultant isochronous data is plotted on the stress - strain axes and represents a family of stress - strain curves for different times of loading.

(3) Extrapolation of creep test data to characterise the long term performance of the composite is best achieved by fitting a mathematical equation to the available short term test data. The creep of polymers can be modelled by a power law function identified by the linear nature of the data when plotted on a logarithmic axes.

P(4) The load level used to induce creep should not exceed 50% of the ultimate tensile strength. The temperatures used for testing should not exceed 20°C below the Tg of the matrix.

4.12.3 Rupture

4.12.3.1 General considerations

(1) In addition to modulus reductions as a consequence of creep, it is possible that strength reductions will occur. These may ultimately lead to a rupture failure.

P(2) The creep rupture behaviour of the composite selected should be checked against Figure 4.14 to ensure that the structure will not fail via a stress rupture mechanism during the required lifetime.

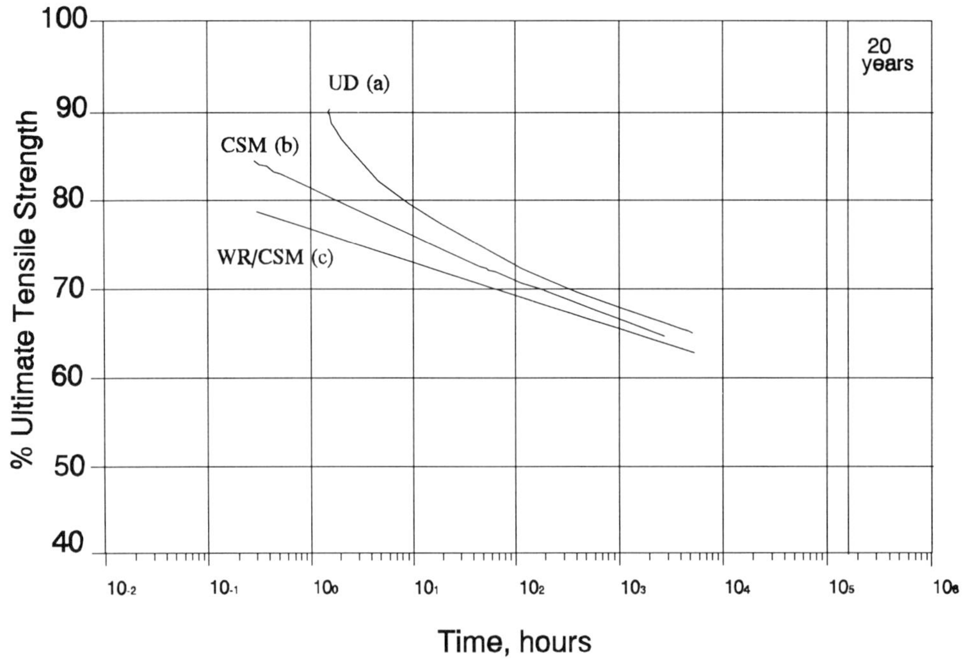

Figure 4.14 Tensile stress rupture of polyester composites in air.
Note: These curves are based on data from References 4.51, 4.60 and 4.61 in the EUROCOMP Handbook.

(3) The stress rupture time does not always represent a gradual decline in strength from the original strength to a value given by the failure load. At low loads which do not cause rupture it is not unusual to obtain residual ultimate strengths close to the original static strength where creep has been limited.

P(4) If the material is loaded in an environment that degrades the matrix or interface, the design must take into account the characteristics of the degraded material in estimating the creep rupture of the composite.

P(5) Testing to generate information on creep rupture time should always be performed in a representative environment.

4.12.3.2 Stress corrosion

(1) If the fibres are attacked by the environment, then a failure via stress corrosion may occur. This mode of failure is predominantly a phenomenon of failure in acidic environments.

P(2) Stress rupture under the appropriate environment should be considered to check whether there is a stress corrosion interaction.

(3) Creep rupture data for glass fibre composites in acid are less reliable than creep data as the size and shape of the specimen obviously influence time to failure. A direct read-over from coupon to structural data is not straight forward.

(4) A conservative design approach is to identify a maximum strain criterion of 0.2%.

(5) If data can be generated to support reduced partial factors then these can be adopted.

(6) Stress corrosion is usually found at stress raisers.

(7) Surface and barrier layers sometimes crack or, if thermoplastic, suffer from weld failures. These are the regions that initiate stress corrosion cracking.

(8) The use of ECR (or equivalent) acid resistant glass is recommended in acidic environments.

P(9) Brittle chemical resistant resins are not particularly good in resisting stress corrosion attack. Tougher resin systems provide greater resistance to stress corrosion cracking. The chemical and stress corrosion requirements of the matrix should always be considered together.

4.13 FATIGUE

4.13.0 Definitions and notation

N	Number of stress cycles
$N_{\sigma f}$) $N_{\tau f}$)	Number of cycles to failure for normal or shear stress range ($\gamma_{Ff}\ \gamma_{Mf}\ \Delta\sigma$, $\gamma_{Ff}\ \gamma_{Mf}\ \Delta\tau$ respectively) each acting alone (Fatigue life)
n	Number of stress cycles to which the FRP component being designed will be subjected during its design life
γ_{Ff}	Partial safety factor for fatigue actions
γ_{Mf}	Partial safety factor for fatigue strength
$\Delta\sigma$	Nominal normal stress range (= $\sigma_{max} - \sigma_{min}$)
$\Delta\tau$	Nominal shear stress range (= $\tau_{max} - \tau_{min}$)
$\Delta\sigma_R$	Normal fatigue strength
$\Delta\tau_R$	Shear fatigue strength

σ_{max}, τ_{max}	)	Maximum, mean and minimum
σ_m, τ_m	)	values of the fluctuating normal
σ_{min}, τ_{min}	)	and shear stresses in a stress cycle (tension positive)
σ_R		Characteristic normal strength
τ_R		Characteristic shear strength

4.13.1 Fundamental requirements

P(1) A structure shall be designed and constructed in such a way that, with an acceptable level of probability, it is unlikely to fail as a result of fatigue loading or to require repair of damage caused by fatigue.

P(2) The fundamental requirements of design for fatigue shall be as specified:

- in Building Regulations or Building Code requirements
- by the client, or by the designer in consultation with the client, provided that minimum provisions, as specified in the relevant regulations and codes or by the competent authority, are observed.

P(3) All structures made of FRP composites shall be considered for the effects of fatigue loading, in particular the following:

- members supporting lifting appliances or rolling loads
- members subject to repeated stress cycles from vibrating machinery
- members subject to wind-induced oscillations or stress reversals due to wind loading
- members subject to crowd-induced oscillations.

P(4) In determining the fundamental requirements, account shall be taken of the following:

- the use to which the structure, of which the component being designed forms part, will be put
- the consequences of failure, due to fatigue, of the components being designed
- the ease of access for inspection or repair and likely frequency of inspection and maintenance of the components.

P(5) The loadings used for fatigue design shall be those given in the Eurocode for Actions for fatigue loading, unless noted otherwise.

4.13.2 Performance requirements

P(1) The following factors shall be taken into account in the fatigue design of components made of FRP composites:

- fatigue failure criteria
- fatigue life required of the structure or member
- the maximum, mean and minimum values of the fluctuating normal and shear stresses in a stress cycle, due account being taken of stress concentrations arising from geometric effects and construction methods and of dynamic amplification factors.

P(2) One of the following limit states shall be considered:

- serviceability: fibre debonding or resin cracking
- ultimate limit state: failure or excessive local deformation of the component.

P(3) The design shall be checked at the serviceability limit state in aggressive environmental conditions, where the structure is to be used for the retention of fluids, or where otherwise agreed between the client and the designer. Neither transverse fibre debonding nor resin cracking shall occur for the number of stress cycles, n, to which the component will be subjected during its design life. Provided this requirement is complied with, it will not be necessary to check the design for fatigue at the ultimate limit state.

P(4) Where none of the above conditions applies, the design for fatigue shall be checked at the ultimate limit state. For n stress cycles the following criteria shall be met at all points in the FRP composite components subject to fatigue loading:

- for normal stress $\gamma_{Ff}\,\Delta\sigma < \Delta\sigma_R/\gamma_{mf}$ (4.58a)
- for shear stress $\gamma_{Ff}\,\Delta\tau < \Delta\tau_R/\gamma_{mf}$ (4.58b)

in which:

$\Delta\sigma$, $\Delta\sigma_R$, $\Delta\tau$, $\Delta\tau_R$ and n are as defined in 4.13.0.

γ_{Ff} = 1.0 unless otherwise determined or agreed, and

$\gamma_{mf} = \gamma_{m,1}\,\gamma_{m,2}\,\gamma_{m,3}\,\gamma_{m,4}$

where:

$\gamma_{m,1}$, $\gamma_{m,2}$ and $\gamma_{m,3}$ are as defined in Table 2.3 and $\gamma_{m,4}$ in Table 4.19.

- for combined stresses:

  $D_d < 1$

Where D_d is the damage indicator for fatigue (see 2.3.2.1 P(5))

$$= (1/N_{\sigma x,f} + 1/N_{\tau xy,f} + 1/N_{\tau xz,f})\ n \qquad (4.59)$$

(similarly for the y- and z-planes)

and $N_{\sigma f}$, $N_{\tau f}$ and n are as defined in 4.13.0.

P(5) The designer shall ensure by appropriate detailing that, wherever possible, the theoretical point of failure of FRP components subject to fatigue loading would occur away from joints, connections, changes of section and areas of stress concentration.

Table 4.19 Partial safety coefficient, $\gamma_{m,4}$, for fatigue strength.

Inspection and access	*"Fail-safe" components (see notes below)*	*Non "fail-safe" components (see notes below)*
Component subject to periodic inspection and maintenance. Detail accessible.	1.5	2.0
Component subject to periodic inspection and maintenance. Poor accessibility.	2.0	2.5
Component not subject to periodic inspection and maintenance.	2.5	3.0

Notes: "Fail-safe" structural components are such that local failure of one component does not result in failure of the structure or large sections of the structure.

Non "fail-safe" structural components are such that local failure of one component could lead to failure of the structure or large sections of the structure.

4.13.3 Design methods

(1) Fatigue strengths should, wherever possible, be determined on the basis of tests on samples of the FRP laminate made with similar resin and glass fibre reinforcement of similar type and proportions. The conditions of the test should be appropriate to the loading conditions of the component being designed, and to the environmental conditions.

(2) For checking of design for fatigue at the serviceability limit state, in the absence of suitable test data, Table 4.20 may be used to estimate the limiting normal strain range for transverse fibre debonding not to occur in laminates made with CSM, WR and UD glass reinforcement.

Table 4.20 Normal strain range and number of stress cycles to transverse fibre debonding.

Normal strain range %	*Maximum strain* ± %	*Number of cycles* *N*
0.30	0.20	10^3
0.23	0.15	10^4
0.17	0.13	10^5
0.10	0.07	10^6
0.03	0.02	10^7

(3) For checking of design for fatigue at the ultimate limit state, in the absence of suitable test data, the normal fatigue strength and shear fatigue strength, as a proportion of the corresponding characteristic normal strength and corresponding characteristic shear strength, respectively, of laminates made with CSM, WR and UD glass reinforcement may be estimated as follows:

$$\Delta\sigma_R/\sigma_R = (1.5 - 0.3\log_{10}N)\,(1 - \sigma_m/\sigma_R)/(1 + \sigma_m/\sigma_R) \quad (4.60)$$

$$\Delta\tau_R/\tau_R = (1.5 - 0.3\log_{10}N)\,(1 - \tau_m/\tau_R)/(1 + \tau_m/\tau_R) \quad (4.61)$$

In which $\Delta\sigma_R$, $\Delta\tau_R$, σ_R, τ_R, N, σ_m and τ_m are as defined in 4.13.0,

provided that:

- the FRP composites subject to fatigue loading are used in normal atmospheric conditions or otherwise are protected from deleterious environmental conditions
- the environmental temperature does not exceed 50°C
- the frequency of loading does not exceed 10 Hz.

(4) When designing for fatigue, the designer should pay particular attention to areas where stress concentrations are likely to be present. These may occur at connections, re-entrant corners and points of acute change of direction. The resulting stresses may be far greater than those present in the adjacent plane section and thus significantly reduce the member's resistance to fatigue.

4.14 DESIGN FOR IMPACT

4.14.1 Fundamental requirements

P(1) The fundamental requirements shall be as specified

- in the appropriate regulations
- by the client, or by the designer in consultation with the client, provided that minimum provision, as specified in the relevant regulations or by the competent authority, are observed.

P(2) In determining the fundamental requirements, account shall be taken of the following:

- that, in general, FRP composites exhibit little or no ductile behaviour beyond the point of linear stress - strain behaviour of the material, failure occurring soon after this point has been reached
- the intended use of the structure, of which the components being designed form part
- whether the design impact occurs as a result of a single major event or repeated minor events over a period of time, or some combination of these
- whether the energy to be absorbed from the impact results from a light projectile of high velocity or a heavy projectile of low velocity
- the consequences of failure, perforation, excessive deformation, loss of strength or loss of stiffness of components, including fixings and means of attachment to other parts of the structure, as a result of impact loading
- the effects of possible delamination in the plane of the laminate perpendicular to the direction of impact on the axial buckling strength in the plane of such possible delamination
- the environmental conditions to which the FRP components will be subjected
- the effects of impact damage on components made of FRP composites or on parts of the structure not directly affected by the impact.

(3) Other factors that should be considered include the acceptable levels of damage, having regard to repair or replacement of damaged components and the continuing use of the facility before and while remedial works are carried out.

4.14.2 Performance criteria

P(1) Buildings, structures and components made of FRP composites subject to impact loading shall be designed to withstand a specific design load event selected to simulate possible realistic in-service conditions.

(2) In such an event, limited damage may be allowed to occur, to an extent, as agreed with the client.

P(3) Failure criteria shall be agreed between the designer and client and selected as one or a combination of the following:

Ultimate limit state

- collapse of the element
- structural integrity of the whole structure, i.e. progressive collapse
- loss of strength due to permanent damage, e.g. delamination
- perforation of the element, i.e. full penetration.

Serviceability limit state

- deformation limits for direct load, flexure and shear
- strain limits
- subsequent leakage rates following impact
- penetration limit, (depth of embedment)
- spalling, ejection of material from the front face of the target/element
- scabbing, ejection of material from the back face of the target/element.

(4) Limits on serviceability limit state design criteria may be selected with regard to the probability of the design impact load occurring.

P(5) Impact loads shall be considered in conjunction with dead and imposed loads but need not be considered in conjunction with wind and seismic loads.

(6) When design for impact is being carried out, the actions may be taken as those appropriate to accidental design situations for which γ_{GA}, ψ_1, ψ_2 = 1.0 (see 2.3.2.2).

γ_m may be taken as 1.20.

P(7) Following an impact loading event, the structure shall retain sufficient strength properties to be functional and to allow the facility to be evacuated and/or shut down as appropriate in a safe and controlled manner.

P(8) The design of FRP composites subject to impact loading shall take account of any reduction of impact strength over time due to weathering caused by the prevailing environmental conditions.

4.14.3 Design methods

P(1) Since impact strength data from standard tests cannot be used directly to predict performance of components, impact design shall be verified by product tests.

P(2) Product impact tests shall be designed to simulate as closely as possible the actual in service impact conditions.

P(3) Factors to be considered in product impact tests shall include:

- projectile size, mass and velocity
- projectile "hardness" (relative hardness between target and projectile)
- temperature
- target geometry
- stress concentrations (notches, holes, etc.).

4.15 DESIGN FOR EXPLOSION/ BLAST

4.15.1 Fundamental requirements

P(1) The fundamental requirements shall be as specified

- in the appropriate regulations
- by the client, or by the designer in consultation with the client, provided that minimum provision, as specified in the relevant regulations or by the competent authority, are observed.

P(2) In determining the fundamental requirements, account shall be taken of the following:

- that, in general, FRP composites exhibit little or no ductile behaviour beyond the elastic limit of the material, failure occurring soon after the elastic limit has been reached
- the intended use of the structure of which the components being designed form part
- the consequences of failure, perforation, excessive deformation, loss of strength or loss of stiffness of components, including fixings and means of attachment to other parts of the structure, as a result of explosion/blast
- the effects of explosion/blast on components made of FRP composites or on parts of the structure not directly affected by the explosion/blast.

P(3) Other factors that should be considered include the acceptable levels of damage, having regard to repair or replacement of damaged components and the continuing use of the facility while the remedial works are being carried out.

4.15.2 Performance criteria

P(1) Buildings, structures and components made of FRP composites, subject to explosion/blast, shall be designed to withstand the effects of an explosion/blast which generates a specific over-pressure for a particular length of time. In such an event, damage may be allowed to occur, to an extent agreed by the client and the designer, but the structure shall be sufficiently preserved to be functional after the explosion/blast and allow the facility to be evacuated and/or shut down as appropriate in a safe and controlled manner.

P(2) Blast loads shall be considered in conjunction with dead and imposed loads but need not be considered in conjunction with wind or seismic loads.

(3) The static load equivalent of blast pressures and duration may be used for design provided that account is taken of the dynamic response of the structure and such factors as stiffness, ductile or lack of ductile behaviour, energy absorption, etc.

P(4) When design for explosion/blast is being carried out, the actions may be taken as those appropriate to accidental design situations for which γ_{GA}, ψ_1, ψ_2, = 1.0 (see 2.3.2.2).

γ_m may be taken as 1.20.

4.15.3 Design methods

(1) The design should ensure that whatever assumptions are made about the dynamic response of the structure or components to blast loading, they can be realised in the structure or components as built.

This is particularly the case with a quasi-static approach to design in which the blast loading is attenuated by the presumed dynamic response. Changes in the design at the detail stage or changes in the construction itself that alter the stiffness of the structure will alter not only the distribution of loads through the structure but the nature of the loads themselves. Thus, where FRP composite cladding panels are used on a blast resistant structure, care should be taken in detailing the fixings, particularly along the face parallel to the direction of blast, to ensure that the panels 'rack' freely (i.e. do no lock), thereby inhibiting any attenuation of the blast pressures assumed in the flexible design of the structure.

(2) Under the effect of blast, FRP structures and components can undergo appreciable elastic distortion without suffering major permanent damage. However, should the elastic limit be reached, the lack of an appreciable ductile range in the stress - strain curve for typical FRP laminates is likely to cause failure by direct rupture or tearing, particularly at locations of stress concentration or where free movement of the laminate is prevented, such as at joints or points of connection. In structures which rely on flexible dynamic response to attenuate the blast loading, joints should be

designed:

- to limit stresses and/or strains at these positions and allow attenuation of blast loading by deformation of the more flexible areas
- to allow deformation to occur by movement at mechanical joints with or without energy absorption
- to allow for quasi-ductile action by designing for controlled failure of local components and shedding of load of local components without causing progressive collapse.

(3) In designing for explosion/blast, attention should be given to second order effects, as the ultimate limit state for stability due to excessive deflection or loss of stiffness due to structural damage may be grossly exceeded in the event of explosion/blast.

Particularly in skeletal structures, all compression members or elements should be provided with stiffeners and/or bracings to prevent local or general buckling before the design stresses are developed, unless such design stresses are reduced to take account of such local or general buckling.

4.16 FIRE DESIGN

4.16.1 Fundamental requirements

P(1) The fundamental requirements of fire design shall be as specified:

- in Building Regulations, or Building Code requirements
- by the client, or by the designer in consultation with the client, provided that minimum provisions, as specified in the relevant regulations and codes or by the competent authority, are observed.

P(2) In determining the fundamental requirements, account shall be taken of the following:

- that the resins of which FRP composites are made are organic compounds; the inherent flammability of some resins shall be taken into account
- the intemded use of the structure of which the components being designed form part
- the direct effects of fire on such components, e.g. flammability, heat generation, smoke emission, toxic and noxious fumes
- the consequences of failure, perforation, loss of strength or loss of stiffness of components, including fixings and means of attachment to other parts of the structure, as a result of fire damage
- the effects of fire damage to components made of FRP composites on parts of the structure not directly affected by fire.

(3) Other factors that should be considered include the acceptable levels of damage, having regard to repair or replacement of fire-damaged components and the continuing use of the facility while the remedial works are being carried out.

4.16.2 Performance criteria

P(1) The following factors shall be taken account of in the fire design of components made of FRP composites:

- fire resistance
- ease of ignition
- surface spread-of-flame
- fire propagation
- fire penetration
- loss of strength due to increase in temperature.

P(2) Other factors, related to safety rather than structural performance, that also shall be taken account of include:

- smoke emission
- emission of toxic and noxious fumes.

P(3) The designer shall exercise caution when seeking to demonstrate compliance with fire performance criteria by means of results of small scale tests carried out in a laboratory on samples of the structural material, as these may not adequately reflect the performance of the materials, products or systems of the structural configuration under actual fire conditions.

(4) Structural fire engineering methods may be used to determine fire resistance of structural components made of FRP composites, subject to the approval of the appropriate authority, provided the fire load (that is the quantity of materials available to burn) is low and the possible sources of combustion can be demonstrated to be sufficiently far away from the FRP-composite components not to cause failure within the period specified for fire resistance.

(5) When fire engineering methods are used to determine fire resistance, the actions may be those appropriate to accidental design situations for which

γ_{GA}, ψ_1, ψ_2 = 1 (see 2.3.2.2).

4.16.3 Design methods

(1) In designing components made of FRP composites for fire conditions, active or passive methods may be employed or a combination of these.

Active methods include:

- fire detection and alarm systems
- fire suppression systems
- provision of adequate means of escape, compartmentalisation of buildings, fire stops and fire doors
- restrictions on the use of combustible or flammable materials.

Passive methods include:

- protection of structural members by non-combustible or low fire-hazard materials or constituents or additives
- use of surface coatings
- intumescent surface coatings.

(2) Active methods of fire design are principally involved with fire safety and are applicable to all materials, not only FRP composites.

(3) Passive protection systems may be of the composite type or the fire barrier type. In the composite type, low fire-hazard resins and additives are incorporated in the gelcoat, or gelcoat and lay-up resins used in the FRP composite. With the fire barrier type, fire retardant coatings or casings are used to delay the time necessary to reach the ignition temperature or that at which the mechanical properties of the composite are degraded, and to reduce the spread of flame. They may be heat resistant, flame retardant or insulative or a combination of these.

P(4) Systems and methods adopted in fire design for fire penetration shall provide adequate fire barriers around pipes, services and other perforations through the FRP composite components.

(5) Where a structure is designed to have joints that transfer forces from one member to another, special account shall be taken of the behaviour of such joints. An assessment should be made of the condition of the composite after the specified period of fire resistance, or fire exposure, with particular attention to the effects of any metal connectors. In structurally indeterminate systems, fire damage may alter the relative stiffness of various parts of the structure and result in a redistribution of forces. Account should be taken of the complete or partial yielding of the joints, as this may change the structural action.

P(6) Where any part of a metal fastening may become exposed to heating during a fire, rapid heat conduction will lead to localised fire damage and loss of anchorage. Where this effect is likely to lead to the structural member that is required to have fire resistance, protection of the fastener should be provided by one of the following methods:

- ensuring that every part of the fastener is fully encased in the composite within the section not subject to fire damage
- covering any exposed part of the fastener with a suitable protective material and ensuring that such protective material will remain in position for the required period of fire exposure.

(7) Special account should be taken of the effects of thermal expansion from fire or heat affected areas and of the risk of permanent damage to parts of the structural fabric not directly affected by the fire.

P(8) In developing methods and systems for fire design of FRP composites, the designer shall consider the effects on other design properties of the materials and of the structure. Such design properties shall include:

- mechanical properties of the composites
- durability
- appearance.

4.17 CHEMICAL ATTACK

4.17.1 General considerations

P(1) Chemical attack is the action of a liquid (e.g. solvent) on the structure.

(2) Glass fibres are only moderately attacked by water, and are not attacked by organic solvents. Alkalis can degrade glass fibres, and acids can result in spontaneous cracking of a fibre.

(3) The action of a stress in combination with an environment that is aggressive to the fibres can result in stress corrosion attack (See 4.12.3.2).

(4) In the absence of stress, and where the environment does not attack the fibres, the primary chemical effects involve degradation of the matrix and the interface.

P(5) All solvents are to a greater or lesser extent absorbed by the matrix.

(6) This leads to swelling and can result in an initial small stiffening of the matrix, followed by a general softening and plasticisation. The extent of swelling is dependent on the chemical compatibility between matrix and solvent.

P(7) The matrix resin should be selected on the basis of the compatibility of that system with the expected service environments.

(8) The criteria for selection of a resin are dependant on the service use. Resin suppliers frequently rate the performance of the resins on a qualitative scale. This is inadequate to assess structural, cosmetic and functional performance separately.

P(9) The information supplied by resin suppliers should be used as an initial guide in selection. If the consequences of failure (as defined by the service application) are great, then test data should always be provided to verify resistance to the environment.

4.17.2 Acids

P(1) Dilute mineral acids are not generally aggressive to most resins. Concentrated sulphuric acid and nitric acids can result in oxidation of the matrix.

4.17.3 Alkalis

P(1) Alkalis can cause severe softening of a matrix. Orthophthalic polyester should not be used with concentrated alkalis.

4.17.4 Organic solvents

P(1) Bisphenol-A based polyester resins are generally regarded as very inert in acids and alkalis, but they do suffer from attack in organic solvents such as toluene. Isophthalic polyesters may provide better performance in such environments.

4.18 DESIGN CHECK-LIST

A check-list for the aspects to be considered in design is given in Figure 4.15.

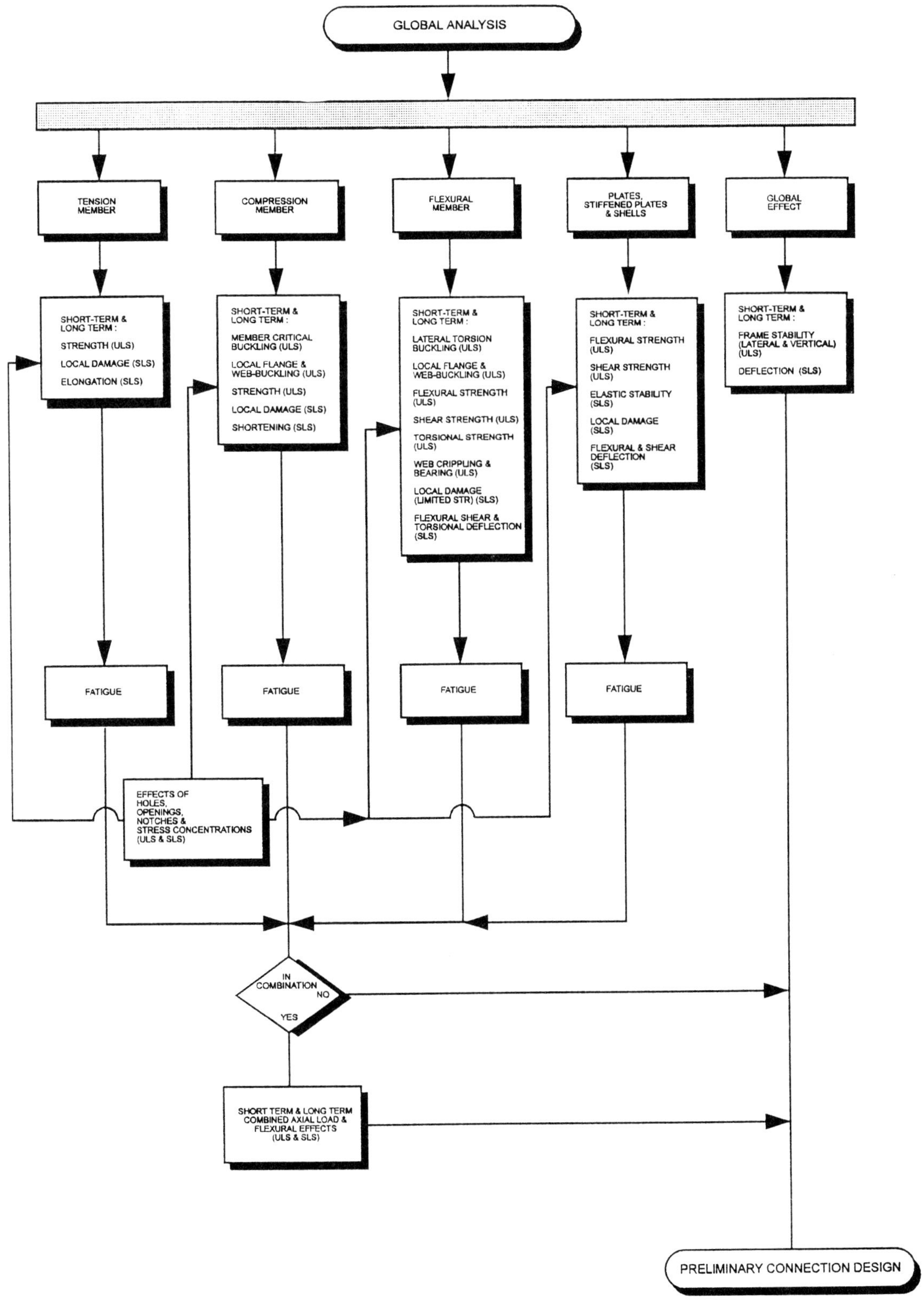

Figure 4.15 Check-list for section/member design in FRP composites.

CONTENTS

5 CONNECTION DESIGN

5 CONNECTION DESIGN

5.1 GENERAL

5.1.1 Definitions

P(1) The definitions specific to connection design in this EUROCOMP Design Code, in addition to those given in 1.4, are listed below.

Fastener connection - A mechanical joint held together by fasteners that are additional to the components being joined. The fastener may be used directly in the transfer of load between the components to be joined (e.g. as in a bolted, riveted or screwed connection) or indirectly (e.g. as in a friction gripped, clamped or strapped connection).

In-plane shear loading - Adherend shear loads that produce shear stresses to the bondline in lap and strap joints.

Joining technique - A classification used within each of the joint categories that characterizes the joint by stating either the method of the load transfer between the joint members or the technique for producing the joint.

Joint category - Classification used on the highest hierarchic level of the connection design to indicate the load transfer mechanism from one member to another.

Joint efficiency - Ratio of joint strength to adherend strength.

Mechanical joint - A joint where two components, one or both of which is/are made of FRP, are held together solely by mechanical means, which may or may not involve mechanical elements, i.e., fasteners, in addition to the components being joined. The joint may be sealed with a sealant bonded to the joint surfaces, but such a sealant is not used for load transfer across the joint.

Non-structural connection - A joint, a failure of which would not endanger life or cause structural or economic damage.

Primary structural connection - A joint that is expected to provide major strength and stiffness to an assembly for the whole service life of the structure. Failure of such a joint would constitute major structural damage and hazard to life.

Secondary structural connection - A joint that contributes some strength and stiffness to an assembly. Any failure due to service loads would not endanger life or cause major structural damage.

Tensile shear loading - Adherend tensile loads that produce shear stresses to the bondline in lap and strap joints.

(2) The terminology and the hierarchic classification system used are illustrated in Figure 5.1.

CONNECTION DESIGN			
Mechanical joints	Bonded joints	Combined joints	JOINT CATEGORIES
Bolted joints (shear loaded) Bolted joints (axially loaded) Riveted joints (shear loaded) Riveted joints (axially loaded) Clamped joints Contact joints (keyed, hooked) Embedded fasteners	Adhesively bonded joints Laminated joints Moulded joints Cast-in joints Bonded insert joints	Bonded-Bolted joints Bonded-Riveted joints	JOINING TECHNIQUES
Lap joints Strap joints Tee joints Angle joints Others	Lap joints Strap joints Scarf joints Butt joints Tee joints Angle joints Others	Lap joints Strap joints Tee joints Others	JOINT CONFIGURATIONS

Figure 5.1 The hierarchic structure for connection design and the terminology used.

5.1.2 Scope

P(1) This chapter covers the design of mechanical, bonded and combined connections, where at least one of the components to be joined is made of FRP.

5.1.3 Joint classification

P(1) Connections are classified into:

- primary structural connections
- secondary structural connections
- non-structural connections.

P(2) A primary structural connection is expected to provide major strength and stiffness to an assembly for the whole service life of the structure. The failure of such a connection has a substantial effect on the performance of the whole structure. This connection category carries the highest requirement for strength, rigidity and durability.

(3) Stressed-skin panels of floors, walls, roofs, box beams and trusses are examples of primary structural applications where the connection is expected to carry static and dynamic design loads without failure.

P(4) A secondary structural connection contributes some strength and stiffness to an assembly, but any failure due to service loading would not endanger life or cause major structural damage. The failure of such a connection would only cause damage to the structural component concerned. Such a failure would be recognizable and easily repaired.

(5) An example of this category is the field-assembly of two modular units. A joint between the units carries loads, but in the event of failure, the modules themselves could withstand loadings without the connection.

P(6) The least critical of all classifications is a non-structural connection, for example joints of decorative panels, which even in the case of failure cause no danger to people or damage to the property.

5.1.4 Joint categories and joining techniques

P(1) Joint categories included in this code are:

- mechanical connections
- bonded connections
- combined connections.

P(2) The following joining techniques are included within each joint category:

- mechanical connections: bolted and riveted joints (shear loaded fasteners), bolted and riveted joints (axially loaded fasteners), clamped joints, threaded joints, contact joints, strap joints, and embedded fasteners
- bonded connections: adhesively bonded joints, laminated joints, moulded joints, bonded insert joints, and cast-in joints.
- combined connections: bonded-bolted joints and bonded-riveted joints.

5.1.5 Joint configurations

(1) Typical joint configurations and loadings within each joining technique are illustrated in Figure 5.2.

5.1.6 Applied forces and moments

P(1) The forces and moments applied to connections at the ultimate limit state shall be determined by global analysis conforming with 2.5.

P(2) These applied forces and moments shall also include:

- second-order effects in structural components; see 2.5.1.4
- the effects of imperfections, see 2.5.1.3.

P(3) The flexibility of joints shall be considered and taken into account in the global analysis when appropriate.

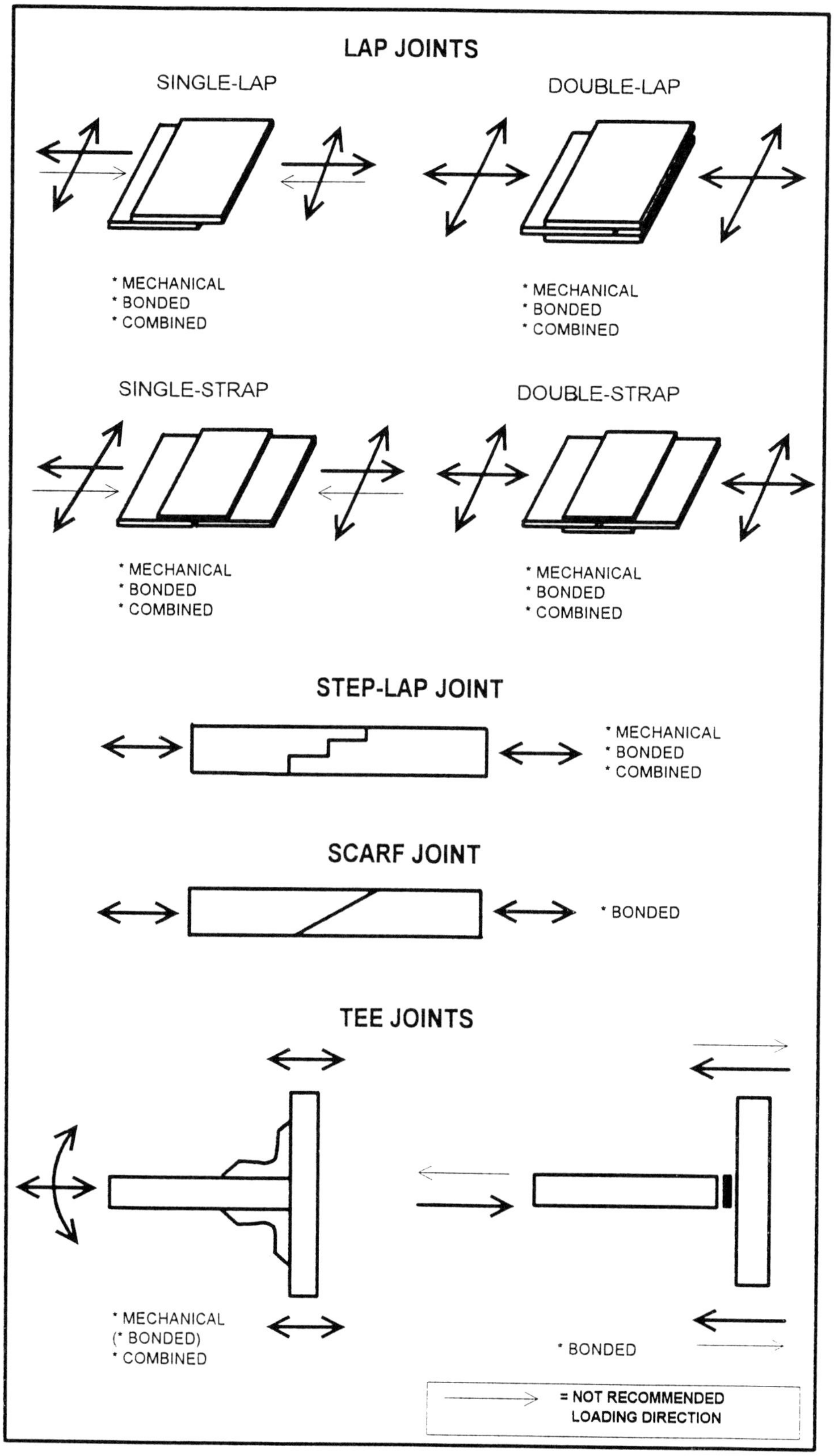

Figure 5.2(a) Typical joint configurations.

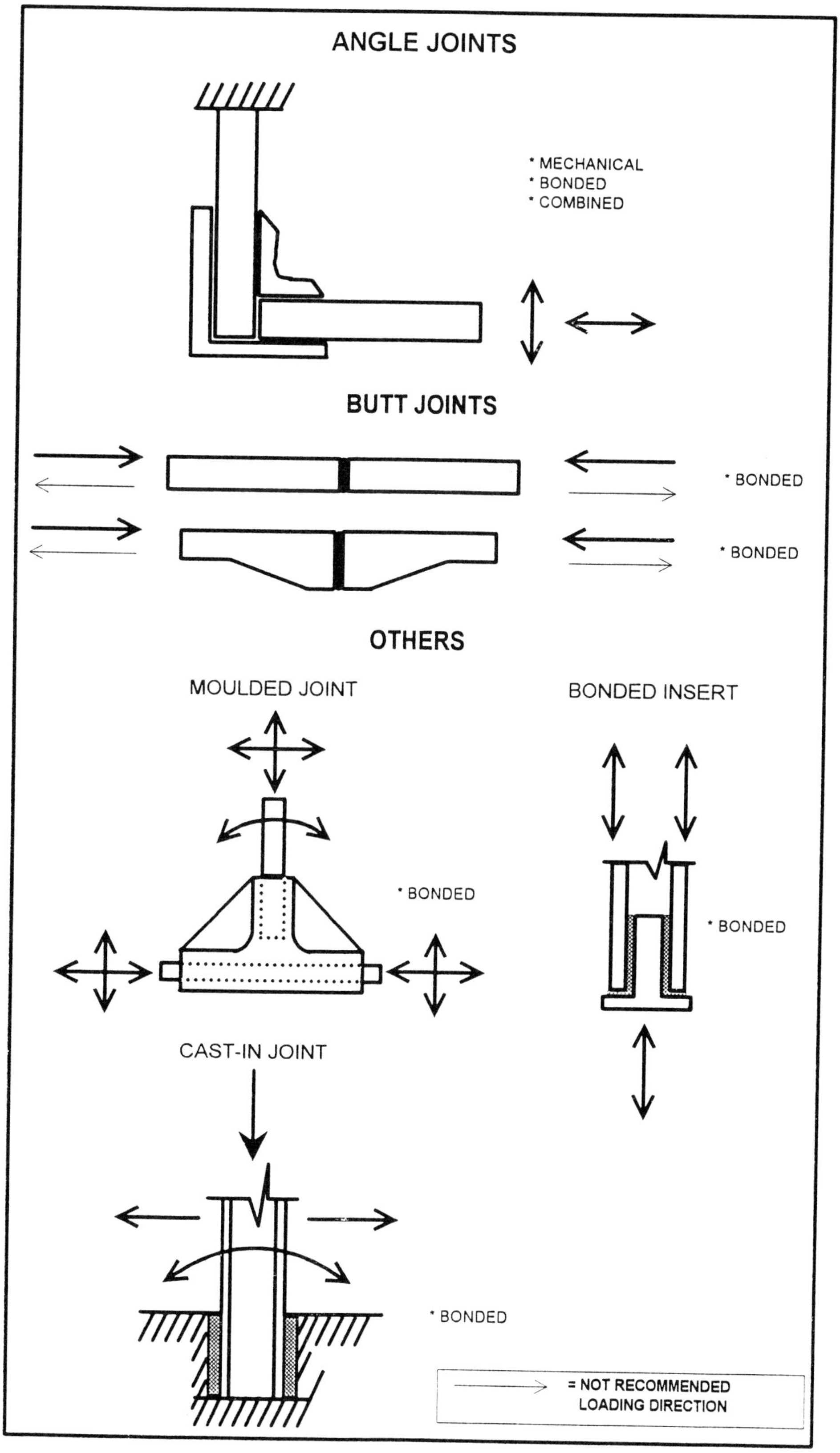

Figure 5.2(b) Typical joint configurations.

(4) The flexibility of joints is affected, for example, by the following factors:

- component (plate) stiffness
- joint configuration.

(5) In addition the flexibility of bolted joints is affected by:

- flexibility of the fasteners
- bolt slip
- bearing of fastener holes.

(6) Similarly the flexibility of bonded joints is affected by:

- flexibility of the adhesive
- bondline thickness
- local stiffeners in the joint area.

5.1.7 Resistance of connections

P(1) All connections shall have a design resistance such that the structure remains effective and is capable of satisfying all the basic design requirements given in Chapter 2.

P(2) Linear-elastic analysis shall generally be used in the design of the connection. Non-linear analysis of the connection may be employed provided that it takes into account the load-deformation characteristics of all components of the connection.

5.1.8 Design approach

(1) In this EUROCOMP Design Code four alternative design approaches are presented for joint design. These are:

1. simplified procedure
2. rigorous procedure
3. design by testing
4. numerical analysis.

(2) Approaches 1 and 2 can only be applied to a limited number of joint configurations.

(3) In the design by testing approach the design is based on testing connections representative of the design case (see 7.3.3.). This approach can be applied to all connections.

(4) Numerical methods (mainly finite element analysis) may be applied in principle in most cases. The practicality of using these methods should be considered separately for each design case. The subject of finite element analysis is not separately addressed for each joint configuration but it is discussed in general terms for each joint category.

(5) For the use of finite element analysis in the design of bolted or riveted connections in shear see 5.2.

(6) For the use of finite element analysis in the design of bonded joints, see 5.3.2.2.

5.1.9 Design requirements

P(1) Connections may be designed by distributing the internal forces and moments in whatever rational way is best, provided that:

(a) the assumed internal forces and moments are in equilibrium with the applied forces and moments,
(b) each element in the connection is capable of resisting the forces or stresses determined according to the procedures stated in this code, and
(c) the deformations implied by this distribution are within the deformation capacity of the mechanical fastener or the adhesive and of the connected parts.

P(2) In addition, the assumed distribution of internal forces shall be realistic with regard to the relative stiffnesses of the joint members and/or the adhesive/fasteners.

P(3) The internal forces seek to follow the path with the greatest rigidity. This path shall be clearly identified and consequently followed throughout the connection design.

P(4) Connections subject to fatigue shall also satisfy the requirements given in 4.13.

5.1.10 Partial safety factors

P(1) Partial safety factors for actions shall be determined as given in Chapter 2.

P(2) Partial safety factors for the material properties of bonded joints shall be determined as given in 2.3.3.2.

P(3) The partial safety factors for mechanical connections shall be as follows:

- for FRP components forming the joint, γ_m shall be as determined from 2.3.3.2
- for metal and other components bolts, rivets, inserts and the like, γ_m shall be in accordance with the corresponding Eurocode or national standard for the material.

P(4) The determination of the partial safety factors for the adhesives is based on the principles given in 2.3.3.2. For the adhesives, γ_m is given by the expression:

$$\gamma_m = \gamma_{m,1}\,\gamma_{m,2}\,\gamma_{m,3}\,\gamma_{m,4} \tag{5.1}$$

where $\gamma_{m,1}$ to $\gamma_{m,4}$ are:

Source of the adhesive properties	$\gamma_{m,1}$
Typical or textbook values (for appropriate adherends)	1.5
Values obtained by testing	1.25

Method of adhesive application	$\gamma_{m,2}$
Manual application, no adhesive thickness control	1.5
Manual application, adhesive thickness controlled	1.25
Established application procedure with repeatable and controlled process parameters	1.0

Type of loading	$\gamma_{m,3}$
Long-term loading	1.5
Short-term loading	1.0

Environmental conditions	$\gamma_{m,4}$
Service conditions outside the adhesive test conditions	2.0
Adhesive properties determined for the service conditions	1.0

P(5) For brittle adhesives (shear strain to failure < 3%), γ_m shall not be taken as less than 1.5.

P(6) For an adhesively bonded connection and a bonded insert joint designed by testing, the following partial material safety factors shall be used:

$\gamma_{m,1}$ = 1.25

$\gamma_{m,2}$ and $\gamma_{m,3}$ shall be taken as defined in P(4) above

$\gamma_{m,4}$ shall be taken as 1.0 if the joint tests have been performed in conditions representative of service conditions and as 2.0 if the test conditions are outside the service conditions.

P(7) For laminated and moulded joints designed by testing, the following partial material safety factors shall be used:

$\gamma_{m,1}$ = 1.25

$\gamma_{m,2}$ and $\gamma_{m,3}$ shall be taken as defined in 2.3.3.2

and

$\gamma_{m,4}$ = 1.0.

P(8) For cast-in joints designed by testing, the following partial material safety factors shall be used:

$\gamma_{m,1}$ = 1.25

$\gamma_{m,2}$ = 2.5

and

$\gamma_{m,3}$ and $\gamma_{m,4}$ shall be taken as defined in P(4) above.

P(9) For connections designed by testing, γ_m shall not be taken as less than 2.0.

5.1.11 Selection of joint category

(1) Selection of joint category usually is determined by:

- loads that need to be transferred, or by the required joint efficiency as a fraction of the strength of the weaker part to be joined
- geometry of the members to be joined
- suitability of the fabrication, considering the component dimensions, manufacturing environment and number of components in a production run
- service environment and the lifetime of the structure
- requirements set for the reliability of joint
- need for disassembly
- need for fluid and weather tightness
- aesthetics
- cost.

(2) Comparison between typical properties of different joint categories is given in Table 5.1. Some of the most commonly considered advantages and disadvantages of the different joint categories are presented in Table 5.2.

(3) The flow chart for the connection design is illustrated in Figure 5.3.

Table 5.1 Characteristics of different joint categories.

	Mechanical	*Bonded*	*Combined*
Stress concentration at joint	high	medium	medium
Strength/weight ratio	low	medium	medium
Seal (water tightness)	no	yes	yes
Thermal insulation	no	yes	no
Electrical insulation	no	yes	no
Aesthetics (smooth joints)	bad	good	bad
Fatigue endurance	bad	good	good
Sensitive to peel loading	no	yes	no
Disassembly	possible	impossible	impossible
Inspection	easy	difficult	difficult
Heat or pressure required	no	yes/no[1]	yes/no[1]
Tooling costs	low	high	low
Time to develop full strength	immediate	long	long

[1] no if cold curing two-part adhesives are used in an appropriate environment (see Chapter 6).

Table 5.2 Typical features of different connections between FRP members.

Advantages	Disadvantages
Mechanical connections	
• Requires no special surface preparation • Can be disassembled • Ease of inspection	• Low strength to stress concentrations • Special practices required in assembly; results in time consuming assembly • Fluid and weather tightness normally requires special gaskets or sealants • Corrosion of metallic fasteners
Bonded connections	
• High joint strength can be achieved • Low part count • Fluid and weather tightness • Potential corrosion problems are minimized • Smooth external surfaces	• Cannot be disassembled • Requires special surface preparation • Difficulty of inspection • Temperature and high humidity can affect joint strength
Combined connections	
• Bolts provide support and pressure during assembly and curing • Growth of bondline defects is hindered by bolts	• Structurally bolts act as backup elements - in an intact joint, bolts carry no load

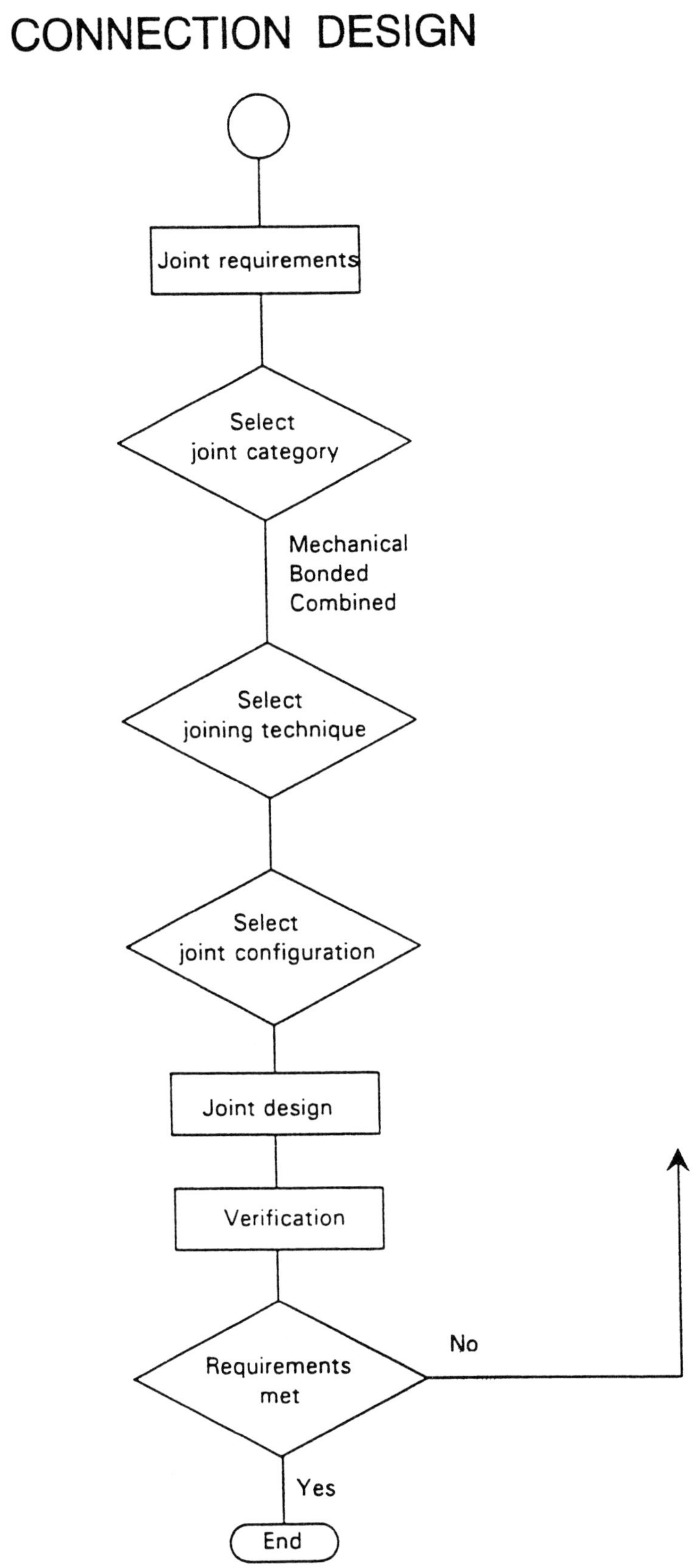

Figure 5.3 Flow chart.

5.2 MECHANICAL JOINTS

5.2.1 Scope

P(1) This section covers the design of mechanical joints where at least one of the primary components to be joined is made of glass FRP. Such joining techniques include fastener connections, friction joints (shear loads), contact joints (direct loads), threaded joints, strap joints and joints incorporating embedded fasteners (see 5.1.4 P(2)).

(2) For the time being, detailed design procedures are given only for connections in which fasteners are loaded in shear or tension; for these joining techniques, a simplified procedure is described (see 5.2.2.4), together with the approach that might be adopted with a more rigorous finite element method.

P(3) The design approach for joining techniques for other mechanical connections shall be by testing. The principles of the design and verification approach shall be as described in 7.3 and the partial safety factors shall be as given in 5.1.10 P(3).

P(4) Bolted and riveted joints are classified according to whether the connecting force in the fastener is acting in shear (see Figure 5.4), tension (see Figure 5.5), or shear and tension combined (see Figure 5.6).

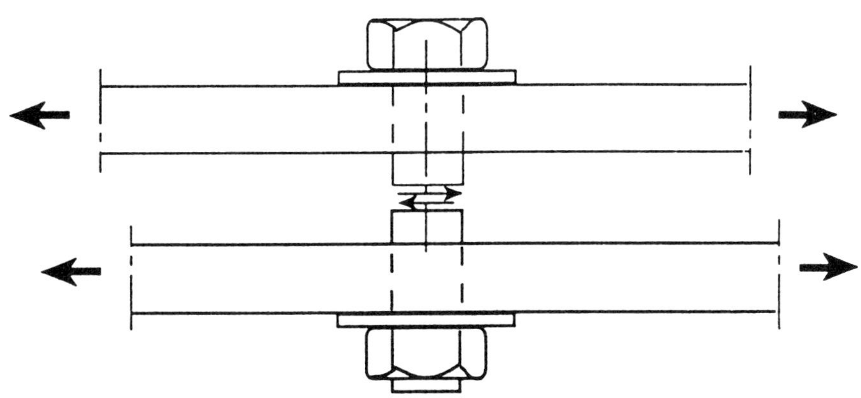

Figure 5.4 Shear loaded fastener installation.

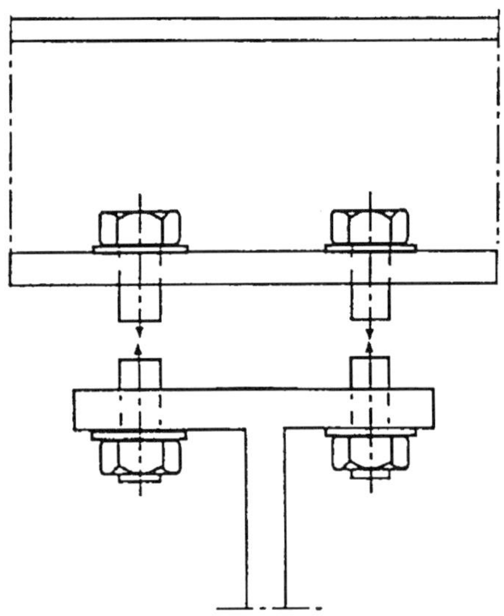

Figure 5.5 Tensile loaded fastener installation.

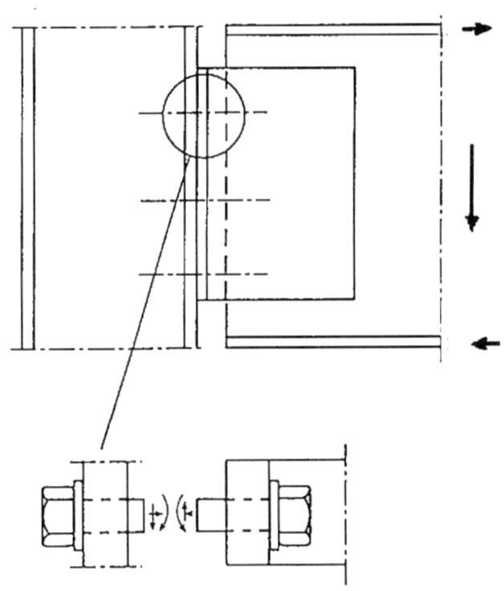

Figure 5.6 Combined loaded fastener installation.

P(5) In the design of bolted and riveted joints, account shall be taken of the following:

(a) Design parameters

- geometry (width, spacings, edge distance, side distance, hole pattern, etc.)
- hole diameter and bolt size
- joint type (single lap, double lap, etc.)
- plate thickness
- loading condition (tensile, compressive, shear, etc.).

(b) Material parameters

- fibre type and form (unidirectional, woven, fabric, etc.)
- resin type
- fibre orientation
- form of construction (e.g. solid laminate, sandwich construction, etc.)
- stacking sequence
- fibre volume fraction
- fastener material.

(c) Fastening parameters

- fastener type (screw, fastener, rivet, etc.)
- clamping force
- washers
- fastener/hole tolerance.

5.2.2 Bolted and riveted joints in shear

5.2.2.0 Notation

See Figures 5.13 to 5.18 and the EUROCOMP Handbook (Definitions).

E_r	Elastic modulus of the radial direction considered
E_x	Elastic modulus in the x-direction
E_y	Elastic modulus in the y-direction
E_ϕ	Elastic modulus in the tangential direction considered
$F_{N,i}$	Direct force in the fastener
F_x	Load on joint in the x-direction
F_y	Load on joint in the y-direction
G_{sn}	Shear modulus in the shear-out plane (s - n) considered
G_{xy}	Shear modulus in the x - y plane
S_b	Average bearing stress on the fasteners given by

$$S_b = F_{N,i}/(d_h t) \tag{5.2}$$

d_b	Fastener shank diameter
d_h	Fastener hole diameter
e	Edge distance (distance from hole centre to the short edge)
s	Side distance (distance from the hole centre to the long edge)
t	Thickness of the laminate considered
V_f	Fibre volume fraction
w	Width of the joint
w_x	Spacing between fastener holes in the x-direction
w_y	Spacing between fastener holes in the y-direction
ν_{xy}	Poisson's ratio for transverse strain in the y-direction when loaded in the x-direction
$\varepsilon_{c,crit}$	Design strain of the glass FRP member in compression
$\varepsilon_{s,crit}$	Design shear strain of the glass FRP member
$\varepsilon_{t,crit}$	Design strain of the glass FRP member in tension
σ_N	Average far field stress
σ_r	Radial stress along the hole boundary
$\sigma_{r,k}$	Characteristic radial strength value
$\sigma_{r,s}$	Calculated radial stress along the hole boundary of actions
$\sigma_{\phi,k}$	Characteristic tangential strength value
σ_ϕ	Tangential stress along the hole boundary
$\sigma_{\phi,s}$	Calculated tangential stress along the hole boundary of actions
τ_{sn}	Shear stress in the shear-out plane
τ_N	Far field shear stress
$\tau_{sn,s}$	Calculated shear stress in the shear-out plane (s - n) of actions
$\tau_{sn,k}$	Characteristic shear strength value

5.2.2.1 Definitions

P(1) Shear loaded fastener connections are divided into two sub-groups depending on the action of the applied load:

- concentrically loaded connections
- eccentrically loaded connections

P(2) In this section, a concentrically loaded connection is one in which the line of action of the applied load passes through the centroid of the fastener group. In a symmetric butt splice (Figures 5.7 and 5.9), symmetry of the shear planes prevents in-plane bending of the plates, whereas in the single lap joint shown in Figure 5.8, in-plane bending of the plates will occur due to the non-alignment of the applied load. Both, however, are examples of concentrically loaded joints.

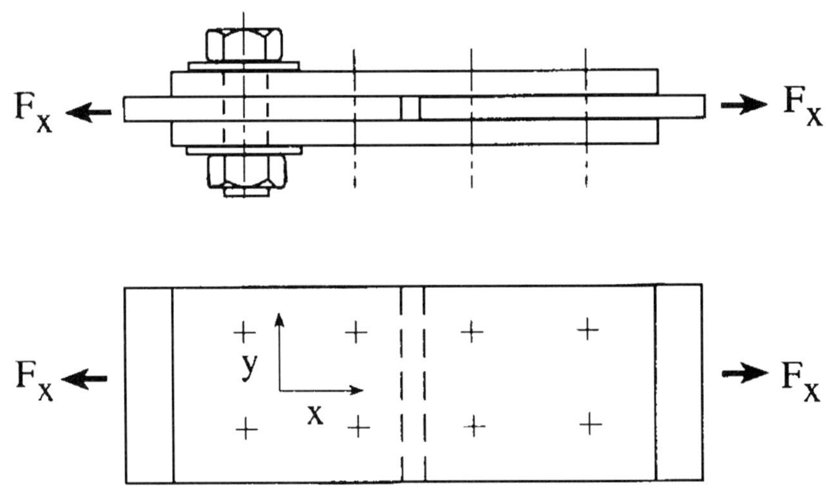

Figure 5.7 A concentrically loaded butt splice.

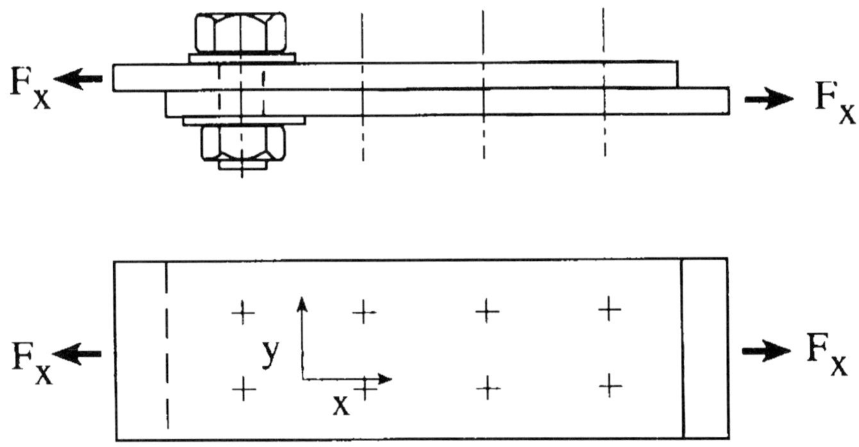

Figure 5.8 A concentrically loaded single-lap joint.

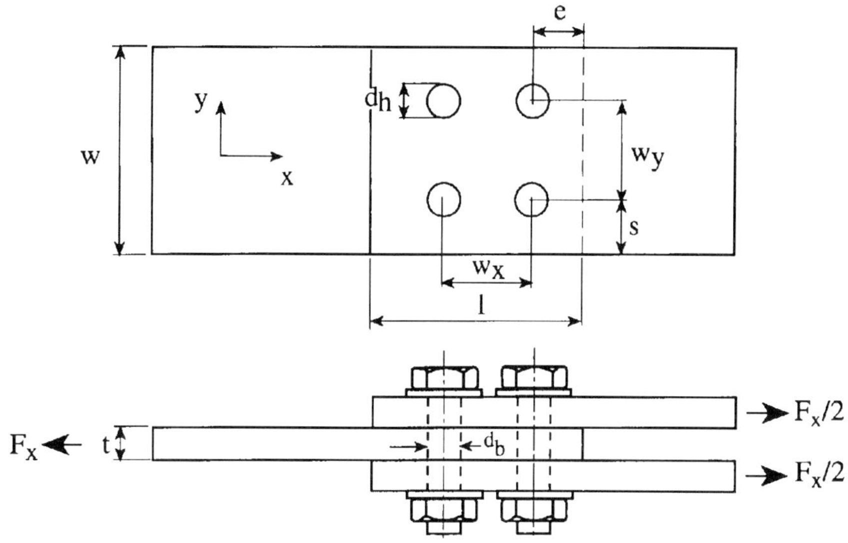

Figure 5.9 Concentrically loaded double - lap joint.

P(3) An eccentrically loaded connection is a connection where the line of action of the applied load does not pass through the centroid of the fastener group. The connection is subjected to a twisting moment in addition to a concentric force. Bracket-type connections, web splices in beams, and standard beam connections are common examples of eccentrically loaded connections (see Figure 5.10).

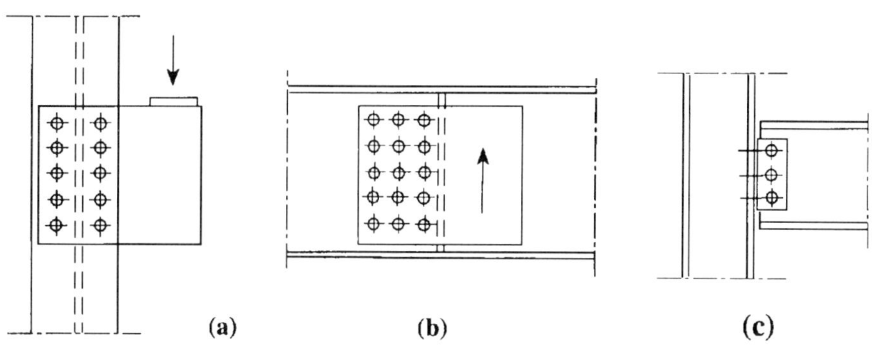

Figure 5.10 Eccentrically loaded connections: (a) bracket type connection, (b) web splice connection and (c) standard beam connection.

(4) Depending on the local arrangement of the connections, the fasteners may act in single or double shear, the former giving rise to out-of-plane bending for both concentrically and eccentrically loaded connections.

P(5) Net-section failure (Figure 5.11(a)) is transverse to the direction of the connecting force, between hole and hole and/or hole and side edges of any one of the components being joined. It is caused primarily by tangential tensile or compressive stresses at the hole edge and, for a joint subjected to uniaxial load, occurs when the ratio of by-pass load to bearing load is high, or when the ratio of hole diameter to plate width (d/w) is high.

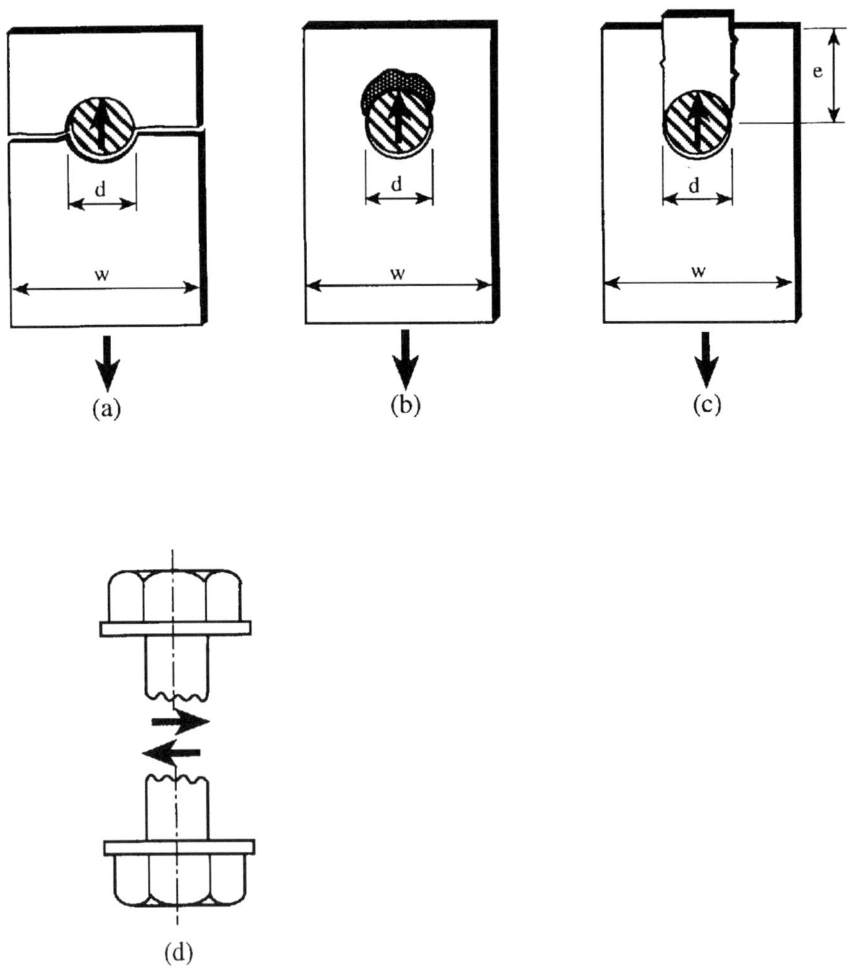

Figure 5.11 Basic failure modes in bolted composite laminates: (a) net-section failure; (b) bearing failure; (c) shear-out failure; (d) bolt shear failure.

P(6) Bearing failure (Figure 5.11(b)) occurs in the material immediately adjacent to the contact area of fastener and laminate and is caused primarily by compressive stresses acting on the hole surface. It is likely to occur when the ratio of d/w is low or when the ratio of by-pass load to bearing load is low. Bearing failure is strongly affected by lateral constraint (clamping force), since lateral constraint prevents delamination of plies.

P(7) Shear-out failure (Figure 5.11(c)) is caused by shear stresses and occurs along shear-out planes on the hole boundary in the principal fastener load direction. Shear-out failure occurs in laminates where the end distance is short. Shear-out failure also occurs in highly orthotropic laminates, such as pultruded laminates, independently of the end distance.

P(8) Bolt shear failure (Figure 5.11(d)) is caused by high shear stress in the fastener. Normally the glass FRP laminate failure occurs before fastener shear failure.

P(9) In pin-loaded joints there is neither any clamping provided by tensile forces in the fasteners nor any lateral constraint provided in the vicinity of the fastener holes.

P(10) Finger-tight joints are those with some clamping and lateral constraint in the vicinity of bolt holes provided by lightly torqued fasteners.

P(11) Torqued fasteners are those with substantial clamping and lateral constraint provided by fasteners tightened to a pre-set torque.

5.2.2.2 Performance requirements

P(1) Either by testing or by calculation the following limit states shall be satisfied:

(a) serviceability

- deflection due to slip of fasteners (within the tolerance of the holes)
- deflection due to excessive deformation of fastener holes
- onset of non-linear load-deflection behaviour of joint under constant load
- onset of increasing deflection under cyclic loading
- separation (at edges) of components or splice plates fastened together
- weather tightness (and/or watertightness) of joint
- fibre debonding or matrix cracking under load or due to assembly techniques
- durability of unsealed edges
- fatigue endurance of laminate and fasteners.

(b) ultimate limit state

- maximum load capacity of complete joint
- static failure of joined parts
- static failure of bolts
- progressive failure of hole edge leading to a permanent hole elongation greater than 4% of the hole diameter
- fatigue endurance of laminate and fasteners.

5.2.2.3 Design requirements

P(1) Glass FRP members that are to be connected by bolts or rivets in shear shall, where the joint is being made, be of solid laminate construction, unless otherwise justified by tests.

(2) Ideally the laminate of the part to be joined should be of balanced symmetrical section with fibre orientations distributed through the thickness of the laminate. Preferably there should be 25% of the fibres in the 0° direction, 25% in the 90° direction and 50% in the ±45° direction. In any case, there should be at least 12.5% of plies in each of the four directions 0°, +45°, −45°, and 90°, the 0° direction being parallel to the load.

It may not always be possible to achieve this, however, particularly with pultruded sections.

(3) In any one connection, there should be not less than two fasteners in a line in the loading direction and all the fasteners in the connection should be of the same diameter.

(4) To achieve maximum strength of multi-row joints, the bearing stress in the most critical region of the joint should be not more than 25% of the design bearing strength and an optimum value of diameter to pitch and side distance should be selected. For guidance, see the EUROCOMP Handbook (5.2.2.3).

P(5) Any load transferred from one member to another by frictional forces between the joined members shall be neglected for the purposes of design, i.e. it should be assumed that all load is transferred by the fasteners.

(6) The joint should be designed, detailed and formed so that the fasteners are tightened to a pre-set torque to provide substantial clamping and lateral restraint around the bolt holes. However, the design strength should be assumed to be that corresponding to finger-tight conditions in which there is little or no lateral restraint. This is to allow for the effects of creep, cyclic loading, fatigue and vibration or some combination of these.

(7) The hole diameter should be not less than the thickness of the thinnest laminate being joined and be not more than one and a half times the thickness ($1.0 \leq d_h/t \leq 1.5$). The clearance of the hole should allow the fastener to be inserted easily, even when all the other fasteners are in place and finger tight, but should not be more than 5% of the fastener diameter. Great care is required, therefore, in forming the holes.

(8) Fasteners should be as tight as possible without causing damage to the laminates of either of the components being joined, particularly in the case of rivet fasteners.

P(9) All bolts should be self-locking or otherwise fitted with locknuts.

P(10) Washers shall be fitted under the head and nut of the fastener and shall have a similar internal diameter to the least diameter of the holes in the laminates through which the fastener passes. The least external dimension of the washer shall be not less than twice the larger or largest diameter of the holes in the laminates through which the fastener passes. The thickness of the washer shall be be sufficient to provide an even surface pressure on the outer laminate. Preferably, the thickness of the washer should be not less than 20% of the thickness of the outermost laminate through which the fastener passes.

(11) The minimum distance between holes, w_x and w_y, centre-to-centre, should be not less than three times the diameter of the hole, ($w_x/d_h \geq 3$, $w_y/d_h \geq 3$) and preferably, about four times the hole diameter.

(12) The shear-out failure mode shall if possible be avoided by selecting an appropriate ply-orientation and by using a minimum value of the e/d ratio.

(13) The minimum end distance, e, centre-to-end, should be not less than three times the diameter of the hole, d_h ($e/d_h \geq 3$), and not less than the side distance, s ($e/d_h \geq s/d_h$).

(14) The ratio of side distance to hole diameter should be not less than half the ratio of transverse pitch to hole diameter ($s/d_h \geq 0.5\ w_y/d_h$).

5.2.2.4 Design methods

P(1) The conditions of equilibrium shall always be fulfilled when determining fastener load distribution, far field stress distribution, and the stress distribution in the vicinity of fastener holes.

P(2) The stiffness properties (elasticity and geometry) of the joined members and fasteners shall, directly or indirectly, be taken into account when determining the fastener load distribution and far field load distribution.

P(3) The failure analysis shall be based on the detailed stress distribution around the fastener holes.

P(4) The failure analysis shall evaluate the following failure modes:

- net-section
- bearing
- shear-out
- fastener shear failure.

(5) In the absence of more rigorous methods or direct applicable test results, the following method of analysis may be used for simple concentrically and eccentrically loaded double lap joints between structural members, of which at least one must be of glass FRP (provided that the preceding design requirements are met).

(a) Determination of fastener shear load distribution

For the concentric loading component of the connection loading, the distribution of load between the rows of fasteners shall be as in Table 5.3.

Table 5.3 Fastener load distribution in multi-row joint (as proportion of average fastener load).

Number of rows		*Row 1*	*Row 2*	*Row 3*	*Row 4*
1	glass FRP/glass FRP	1.0	-	-	-
	glass FRP/metal	1.0	-	-	-
2	glass FRP/glass FRP	1.0	1.0	-	-
	glass FRP/metal	1.15	0.85	-	-
3	glass FRP/glass FRP	1.1	0.8	1.1	-
	glass FRP/metal	1.50	0.85	0.65	-
4	glass FRP/glass FRP	1.2	0.8	0.8	1.2
	glass FRP/metal	1.7	1.0	0.7	0.6
>4	Not recommended				

Note: In the glass FRP/metal joint, Row 1 is the row nearest the end of the glass FRP member.

For the eccentric loading component, the loading should be carried by each fastener in proportion to the distance of each fastener from the centroid of the fasteners group.

(b) Determination of by-pass load distribution

Resolve the resultant force in each fastener due to concentric loading and due to eccentric loading into each of the six load cases illustrated in Figure 5.12.

From consideration of each of the six load cases, calculate:

σ_N, s_b and τ_N

(c) Determination of normalised stress distribution around the fastener hole

From appropriate analytical formulae or from the graphs in Figure 5.13 or 5.14 estimate the distribution of σ_r. Figure 5.13 shall be used if the action is tensile and Figure 5.14 shall be used if the action is compressive.

From appropriate analytical formulae or from the graphs in Figures 5.13 to 5.18, estimate the combined distribution of σ_ϕ and τ_{xy}

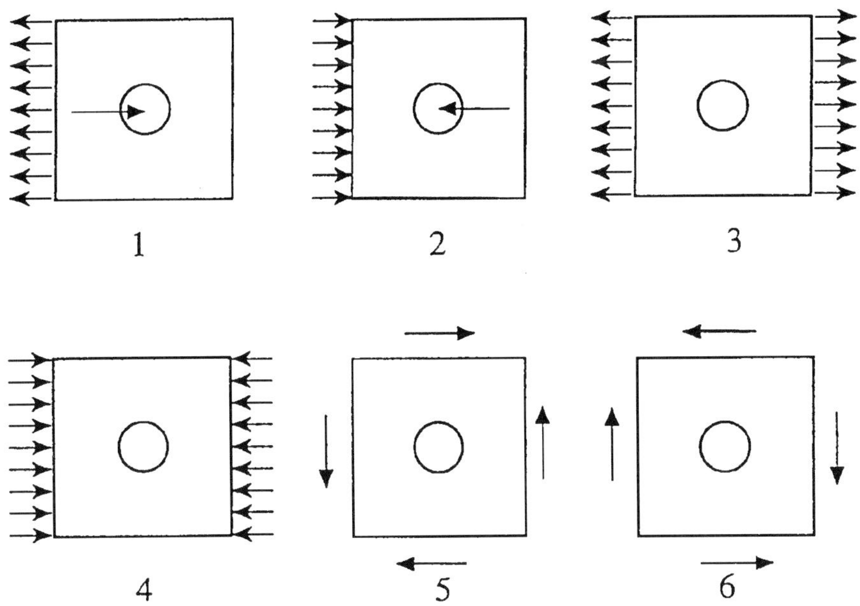

Figure 5.12 Basic load cases in bolted composite laminates.

For single-lap joints, all stresses as calculated above should be increased by 100% ($C_{m,1}$ = 2.0). For double-lap joints and similar joints which may reasonably be assumed to be symmetrically loaded, only σ_r, as calculated above, should be increased by 20% ($C_{m,1}$ = 1.2). These increases are to allow for non-uniform stress distributions through the thickness of the laminate due to non-symmetry, bolt-bending and loss of lateral restraint.

Where the ratio of the distance between holes centre-to-centre and the diameter of the fastener hole is less than 4, the tangential stress σ_ϕ shall be further modified by a factor $C_{m,3}$ obtained from Figure 5.20.

(d) Compliance check

Net-section failure is evaluated around the hole circumference at a sufficient number of locations (Figure 5.19(a)). The most critical point defines the location of failure. Failure at any of the evaluated points is assumed to have occurred when the tangential stress $\sigma_{\phi,s}$ at the hole edge reaches the characteristic strength $\sigma_{\phi,k}$ of the laminate in the direction considered. Both tensile and compressive net-section failures are considered. Hence, characteristic strengths (in tension and compression) are required to evaluate net-section failure.

Bearing failure is evaluated around the loaded part of the hole circumference at a sufficient number of locations (Figure 5.19(b)). Bearing failure occurs at any of the selected hole boundary points when the radial compressive stress $\sigma_{r,s}$ reaches the characteristic compressive strength $\sigma_{r,k}$

of the laminate at the hole edge.

Shear-out failure is evaluated along shear-out planes (s - n planes) (Figure 5.19(c)). It is assumed that failure occurs when the maximum shear stress $\tau_{sn,s}$ along the shear-out plane reaches the characteristic shear strength $\tau_{sn,k}$ of the laminate.

Verify that:

$$\gamma_F\, C_m\, (\sigma_r,\ \sigma_\phi,\ \tau_{sn})_s \leq (\sigma_r,\ \sigma_\phi,\ \tau_{sn})_k/\gamma_m \qquad (5.3)$$

where:

C_m = $C_{m,1}\, C_{m,3}$
$()_s$ = calculated values of actions
$()_k$ = corresponding characteristic values of strengths determined from section 4.0
γ_m = partial materials safety factor determined in accordance with 5.1.10 P(2), with $\gamma_{m,1}$ taken as 1.5.

If the constitutive properties of the laminate, i.e. E_x, E_y, G_{xy}, ν_{xy}, are known, then $\sigma_{r,k}$, $\sigma_{\phi,k}$, $\tau_{xy,k}$ may be predicted from the following relationships:

$\sigma_{r,k}$ = $E_r\, \varepsilon_{c,crit}$ for bearing

$\sigma_{\phi,k}$ = $E_\phi\, \varepsilon_{t,crit}$ or $E_\phi\, \varepsilon_{c,crit}$ for net section (tension and compression)

$\tau_{sn,k}$ = $G_{sn}\, \varepsilon_{s,crit}$ for shear out

where:

E_r is the radial elastic modulus given by:

$$\frac{1}{E_r} = \frac{1}{E_x}\cos^4\phi + \left(\frac{1}{G_{xy}} - \frac{2\nu_{xy}}{E_x}\right)\sin^2\phi\cos^2\phi + \frac{1}{E_y}\sin^4\phi \qquad (5.4)$$

E_ϕ is the tangential elastic modulus given by:

$$\frac{1}{E_\phi} = \frac{1}{E_x}\sin^4\phi + \left(\frac{1}{G_{xy}} - \frac{2\nu_{xy}}{E_x}\right)\sin^2\phi\cos^2\phi + \frac{1}{E_y}\cos^4\phi \qquad (5.5)$$

G_{sn} is the shear modulus in the shear out plane given by:

$$\frac{1}{G_{sn}} = \frac{1}{G_{xy}} + \left(\frac{1+\nu_{xy}}{E_x} + \frac{1+\nu_{yx}}{E_y} - \frac{1}{G_{xy}}\right)\sin^2 2\phi \qquad (5.6)$$

$\varepsilon_{t,crit}$ is the design strain of the laminate in tension (which may be taken as 0.001 in exposed polluted conditions and 0.0025 elsewhere, in the absence of test data)

$\varepsilon_{c,crit}$ is the design strain of the laminate in compression (which may be taken as 0.002 in all conditions)

$\varepsilon_{s,crit}$ is the design strain of the laminate in shear (which may be taken as 0.002 in exposed, polluted conditions and 0.004 elsewhere).

- where the above predictions are used, $\gamma_{m,1}$ should be taken as 2.0 (see 5.1.10).

Check the strength of the fasteners for the maximum force.

(7) Where the design requirements for the application of the simplified method described above cannot be met or, for reasons of economic design of a large number of joints, a more accurate method of analysis is required, advanced numerical methods of analysis using finite element methods (FEM) may be used.

One such method (FE-Code BOLTIC) is described in the EUROCOMP Handbook (5.2.2.4).

Stress charts, basic load case 1

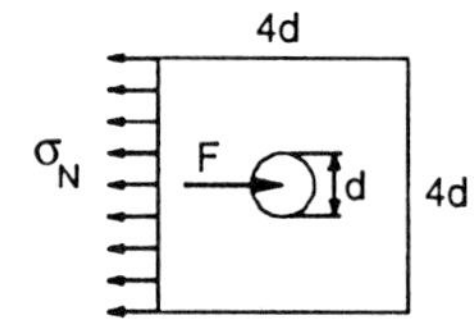

- - - - - Laminate A1 and A2, uni-directional, 0°

– – – – Laminate B1 and B2, uni-directional, 90°

——— Laminate C1 and C2, quasi-isotropic

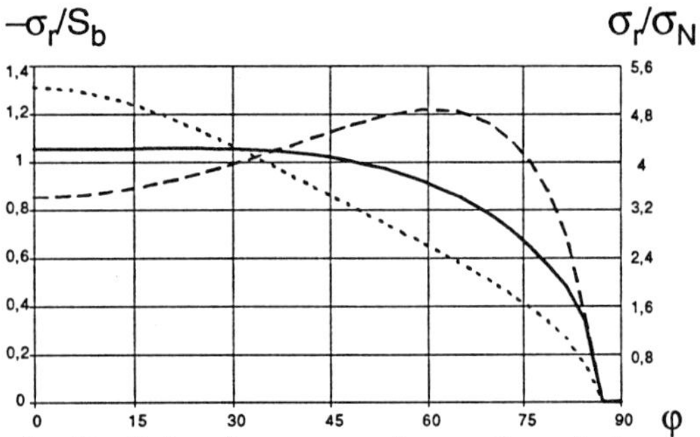

a) Radial stresses along the hole boundary, v_f=60% and v_f=35%

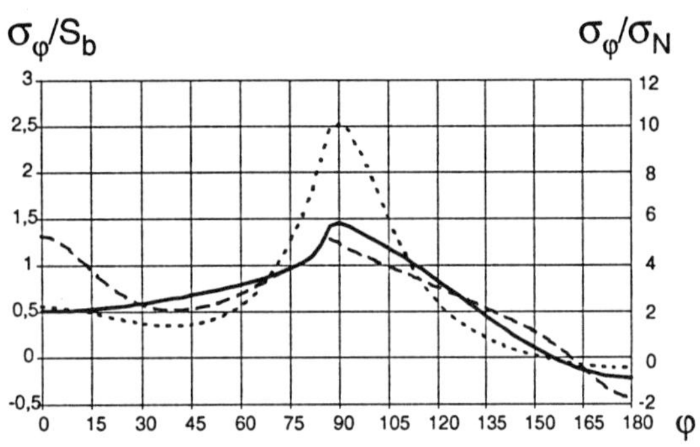

b) Tangential stresses along the hole boundary, v_f=60%

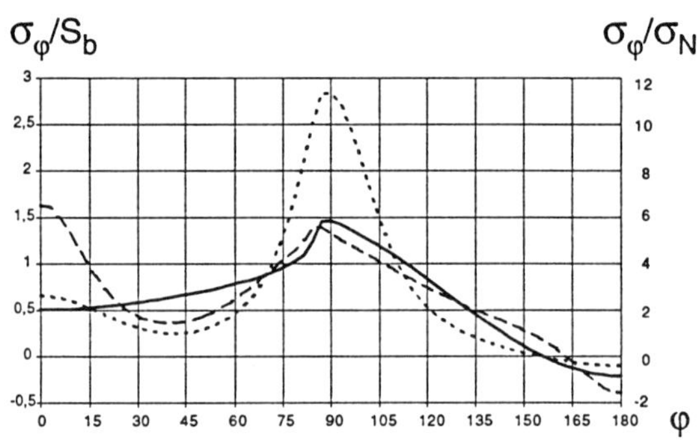

c) Tangential stresses along the hole boundary, v_f=35%

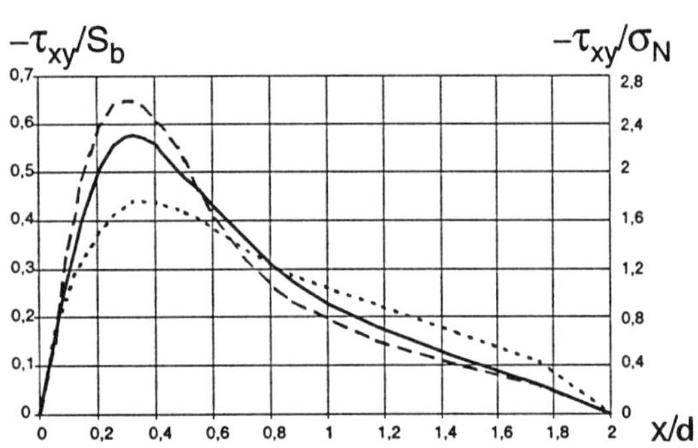

d) Shear stresses in the shear-out plane, v_f=60%

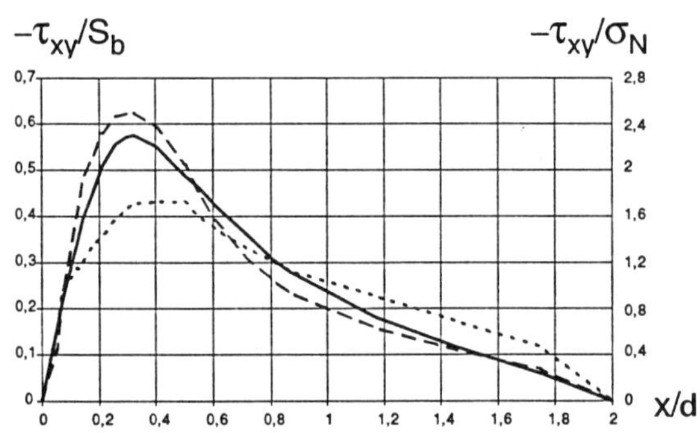

e) Shear stresses in the shear-out plane, v_f=35%

Figure 5.13 Stress distribution around bolt holes for basic load case 1. Stress normalised by nominal bearing stress S_b = |F/dt|.

Stress charts, basic load case 2

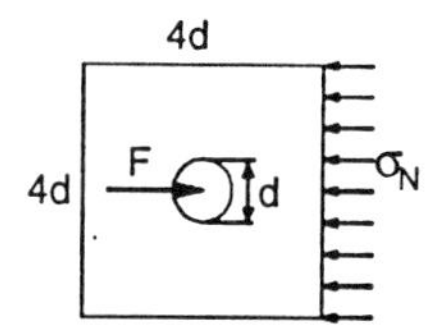

- - - - Laminate A1 and A2, uni-directional, 0°

– – – – Laminate B1 and B2, uni-directional, 90°

——— Laminate C1 and C2, quasi-isotropic

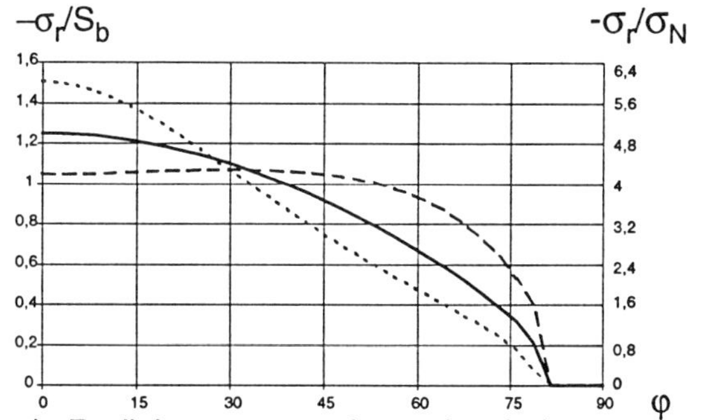

a) Radial stresses along the hole boundary, v_f=60% and v_f=35%

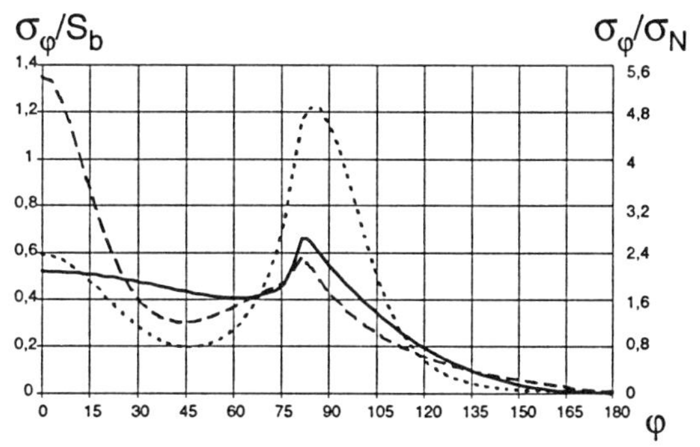

b) Tangential stresses along the hole boundary, v_f=60%

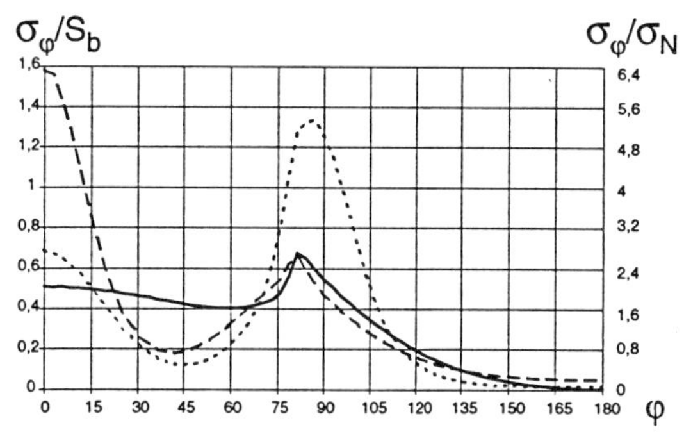

c) Tangential stresses along the hole boundary, v_f=35%

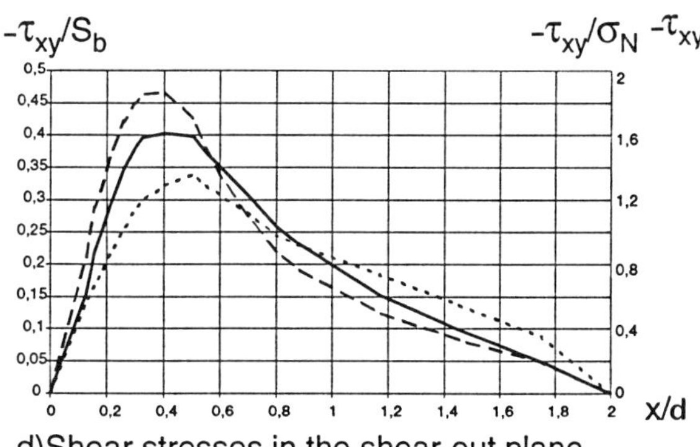

d)Shear stresses in the shear-out plane, v_f=60%

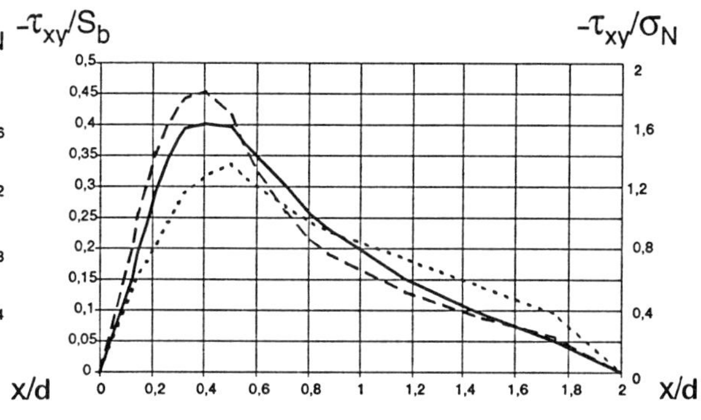

e)Shear stresses in the shear-out plane, v_f=35%

Figure 5.14 Stress distribution around bolt holes for basic load case 2. Stress normalised by nominal bearing stress S_b |F/dt|.

Stress charts, basic load case 3

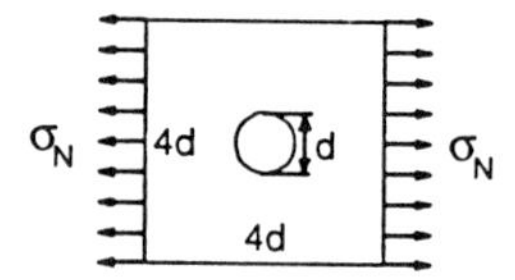

- - - - - Laminate A1 and A2, uni-directional, 0°

– – – – – Laminate B1 and B2, uni-directional, 90°

——— Laminate C1 and C2, quasi-isotropic

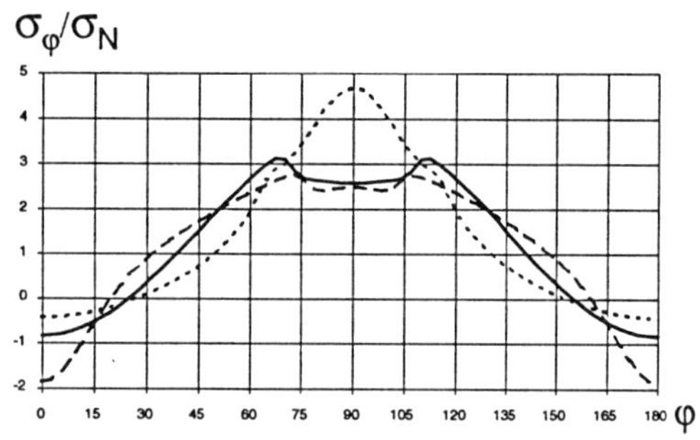

a) Tangential stresses along the hole boundary, v_f=60%

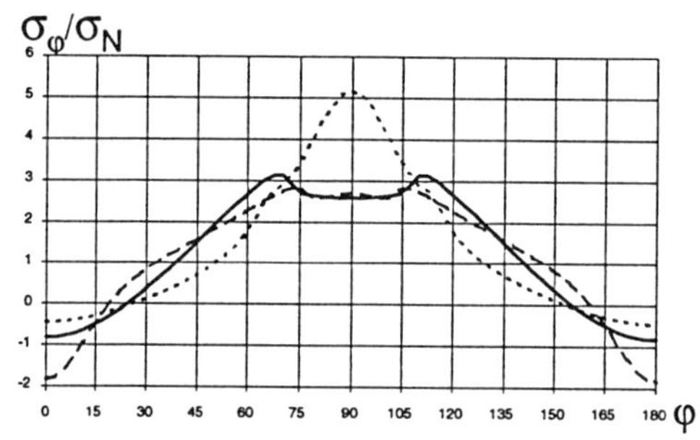

b) Tangential stresses along the hole boundary, v_f=35%

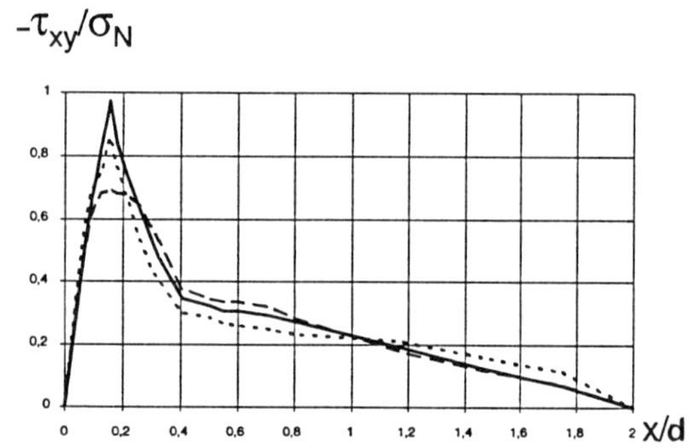

c) Shear stresses in the shear-out plane, v_f=60%

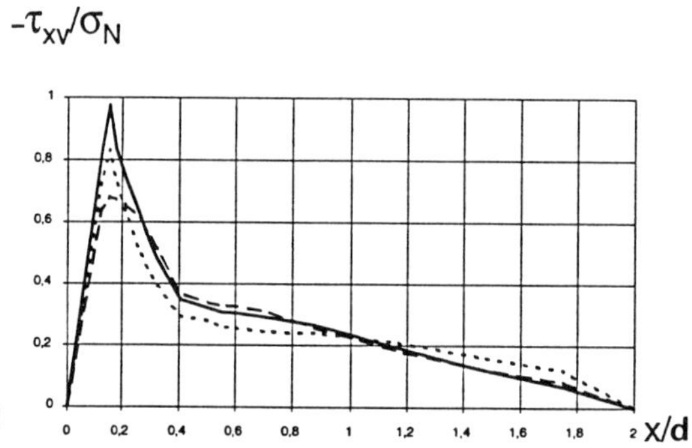

d) Shear stresses in the shear-out plane, v_f=35%

Figure 5.15 Stress distribution around bolt holes for basic load case 3. Stress normalised by far-field stress σ_N.

Stress charts, basic load case 4

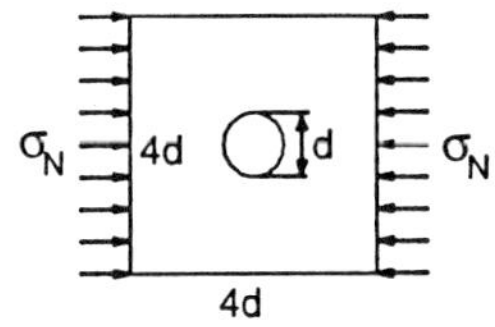

- - - - - Laminate A1 and A2, uni-directional, 0°
– – – – – Laminate B1 and B2, uni-directional, 90°
——— Laminate C1 and C2, quasi-isotropic

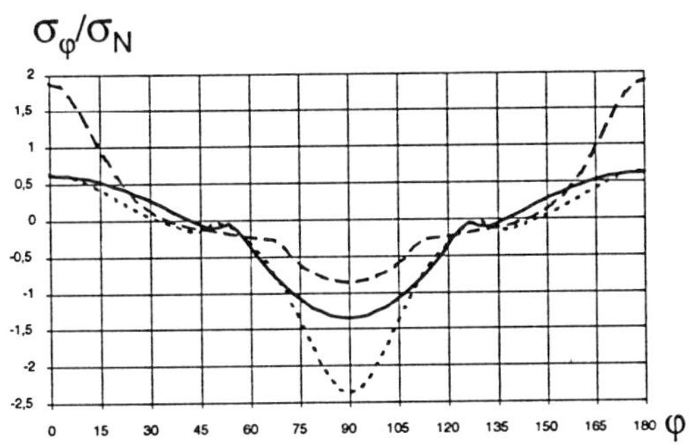

a) Tangential stresses along the hole boundary, v_f=60%

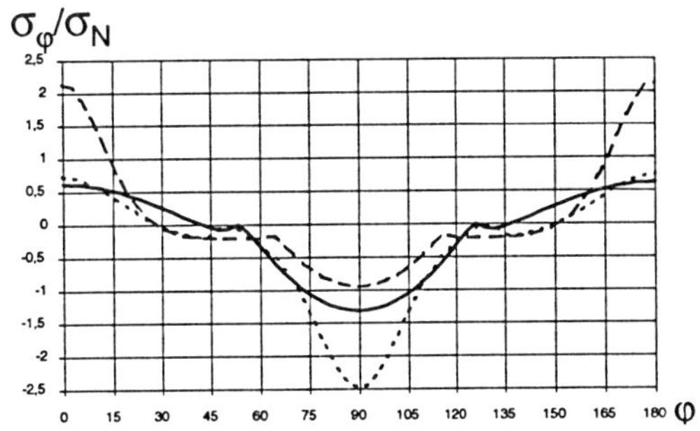

b) Tangential stresses along the hole boundary, v_f=35%

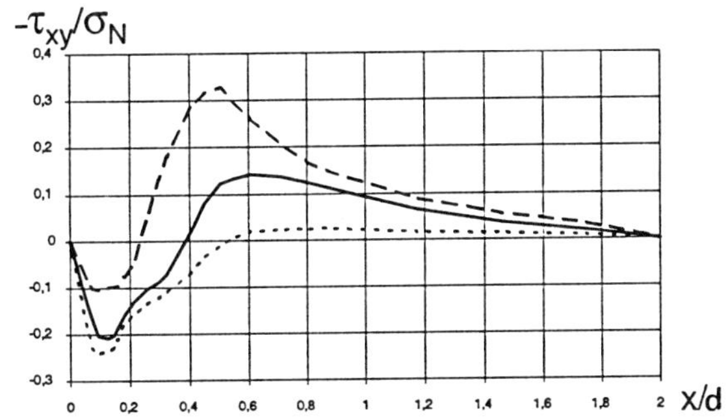

c) Shear stresses in the shear-out plane, v_f=60% and v_f=35%

Figure 5.16 Stress distribution around bolt holes for basic load case 4. Stress normalised by far-field stress σ_N.

Stress charts, basic load case 5

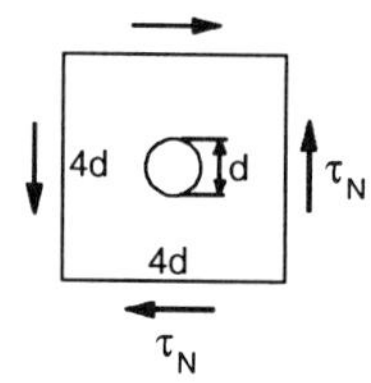

- - - - -	Laminate A1 and A2, uni-directional, 0°
– – – – –	Laminate B1 and B2, uni-directional, 90°
———	Laminate C1 and C2, quasi-isotropic

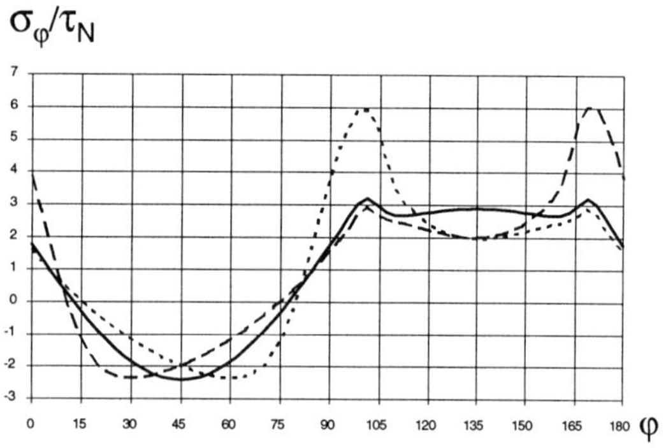

a) Tangential stresses along the hole boundary, v_f=60%

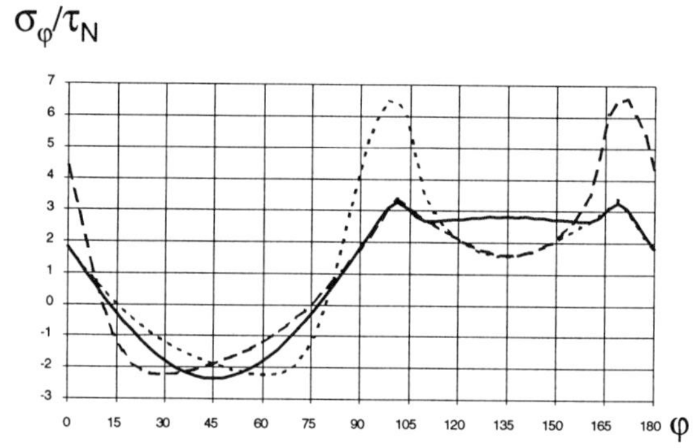

b) Tangential stresses along the hole boundary, v_f=35%

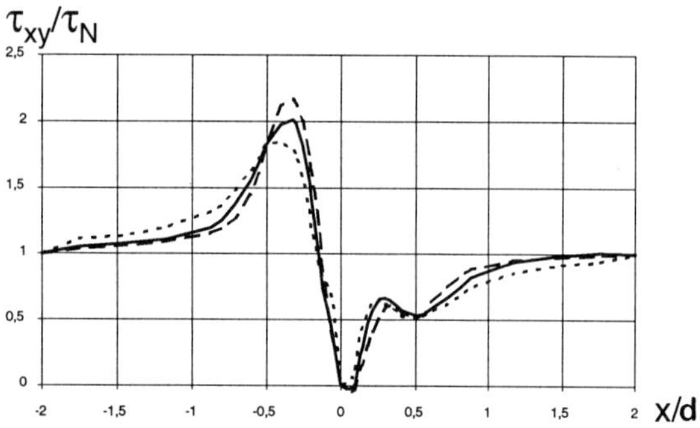

c) Shear stresses in the shear-out plane, v_f=60% and v_f=35%

Figure 5.17 Stress distribution around bolt holes for basic load case 5. Stress normalised by nominal far-field stress $|\tau_N|$.

Stress charts, basic load case 6

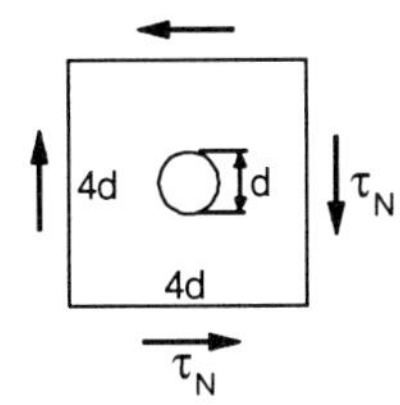

- - - - -	Laminate A1 and A2, uni-directional, 0°
– – – – –	Laminate B1 and B2, uni-directional, 90°
———	Laminate C1 and C2, quasi-isotropic

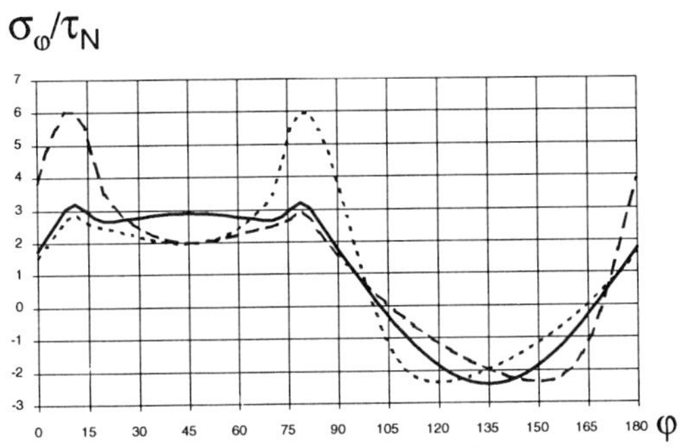

a) Tangential stresses along the hole boundary, v_f=60%

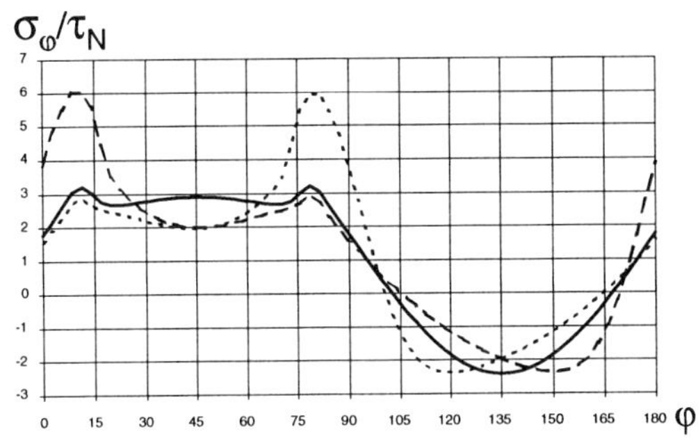

b) Tangential stresses along the hole boundary, v_f=35%

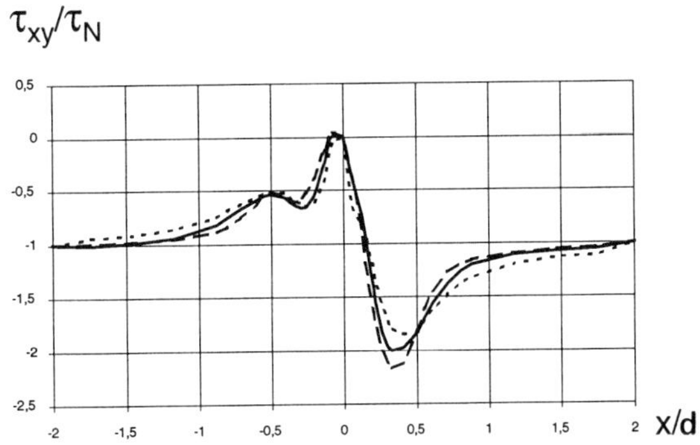

c) Shear stresses in the shear-out plane, v_f=60% and v_f=35%

Figure 5.18 Stress distribution around bolt holes for basic load case 6. Stress normalised by nominal far-field stress $|\tau_N|$.

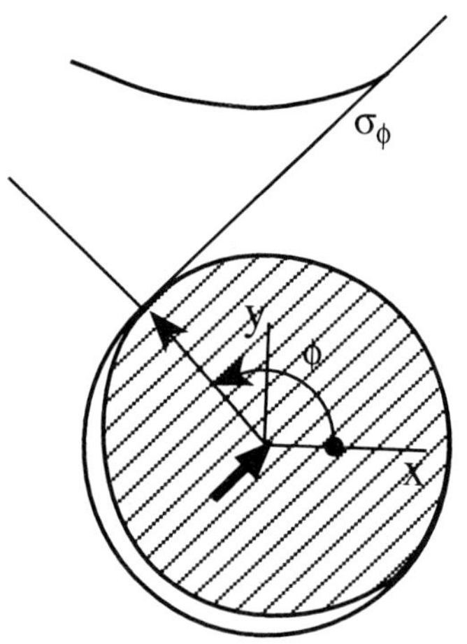

Figure 5.19(a) Evaluation of net-section bearing failure around the hole circumference.

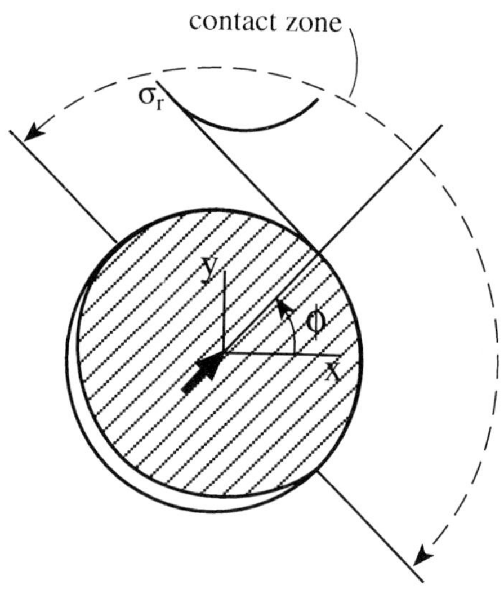

Figure 5.19(b) Evaluation of failure around the loaded half of the hole circumference.

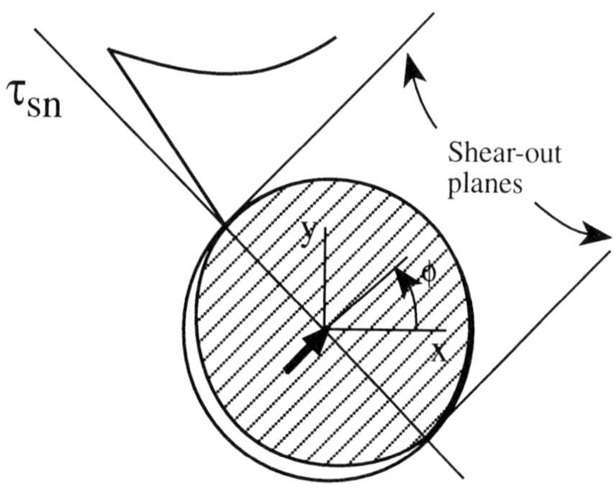

Figure 5.19(c) Calculation of shear-out failure along shear-out planes (s - n planes).

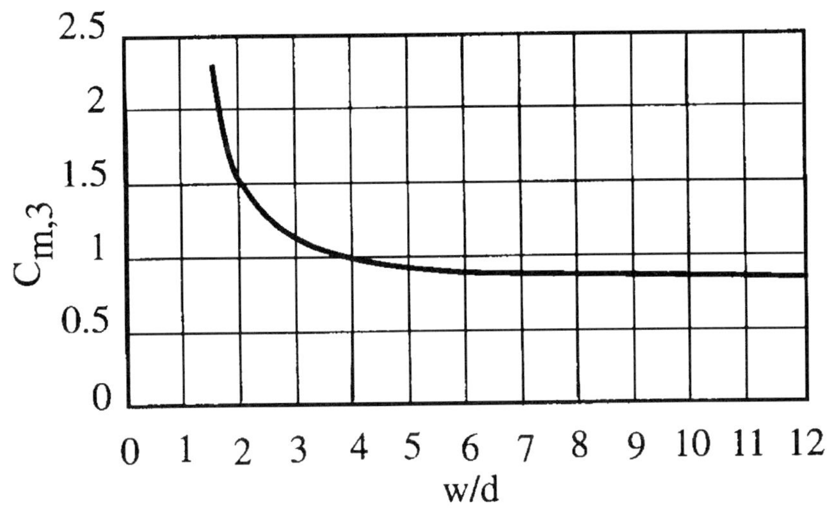

Figure 5.20 Correction factor $C_{m,3}$ for σ_ϕ.

5.2.2.5 Experimental data

(1) Within the EUROCOMP Programme, data for three different glass FRP laminate configurations were generated from two different material systems. The data are listed in the EUROCOMP Handbook.

5.2.3 Bolted and riveted joints in tension

5.2.3.0 Notation

- D Diameter of the head of the fastener
- D_w Diameter of the washer under the head
- $F_{bolt,k}$ Tensile force in the fastener under consideration
- τ_{xz} Average shear stress through the thickness of the laminate for a tension loaded fastener

5.2.3.1 Definitions

P(1) Pull-out failure occurs wholly within the laminate and is a through-thickness shear failure on the periphery of the fastener head or of the washer (see Figures 5.21 and 5.22).

P(2) Flexural failure occurs within the end-plate (or end-plates) being connected, as a result of either single or double curvature bending of the flange (see Figures 5.23(a) and (b)).

P(3) Fastener failure is a simple tension failure (see Figure 5.24).

5.2.3.2 Performance requirements

P(1) Either by testing or by calculation, the serviceability and ultimate limited states defined in 5.2.2.2 shall be satisfied.

5.2.3.3 Design requirements

(1) The design requirements should be generally in accordance with 5.2.2.3, as appropriate.

5.2.3.4 Design methods

P(1) The conditions of equilibrium shall always be fulfilled when determining fastener load distribution, the shear stress distribution in the vicinity of fastener holes and the flexural stresses in the end-plate.

P(2) The stiffness properties (elasticity and geometry) of the joined members and fasteners shall directly or indirectly be taken into account when determining the fastener load distribution and the flexural stresses in the end-plate.

P(3) In general, the fastener load distribution shall be determined from the equilibrium conditions and the genetic compatibility of the joint in conjunction with a constitutive relation for the fasteners.

P(4) The failure analysis shall be based on the average pull-out stress distribution in the vicinity of the bolt holes and on the flexural stresses elsewhere in the end-plate.

P(5) The failure analysis shall evaluate the following failure modes:

- pull-out failure in the end-plate around the fastener
- flexural failure of the end-plate due to both single curvature and double curvature bending
- fastener failure.

(6) For pull-out failure, it should be verified that:

$$\gamma_f C_m \ \tau_{xz,s} \geq \tau_{xz,k}/\gamma_m \tag{5.6}$$

where

C_m = 2 unless otherwise justified
$\tau_{xz,s}$ = $F_{bolt,k}/(\pi Dt)$ (no washer)
$\tau_{xz,s}$ = $F_{bolt,k}/(\pi D_w t)$ (with washer)
$\tau_{xz,k}$ = through laminate-thickness shear strength.

(7) For pull-out failure and fastener failure, the prying force, Q, should, in the absence of rigorous analysis, be taken as 0.5 F_{bolt} (see Figure 5.23).

(8) The fastener strength should be calculated in accordance with the appropriate design codes for the fastener material.

(9) For flexural failure of the end-plate, the prying force, Q, should, in the absence of rigorous analysis, be taken as zero or 0.5 F_{bolt}, whichever gives the worse result at the location where the flexural stresses are being calculated.

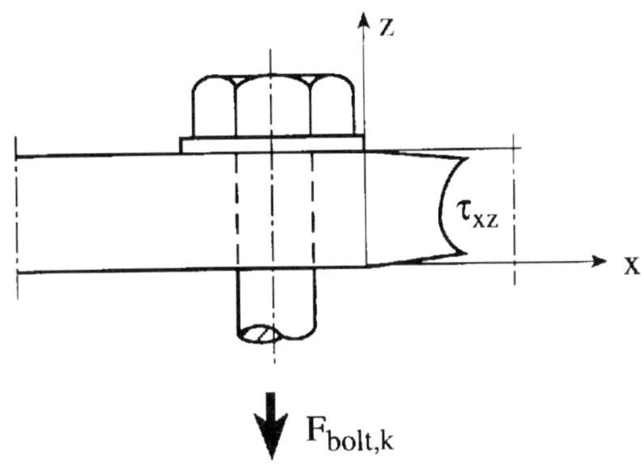

Figure 5.21 Shear stress distribution (τ_{xz}) through the thickness of the glass FRP laminate due to a tension loaded stresses.

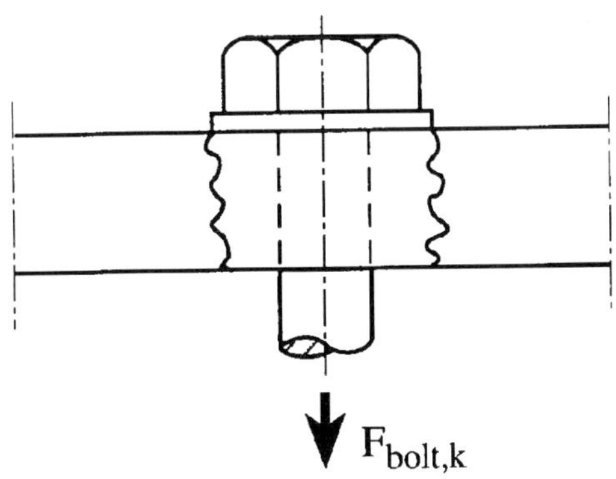

Figure 5.22 Pull-out failure caused by through-thickness stresses.

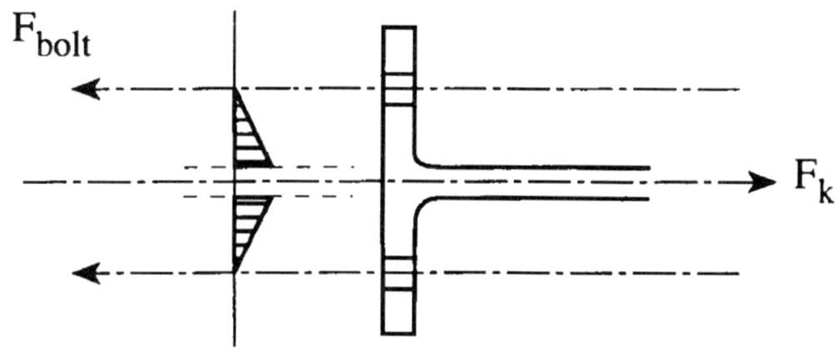

Figure 5.23(a) Single curvature bending. (typical) $F_{bolt} = 0.5F_k$

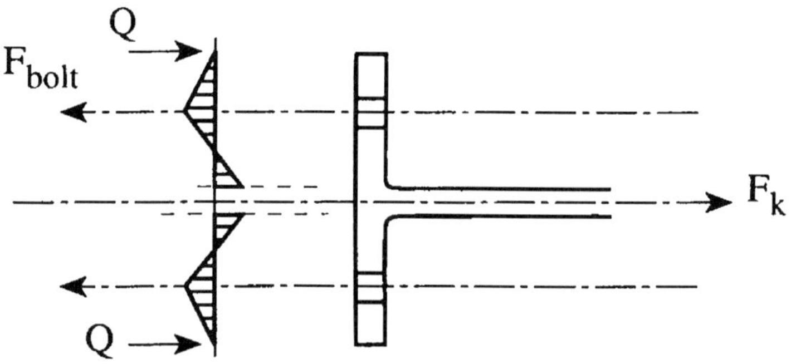

Figure 5.23(b) Double curvature bending (typical) $F_{bolt} = 0.5F_k + Q$

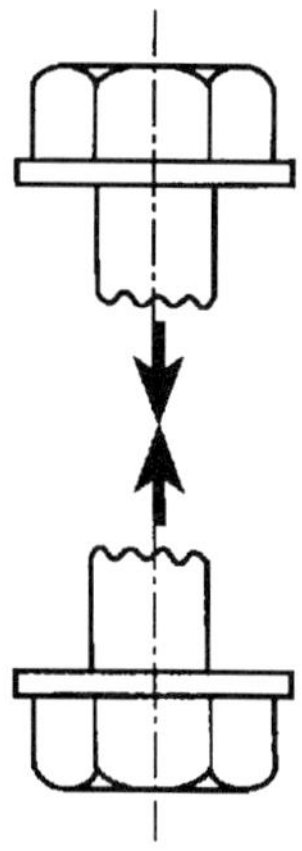

Figure 5.24 Bolt tensile failure.

5.3 BONDED JOINTS

5.3.0 Notation

C_m	Coefficient for design method inaccuracy
E	Adherend tensile/compressive modulus (N/mm^2)
E_a	Adhesive tensile/compressive modulus (N/mm^2)
E_a'	Effective adhesive transverse modulus (N/mm^2)
F	Force or deformation (see 2.2.2.1)
G	Adherend shear modulus (N/mm^2)
G_a	Adhesive shear modulus (N/mm^2)
L	Length of overlap (bond length) (m)
N	Ultimate load on joint per unit width (kN/m)
P	Applied load on joint per unit width (kN/m)
R	Joint resistance (kN/m)
S	Applied shear load on joint per unit width (kN/m)
S_d	Design value of an internal force or moment or strain or deformation
b	Joint width (m)
c	Half-length of overlap (m)
k	Bending moment factor, constant in tee and angle joints
k'	Rotation factor
t	Adherend thickness (m)
t_a	Adhesive layer thickness (m)
β	Parameter for overlap
γ_e	Elastic adhesive shear strain
γ_f	Partial safety factor for the load
γ_m	Partial material safety factor
γ_p	Plastic adhesive shear strain
λ	Shape factor of adhesive elastic shear stress distribution (1/m)
σ	Average adherend tensile stress (N/mm^2)
σ_z	Adherend through-thickness tensile strength (N/mm^2)
$\sigma_{z\ allowable}$	Allowable adherend through-thickness tensile stress (N/mm^2)
$\sigma_{o\ max}$	Maximum adhesive peel (tensile) stress (N/mm^2)
$\sigma_{o\ allowable}$	Allowable adhesive peel (tensile) stress (N/mm^2)
τ_m	Average adhesive shear stress (N/mm^2)
τ_p	Plastic adhesive shear stress (N/mm^2)
$\tau_{o\ allowable}$	Allowable adhesive shear stress (N/mm^2)
$\tau_{o\ max}$	Maximum adhesive shear stress (N/mm^2)
ν	Adherend Poisson's ratio

Subscripts

BA	Bonding angle in tee and angle joints
OL	Overlap in tee and angle joints
Rd	Design resistance (see 2.2.3.2)
Ru	Temporary maximum resistance
Sd	Design force
a	Adhesive
d	Design

i	Double-lap joint inner adherend, or double-strap joint adherend
k	Characteristic value (see 2.2.3.1)
o	Double-lap joint outer adherends, or double-strap joint straps
t	Tensile loading
u	Ultimate
1, 2	Single-lap joint: adherends 1 and 2 single-strap joint: adherend and strap

5.3.1 General

P(1) Definitions for the joining techniques used for bonded joints in this EUROCOMP Design Code are given in 1.4.2 and 5.1.

P(2) The design of any bonded joint shall satisfy the following conditions:

- allowable shear stress of the adhesive is not exceeded
- allowable tensile (peel) stress of the adhesive is not exceeded
- allowable through-thickness tensile stress of the adherend is not exceeded.

(3) In addition, the allowable in-plane shear stress of the adherend should not be exceeded. This condition is not included with the design conditions, as in practice it may only be verified by testing. With the joints and structures within the scope of this EUROCOMP Design Code, typically one or more of the three design conditions given above will become critical before the in-plane shear stress limit in any of the adherends is exceeded.

5.3.1.1 Stresses and strains

P(1) The four main loading modes of bonded joints are (see Figure 5.25):

- peel loads produced by out-of-plane loads acting on a thin adherend
- shear stresses produced by tensile, torsional or pure shear loads imposed on adherends
- tensile stresses produced by out-of-plane tensile loads
- cleavage loads produced by out-of-plane tensile loads acting on stiff and thick adherends at the end of the joint.

(2) Typically, a bonded joint is simultaneously loaded by several of these load components.

(3) Adhesive layers of bonded joints should primarily be stressed in shear or compression. Tensile, cleavage and peel loads should be avoided, or their effect evaluated with great care.

(4) Where non-linear behaviour of adherends or adhesive is expected, the strains (deformations) should also be considered.

5.3.1.2 Failure modes

P(1) A bonded joint has the following three primary failure modes:

- adhesive failure
- cohesive failure of adhesive
- cohesive failure of adherend.

(2) Adhesive failure is a rupture of an adhesive bond, such that the separation is at the adhesive - adherend interface. This failure is mainly due to a material mismatch or inadequate surface treatment. Adhesive failure should be avoided.

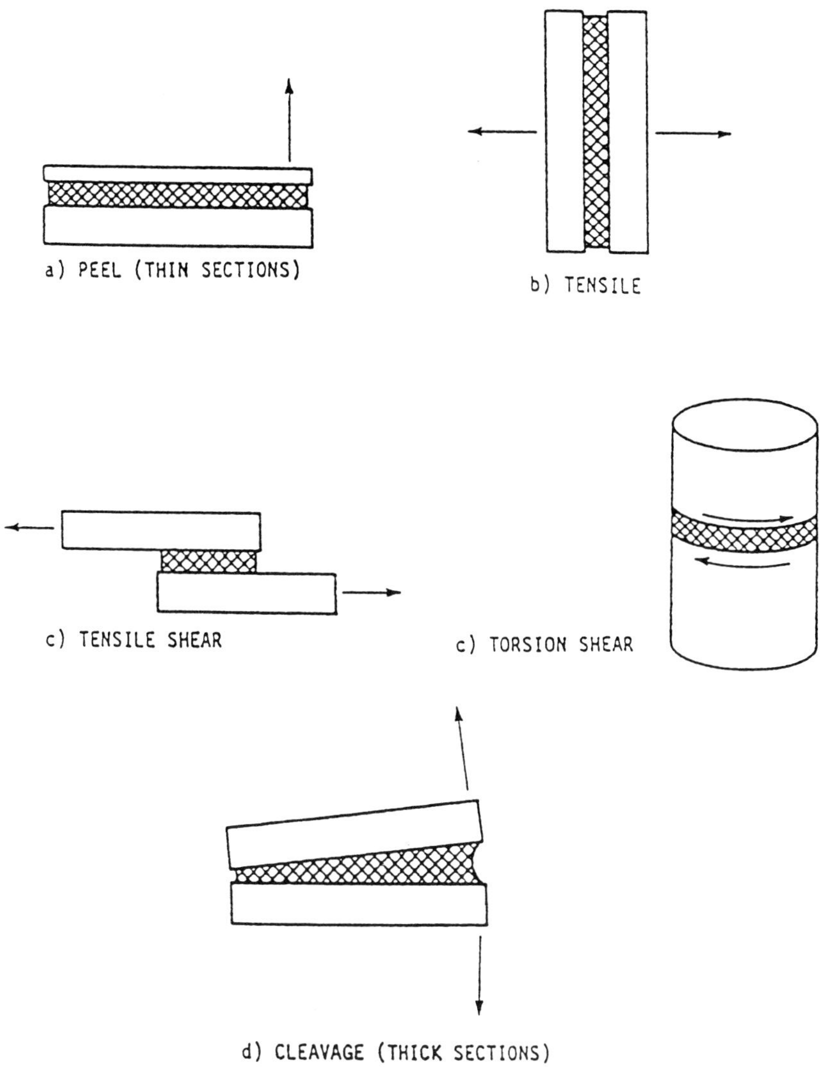

Figure 5.25 Loading modes or types of stresses (see reference 5.21, EUROCOMP Handbook).

(3) Cohesive failure of the adhesive takes place when the adhesive fails due to loads exceeding the adhesive strength.

(4) Cohesive failure of the adherend takes place when the adherend fails due to loads in excess of the adherend strength. In laminate structures the failure typically initiates from the matrix between the laminae as a result of out-of-plane loads or interlaminar shear loads. Other types of failure initiation are also possible, especially for FRP composites that do not have a layered structure.

(5) Typical locations of possible failure initiation and critical strengths are indicated in Figure 5.26.

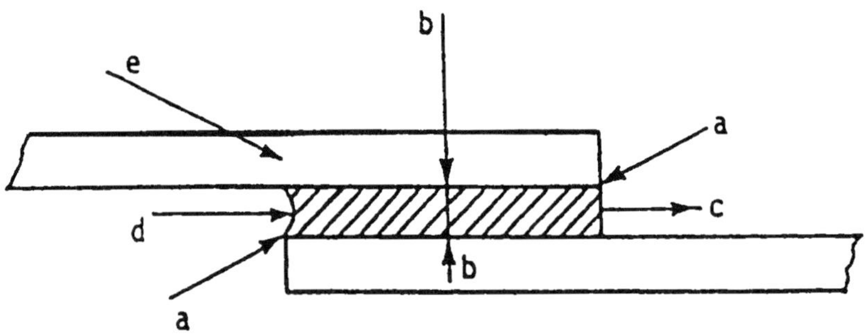

Figure 5.26 Areas of failure initiation and critical strengths (see reference 5.21, EUROCOMP Handbook).

(6) Single- and double-lap and single- and double-strap joints have stress concentrations at the ends of the overlap when loaded with in-plane loads. These are the most probable locations for the failure initiation. A typical shear stress distribution is shown in Figure 5.27.

(7) The predominant failure mode depends upon the lap length, adhesive thickness and fibre orientation in the layer adjacent to the adhesive.

5.3.1.3 Effect of joint geometry on joint strength

(1) Within a certain joint configuration the joint strength can be affected significantly by the joint geometry.

(2) In the bonded joint design the most basic problems are the unavoidable shear stress concentrations and the inherent eccentricity of the forces causing peel stresses both in the adhesive and in the adherends. At the ends of the overlap both the peel and shear stresses reach their maximum values, resulting in reduced load-bearing capacity of the joint, see Figure 5.28.

(3) The effects of the eccentricity are the greatest in lap and strap joints, particularly in a single-lap joint.

(4) Owing to the fact that the joint failure is triggered by the practically constant peel stresses induced by the eccentricity and by shear stress peaks at the ends of the overlap, the static load-bearing capacity of a bonded lap or strap joint cannot be increased significantly by increasing the lap length beyond a certain minimum.

P(5) However, the lap length shall be long enough to provide a moderately loaded adhesive region in the middle of the joint to resist creep deformations of the adhesive, see Figure 5.27.

(6) The eccentricity should always be minimized within the limits dictated by other factors.

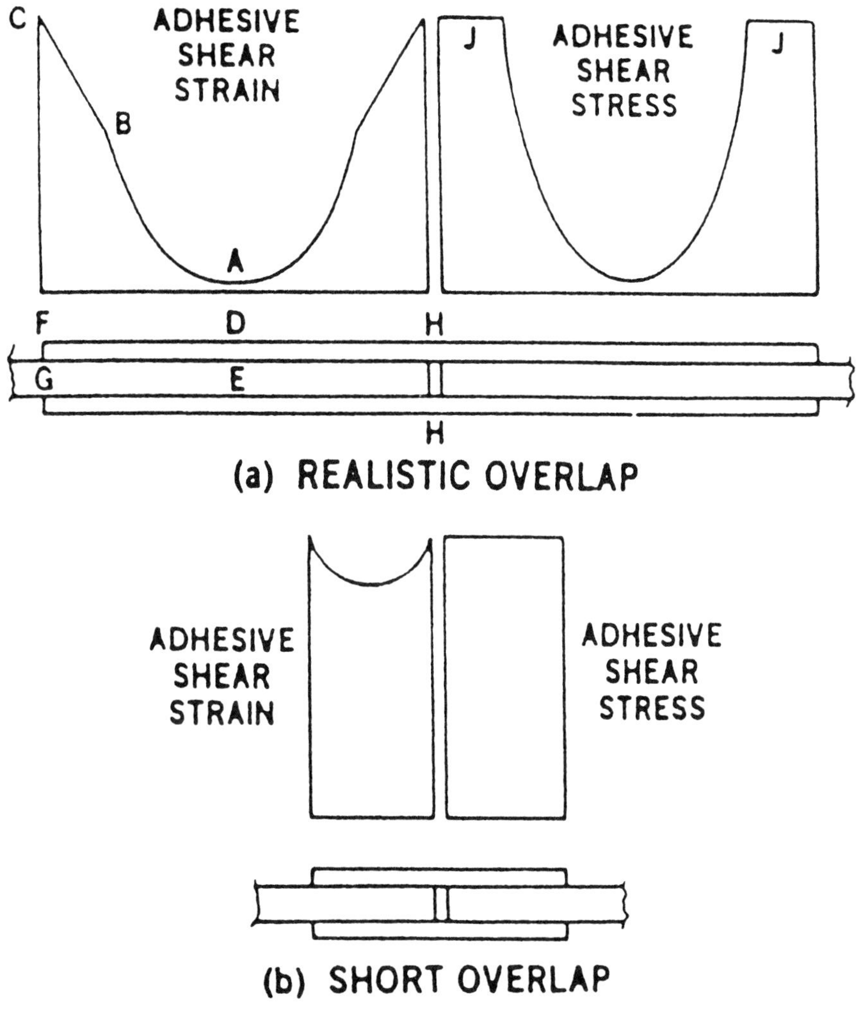

Figure 5.27 A typical adhesive shear stress distribution in a lap joint according to elastic-plastic model (see reference 5.24, EUROCOMP Handbook).

(7) The effects of peel stresses induced by eccentricity may be reduced by the following methods:

- increasing the adherend stiffness without increasing its thickness
- increasing the lap length
- tapering the ends of the adherends
- using adhesive fillets.

(8) The effects of eccentricity are also reduced when a double-lap or a double-strap configuration is used instead of a single-lap or a single-strap configuration.

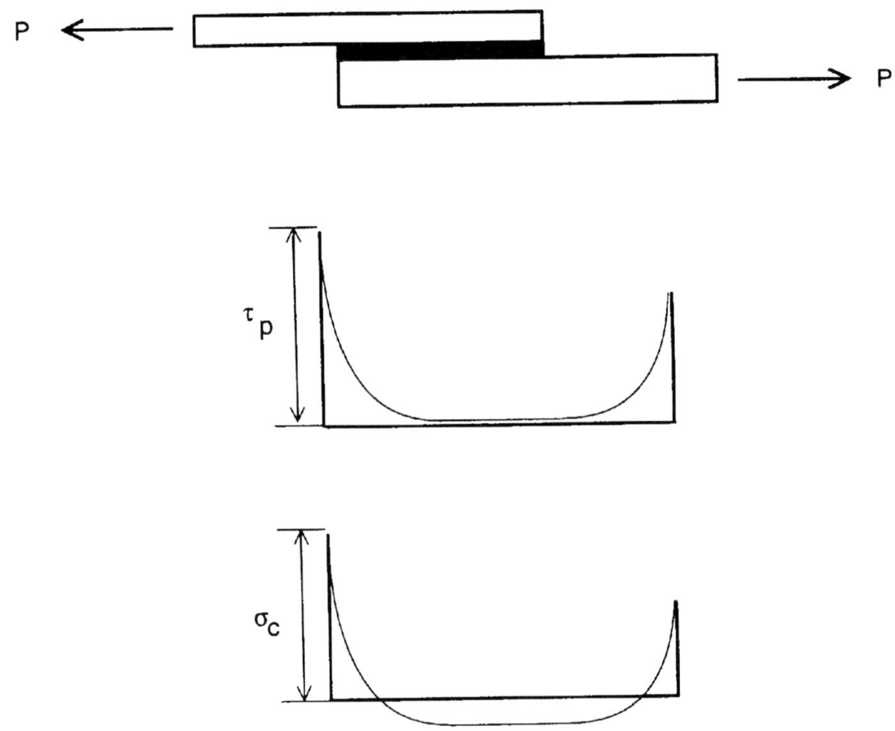

Figure 5.28 Typical shear and peel stress distributions in an unbalanced single-lap joint.

(9) The use of adhesive fillets and tapering the adherend ends may significantly increase the load-bearing capacity of the joint by reducing stress concentrations at the ends of the overlap, as illustrated in Figure 5.29. The effect of these methods is based on the increased flexibility of the joint ends, resulting in more favourable stress distribution within the overlap. Full utilization of these methods requires the use of a ductile adhesive, see 5.3.4.3.

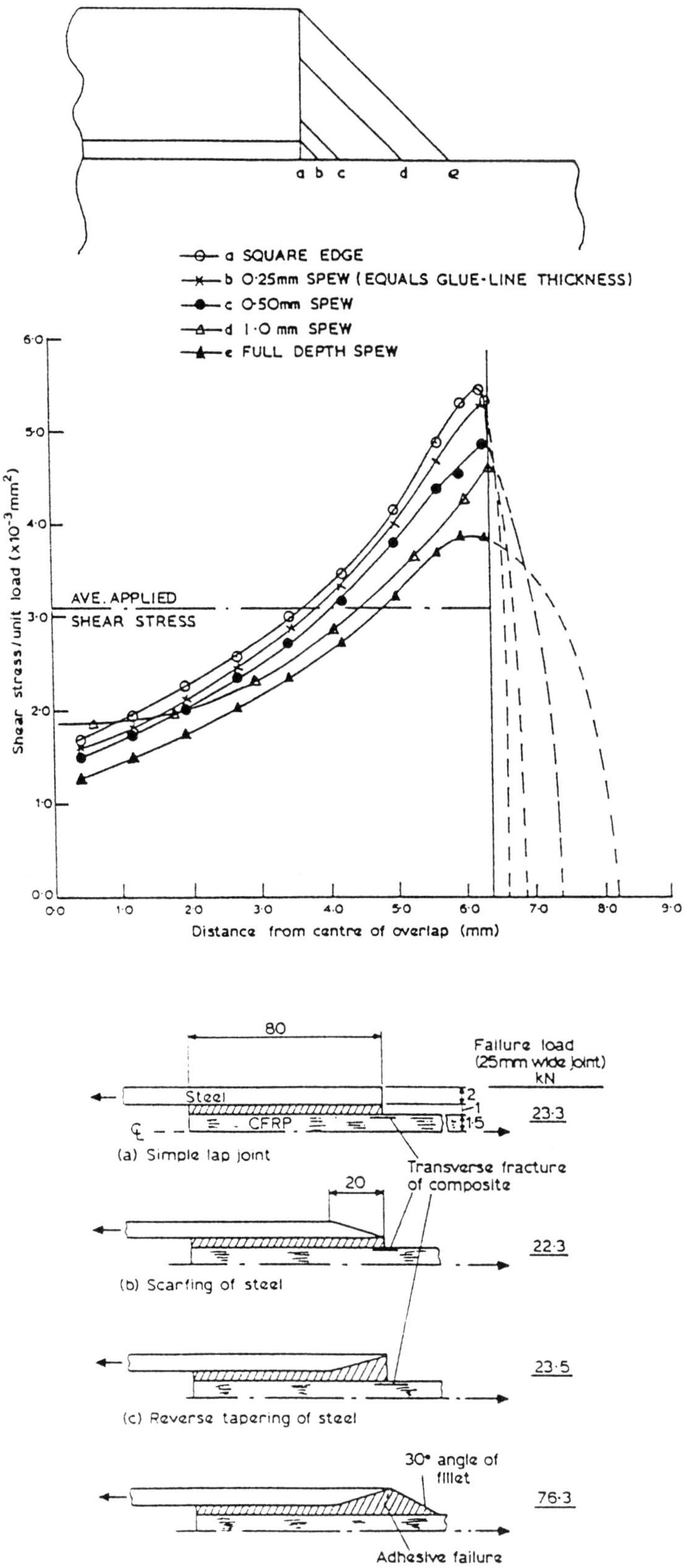

Figure 5.29 Reducing peak shear stresses by using adhesive fillets and adhered tapering on steel/carbon adherends (see reference 5.17, EUROCOMP Handbook).

(10) Further improvements in the load bearing capacity and joint efficiency of bonded joints can be achieved by using stepped-lap or scarfed configurations where the stress concentrations are reduced significantly and the peel stresses minimised or eliminated.

5.3.1.4 Adherends

P(1) Special attention shall be paid to the through-thickness properties of the adherends. Delamination due to peel and interlaminar shear stresses shall be considered.

P(2) The adherend bond surface in primary structural connections shall consist of fabrics, woven rovings or uni-directional reinforcements where at least half of the fibres are aligned parallel to the primary loading direction. No random mats shall be allowed on the bond surface in primary structural connections.

(3) For other than primary structural connections, random mats may be used on the bond surfaces.

5.3.1.5 Surface treatments

P(1) FRP components shall be stored and preconditioned prior to bonding as specified by the adhesive manufacturer or the component manufacturer.

P(2) All the bond surfaces shall be properly treated prior to bonding to achieve an adequate joint strength and to avoid adhesive failure.

P(3) Bond surfaces of FRP composites shall be prepared prior to bonding as follows:

1. Degrease with solvent using a clean absorbent material which does not itself contaminate the surface.

2. Abrade with medium grit abrasive paper or an appropriate grit blaster (when using abrasive paper care should be taken to select one with a suitable binder).

3. Repeat degreasing.

P(4) In grit blasting, the grit size and type and the blast pressure shall be adjusted so as not to damage the outer fibres.

(5) The use of a peel ply on the bond surface during the component manufacture is recommended to prevent surface contamination.

(6) When the laminate surface plies are of coarse fabrics or woven rovings (high areal weight, high tex number) a peel ply should be used on the bond surface. Alternatively, reinforcement plies with finer surface texture may be used on the bond surface.

P(7) Bond surfaces having peel plies shall be prepared prior to bonding as follows:

1. Remove the peel ply

2a. For other than primary structural connections:

Check that the bond surface has no contamination. If any contamination is observed or suspected, treat the surface according to P(3). Otherwise no further treatment is required.

2b. For primary structural connections:

Perform the surface treatment as stated in P(3) above.

P(8) Bond surfaces of metallic materials shall be prepared for bonding according to the instructions of the adhesive manufacturer or the component manufacturer.

P(9) Pretreated surfaces shall be bonded together immediately after surface treatment. If that is not possible, pretreated surfaces shall be protected according to the instructions of the adhesive manufacturer or the component manufacturer.

P(10) The maximum time allowed between the surface treatment and bonding shall be specified by the designer in accordance with the instructions of the material manufacturer.

P(11) Environmental conditions during bonding shall be those specified by the adhesive manufacturer or the component manufacturer.

5.3.2 Design principles

5.3.2.1 General principles

P(1) The loads for the bonded joint analysis shall be obtained from the global analysis of the structure. The bonded joint design ensures that the bondline is capable of transferring the applied loads between the joint members and that the joints members (adherends) are capable of with-standing the joint-induced internal loadings. The basic strength of the components to be joined under the applied external loads shall be evaluated as a part of the component design process (Chapter 4).

(2) The design of bonded joints may be based on:

- analytical models for plate-to-plate connections
- design guidelines supplemented by testing
- finite element analysis.

Only design methods which use closed form analytical models for plate-to-plate connections are presented in detail in this EUROCOMP Design Code.

P(3) Other than plate-to-plate connections should be decomposed to plate-to-plate connections using standard engineering practises. Particular care should be exercised when decomposing primary structural connections. When the internal distribution of loadings is not sufficiently understood, the design should be verified by testing. Finite element analysis may also be used to enhance the understanding of the load distribution.

(4) Guidelines for performing a finite element analysis of bonded joints are given in 5.3.2.2.

P(5) The design of bonded joints is based on the assumption of a perfect bond between the adhesive and adherends, i.e. cohesive failure in the adhesive or adherend always occurs before adhesive failure at the interface.

(6) The perfect bond assumption may become invalid, owing to:

- chemical incompatibility of the adhesive and adherends
- inadequate surface treatment
- environmental factors, i.e. temperature and humidity during bonding
- other bondline defects.

(7) Where there is any doubt about the quality of the adhesion, it is recommended that the joint be tested using samples identical to the actual joint in order to verify the failure mode.

(8) Depending on the results of the tests, the following actions should be taken:

- if the failure mode is not an adhesive failure, the assumption of a perfect bond can be used
- if an adhesive failure is encountered:

  (a) Improved surface treatment, another adhesive or another joint configuration is selected.

  (b) The strength of the modified joint is tested. If the test results prove the joint strength to be adequate but are based on a limited number of samples or the samples have been tested to other than ultimate conditions, only non-structural or secondary structural connections can be dimensioned using the procedures of this EUROCOMP Design Code. Where a complete set of tests is performed and the tests indicate that no adhesive failure is encountered, primary structural connections can be dimensioned using the procedures given in this EUROCOMP Design Code.

(9) Where there is any doubt about the quality of the adhesion but performing tests is considered impossible or impractical, only non-structural connections may be designed using the procedures of this EUROCOMP Design Code.

P(10) The design of bonded joints shall be based on the practical and achievable tolerances of manufacture and installation (see Chapter 6).

P(11) Should the design call for tolerances or accuracy beyond standard practice, the designer shall issue clear instructions, and the quality control during the manufacture shall be arranged accordingly.

5.3.2.2 Finite element analysis

(1) Finite element (FE) analysis using a global three-dimensional model (as defined in the EUROCOMP Handbook, 5.3.2.2) may be employed to assess the following:

- type of loadings to which the joint is subjected
- possible stiffness mismatch between the joint members
- possible large displacements or strains
- the magnitude of peel loadings in relation to other loads
- potential problem areas
- any other characteristic feature related to the overall behaviour of the joint.

P(2) The global model shall not be used directly for any detailed analysis or to verify the load-bearing capacity, as the results may be highly misleading or erroneous.

(3) It may be possible to study the critical details identified in the global analysis further by using a more accurate two-dimensional or axisymmetric model of the area of interest. The model should follow the guidelines given in the EUROCOMP Handbook.

(4) Any FE analyses of a bonded joint should be used primarily to investigate the behaviour of the joint or to improve the joint performance, not to establish the exact level of the load-bearing capacity.

P(5) The flexibility of the joint area, and particularly that of the bondline, in relation to the global structure, shall be considered and taken into account regardless of whether or not it has been incorporated into the FE model.

P(6) The level of reliability of the input data, particularly loads, geometry and material data, shall be appropriate to the accuracy of the analysis and to that required and expected from the results.

5.3.3 Selection of joining techniques

(1) The possible joining techniques for bonded joints covered in this EUROCOMP Design Code are the following:

- adhesively bonded joint
- laminated joint
- moulded joint
- bonded insert joint
- cast-in joint.

(2) For laminated sheets recommended joining techniques are:

- adhesive bonding
- laminating.

(3) For laminated sheets with a sandwich structure recommended joining techniques are:

- adhesive bonding
- laminating
- cast-in joining
- use of bonded inserts.

Sandwich components should be joined using connecting sections or doublers, see Figure 5.30.

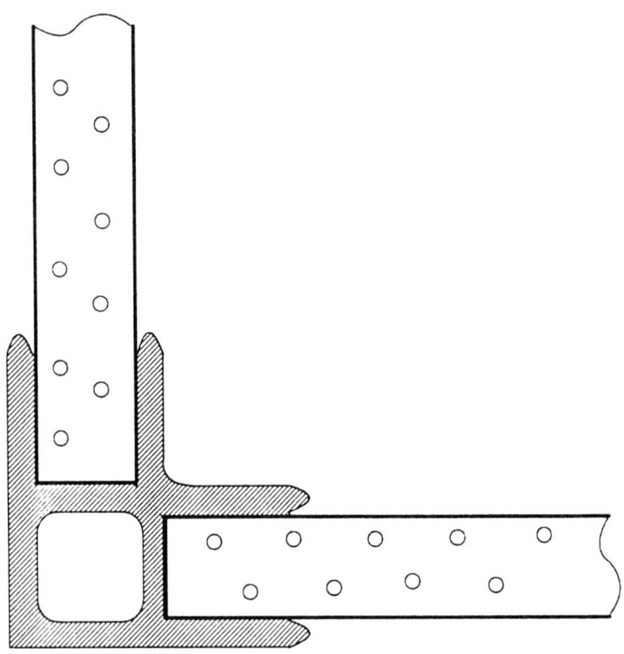

Figure 5.30 Angle joint of sandwich components.

(4) For pultruded sections recommended joining techniques (see Figure 5.31) are:

(a) intersecting sections:

- laminating
- bonded doublers
- moulding.

(b) axially connected sections:

- bonded doublers
- laminated doublers
- bonded or laminated scarf joints
- use of bonded inserts.

(c) basement/section connections:

- cast-in joining
- use of bonded inserts.

5.3.4 Adhesives

5.3.4.1 Classification and characterisation of adhesives

(1) Adhesives may be categorised in several ways. Typically, the categorisation is based on one of the following factors:

- type of adhesive
- curing process activation
- curing process requirements
- form of adhesive.

(2) The primary categorisation is based on the polymer type, i.e. whether the adhesive has a thermoset or thermoplastic base. Thermoset adhesives are infusible and insoluble after curing, while thermoplastic adhesives remain fusible and soluble and soften when heated.

(3) The most common thermoset adhesives are epoxies, phenolics and thermoset polyurethanes. The most widely used thermoplastic adhesives are acrylics (including anaerobics, hot melts, cyanoacrylates) and thermoplastic polyurethanes. A brief description of some adhesives is given in the EUROCOMP Handbook, 5.3.4.

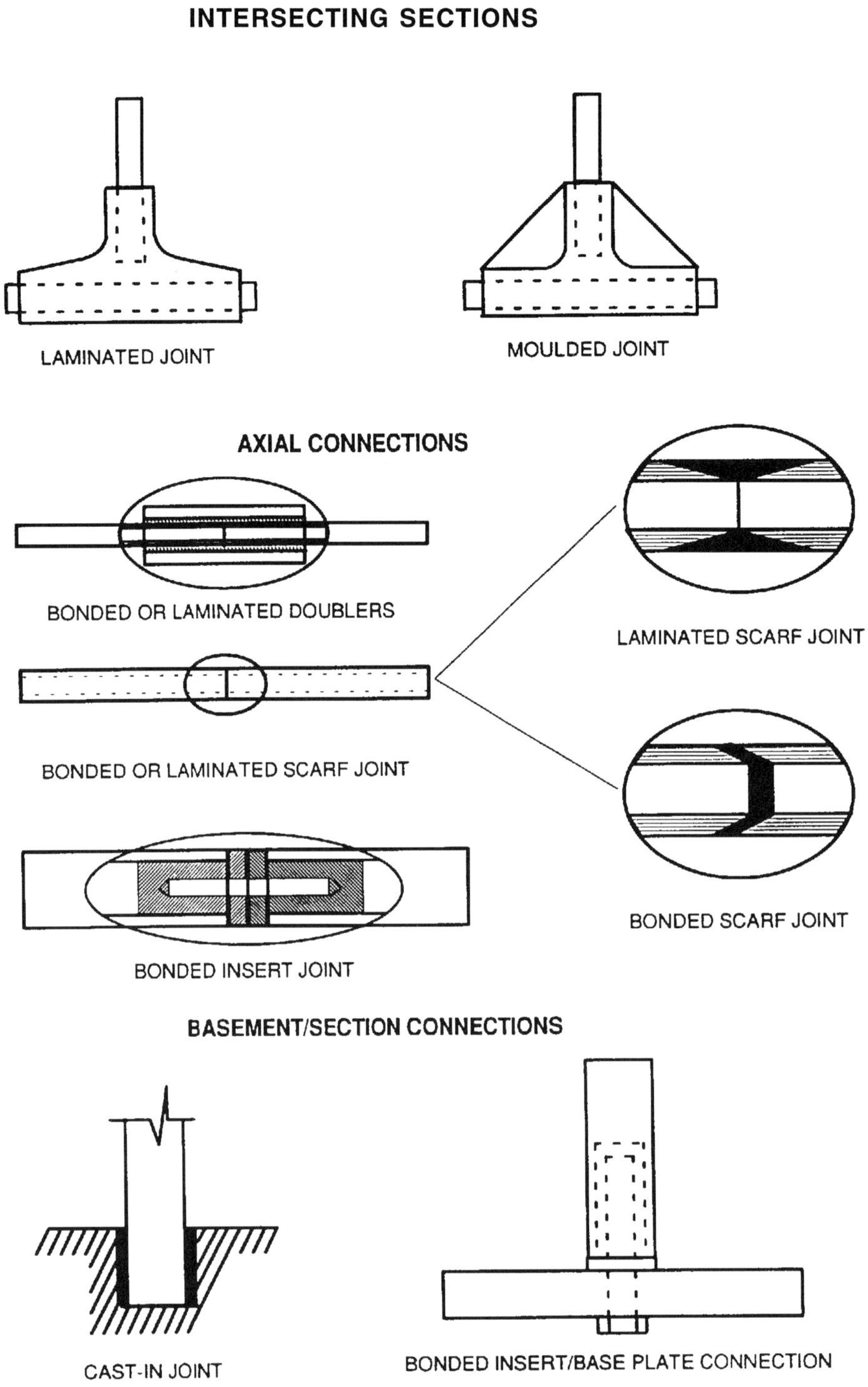

Figure 5.31 Joining techniques for pultruded sections.

(4) Adhesives may also be categorised according to curing process activation. The following methods are available:

- chemical activation
- solvent activation

- heat activation
- other activation.

(5) When the categorisation is based on curing process requirements, the following are typically considered:

- curing temperature and cycle
- curing pressure and cycle
- post-curing.

(6) The most common forms of adhesive are:

- paste
- liquid
- film.

5.3.4.2 Adhesive mechanical properties

P(1) The most important mechanical properties of adhesives that shall be considered in structural applications are:

- shear modulus
- shear strength
- maximum shear strain
- tensile modulus
- tensile (peel) strength.

(2) Material properties of the adhesives should be obtained from the adhesive manufacturers or by testing. For appropriate test methods, see EUROCOMP Handbook 7.3.3. Where no other data are available the values given in Table 5.4 may be used.

(3) Environmental factors, such as temperature, moisture and chemicals, may affect adhesive mechanical properties considerably.

P(4) Adhesives have either ductile or brittle behaviour, see Figure 5.32. This shall always be considered when the joint is subjected to any of the following loadings:

- peel loads
- impact loads
- creep or other long-term loads
- environmental loads.

(5) Adhesives creep under constant load even at room temperature and especially at elevated temperatures. Absorbed humidity increases the creep.

(6) In general, thermoset adhesives have better creep resistance than thermoplastic adhesives, which are susceptible to creep even at room temperature.

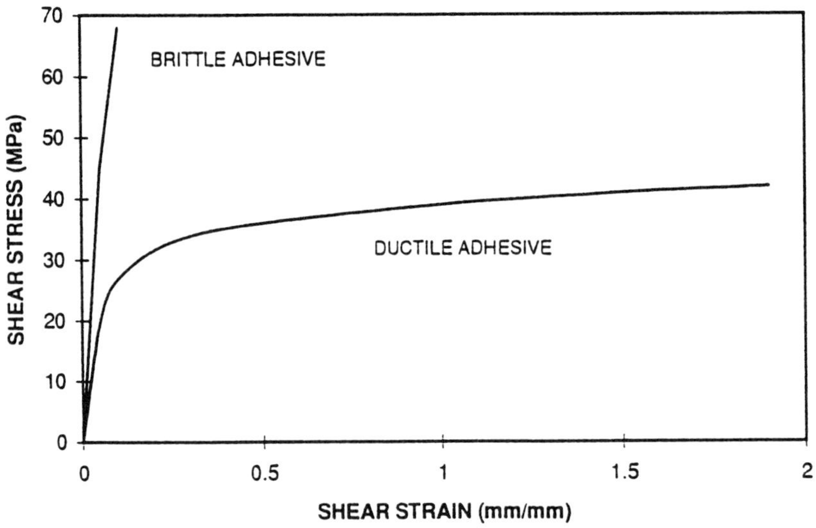

Figure 5.32 Typical brittle and ductile adhesive behaviour.

5.3.4.3 Adhesive selection

(1) Adhesive selection should be based on previous experience or on a specific selection process.

(2) Preliminary adhesive selection should be performed using any unbiased method which includes all the factors required for a reliable selection procedure. Guidelines of applicable selection procedures are given in the EUROCOMP Handbook.

(3) Currently any selection process can do no more than suggest one or more generic types of adhesives that are worthy of more detailed examination.

(4) A detailed selection within the most promising groups of adhesives can be based on information provided in this code, in published adhesive data or in data sheets of adhesives, manufacturers.

P(5) Whenever necessary, tests shall be performed to verify the adhesive material data.

(6) Factors to be considered in adhesive selection are:

- adherend materials
- environmental factors
- applied loading
- joint geometry restrictions
- bonding and curing processes
- cost
- special requirements, including health and safety.

Table 5.4 Characteristic mechanical properties of certain adhesives.

Adhesive	Description	G_a (N/ mm^2)	E_a (N/ mm^2)	τ_{ult} (N/ mm^2)	$\tau_{o,k}$ (N/ mm^2)	ε_{ult} (N/ mm^2)	□
Araldit AV138 (Ciba Geigy)	Epoxy 2-part paste		9160	18.4		17.7	5.3.27
EC 2216 (3M)	Epoxy modified 2-part paste	270[1]		21.3		20.7	5.3.28 5.3.27
Araldit 2015 (Ciba Geigy)#	Epoxy 2-part paste	670[2]		21.0[2]	14.2[2]		5.3.26
DP 460 (3M)#	Epoxy 2-part paste			32.9[2]	28.1[2]		5.3.26
Foss Than 2 K 1897 (Casco Nobel)#	Polyurethane 2-part paste	145[2]		11.8[2]	1.6[2]		5.3.26

\# Results from EUROCOMP test programme at Helsinki University of Technology
[1]) Napkin ring torsion tests
[2]) Thick adherend test
□ Sections in EUROCOMP Handbook

P(7) The material compatibility between the adhesive and the adherends shall be checked in cooperation with the adhesive manufacturer or by testing, as the possible material incompatibility may significantly reduce the adhesion.

(8) When the adherend materials have dissimilar stiffnesses or coefficients of linear thermal expansion, a ductile adhesive is recommended.

(9) Where possible, ductile adhesives should be selected in preference to brittle adhesives, see Figure 5.32.

5.3.5 Adhesively bonded joints

5.3.5.1 General

P(1) These design rules are applicable to static loading conditions only.

(2) Material properties of the adherends may be obtained, for example, from 4.11 or by testing.

5.3.5.2 Selection of joint configuration

(1) For in-plane loaded plate-to-plate joints the recommended joint configurations and allowable loading conditions are presented in Figure 5.33.

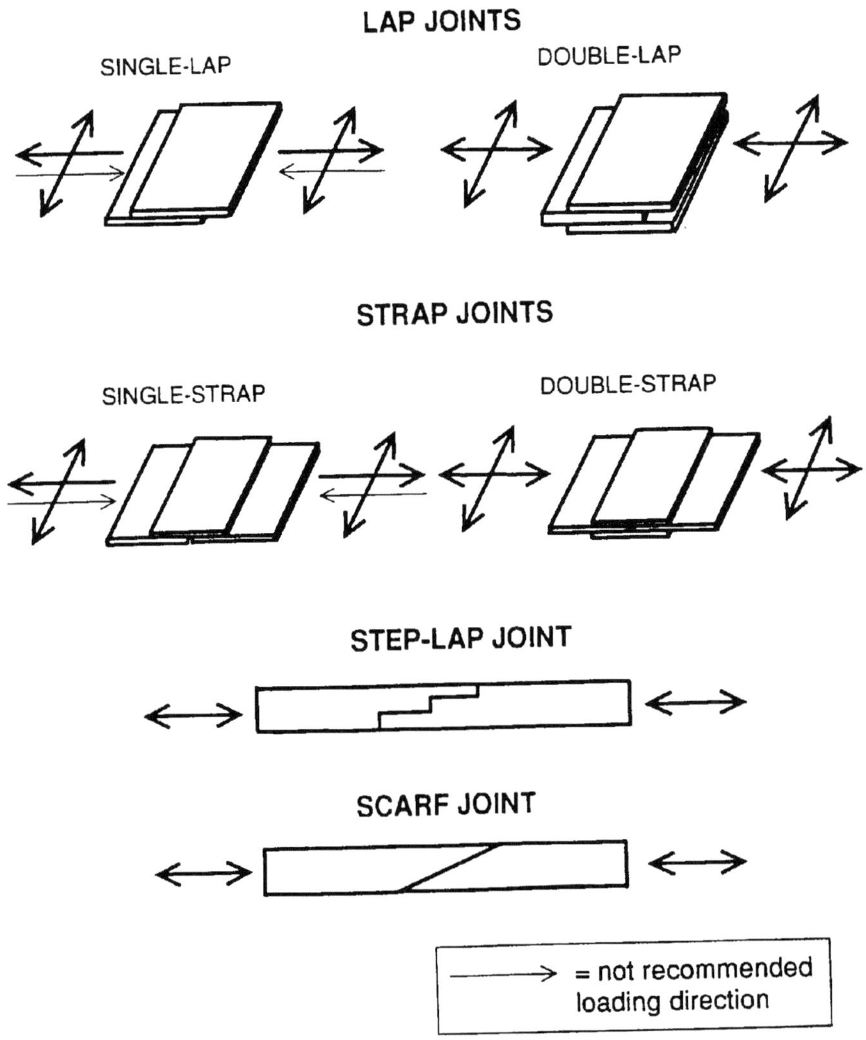

Figure 5.33 Recommended joint configurations for in-plane loaded plate-to-plate connections.

(2) The relative joint strength of various axially loaded joint configurations for plate-to-plate connections as a function of the adherend thickness is presented in Figure 5.34.

(3) For angle and tee-joints the recommended joint geometries and allowable loading conditions are presented in Figure 5.35.

(4) For laminate thicknesses greater than 5 mm, joint configurations other than single-lap or double-lap are recommended, as with greater thicknesses these configurations become inefficient.

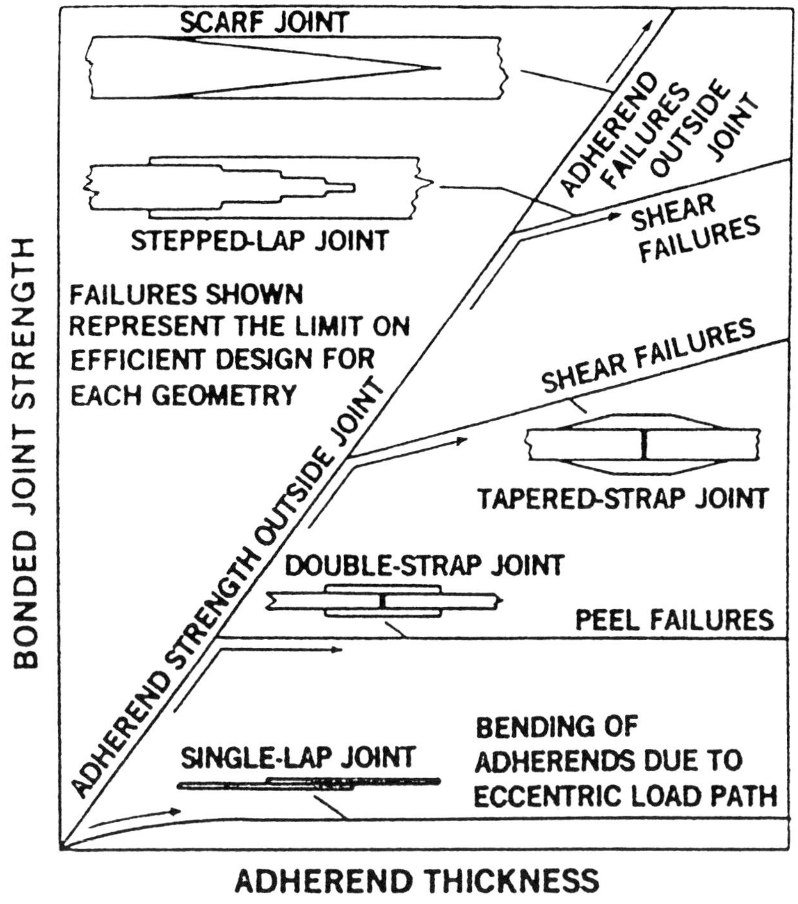

Figure 5.34 Relative joint strength of various joint configurations (see Reference 5.24, EUROCOMP Handbook).

P(5) In the case of laminate thicknesses below 5 mm the use of scarf and step-lap configurations shall be considered with care, as producing the required shapes on the adherends may become complicated. Also the bonding may become impossible if adequate jigging cannot be provided, see 5.3.5.11.

5.3.5.3 Design of lap and strap joints

P(1) Two design methods are given for the design of lap and strap joints: a simplified procedure and a rigorous procedure (5.3.5.4 and 5.3.5.5 respectively). In both procedures lap and strap joints are treated identically. In the design of strap joints the strap shall also be treated as one of the adherends.

P(2) The simplified design procedure is based on experimental test results produced by standardised lap shear tests or by corresponding test arrangements. An additional coefficient is introduced to take into account the inaccuracy of the design method.

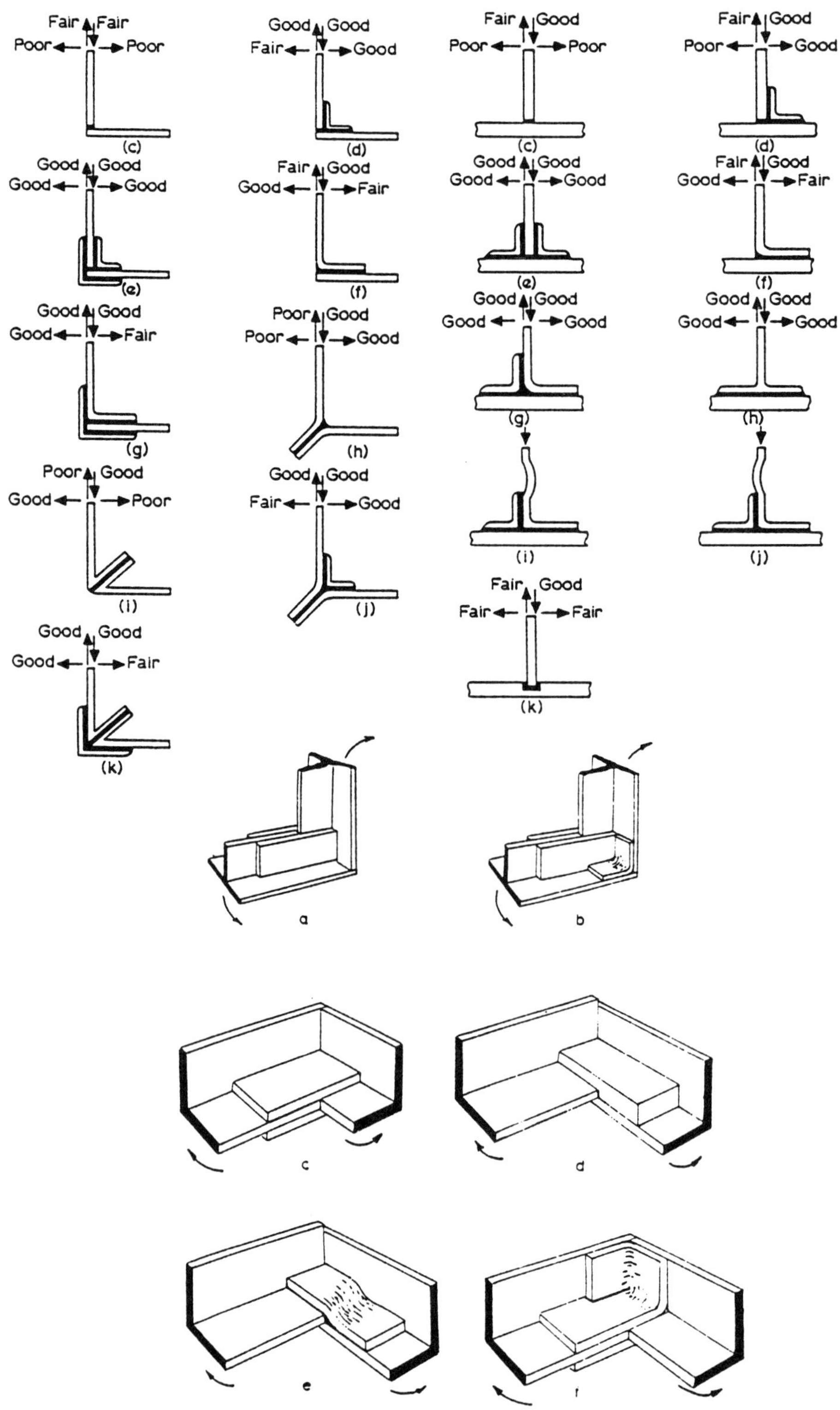

Figure 5.35 Recommended joint arrangements and loadings for tee and angle joints (see reference 5.17, EUROCOMP Handbook).

(3) The simplified procedure may be used for the preliminary design of all connections and for the final design of secondary and non-structural connections.

P(4) The rigorous procedure is based on fully elastic adhesive behaviour and closed-form analytical models. Adhesive plastic behaviour may also be utilised when determining the temporary maximum joint resistance.

P(5) The rigorous procedure shall be used for the final design of primary structural connections.

(6) The use of the rigorous procedure leads to more effective designs with higher load carrying capacities than the use of the simplified procedure due to the built-in conservatism in the latter method. Therefore, the use of the rigorous procedure is recommended for all structural connections.

(7) The models used in the rigorous procedure have been developed for lap joints. The use of the same models for the corresponding strap joints brings added conservatism to the design.

5.3.5.4 Simplified design procedure for lap and strap joints

P(1) The simplified procedure shall only be used for the preliminary design of primary structural connections or for the final design of secondary and non-structural connections, that are not critically loaded.

P(2) The simplified procedure shall only be used for tensile loaded joints.

(3) The simplified design procedure for lap joints is based on the experimental results from standardised lap shear tests such as ISO 4587-1979, ASTM D 3163 or corresponding test arrangements.

(4) The following properties of the lap shear specimen and the actual joint should be identical within reasonable accuracy:

- adhesive
- joint configuration
- adherend tensile stiffness
- adherend flexural stiffness
- adherend thicknesses
- environmental conditions
- preconditioning
- joint preparation
- curing.

(5) The design proceeds as follows:

Step 1

Define the ultimate lap shear load using existing data or by performing experiments. Express the ultimate load as load per unit width (N_u).

Step 2

Define the actual joint lap length to be twice the lap length of the one used in the test specimen. However, a minimum lap length of 50 mm shall be used.

Step 3

Calculate the maximum joint resistance $N_{t,Rd}$ of the actual joint as follows:

$$N_{t,Rd} = \frac{N_u}{\gamma_m} \tag{5.8}$$

where γ_m is partial material safety factor, see 5.1.10.

Step 4

Investigate the magnitude of the design tensile load $N_{t,Sd}$ with respect to the joint resistance as follows:

$$N_{t,Sd} = F_k \, \gamma_f \, C_m < N_{t,Rd} \quad \text{or} \quad N_{t,Sk} = S_k \, \gamma_f \, C_m < N_{t,Rd} \tag{5.9}$$

where γ_f is the partial coefficient for the load, see 2.2.2.4, and C_m has the value of 10.0.

Step 5

Consider whether the joint should be analysed according to the rigorous procedure given in 5.3.5.5.

5.3.5.5 Rigorous design procedure for lap and strap joints

P(1) The following conditions shall be satisfied:

- for the adhesive the maximum adhesive shear stress shall be equal to or less than the maximum allowable adhesive shear stress

$$\tau_{0\,max} \leq \tau_{0\,allowable}$$

- for the adhesive the maximum tensile (peel) stress shall be equal to or less than the maximum allowable adhesive tensile stress

$$\sigma_{0\,max} \leq \sigma_{0\,allowable}$$

- for the adherend the maximum tensile stress in the through-thickness direction shall be equal to or less than the maximum allowable adherend through-thickness tensile stress

$$\sigma_{0\,max} \leq \sigma_{z\,allowable}$$

(2) The material properties required for the rigorous design procedure are listed in Table 5.5.

P(3) The characteristic adhesive shear strength $\tau_{0,k}$ shall be taken equal to the shear stress in the adhesive at the elastic limit τ_e if not specified otherwise. See also Figure 5.34 of the EUROCOMP Handbook.

(4) The flow chart for the rigorous design procedure is shown in Figure 5.36.

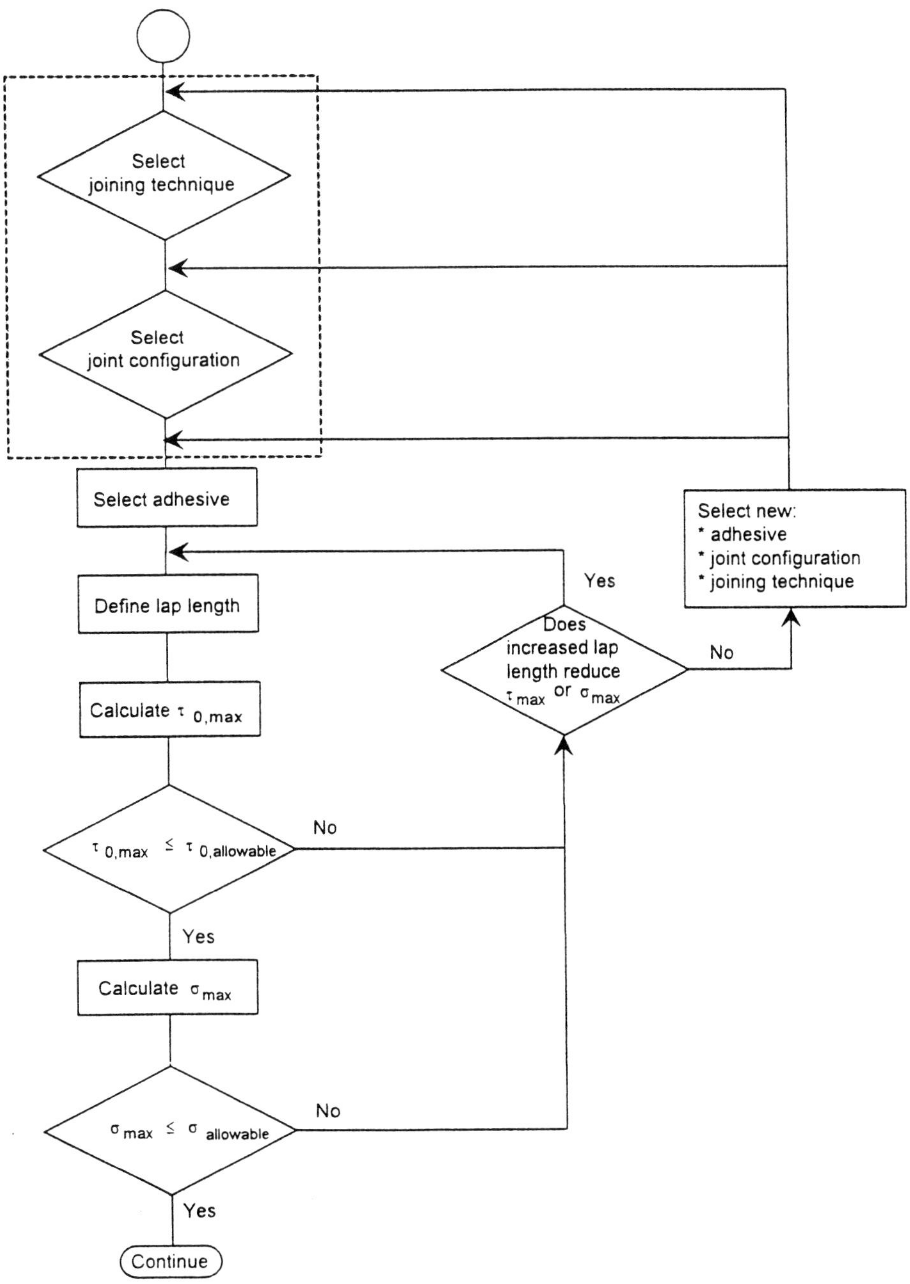

Figure 5.36 Rigorous design procedure for lap joints.

Table 5.5 List of material properties required for the rigorous design procedure.

Tensile and compressive shear loading	
G_a	adhesive shear modulus
E_a	adhesive elastic tensile or compressive modulus
E_a'	effective adhesive transverse modulus
$\tau_{0,k}$	characteristic adhesive shear strength
$\sigma_{0,k}$	characteristic adhesive tensile or compressive strength
E	adherend tensile or compressive modulus
ν	adherend Poisson's ratio
$\sigma_{z,k}$	characteristic adherend through-thickness tensile strength
Theoretical maximum joint strength (in addition to above mentioned)	
τ_p	plastic adhesive shear stress
γ_e	elastic adhesive shear strain
γ_p	plastic adhesive shear strain
In-plane shear loading (in addition to tensile shear loading)	
G_i, G_o	adherend shear modulus

(a) Tensile shear loading

P(1) This design procedure shall be applied only to tensile shear loaded single- and double-lap and single- and double-strap joints.

(2) The design proceeds as follows:

Step 1

Lap length (L = 2c) is obtained using Figure 5.37 where parameter β/t is defined as follows:

$$\frac{\beta}{t} = \sqrt{\frac{8\,G_a t}{E\,t_a}} \qquad (5.10)$$

where

G_a = adhesive shear modulus
E = adherend Young's modulus = E_1 = E_2
t = minimum adherend thickness
t_a = adhesive layer thickness

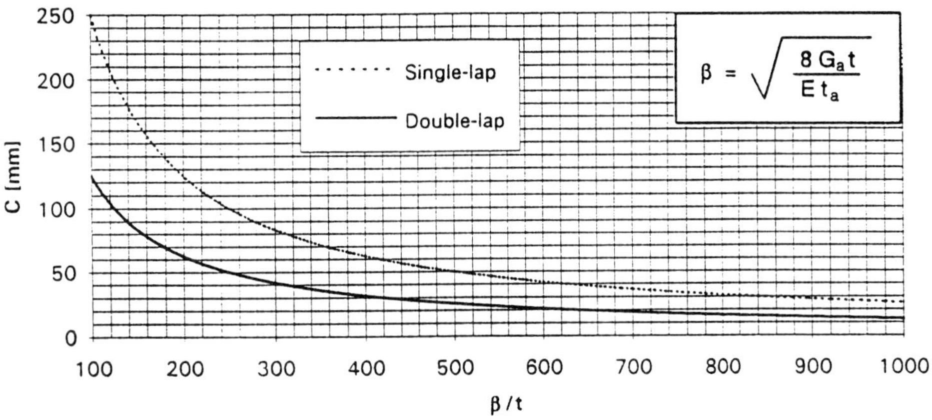

Figure 5.37 Half lap length (c) as a function of parameter β/t.

Step 2a

For single-lap and single-strap joints calculate the maximum adhesive shear stress:

$$\tau_{0\,max} = \frac{\sigma}{8}\,(1 + 3k)\sqrt{8\,\frac{G_a t}{E t_a}} \tag{5.11}$$

where

$$\sigma = \frac{P_k d\,\gamma_f}{t}$$

$$k = \frac{\cosh(u_2 c)\;\sinh(u_1 L)}{\sinh(u_1 L)\;\cosh(u_2 c) + 2\sqrt{2}\,\cosh(u_1 L)\;\sinh(u_2 c)}$$

$$u_1 = 2\sqrt{2}\,u_2$$

$$u_2 = \frac{1}{\sqrt{2}\,t}\sqrt{3(1 - \nu^2)\,\frac{\sigma}{E}}$$

and

P_k = characteristic load per unit width, F_k or S_k
ν = adherend Poisson's ratio.

Step 2b

For double-lap and double-strap joints calculate the maximum adhesive shear stress (for notations see Figure 5.38):

$$\tau_{0\,max} = \frac{\lambda P_k \gamma_f}{4}\left[\frac{\cosh(\lambda c)}{\sinh(\lambda c)} + \Omega\,\frac{\sinh(\lambda c)}{\cosh(\lambda c)}\right] \qquad (5.12)$$

where

$$\lambda^2 = \frac{G_a}{t_a}\left(\frac{1}{E_o t_o} + \frac{2}{E_i t_i}\right)$$

and

P_k = characteristic load per unit width
Ω = the greater of $(1-\Psi)/(1+\Psi)$ or $(\Psi-1)/(1+\Psi)$

where

$$\Psi = \frac{E_i t_i}{2 E_o t_o}$$

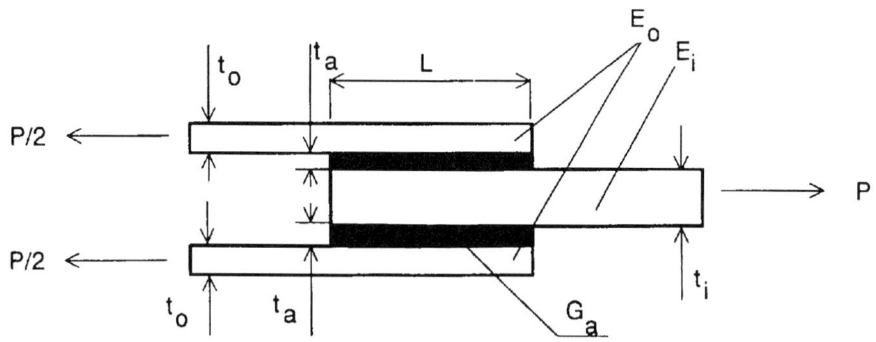

Figure 5.38 Double-lap joint notations.

Step 3

Investigate the magnitude of the adhesive shear stress maximum value as follows:

$$\tau_{0\,max} \leq \tau_{0\,allowable} \qquad (5.13)$$

where

$$\tau_{0\,allowable} = \frac{\tau_{0,k}}{\gamma_m}$$

If the condition given in Equation (5.13) is satisfied, go to step 4. If the condition is not satisfied, act as indicated in Figure 5.36.

Step 4a

For single-lap and single-strap joints calculate the maximum adhesive peel stress:

1. Calculate the value of λ as follows:

$$\lambda = \frac{c}{t}\left(\frac{6\,E_a\,t}{E\,t_a}\right)^{0.25} \tag{5.14}$$

2. If $\lambda > 2.5$, then:

$$\frac{\sigma_{0\,max}}{\sigma} = \frac{k}{2}\sqrt{6\,\frac{E_a}{E}\,\frac{t}{t_a}} + k'\,\frac{t}{c}\sqrt{6\,\frac{E_a}{E}\,\frac{t}{t_a}} \tag{5.15}$$

$$\frac{\sigma_{0\,max}}{\sigma}\left(\frac{c}{t}\right)^2 = \lambda^2\,\frac{k}{2}\,\frac{\sinh(2\lambda) - \sin(2\lambda)}{\sinh(2\lambda) + \sin(2\lambda)} - \lambda k'\,\frac{\cosh(2\lambda) + \cos(2\lambda)}{\sinh(2\lambda) + \sin(2\lambda)} \tag{5.16}$$

where

$$k' = k\,\frac{c}{t}\sqrt{3(1 - \nu^2)\,\frac{\sigma}{E}}$$

and others as above.

Step 4b

For double-lap and double-strap joints calculate the maximum adhesive peel stress:

$$\sigma_{0\,max} = \tau_{0\,max}\left(\frac{3\,E_a'(1 - \nu^2)\,t_o}{E_o\,t_a}\right)^{0.25} \tag{5.17}$$

where

E_a' = effective transverse modulus of the adhesive

ν = Poisson's ratio of the outer adherend.

Step 5

Investigate the magnitude of the adhesive peel stress with respect to the allowable adhesive tensile strength and the allowable adherend through-thickness tensile strength as follows:

$$\sigma_{0\,max} \le \sigma_{0\,allowable} \tag{5.18}$$

and

$$\sigma_{0\,max} \le \sigma_{z\,allowable} \tag{5.19}$$

where

$$\sigma_{0\,allowable} = \frac{\sigma_{0,k}}{\gamma_m}$$

where γ_m is the adhesive partial material safety factor (see 5.1.10)

and

$$\sigma_{z\,allowable} = \frac{\sigma_{z,k}}{\gamma_m}$$

where γ_m is the adherend partial material safety factor (see 2.3.3.2).

If the conditions given in Equations (5.18) and (5.19) are satisfied, the design procedure is completed. If either of the conditions given in Equations (5.18) and (5.19) is not satisfied, act as indicated in Figure 5.36.

P(3) The theoretical temporary maximum joint resistance for double-lap and double-strap joints with long overlaps shall be specified by the lesser of the values given by the following pair of equations:

$$\begin{aligned} N_{Ru} &= \sqrt{2\,\tau_p t_a \left(\frac{\gamma_e}{2} + \gamma_p\right) 2\,E_i t_i \left(1 + \frac{E_i t_i}{2\,E_o t_o}\right)} \\ N_{Ru} &= \sqrt{2\,\tau_p t_a \left(\frac{\gamma_e}{2} + \gamma_p\right) 4\,E_o t_o \left(1 + \frac{2\,E_o t_o}{E_i t_i}\right)} \end{aligned} \tag{5.20}$$

where

τ_p = plastic adhesive shear stress in the elastic-plastic adhesive model

γ_e = elastic adhesive shear strain in the elastic-plastic adhesive model

γ_p = plastic adhesive shear strain in the elastic-plastic adhesive model.

P(4) If there is any doubt that the adherend allowable stress or strain may be exceeded due to combined in-plane loading and eccentricity-induced local bending, the possible failure of the adherend shall be investigated.

(b) Compressive loading

P(1) Compressive loading shall not be applied to unsupported single-lap and single-strap joints.

P(2) Compressively loaded double-lap, double-strap, supported single-lap, and supported single-strap joints shall be designed as the corresponding tensile loaded joint in respect of adhesive shear stresses. The load shall be taken as negative.

(3) In compressive loading, peel stresses are negligible and need not be calculated.

(c) In-plane shear loading

P(1) The notation to be used for an in-plane shear loaded double-lap joint is given in Figure 5.39.

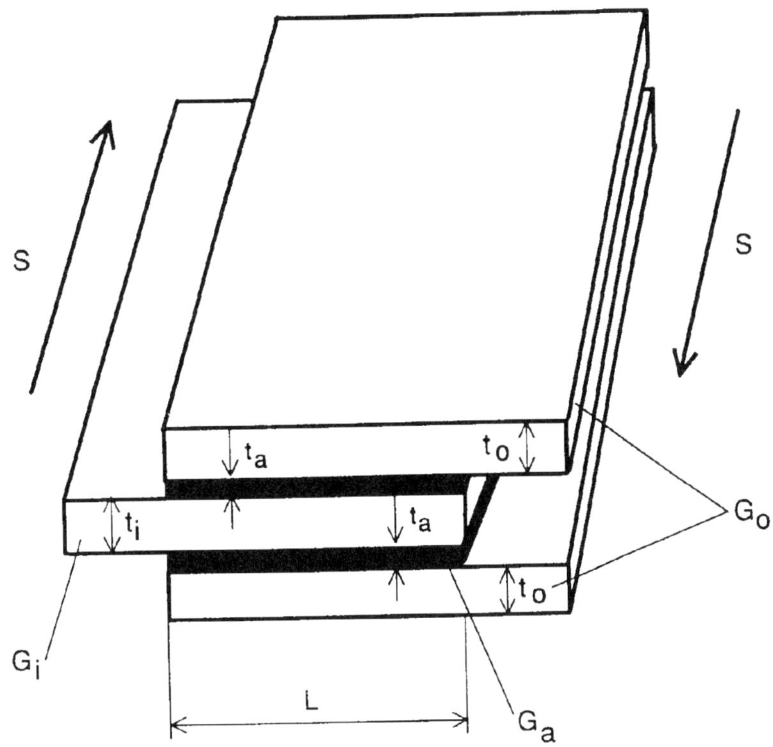

Figure 5.39 In-plane shear loaded double-lap joint.

P(2) In-plane loaded single-lap and single-strap joints shall be designed using the corresponding equations for tensile shear loaded joints and using the parameter conversions indicated in Table 5.6.

P(3) In-plane shear loaded double-lap and double-strap joints shall be designed using the corresponding equations for tensile shear loaded joints and using the parameter conversions of Table 5.6.

Table 5.6 Equivalent parameters for tensile shear and in-plane shear loadings.

Tensile shear loading	*In-plane shear loading*
τ_0	τ_0
$\gamma, \gamma_e, \gamma_p$	$\gamma, \gamma_e, \gamma_p$
G_a, t_a	G_a, t_a
E, E_i, E_o	G, G_i, G_o
t, t_i, t_o	t, t_i, t_o
L	L

(d) Combined loading

P(1) When lap and strap joints are loaded simultaneously by more than one of the above listed loadings, the joint resistance shall be determined using the maximum strain failure criterion applied to the resultant shear strain vector in the adhesive.

P(2) If there is any possibility that the adherend allowable stress or strain may be exceeded due to combined in-plane loading and eccentricity- induced local bending, the possible failure of the adherend shall be investigated.

5.3.5.6 Design of scarf joints

P(1) If the scarf angle $\theta < 20°$ (see Figure 5.40), the joint resistance shall be calculated as follows:

$$N_{Rd} = \frac{\tau_{0\ \text{allowable}}\ t}{\cos\theta\ \sin\theta} \qquad (5.21)$$

where

$\tau_{0\ \text{allowable}}$ = allowable adhesive shear strength.

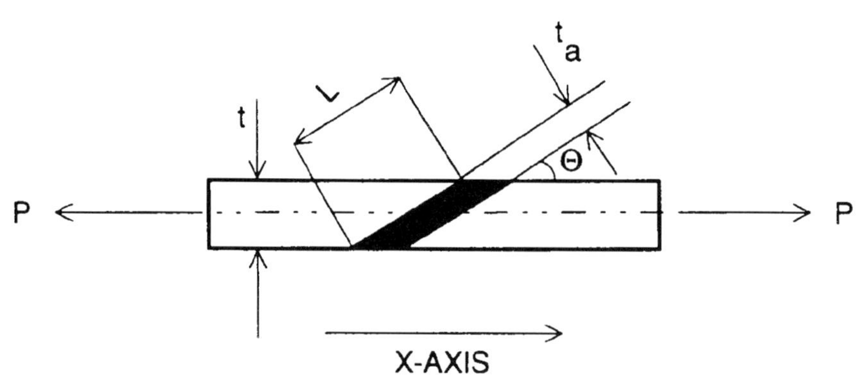

Figure 5.40 Scarf joint notation.

P(2) The joint shall satisfy the following criterion:

$$P_k \gamma_f \leq N_{Rd} \tag{5.22}$$

(3) If the condition given in Equation (5.22) is not satisfied, the following approaches may be taken:

- reduce the scarf angle and recalculate the joint strength
- consider the use of another joint configuration or joining technique.

(4) At very small scarf angles the joint resistance obtained from Equation 5.21 exceeds the component resistance, indicating that the design has become component critical instead of being joint critical.

(5) When designing joints with small scarf angles, it shall be ensured that these scarfs can be manufactured. It is recommended that 2° be considered as an absolute practical minimum, but typically even greater angles require special tooling and jigging.

P(6) If the scarf angle $\theta \geq 20°$, the joint shall be considered to be a butt joint and treated accordingly.

P(7) Knife edges are not allowed in scarf joints. The adherend ends should have a minimum thickness of 1.0 mm.

5.3.5.7 Design of butt joints

P(1) Butt joints shall not be used in load-bearing connections, owing to their inferior resistance to bending loads.

P(2) Instead of butt joints a strap, scarf or lap joint configuration shall be used (see Figure 5.41). The joint design is then undertaken according to the procedures for strap or scarf joints respectively. When using strap configurations the adherend ends shall also be bonded.

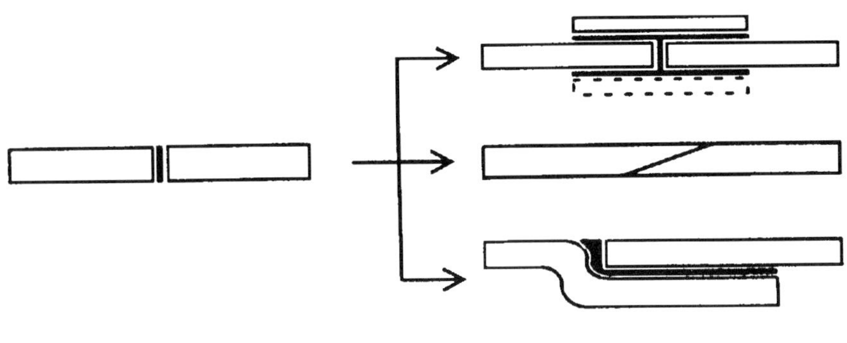

Figure 5.41 Replacing a butt joint configuration using other configurations.

P(3) Elastic adhesives or sealants may be used in non-structural butt joint connections for sealing. The adhesive or sealant shall not be expected to carry any load.

(4) When the joint is supported against bending loads, the butt joint configuration can be used in load-bearing non-critical connections.

P(5) The design of supported butt joints shall be based on testing.

5.3.5.8 Design of step-lap joints

(1) Notation and a typical geometry for a stepped lap joint are given in Figure 5.42.

P(2) From one to three steps the joint shall be analysed as a single-lap joint taking each lap as a single-lap of thickness t_i. For each step:

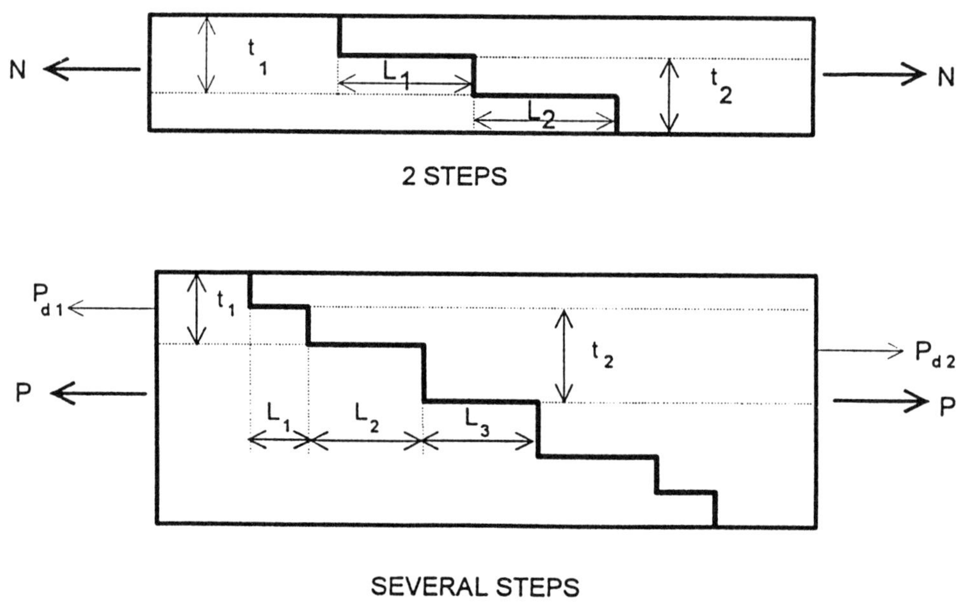

Figure 5.42 Step-lap joint notation.

P_d = P_{d1}, P_{d2} or P_{d3}
L = L_1, L_2 or L_3
t_a = t_{a1}, t_{a2} or t_{a3}

P(3) For four or more steps the analysis of a double-lap joint shall be carried out by taking the step thickness as follows:

$$t_{o,DL} = t_{i,ST} \quad \text{and} \quad t_{i,DL} = 2\,t_{i,ST}$$

where subscripts DL and ST refer to double-lap and step-lap joint respectively.

P(4) The step loads P_{di} shall be determined by using any generally approved method. The following condition shall always be satisfied:

$$\sum_{i=1}^{n} P_{di} = P_d \tag{5.23}$$

where n is the number of steps.

(5) If the stiffness of any single step differs significantly from that of the other steps, this should be taken into account when distributing the external loads into the individual step loads.

P(6) The analysis stated in either P(2) or P(3) shall be repeated for each step. However, if the step loads and the step geometries are identical in each step, the analysis may be performed for one step only.

(7) The thickness and the L/t ratio of the outer steps should be less than or equal to the respective values of any inner step.

(8) Increasing the length of any or all of the steps does not increase the joint strength indefinitely. A more effective method is to increase the number of steps.

(9) As step-lap joints have similar features to lap and double-lap joints, every design aspect that has been indicated in the design of lap joints should also be considered in the design of step-lap joints. In step-lap joints factors such as fibre orientation on the bond surface still have a major effect on the joint strength (see 5.3.1.4), while tapering the adherend ends becomes irrelevant.

5.3.5.9 Design of angle joints

(1) A typical bonded angle joint is presented in Figure 5.43. The joint is achieved by bonding two angle sections (i.e. bonding angles) to the FRP members meeting at the joint.

P(2) When joining sections to panels or to other sections, both the web and the flange parts of the members shall be joined by using bonding angles or straps, as appropriate.

(3) Factors affecting the performance of the joint are:

- lay-ups of the adherend members
- type of adhesive
- bondline thicknesses
- lay-up, thickness and length of the overlap of the bonding angles.

P(4) The choice of the design variables shall be such that the joint is capable of transferring the design loads between the members. Assuming the bonding angle sections to be made of similar material to the members to

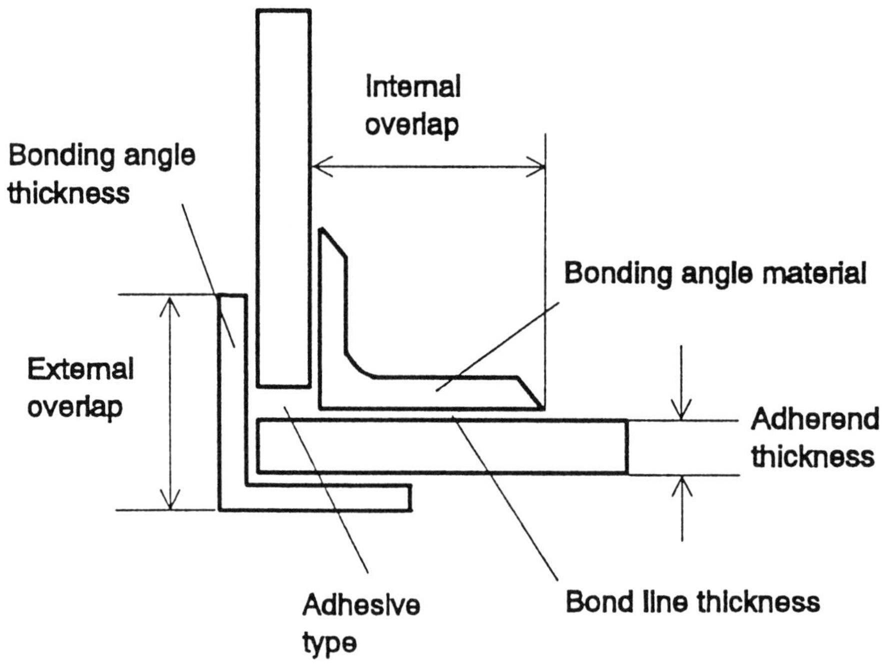

Figure 5.43 Design variable of a typical bonded angle joint.

be joined, the initial estimation of the joint parameters may be made as follows:

$$t_{BA} = t/2$$

$$L_{OL} \geq 10\,t$$

where:

t_{BA} = thickness of bonding angle
t = thickness of the thicker of the two adherends
L_{OL} = length of overlap of the bonding angle.

(5) Potential failure modes are:

- peeling of the ends of the bonding angle from the adherend members
- failure in the adhesive layer due to shear
- failure in the adherend members; this mode, however, should be designed out by suitable proportioning.

P(6) The integrity and sufficiency of the joint shall be verified by testing or be supported by relevant performance or test data.

(7) For primary structural connections this should be supplemented by finite element analysis.

P(8) The possible loading and boundary conditions to be used in the testing are illustrated in Figure 5.44. The exact choice will be dependent on the requirements imposed on the joint by its location in the overall structure.

(9) The finite element analysis shall be such as to enable a detailed study of the response patterns in the bonding angles, adhesive layers and adherend members. Particular items of interest are:

- displacement at the point of application of the load
- peel/cleavage stresses at the ends of the bonding angles
- shear stresses in the adhesive layer
- in-plane and through-thickness stresses in the adherends.

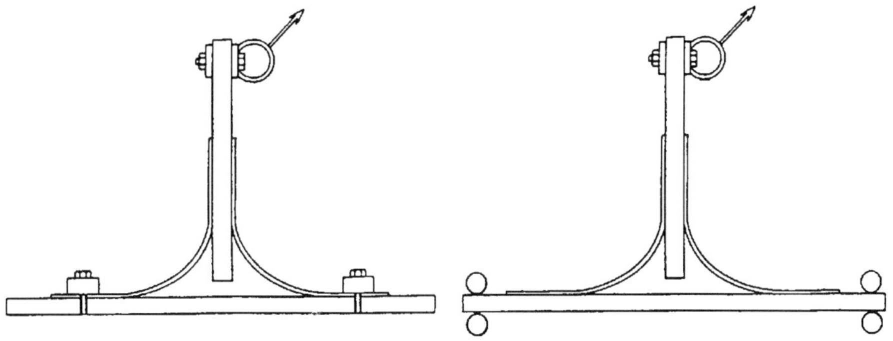

Figure 5.44 Possible loading and boundary conditions for testing of tee joint.

5.3.5.10 Design of tee joints

(1) A typical bonded tee joint is presented in Figure 5.45. The joint is achieved by bonding angle sections (i.e. bonding angles) on both sides of the leg of the tee.

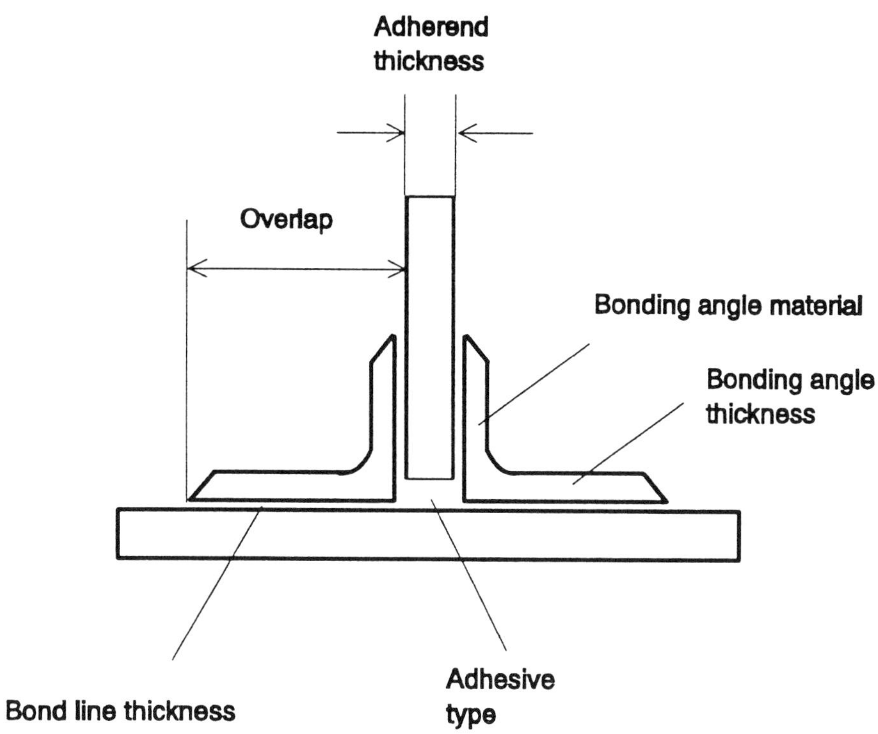

Figure 5.45 Design variable of a typical bonded tee joint.

P(2) When joining sections to panels or to other sections, both the web and flange parts of the leg shall be joined by using bonding angles or straps, as appropriate.

(3) Factors affecting the performance of the joint are the lay-ups of the adherend members, type of adhesive, bondline thicknesses, and the lay-up, thickness and length of the overlap of the bonding angles.

P(4) The choice of the design variables shall be such that the joint is capable of transferring the design loads between the adherend panels. Assuming the bonding angle sections to be made of similar material to the members to be joined, the initial estimation of the joint parameters may be made as follows:

$$t_{BA} = t/2$$

$$L_{OL} \geq 10\,t$$

where:

t_{BA} = thickness of bonding angle
t = thickness of the thicker of the two adherends
L_{OL} = length of overlap of the bonding angle.

(5) Potential failure modes of the joint are:

- peeling of the ends of the bonding angle from the adherend members
- failure in the adhesive layer due to shear
- failure in the adherend members; this mode, however, should be designed out by suitable proportioning.

P(6) The integrity and sufficiency of the joint shall be verified by testing or be supported by relevant performance or test data.

(7) For primary structural connections this should be supplemented by finite element modelling. Such modelling shall enable a detailed study of the internal response states in the adherend members, adhesive layer and bonding angles. Parameters of particular interest are:

- displacement at the point of application of the load
- peel/cleavage stresses at the ends of the bonding angles
- shear stresses in the adhesive layer
- in-plane and through-thickness stresses in the adherends.

P(8) The possible loading and boundary conditions to be used in the testing and numerical modelling are similar to those adopted for laminated tee joints, illustrated in Figure 5.46.

5.3.5.11 Manufacturing aspects

P(1) Only properly trained and skilled personnel may perform adhesive bonding.

P(2) Components to be joined shall be preconditioned according to the requirements of the designer and/or the adhesive manufacturer. Preconditioning may, for example, include drying of the components before bonding.

P(3) Adequate surface treatments have to be carried out on all bond surfaces, see 5.3.1.5.

P(4) The bond surfaces of the components to be joined shall match each other within the specified tolerances given for the adhesive layer thickness.

P(5) Environmental conditions (temperature, humidity) during bonding shall be within the limits defined by the designer or the adhesive manufacturer. The conditions shall be monitored.

P(6) Components to be joined shall be correctly supported or clamped during the bonding process to ensure that required geometrical tolerances are met, see also 5.3.2.1.

P(7) When using strap configurations the butt surfaces of the adherends shall also be bonded.

P(8) The steps of step-lap joints may be formed during the laminate manufacture or afterwards by machining. A suitable method shall be chosen in respect of the requirements of 5.3.1.3 and in 5.3.5.8.

P(9) If the tapering of a scarf joint or the steps of a step-lap joint are to be machined to a cured adherend having a thickness of less than 5 mm, proper jigging during machining shall be arranged and tooling appropriate to the precision shall be used.

P(10) If scarf joints or step-lap joints are used with adherend thicknesses of less than 5 mm, proper jigging shall be used during bonding to guarantee an adequate bondline quality.

P(11) In adhesively bonded angle joints the gap between the butt end and the base surface shall be filled with fillet resin. The volume of the resin shall be such as to give the appropriate gap and radius as specified in the design. The method of application shall be such as to ensure the fillet is reasonably void free.

P(12) The joint shall be cured so that it reaches the level of temperature resistance required from the whole structure or component.

P(13) It shall be verified that the cure cycle required for the joint does not damage the components to be joined.

5.3.6 Laminated joints

5.3.6.1 General

P(1) The preliminary design shall be performed using the corresponding procedures for adhesively bonded joints (see 5.3.5) by taking the resin layer as an adhesive.

(2) Material properties for the members to be joined may be obtained, for example, from 4.11 or by testing.

P(3) The resin layer (bondline) thickness should always be minimised. In laminated joints the bondline thickness is difficult or impossible to define accurately. For the design, it shall be estimated using a realistic value. However, the design value for the bondline thickness shall never be taken as less than 0.1 mm.

P(4) The bond surfaces shall be treated prior to bonding as stated in 5.3.1.5.

P(5) An adequate number of component tests shall be performed to verify the preliminary design, see 7.3.2.

5.3.6.2 Joint configurations and applied loads

(1) Strap and tee joint configurations are the most frequently used laminated joints.

P(2) Special attention shall be paid to minimising the peel loads because of the brittle nature of lamination resins, see 5.3.1.3.

5.3.6.3 Design of lap and strap joints

P(1) Laminated lap and strap joints may be designed according to the simplified and rigorous design procedures for adhesively bonded joints. In the case of primary structural connections the design shall be verified by testing.

P(2) The joint geometry shall be determined following similar principles to those for adhesively bonded joints (see also 5.3.1.3).

P(3) When the thickness of the matrix resin layer acting as an adhesive layer cannot be defined, a representative estimate of the thickness value shall be given by the designer, see 5.3.6.1.

5.3.6.4 Design of scarf joints

(1) A typical laminated scarf joint is shown in Figure 5.46. The scarf angle should be equal to or less than 10°.

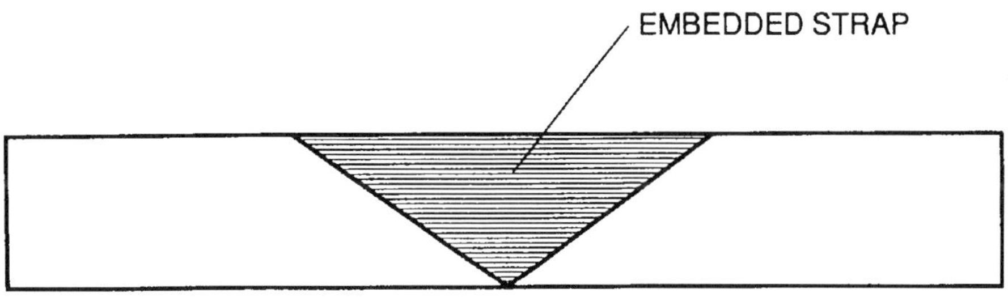

Figure 5.46 A typical laminated scarf joint.

(2) Laminated scarf joints may be designed according to the design procedures for adhesively bonded scarf joints. In the case of primary structural connections the design shall be verified by testing.

P(3) When the thickness of the matrix resin layer acting as an adhesive layer cannot be defined, a representative estimate of the thickness value shall be given by the designer, see 5.3.6.1.

P(4) When designing the joint shown in Figure 5.46 the embedded strap shall be considered as one of the adherends.

5.3.6.5 Design of step-lap joints

(1) A typical laminated step-lap joint is shown in Figure 5.47.

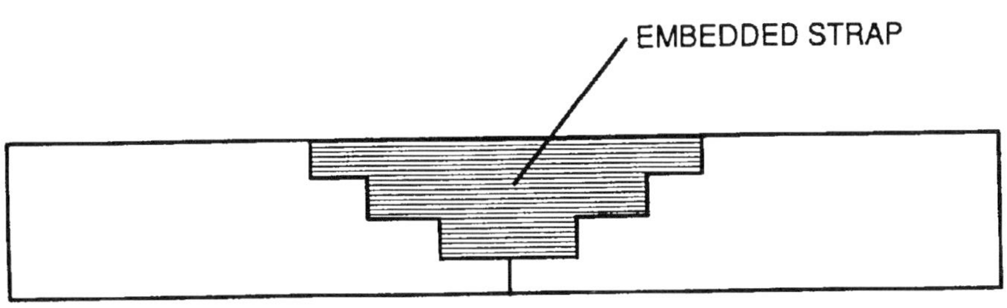

Figure 5.47 A typical laminated step-lap joint.

P(2) Laminated step-lap joints may be designed according to the design procedures for adhesively bonded joints. In the case of primary structural connections the design shall be verified by testing.

(3) The step surfaces should coincide as accurately as possible with the layer interfaces of the adherends (see Figure 5.48).

P(4) When the thickness of the matrix resin layer acting as an adhesive layer cannot be defined, a representative estimate of the thickness value shall be given by the designer, see 5.3.6.1.

P(5) When designing the joint shown in Figure 5.47 the embedded strap shall be considered as one of the adherends.

5.3.6.6 Design of angle joints

P(1) A typical laminated angle joint is shown in Figure 5.49, which also indicates the relevant design variables. The joint is achieved by running boundary angles on both the faces of the two members meeting at the joint. Each boundary angle comprises reinforcement plies laminated over the pre-existing base members with filleting resin in the corner.

P(2) When joining sections to panels or to other sections, both the web and the flange parts of the members shall be joined by bonding using boundary angles or straps, as appropriate.

(3) The potential failure modes are:

- peeling of the ends of the boundary angles from either of the two members meeting at the joint
- delamination of the plies within the boundary angles
- debonding of the boundary angle from the filler fillets
- cracking within the filler fillets.

P(4) The choice of design variables shall be such that the joint is capable of transferring loads between the two members adequately. The choice of length of the overlap of the boundary angles shall be carefully made to ensure smooth transition of load from one member to the other.

(5) It is recommended that the boundary angle consists of CSM and woven roving. An initial choice of key parameters, the thickness of the CSM plies (t_{CSM}), thickness of the woven roving (t_{WR}) and the radius of the boundary angle (r), may be made as follows:

$$t_{BA} = t_{CSM} + t_{WR}$$

where

$$t_{CSM} = (t - t_{WR} - 2)/2$$

and

t_{WR} = 4 mm for joints transmitting structural loads
t_{WR} = 2 mm for joints not carrying significant loads
t = greater of the thicknesses of the two members to be joined

and the boundary angle radius

$$r = kt$$

where

k = a value between 1 and 2.

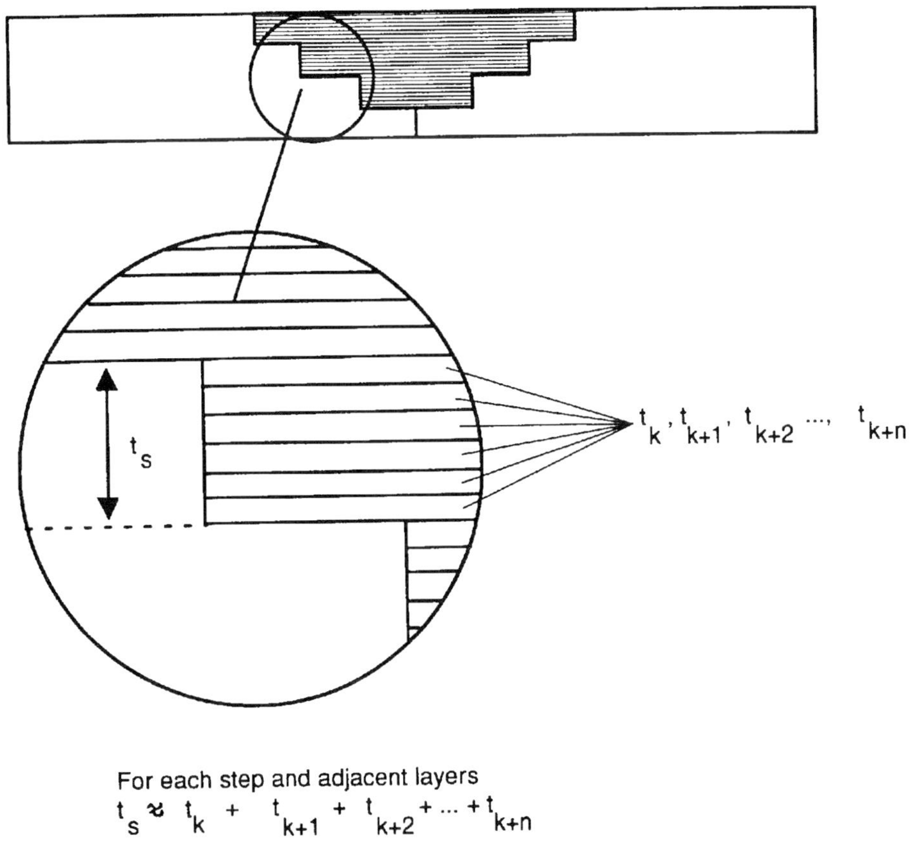

Figure 5.48 Matching the step height with layer thicknesses.

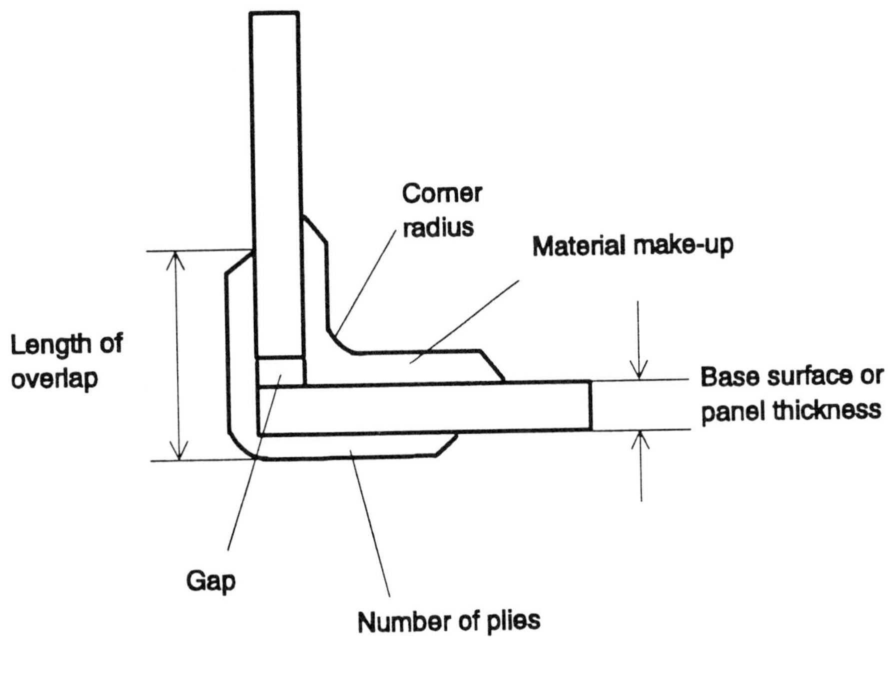

Figure 5.49 Design variables of a typical laminated angle joint.

P(6) However, r shall not be taken less than 5 mm for $t < 5$ mm and less than 10 mm for $t \geq 5$ mm.

(7) Other bi-directional reinforcements in place of woven roving may also be considered provided the thickness requirements are met.

P(8) The minimum overlap of the boundary angle along the members should be on both sides 100 mm or 10*t*, whichever is greater. These values, however, should be verified by detailed numerical analysis and/or experimental testing as detailed below.

P(9) The integrity and sufficiency of the joint shall be checked by testing the specimen. Ideally, for primary structural connections, this should be supplemented by finite element analysis.

P(10) The possible loading and boundary conditions to be used in the testing are illustrated in Figure 5.44. The exact choice of loading mechanisms and holding devices depend on the requirements of the joint imposed by its location within the overall structure.

P(11) The finite element analysis shall be such as to enable a study of the detailed, internal response patterns in the boundary angles and filler region. Parameters of particular interest are:

- the displacement at the point of application of the load
- in-plane and through-thickness stresses in the plies forming the boundary angle
- principal stresses in the filler fillet region.

(12) There a two basic ways to construct the boundary angle, as indicated in Figure 5.50 (see also 5.3.6.8). The relative merits of the two arrangements should be ascertained after analysing the exact design requirements. Validation shall be carried out as outlined above.

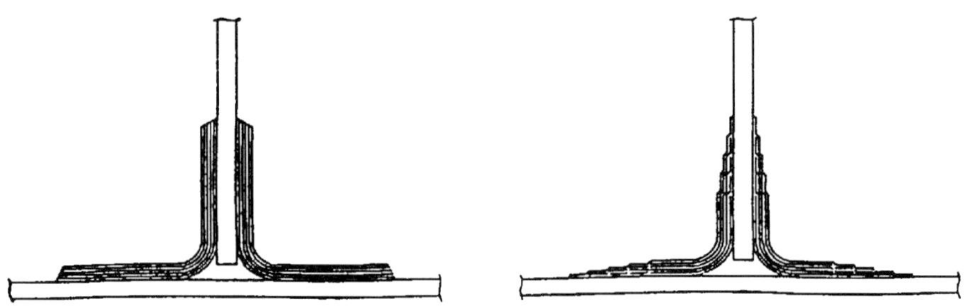

a) b)

Figure 5.50 End shape of overlaminates.

5.3.6.7 Design of tee joints

(1) A typical laminated tee joint is shown in Figure 5.51, which also indicates the relevant design variables. The joint is achieved by running a boundary angle, comprising reinforcement plies laminated over a radiused resin fillet, along both sides of the leg of the tee.

P(2) When joining sections to panels or to other sections, both the web and the flange parts of the leg shall be joined by bonding using boundary angles or straps, as appropriate.

(3) The potential failure modes are:

- peeling of the ends of the boundary angles from the leg or top of the tee
- delamination of plies within the boundary angles
- debonding of the boundary angles from the filler fillets
- cracking within the filler fillets.

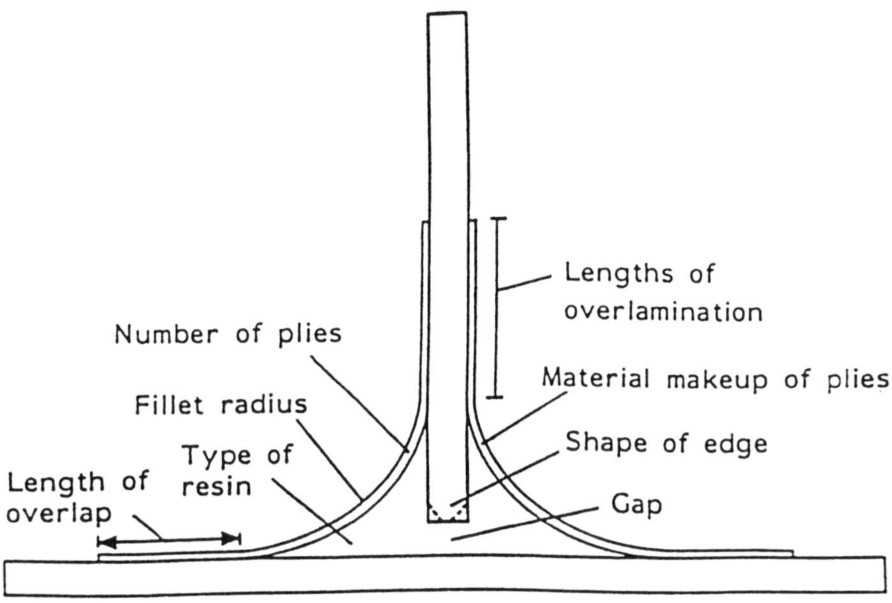

Figure 5.51 Design variables of a typical laminated tee joint.

P(4) The choice of the design variables, especially the boundary angle thickness and radius, shall be such as to be adequately strong and stiff and capable of transmitting the applied loads. The length of overlap of the boundary angle on the leg and flange panels shall be carefully determined to ensure smooth transition of load.

(5) It is recommended that the boundary angle consists of CSM and woven roving. An initial choice of key parameters, the thickness of the CSM plies (t_{CSM}), thickness of the woven roving (t_{WR}) and the radius of the boundary angle (r), may be made as follows:

$$t_{BA} = t_{CSM} + t_{WR}$$

where

$$t_{CSM} = k_1(t - t_{WR})$$

and

t_{WR} = 4 mm for joints transmitting structural loads
t_{WR} = 2 mm for joints not carrying significant loads
t = greater of the thicknesses of the two members to be joined
k_1 = 0.5 to 0.67

and the boundary angle radius

$$r = k_2 t$$

where
k_2 = a value between 1 and 4.

P(6) However, r shall not be taken less than 5 mm for $t < 5$ mm and less than 10 mm for $t \geq 5$ mm.

(7) Other bi-directional reinforcements in place of woven roving may also be considered provided the thickness requirements are met.

P(8) The minimum overlap of the boundary angle along the leg and top of the tee should be about 150 mm or 15*t*, whichever is greater. These values, however, should be verified by detailed numerical analysis and/or experimental testing as detailed below.

P(9) The integrity and sufficiency of the joint shall be checked, by using experimental testing or finite element modelling, or preferably both, of a slice of the joint.

(10) The possible loading and boundary conditions to be used in the testing are illustrated in Figure 5.44. The exact choice of the loading mechanisms and holding devices will depend on the location of the joint within the overall structure.

(11) The finite element analysis shall be such as to enable a study of the detailed, internal response states in the boundary angles and filler region. Parameters of particular interest are:

- the displacement at the tip of the leg in the direction of the load
- in-plane and through-thickness stresses in the plies forming the boundary angle
- the principal stress in the filler fillet region.

(12) There are two basic ways to construct the boundary angle, as indicated in Figure 5.50 (see also 5.3.6.8). The relative merits of the two arrangements should be ascertained after analysing the exact design requirements. Validation of the adequacy shall be done as outlined above.

5.3.6.8 Manufacturing aspects

P(1) The same requirements of quality and supervision shall be applied to the fabrication of laminated joints as to the manufacture of laminated structures and components.

P(2) Components to be joined shall be preconditioned according to the requirements of the designer and/or the resin manufacturer. Preconditioning may include, for example, drying of the components before bonding.

P(3) Adequate surface treatments shall be performed on all joint surfaces, see 5.3.1.5.

P(4) When the joint is laminated on components which are not fully cured, i.e. which have been laminated recently, surface treatments need not be performed, provided that the joint surfaces are kept free from any contaminants. The designer shall specify the maximum allowable time between the lamination of the components and the joint (according to the instructions given by the resin manufacturer).

P(5) Environmental conditions (temperature, humidity) during laminating shall be within the limits defined by the designer or the resin manufacturer. The conditions shall be monitored.

P(6) Components to be joined shall be correctly supported or clamped during the lamination process to ensure that the required geometrical tolerances are met.

P(7) If the tapering of a scarf joint or the steps of a step-lap joint are to be machined to the cured adherend having a thickness of less than 5 mm, proper jigging during machining shall be arranged and tooling appropriate to the accuracy shall be used.

P(8) If scarf joints or step-lap joints are used with adherend thicknesses of less than 5 mm, proper jigging shall be used during lamination to guarantee an adequate bondline quality.

P(9) The steps of step-lap joints may be formed during the laminate manufacture or afterwards by machining. The methods shall comply with 5.3.1.3, 5.3.5.8, 5.3.5.11 and 5.3.6.5.

(10) During the lamination process the excessive use of resin at the bondline should be avoided.

P(11) In angle joints the gap between the butt end and the base surface shall be filled with fillet resin. The volume of the resin shall be such as to give the appropriate gap and radius, as specified in the design. The method of application shall be such as to ensure the fillet is reasonably void free.

P(12) In tee joints the gap between the leg and flange panels shall be filled with filleting resin. The quantity of the resin shall also be adequate to provide the required radius, as specified in the design. The method of application shall be such as to ensure the fillet is reasonably void free.

P(13) In angle joints the overlaminating shall be commenced at an appropriate stage of the cure of the filleting resin, as stated by the resin manufacturer. The actual lamination of the boundary angle plies shall be performed in accordance with standard practice (see 6.2).

P(14) In angle and tee joints the boundary angle may be constructed in two ways. The reinforcement lengths for the individual plies forming the overlaminate may be cut either to be shorter in each successive ply, in which case the end shape is as shown in Figure 5.50(a), or to be equal or marginally longer, in which case the resulting arrangement is as shown in Figure 5.50(b).

P(15) The laminated joint shall be cured so that the joint reaches the level of temperature resistance required from the whole structure or component.

P(16) It shall be verified that the cure cycle required for the joint does not damage the components to be joined.

5.3.7 Moulded joints

5.3.7.1 General

(1) In this EUROCOMP Design Code moulded joints are formed by moulding an external composite member on to the members to be joined, thus forming a bonded joint.

(2) The possible moulding processes include:

- resin transfer moulding (RTM)
- sheet moulding compound (SMC) process
- injection moulding of fibre reinforced thermoplastics.

5.3.7.2 Joint configurations and applied loads

(1) Moulded joints are typically used in frames or in truss or lattice type structures.

(2) The members to be joined are typically pultruded sections having closed cross-sections.

(3) The possible applications for moulded joints range from highly loaded structures using SMC or RTM joint members to lightly loaded ones utilizing injection moulded joint members.

(4) In a typical loading situation one of the members to be joined is loaded in axial tension, thus loading the actual joint primarily in shear.

5.3.7.3 Design

(1) In principle, thermosetting moulded joints (RTM, SMC) are related to laminated double-lap joints, enabling similar types of design procedures to be used for preliminary dimensioning.

P(2) These procedures shall not be used for the final design of load, carrying structures owing to certain basic differences between these joining techniques.

(3) As a preliminary estimate, the following overlap lengths for the moulded members (over the members to be joined) may be given:

(a) for members having tubular or other closed cross-section, the overlap length should be the greater of the following:

$$L_{OL} = 1.5\,D$$

$$L_{OL} = 25\,t$$

where:

D = outer diameter of a tube, or for other than tubular sections, the square root of the area inside the outer surface of the cross-section

t = section wall thickness.

(b) for other members with other cross-sections:

$$L_{OL} = 25\,t$$

where:

t = section wall thickness.

(4) Methods to reduce peel loads and peak stresses described in 5.3.1.3 should be applied when practical.

P(5) Injection moulded reinforced thermoplastic joints shall only be used in lightly loaded non-structural connections.

P(6) The load transfer to injection moulded thermoplastic joint members shall be based on mechanical interlocking and not on adhesion, unless additional surface treatments specific to thermoplastic materials are applied.

P(7) The final design of moulded joints shall be based on testing equivalent joints.

5.3.7.4 Manufacturing aspects

P(1) Those manufacturing aspects given in 5.3.5.11 and 5.3.6.8 applicable to moulded joints shall be considered.

P(2) Components to be joined shall be correctly supported or jigged during the moulding process.

P(3) The properties of the moulded members and joints are sensitive to the process settings and variations. The correct moulding parameters shall be determined experimentally for each moulding application and, subsequently, be monitored throughout the moulding process. The parameters that need to be adjusted and balanced against each other typically include the following:

(a) for SMC joints:

- amount of moulding material
- moulding pressure
- moulding temperature
- moulding time.

(b) for RTM joints:

- injection pressure and time
- fibre volume
- reinforcement lay-out.

(c) for injection moulded joints:

- matrix material
- shrinkage vs. tolerances
- locations of weldlines
- process settings.

P(4) The moulding area shall be properly sealed in order to prevent material escape during the moulding process. The locations to be sealed are typically the outer edges of the mould and the open ends of the sections to be joined.

P(5) The use of internal release agents in the moulding material shall be avoided or minimised. When such additives are used the level of the adhesion shall be verified by testing.

5.3.8 Bonded insert joints

5.3.8.1 General

(1) In this EUROCOMP Design Code bonded inserts are components adhesively bonded to a composite member, enabling later mechanical attachment of this member to another member through the bonded insert, see Figure 5.31.

(2) This EUROCOMP Design Code only addresses the design of the bonded joint, not the later mechanical attachment.

(3) There is a basic difference between the bonded inserts, which rely solely on adhesion for their attachment to the parent component, and the embedded inserts, where the attachment to the composite member is based on mechanical interlocking or entrapment, although the latter are often secured by bonding for convenience.

(4) Bonded inserts are typically of metallic materials.

5.3.8.2 Joint configurations and applied loads

(1) Bonded inserts are typically used with pultruded sections or with similar types of components having closed cross-sections.

(2) In sections the inserts may be bonded on to either the outer or inner surface.

(3) Bonded insert joints should be loaded primarily in axial tension or compression.

(4) The maximum possible axial loads applied to a bonded insert joint are, with careful design, of the same magnitude as the ultimate strength of the composite member/section.

(5) Bonded inserts are also used in sandwich structures, most typically in panels.

(6) In sandwich structures the loads are typically directed to the bonded insert joint through an axially loaded mechanical fastener. This causes out-of-plane loadings to the sandwich component.

(7) If the insert is loaded in the in-plane direction of the sandwich component, the connection is defined to be mechanically interlocked and shall be designed as an embedded insert joint.

5.3.8.3 Design

(1) If the insert is bonded on one side of the section, i.e. inside or outside, the preliminary design of a bonded insert joint may be performed using the procedures given for the design of double-lap joints by assuming, for

example, the lap joint to be circular. If the insert is bonded to both inner and outer surfaces of the section or separate inserts are used on both sides, both bondlines may be considered as separate double-lap joints.

P(2) In sandwich structures the properties of the core surrounding the insert have a major effect on the total load-bearing potential of the bonded insert joint. Therefore the designer shall verify that the core material is capable of carrying the loads transferred from the insert.

(3) The methods to reduce joint peak stresses described in 5.3.1.3 should be applied to the bonded insert joint design when practical.

P(4) The final design shall always be based on testing of equivalent structures or components.

5.3.8.4 Manufacturing aspects

P(1) Those manufacturing aspects given in 5.3.5.11 applicable to bonded insert joints shall be considered.

P(2) The designer shall be satisified that the whole bond area is still covered by the adhesive after the installation of the insert.

5.3.9 Cast-in joints

5.3.9.1 General

(1) In this EUROCOMP Design Code cast-in joints are defined as joints where a composite member is embedded in another member and the joint is made by casting adhesive material between the two members.

(2) The joint is defined as a cast-in joint only if the load transfer is at least partly based on the adhesion between the casting adhesive and the two members. Typically the joint relies to some extent, or even primarily, on geometrical interlocking.

5.3.9.2 Joint configurations and applied loads

(1) Cast-in joints are used to join intersecting members.

(2) The members to be joined may be, for example, large, flat panel type structures or sections.

(3) Cast-in joints are typically loaded by compression and bending. Tensile loads should be avoided as the joint length (overlap) and the joint quality would typically be insufficient.

5.3.9.3 Design

P(1) The strength of a cast-in joint shall be determined only when the adhesive/resin failure leads to total failure of the structure or when the serviceability limit is exceeded as a result of the adhesive/resin failure. In both of these cases the strength of cast-in joints shall always be verified by testing equivalent structures.

P(2) The adhesive/resin layer thickness shall not exceed the maximum value given by the adhesive/resin manufacturer because of the temperature peak produced by the exothermal cure process of adhesives/resins.

5.3.9.4 Manufacturing aspects

P(1) Those manufacturing aspects given in 5.3.5.11 and in 5.3.6.8 applicable to cast-in joints shall be considered.

(2) The use of fillers in the adhesive/resin is recommended when the adhesive/resin layer thickness is high.

(3) The use of resins and adhesives that have low cure shrinkage is highly recommended.

5.3.10 Defects and quality control

(1) The most common bonding defects in all bonded joints are:

- inadequate adhesion
- debonding
- flaws
- porosity
- undercured adhesive/resin.

P(2) Possible defects shall be identified using a proper inspection method.

P(3) Actions to be taken on defective bonds shall be based on the importance of the defects and on the consequences of the bond failure.

P(4) An adequate quality control system is required to guarantee that the joint is manufactured according to the design assumptions.

P(5) When critical joints are manufactured a casting shall be made for quality control purposes from each adhesive batch applied. The casting shall be cured identically to the actual bond. The curing of the adhesive/resin shall be verified by an appropriate test method, such as a hardness test, mechanical testing or chemical analysis, see Chapter 7.

(6) It is strongly recommended that lap-joint samples are prepared and tested to control the bond quality (for test methods see Chapter 7). The samples shall have identical surface treatments to the actual bond surfaces and shall be cured identically to the actual bond.

(8) Laminated joints are often manufactured under on-site conditions which may increase the number of defects or even be the primary reason for these defects. The on-site conditions may also introduce types of defects

other than those listed in (1). This should be considered separately for each application.

P(9) In moulded joints special attention shall be paid to the defects that may result from the moulding process, such as:

- inadequate adhesion caused by the use of internal release agents or thermoplastic moulding materials
- porosity, voids or debonding due to inadequate filling of the mould
- damage to the components to be joined due to excessive moulding pressure.

P(10) In bonded-insert joints special attention shall be paid to inadequate coverage of the bond area by the adhesive, which often results from the installation of the insert.

5.3.11 Repairability

P(1) Before any repair is undertaken the reason for the failure shall be investigated. If the primary reason for the failure is any of the following, the component or the joint shall be redesigned:

- erroneous or inaccurate design
- material properties outside the specified ranges
- applied loads outside the design ranges
- service conditions outside the specified ranges.

P(2) If the original design has been found to be satisfactory, the failure shall be investigated to determine whether it is due to insufficient manufacturing requirements or inaccurate manufacturing specification. If either of these is found to be the primary reason for the failure, it shall be taken into account in the instructions for the repair or in the manufacture of the replacement member(s).

P(3) When the FRP structure to be repaired may have been exposed to conditions in which it may have absorbed more humidity than under typical in-doors atmospheric conditions, the structure shall be dried prior to bonding.

(4) Localised adhesive failure may be repaired by using mechanical fasteners. This may be supplemented by injecting adhesive or resin into any area of disbond or any cavity.

(5) When an adhesively bonded or laminated joint has failed due to adhesive/resin failure and none of the joint members is damaged, the joint may be reassembled using adhesive bonding with proper structural adhesive. Surfaces shall be treated according to 5.3.1.5 prior to bonding.

(6) When a moulded joint has failed due to adhesive failure and none of the members is damaged, the joint may be reassembled using adhesive bonding. Surfaces shall be treated according to 5.3.1.5 prior to bonding.

P(7) When a bonded insert has separated from a composite section, the damaged part of the structure shall be replaced by an intact structure.

P(8) When a bonded insert has separated from a sandwich structure and the structure is to be repaired, the damaged part of the structure shall be rebuilt.

(9) A cast-in joint may be repaired using one of the following methods:

- the joint is disassembled, repaired and bonded; this is provided that disassembling does not damage the joint members
- the adhesive is removed from the joint by mechanical methods without disassembly and the joint is re-cast; this requires verification of achieving a sufficient level of adhesion in regard to the load-bearing requirements
- external doublers are bonded or laminated onto the outer surfaces of the joint members; in this case the intersecting member shall be fully supported to carry the compression at the base of the joint and the space between the joint members and the doublers shall be filled with an adhesive.

5.4 COMBINED JOINTS

5.4.1 Bonded-bolted joints

P(1) In an intact bonded-bolted joint only the bond or the bolts is assumed to carry the joint load. No load sharing between the bond and the bolts is taken into account.

(2) Structural adhesives provide a much stiffer load path than bolts. Thus, the adhesive carries practically all of the joint load.

P(3) Where structural adhesives are used, bonded-bolted joints shall be designed according to the design procedure of the corresponding adhesively bonded joint.

P(4) Where elastomeric adhesives or sealants are used, often with a relatively thick bondline, bonded-bolted joints shall be designed according to the design procedure of the corresponding bolted joint.

(5) The combined use of bonding and bolting may be justified in the following cases:

- bolts can prevent manufacturing defects and service-induced bondline damage from spreading.
- in certain cases, bonded-bolted joints may be used in such a way

that, for example, the bonded joint satisfies the serviceability limit state requirements and the bolted joint satisfies the ultimate limit state requirements (see EUROCOMP Handbook 5.4).

- bolts can provide the required clamping pressure during the bonding process.

5.4.2 Bonded-riveted joints

P(1) Bonded-riveted joints are, from the design point of view, identical to bonded-bolted joints and shall be dealt with accordingly.

P(2) When determining the load transfer mechanism, i.e. whether the loads are transmitted through the bondline or through the rivets, special attention shall be paid to the type of adhesive used. See clauses P(3) and P(4) of 5.4.1.

CONTENTS

6 CONSTRUCTION AND WORKMANSHIP

6 CONSTRUCTION AND WORKMANSHIP

6.1 OBJECTIVES

6.1.1 Design and co-ordination

P(1) The designer, manufacturer, erector and finisher, and others involved in the work processes shall ensure that the designer's intentions and any assumptions made in the design are realised or are otherwise properly taken account of in the completed structure.

P(2) All FRP units (including fixings and joints) shall be designed and constructed to ensure compliance with requirements for accuracy in manufacture and erection, and to accommodate permissible deviations in the building structure. Provision shall be made for adjustments, if required.

P(3) Structural FRP components shall be fully designed and detailed prior to manufacture, except where full-scale testing or testing of prototypes forms part of the design process (see Chapter 7). In all cases, manufacture or fabrication of FRP components shall take account of the methods of erection and completion of the work so as to ensure co-ordination of the FRP work with that of related building elements.

P(4) Any proposed departure from the designer's requirements shall be clearly indicated and supported by all relevant information.

6.1.2 Proposed method of fabrication and erection

P(1) The manufacturer shall prepare in conjunction with the designer or otherwise for his approval:

- a full description of the construction, including materials, method of manufacture, finishes, nature and disposition of stiffening members, type(s) and locations of fixings, jointing materials and methods. Any proposed departure from the specified requirements must be clearly indicated
- a statement of the relevant properties of the proposed construction, together with independent test certificates where appropriate
- a representative sample(s) of the proposed FRP if required
- the estimated weight of each type and size of unit, including accessories and components incorporated in the composite at the place of manufacture
- details of design verification, quality control and other testing procedures
- methods of transportation, erection and assembly and details of transportation, erection and assembly requirements.

(2) With the methods and materials specified or otherwise available, consideration should be given to:

- the maximum size that can safely, effectively and economically be manufactured, cured, transported, stored, erected and fixed, allowing for temporary holding in place while permanent connections are made
- standardisation of FRP units and accessories, making maximum reuse of dies, forms, moulds, etc.

6.1.3 Production of mock-ups and samples

P(1) Where required, the FRP manufacturer shall produce design samples, full size or scaled mock-ups for design verification, inspection and approval, routine testing or compliance testing (see Chapter 7). The exact purpose(s) of the samples and/or mock-up shall be agreed between the designer and manufacturer. This is to ensure that only those characteristics or properties that are intended to be approved are actually approved.

P(2) The description of the samples and/or mock-up shall include: scale, material, size and accessories or features to be incorporated, e.g.:

- fixings
- jointing details
- finish (colour and surface texture).

P(3) The purpose of the sample or mock-up shall be stated, e.g.:

- to test the strength of the unit
- to test the stiffness of the unit
- to test the appearance and/or performance of finish and details, including colour
- to test for durability
- to prove methods of handling, assembling, fixing and jointing
- to submit for fire-testing
- to test for manufacturing quality control purposes.

6.2 MANUFACTURE AND FABRICATION

6.2.1 Quality of work

P(1) FRP units shall be carefully and accurately manufactured to ensure compliance with design and performance requirements, using materials and workmanship appropriate for the purpose.

P(2) All materials must be compatible with each other, and must be stored and used in accordance with the manufacturers' recommendations. Resins must be used as supplied and not mixed with other materials; fillers and additives may be used only where approved.

P(3) The standard of finish shall be appropriate to the end use and position in the building.

P(4) The manufacturer shall take all necessary and reasonable steps to ensure that defects such as wrinkling, spotting, striations, fibre patterning, fish eyes, blisters, crazing, dry patches and uneven or inconsistent colour do not occur.

6.2.2 Workshop conditions

P(1) Workshops where FRP units are manufactured must be at or about normal room temperature, dry, clean, well ventilated and well lit.

P(2) Moulds, tools and other equipment shall be clean, dry and at a similar temperature to the ambient temperature of the workshop.

P(3) Moulding work shall not be carried out when the temperature of the workshop, equipment moulds or materials is below that stipulated by the resin manufacturer or below 17°C or below the dew-point temperature.

P(4) Adequate time shall be allowed for the temperature of materials and equipment transferred from elsewhere to acclimatise to the workshop temperature and humidity conditions.

6.2.3 Manufacturing accuracy

P(1) Unless agreed otherwise between the designer and manufacturer, finished dimensions of completed units shall be such that:

- the FRP assembly, when erected, complies with 6.3.4
- all sizes fall within the permissible deviations given in Table 6.1
- at least two thirds of sizes fall within one-third of the permissible deviations given in Table 6.1.

6.2.4 Gelcoats to external surfaces (where used)

(1) All FRP units of one colour should have gelcoats from the same colour batch of resin.

P(2) Unless otherwise agreed, the overall thickness of single gelcoats shall be 500 microns (nominal) and nowhere be less than 400 microns nor more than 600 microns.

P(3) Double gelcoats must be approved and shall have thickness limits in accordance with the resin manufacturer's recommendations.

Table 6.1 Permissible deviations in FRP units.

	Overall dimension involved (m)				
	Up to 2	*2-3*	*3-4.5*	*4.5-6*	*6+*
Width and height: deviation from designed dimension (mm)	±3	±4	±5	±6	±7
Straightness of edges: deviation from intended line, any variation to be evenly distributed with no sudden bends or irregularities (mm)	±3	±4	±5	±6	±7
Squareness: taking the longer of 2 sides at any corner as a base line, the deviation of the shorter side from the perpendicular (mm). Dimension involved is at the far end of the shorter side.	±3	±4	±5	±6	±7
Flatness: deviation under a 1 m straight edge placed anywhere on a flat panel surface: 3mm.					

(Measurements taken at an ambient temperature of 20-22°C)

6.2.5 Laminating

P(1) Random reinforcement shall be distributed uniformly, and non-random reinforcement shall be correctly positioned and aligned.

P(2) The fibre reinforcement shall be fully wetted out by the resin.

(3) To assist wetting out of woven fabric reinforcement, a layer of chopped strand mat may be placed between each layer of fabric.

P(4) There shall be a good overall bond between all gel coats and all layers of laminate.

P(5) The FRP shall be well consolidated and reasonably free from air voids.

(6) The allowable air voids content should be agreed upon between the designer and the manufacturer, but, in general, it should not exceed 5% by volume.

P(7) All core materials, ties, ribs, fixings and accessories shall be properly bonded to the FRP over the full contact surface area.

(8) Unless otherwise agreed, a flow coat should be applied to all surfaces of the finished units that are not gel coated and to all cut edges, holes, etc., to protect the glass fibre from penetration of moisture.

6.2.6 Curing

P(1) FRP units shall be adequately cured in accordance with the resin manufacturer's recommendations, to fulfil service temperature requirements.

P(2) All components for use inside buildings shall be fully cured so that no toxic or noxious vapour (e.g. styrene) are emitted during service.

P(3) The methods of curing shall be agreed upon between the designer and the manufacturer.

(4) Unless otherwise agreed, curing should be carried out at a temperature not less than 50°C for not less than 8 hours.

P(5) Care shall be taken to ensure that units are not distorted while being cured.

6.2.7 Records

P(1) Complete records shall be kept for each unit, including details in accordance with 8.7 *Control of component delivery.*

(2) Records should be available for inspection on request.

6.3 DELIVERY AND ERECTION

6.3.1 Information to be provided

P(1) The FRP manufacturer shall provide

- clear and comprehensive instructions for delivery and erection of FRP units and shall ensure that they are understood by the site operatives
- adequate site supervision by a suitably skilled person experienced in the erection of composite units.

6.3.2 Preparation of FRP units for transportation

(1) Panels should be cleaned of all traces of release agent, fibres, dust and other foreign matter.

(2) Temporary protection of the surface, flange faces and raised areas should be provided to prevent damage and contamination likely to be met during transport and storage of the panels.

(3) Special restrictions on angle of lift, minimum number of lifting positions, etc., should be clearly labelled on the panel, and written details should be provided. The handling technique and equipment required should be stated.

6.3.3 Protection

P(1) Protection shall be provided to prevent mechanical damage and disfigurement. Units should be separated during transport and storage to prevent chaffing. All slings, ropes, bearers, ladders, etc., should be padded.

P(2) Units shall be supported as necessary so that they do not bow, twist or distort.

P(3) Units shall be protected from the weather. Surfaces not having a weathering gel coat must not have prolonged exposure to direct sunlight or water.

(4) Where units are covered with plastic sheeting, adhesive tape should not be stuck on exposed surfaces.

(5) Fixing and jointing materials should be stored under suitable conditions as instructed or recommended by the manufacturer.

(6) Units should not be delivered to site that cannot be erected immediately or unloaded into a suitable well protected storage area.

(7) Items shall be stored on a level surface free from sharp protrusions and supported to prevent local damage.

6.3.4 Accuracy of erection

(1) Any supporting structure including any fixing inserts should be surveyed before commencing erection of the FRP units.

(2) Units should be erected using temporary spacers to suit the survey results and ensure consistent spacing.

P(3) The relative positions of members or units to be joined shall be such as to ensure that the joints perform as intended structurally, so that, where required, they may be properly sealed or otherwise made weathertight.

P(4) The finished work must have a satisfactory appearance and be square, regular, true to line, level and plane with a satisfactory fit at all junctions and be within tolerances agreed with the designer.

6.4 CONNECTIONS

6.4.1 Bolted connections

P(1) Bolts, fixings and metal inserts shall be of a suitable type of stainless steel, non-ferrous metal or FRP and shall be such as to avoid bi-metallic corrosion.

P(2) Bolted connections shall be formed in such a way as to ensure that the load on the connection is properly distributed between the bolts without damage to the parts being joined, either during the forming of the joint or in service.

P(3) The methods of making the joint, forming the holes and fixing and tightening the bolts shall be agreed between the FRP designer and the manufacturer.

P(4) Joints that are to accommodate movement between the units to be connected and rigidly bolted joints shall be clearly identified by the designer.

P(5) Holes for structural connections using bolts loaded in shear shall be formed so that each bolt is a close fit in the mating holes of the parts to be joined and so that no loads due to misaligned holes are imparted to either of the FRP laminates being joined.

(6) Groups of such holes may be formed on site by clamping together the parts to be joined and drilling through both laminates or by forming the holes by suitable means at the place of manufacture and checking that they are aligned before the units leave the works. It is unlikely that separately drilling the parts to be joined, whether or not using a template, will produce alignments of sufficient accuracy for bolted connections involving more than four bolts loaded in shear.

(7) Groups of holes for non-structural or lightly loaded connections or where the bolting is used in conjunction with adhesives may, with the approval of the designer, be formed separately in the components to be joined, using a template. Such holes (made separately in the parts to be joined) should be 2 mm greater in diameter than the bolt(s) used for the connection.

(8) Holes may be formed using a diamond-tipped or tungsten carbide drill, by turning and milling or, for laminate thicknesses less than 4 mm, by bit-by-bit punching with a concave punching tool.

P(9) In all cases, components being drilled shall be properly supported at the back, using a suitable backing piece so as to avoid splitting or cracking of the laminate from the drilling.

P(10) Bolts shall be tightened to a pre-set torque. Care shall be taken not to overtighten the bolts. Tightening of the bolts shall not cause crushing of the laminate or restrain fixings intended to permit lateral movement.

(11) All drilled or punched holes not moulded should be sealed to protect the fibre reinforcement from moisture penetration.

P(12) The use of drifts or bars to align holes for the insertion of bolts shall not be permitted.

6.4.2 Adhesively bonded and laminated connections

P(1) Dimensions of joint, joint thickness, preparation, application and curing shall be as specified by the designer and in accordance with the resin manufacturer's recommendations.

P(2) The surfaces to be joined must be thoroughly clean, dry and free from frost, oil and release agent.

(3) Surfaces should be abraded and primed as required and as specified.

(4) Where exposed to view, edges of joints should be masked with tape before priming which should be removed immediately after sealing.

(5) This section also applies to bolted-bonded connections where structural adhesive is used, and laminated joint connections, as appropriate, unless otherwise specified by the designer.

6.4.3 Final Fixing

P(1) Units to be joined should be checked for position and alignment before final assembly.

6.5 SEALANT JOINTS

P(1) Sealant, backing strip, bond breaker and primer shall be suitable for the type, size, expected movement and location of the joint to be sealed and for fire rating.

P(2) Backing strip, bond breaker and primer shall be of make and type recommended by the sealant manufacturer.

P(3) Preparation, depth of sealant and application shall be strictly in accordance with the sealant manufacturer's recommendations.

P(4) Joint surfaces must be thoroughly clean, dry and free from frost, oil and release agent. Surfaces shall be finely abraded and primed as appropriate. Where exposed to view, edges of joints shall be masked with tape before priming, which shall be removed immediately after sealing.

P(5) Sealant shall be applied ensuring maximum adhesion to the sides of the joint and a neat, smooth and clean finish.

6.6 REPAIR OF LOCAL DAMAGE

P(1) Repairs shall be carried out only in accordance with a procedure agreed by the designer and the fabricator and in accordance with the manufacturer's recommendations.

P(2) Repairs shall not be carried out on badly damaged components or where the proposed repair will impair appearance, performance or durability.

(3) Where appearance is not a factor, repairs may be carried out on damaged units provided such repairs can be justified by tests.

P(4) Where repairs are carried out:

- the surrounding laminate shall be suitably roughened, if necessary, using both abrasion and a suitable chemical primer
- material compatible with that used in the manufacture of the FRP units shall be used for the repair
- the repair shall be adequately cured, using suitable portable heaters for site curing, where necessary.

6.7 MAINTENANCE

6.7.1 Inspection

(1) On completion of the erection work, and at predetermined intervals, all FRP components should be given a full visual inspection. This should be to check for signs of damage during erection and for any environmental deterioration.

6.7.2 Washing

(1) When and where required, surfaces of FRP should be cleaned with detergent and water using a stiff bristle brush or nylon pad followed by rinsing with clean water.

P(2) Wire wool and coarse abrasives shall not be used.

P(3) Cleaning agents shall be approved by the resin manufacturer.

6.7.3 Painting

(1) Faded, scratched or lightly damaged surfaces may be improved or repaired using a suitable brush or spray-applied coating.

P(2) Paint or other coating systems used on FRP surfaces shall be compatible with the FRP resins and shall be approved by the resin manufacturer.

P(3) Common paint strippers containing methylene dichloride shall not be used nor shall burning off be performed under any circumstances.

(4) In non-structural components: after washing as in 6.7.2 above, surface cracks up to about 0.5 mm wide may be filled using an approved brush-applied solvent-free epoxide solution extending at least 50 mm beyond the visible ends of the cracks. This should be wet-abraded before over-painting.

(5) Suitable paint systems are:

- non-yellowing (aliphatic) two-part polyurethane paints
- silicone alkyd enamels.

applied strictly in accordance with the resin and paint manufacturers' recommendations.

6.8 HEALTH AND SAFETY

P(1) The fundamental requirements shall be as specified in the appropriate regulations.

P(2) Reference must be made to EC Directives on the Control of Substances Hazardous to Health, e.g. "Protection of workers from the risks related to chemical, physical and biological agents at work" (1980) and to the Environmental Protection Act for emission control regulations for volatile organic compounds.

(3) Particular points requiring attention are:

- provision of adequate ventilation
- possible eye and skin irritation from resin, glass and certain reagents and curing agents
- need for protective clothing, gloves, goggles and face masks when working with FRP materials
- use of barrier creams
- need for stringent personal hygiene measures
- toxicity of unreacted isocyanate components in the spray application of two-part polyurethane paints.

CONTENTS

7 TESTING

7 TESTING

7.1 GENERAL

P(1) Testing shall be carried out as part of the general quality control process outlined in Chapter 8 to ensure that the finished structure conforms with the specification.

P(2) Testing shall be carried out where there is insufficient knowledge of the properties of the material. The resulting data may then be used in design with reduced partial safety factors (see 2.3.3), and in the design of connections (see Chapter 5 and 7.3.3).

P(3) Testing may also be carried out as part of the development of new materials, elements or structures, as follows:

- for newly designed elements using newly designed laminates
- for newly designed elements using established and proven laminates
- for structures, or parts thereof, particularly when subjected to unusual loadings or environments.

P(4) The amount and type of testing that is to be carried out should be agreed between the client and the designer at the start of the project.

(5) All testing should be carried out in Approved Laboratories in accordance with International Standards and under approved quality schemes, such as NAMAS or ISO 9000. Where this is not possible, the testing procedure should be agreed between the client and the designer. Traceability of the materials should be ensured in all cases.

(6) Compliance testing is described in 7.2 and covers all stages of the production process, from the selection of raw materials, through the manufacturing process, to fabrication and the erection of the finished structure.

(7) Design and verification testing is described in 7.3 and covers the development process, from the laminate, through components and connections, to the finished structure.

(8) The amount of testing that will be required will depend on the data already available. For example, data sheets may be available for standard structural profiles which contain all the necessary information for the specifier and the designer. Figure 7.1 illustrates the design and construction process and indicates areas in which testing may be required.

(9) All testing should be carried out in accordance with the relevant ISO Standard or Euronorm. Where such standards do not exist, national or internationally accepted standards should be used.

7.2 COMPLIANCE TESTING

7.2.1 During production

P(1) As part of the manufacturer's overall quality control process, tests shall be carried out on constituent materials (resins, fibres, fillers, etc.) to ensure uniformity and conformity.

P(2) Testing shall consist of visual checks on the basic materials and on the finished product, and may include physical tests on the finished element.

P(3) Visual or other tests should be carried out on mats, woven rovings and multiaxial fabrics to ensure uniformity and conformity.

P(4) Tests shall also be carried out on laminates (using coupons either specially manufactured or cut from large units) to ensure uniformity and conformity of the production process.

P(5) When the manufacturer makes a standard range of FRP products, for example pultruded sections, tests such as load-deflection tests and dimensional checks shall be carried out on fabricated elements.

P(6) In addition, the FRP manufacturer may be required to carry out ultimate load tests on standard fabricated elements.

P(7) The type of testing required to ensure compliance will depend upon the intended use for the product and shall be agreed by all parties before production begins.

P(8) All fibres, resins, catalysts, fillers, etc. should be in accordance with the relevant ISO specifications or Euronorms and should be supplied with a certificate of conformity.

(9) Careful control of the amount and location of the fibre in the product is essential. The supply of fibres should be monitored on a regular and agreed basis. For continuous processes, this should be at least once per hour. For batch processes, where practicable, the fibres should be divided into packages such that one package is used for each finished unit.

(10) The amount and location of the fibres should be checked by testing samples at an agreed rate. For pultrusion, one sample should be taken from every 2000 m of finished product or once every day, whichever is the more frequent.

For hand lay-up processes, trial pieces should be made and tested. These may be either sacrificial parts of the unit that can be removed for testing or specially prepared samples made at the same time and by the same process. In both cases care should be taken to ensure that the sample is representative of the material in the finished unit.

The samples should be weighed to check the density of the product.

The samples should be tested in direct tension, to check the fibre content, and in bending, to check the fibre location.

Finally, the samples may be burned to remove the resin so that the amount, location, type and orientation of the fibres may be checked visually. The reinforcement layers may also be weighed in order to check their areal weights of reinforcement.

The tolerances for the amount of fibres, determined on the basis of cross-sectional area should be agreed prior to production.

(11) A visual check on the finished product will not only be a measure of the quality but also a check on the production process. Defects such as pin-holes and blisters in the resin should be checked against an agreed Standard and, where possible, adjustments made to the production process, such as the pull speed in pultrusion, to eliminate the problem subsequently.

(12) The dimensions of the finished product should be checked. For continuous or batch processes this should be at least once every hour during production. Pultrusion is basically a highly stable production process, with constant overall dimensions for the material being produced. However, internal formers may move during the production run. A visual check should be made on all cut ends. In addition, the samples used to test for strength, see (10) above, should be checked for dimensional accuracy.

For hand lay-up units, the thickness may be checked on the test piece, provided that this is representative of the general thickness. Otherwise the thickness should be measured on the unit at agreed locations. The tolerance for the dimensions of the unit, and of cast-in holes and fixings, are dealt with in 7.2.2.

(13) Additional specific tests, such as resistance to local impact, surface hardness, etc., may be required and should be carried out at an agreed rate and in an agreed manner.

(14) Where the product is a standard item, for example a standard pultruded section, the test samples in (10) should be used to check conformity with the properties specified by the manufacturer. Typical specified properties might include:

- elastic modulus
- shear modulus
- tensile strength

- compressive strength
- flexural strength
- interlaminar shear strength.

The test data should be used to obtain characteristic values for the properties. Where insufficient data exist, a nominal value may be obtained (see 2.2.3).

Testing methods should be in accordance with 7.3.2.

7.2.2 On site

P(1) As part of the constructor's overall quality control process, dimensional checks shall be carried out on the units delivered to site. These shall include checks on the locations of fixings, holes, etc.

P(2) In addition, tests may be required on connections made on site.

(3) The testing for dimensional accuracy should be carried out in a manner agreed by all concerned.

(4) The rate of testing should be agreed, but not less than 5% of the components delivered to site should be checked for dimensional accuracy.

(5) Allowable tolerances for components should be in accordance with Table 6.1, unless agreed otherwise between the designer and the manufacturer .

(6) The allowable tolerance for the positions of groups of holes and fixings to be used for structural connections should be ±5 mm, unless agreed otherwise between the designer and the manufacturer.

The allowable tolerances for the positions of groups of holes and fixings to be used for non-structural connections, or for other purposes, should be as agreed between the designer and the manufacturer.

(7) At least 10% of components should be checked for visual defects, against an agreed Standard. The Specification should state the agreed actions to be taken in the event of components having more than the allowable number of visible defects.

(8) The rate of testing of connections should be agreed in advance, but should not be less than 5% of a particular type of connection. Testing will generally be to a factored service load, in accordance with 7.3.3.

(9) Where a connection or type of connection is particularly critical to the behaviour of the finished structure, compliance tests may be carried out on specially fabricated connections that will not form part of the completed structure. Care should be taken that the workmanship is representative of that in the completed structure. Testing of the units may be to a factored

ultimate load or to collapse, in accordance with 7.3.3. The number of units to be tested and the acceptance criteria shall be agreed by all concerned.

7.3 TESTING FOR DESIGN AND VERIFICATION

7.3.1 Laminates

P(1) Testing shall be carried out where there is insufficient knowledge of the properties of the material. The resulting data may then be used in design with reduced partial safety factors (see 2.3.3), and in the design of connections (see Chapter 5 and 7.3.3).

P(2) Testing of laminates may be required:

- as part of the development of a new manufacturing process or technique
- when new materials, or combinations of materials are being used in an already developed manufacturing process
- when new materials are being developed for specific purposes, for example fire resistance.

P(3) Testing shall consist of loading the element, or a representative sample, to determine specified mechanical properties, under static, dynamic or long-term loads.

(4) Testing may also consist of environmental testing, to consider effects such as temperature and humidity changes and a corrosive atmosphere.

P(5) Prior to the execution of the tests, a test plan shall be drawn up by the Client and the testing organisation. Wherever possible this should be in accordance with agreed International Standards, or failing that National Standards, which will specify the number, preparation and testing of the specimens and the analysis and interpretation of the results.

(6) The use of International Standards will ensure the uniformity of the test method used and hence enable direct comparison with the results of other programmes. Table 8.1 lists ISO standards for production and construction control.

(7) Where no suitable ISO or Euronorm test methods exist, other National or Internationally accepted tests, such as ASTM, may be used.

(8) Testing should be carried out at a temperature appropriate to that in the finished structure.

(9) Because different test methods may give different values for a particular property, the method used should always be stated when reporting results. In addition, full details of the material's composition should be recorded to aid traceability.

(10) Where particular circumstances are to be tested, such as stressed laminates in an aggressive environment, standard tests are unlikely to exist. Suitable methods must be agreed by all concerned with the testing. This particularly applies when laminate testing is carried out as part of the testing of a complete structure; see 7.3.5.

(11) The number of tests to be carried out will be stated in any standard test method. When agreed non-standard tests are used, at least 5 specimens representative of the batch should be tested. If the measured behaviour of individual test specimens varies significantly from the mean value, additional specimens should be tested so that statistical methods may be used to determine a characteristic response.

(12) When analytical work is used to widen the range of parameters considered during testing, the range of properties of the resin and of the fibres should be obtained from the suppliers.

(13) Accelerated testing should only be used when there is sufficient correlation with testing under normal conditions for an accurate evaluation of the test results to be made.

7.3.2 Components

P(1) Testing of components may be required:

- as part of the development of a new manufacturing process or technique
- when new materials, or combinations of materials, are being used in an already developed manufacturing process or technique.

P(2) Testing may also be required when the material properties are known but the structural behaviour of the component is not adequately understood.

P(3) Testing may consist of loading the element, either statically or dynamically, to determine its mechanical properties. Alternatively, environmental loads, such as temperature or a corrosive atmosphere, may be used to determine the durability of the component.

P(4) Prior to the execution of the tests, a test plan shall be drawn up by the designer and the testing organisation. This shall contain the objectives of the tests and all the instructions and other specifications necessary for the selection or production of the test specimens, the execution of the tests, including the number of specimens to be tested, and the evaluation of the results.

P(5) Where possible, testing should be in accordance with International Standards or Euronorms or, failing that, National Standards.

(6) The loading arrangement for structural tests should, where possible, be kept as simple as possible to aid interpretation of the test results. For example components should be loaded in pure bending or in axial compression.

(7) Where component testing is carried out in conjunction with the testing of an assembly or a complete structure, as described in 7.3.4 and 7.3.5, it may be possible to use the appropriate loading pattern and boundary conditions.

(8) Similarly, the environmental conditions, particularly the temperature, should, as far as is appropriate, represent those for the assembly or the complete structure.

(9) Unless otherwise stated in the International or National Standard, a minimum of three components should be tested under any given loading configuration. If the measured behaviour of individual components varies significantly from the mean value, additional components should be tested so that statistical methods may be used to determine a characteristic response.

(10) To evaluate the experimental behaviour, relevant material properties should be obtained from suitable specimens cut from lightly stressed parts of the component. Failing this, representative elements may be used, for example untested units made by the same process and using the same materials. Finally, relevant data may be provided by the manufacturers when there is sufficient experience of the material and manufacturing process. With standard materials the third option is likely to give the most realistic values.

(11) Accelerated environmental testing should only be used when there is sufficient correlation with testing under true environmental conditions for an accurate evaluation of the test results to be made.

(12) When environmental testing is carried out, the component should be subjected either to a specified level of stress or, if the work is part of the testing of an assembly or a complete structure, to loads appropriate to its in-service conditions.

(13) The components should be loaded to the design service load or to failure. Loading to the design ultimate load may be relevant for compliance testing, see 7.2.2, but gives no information on the safety factors associated with collapse.

7.3.3 Connections

P(1) Testing of connections may be required

- as part of the verification process and for cases where no calculations are accepted without verification
- as part of the development of a new connection technique
- when new materials are being used in an already developed connection detail
- to obtain the full benefit of details that can not be taken into account in the design methods
- when connections have been formed under conditions that do not conform with the specification.

P(2) Testing shall consist of loading the connection, either statically or dynamically, to determine its mechanical properties and/or behaviour. Alternatively the effects of environmental loads, such as temperature cycling or a corrosive atmosphere, may be used to determine the durability of the connection.

P(3) Prior to the execution of the tests, a test plan shall be drawn up by the designer and the testing organisation. This shall contain the objectives of the tests and all the instructions and other specifications necessary for making the connections, the execution of the tests, including the number of specimens to be tested, and the evaluation of the results.

P(4) All the connections that are to be tested shall be constructed in accordance with the agreed specification under conditions similar to those in practice.

(5) When the connections are tested as part of the testing of an assembly or a complete structure, the loading and support systems should represent, as accurately as possible, those experienced in practice.

(6) Where the connections are tested as part of a development programme, the significant mode of loading should be determined and the loading arrangement should be kept as simple as possible to aid interpretation of the test data.

(7) Similarly the environmental conditions, particularly the temperature, should, as far as is appropriate, represent those in the complete structure.

(8) When environmental testing is carried out, the connection should be subjected either to a specified loading or, if the work is part of the testing of an assembly or a complete structure (as described in 7.3.5) to loads appropriate to its in-service conditions.

(9) A minimum of five representative sample connections should be tested under any given loading configuration. If the measured behaviour of individual components varies significantly from the mean value or the failure mode differs from that assumed in design, additional connections

should be tested so that statistical methods may be used to determine a characteristic response.

(10) The connections should be loaded to the design service load, or a factored service load (as described in 7.3.5.3) where appropriate, or to failure. Loading to the design ultimate load may be relevant for compliance testing, see 7.2.2, but gives no information on the safety factors associated with collapse. However, it will confirm that premature failure, due to an unexpected failure mode, will not occur.

(11) To evaluate the experimental behaviour of the connection, relevant material properties should be obtained from suitable specimens, cut from parts of the component well away from the connection. Failing this, representative elements such as untested units, made by the same process and using the same materials, may be used. Finally, relevant data may be provided by the manufacturer when there is sufficient experience of the materials and manufacturing process. Where standard components are joined by the connections, the third option is likely to give the most realistic values.

(12) Similarly when the connection is bonded, or bolted-bonded, information should be obtained on the properties of the adhesive used.

7.3.4 Assemblies

P(1) Testing of assemblies may be required when:

- the pattern of loading differs from that in the final structure, such as stresses induced during transport
- the response differs from that in the final structure, for example when stability is achieved only after the assembly is connected to other parts of the structure.

P(2) Prior to the execution of the tests, a test plan shall be draw up by the designer and the testing organisation. This shall contain the objectives of the tests and all the instructions for executing the tests, including the number of specimens to be tested and the evaluation of the results.

(3) Testing of assemblies may be undertaken as a limited part of testing a complete structure. Thus the recommendations in 7.3.5 generally apply.

(4) The loadings to be applied are those appropriate to the assembly only and not those that will be applied in the completed structure. Thus only the static, short-term response to the service loads should be considered. The ultimate load capacity will generally be meaningless.

(5) When analysing the behaviour, the effect of any excessive deformation on the subsequent behaviour of the complete structure should be considered carefully.

7.3.5 Structures

7.3.5.1 Purpose of testing

P(1) Testing of the complete structure, or part of the structure, may be required when:

- the structural behaviour is not accurately reflected by the design assumptions
- the behaviour of the joints between component members is not adequately understood
- the structure is subjected to loadings that are not adequately covered in the design, such as dynamic or seismic effects.

P(2) Testing may also be required where:

- the compliance procedures in 7.2 indicate that the materials may be sub-standard
- supervision and inspection procedures indicate poor workmanship on site, producing construction outside the specification
- there are visible defects, particularly at critical sections of structural members or at critical joints
- a check is required on the quality of the construction.

P(3) Testing may be required when the finished structure differs significantly from that specified, for example following the repair of damaged units or the use of non-standard connections.

P(4) Testing may also be undertaken where the rules for design by calculation given in this EUROCOMP Design Code would lead to uneconomic results. However, the conservative assumptions in the specified calculation models, which are intended to account for unfavourable influences not explicitly considered in the models, shall not be bypassed.

(5) Because circumstances vary greatly, the requirement for testing and the intended objectives should be agreed in advance by all concerned.

7.3.5.2 Load tests to substantiate design

P(1) The experimental assessment shall be based on tentative calculation models, which may be incomplete but which relate one or several relevant variables to the structural behaviour under consideration, such that basic tendencies are adequately predicted. The experimental assessment shall then be confined to the evaluation of correction terms in the tentative calculation model.

P(2) If the prediction of the relevant calculation models, or the failure mode to be expected in the tests, is extremely doubtful, the test plan shall be developed on the basis of accompanying pilot tests.

P(3) Prior to the execution of tests, a test plan shall be drawn up by the designer and the testing organisation. This shall contain the objective of the tests and all the instructions and other specifications necessary for the selection or production of the test specimens, the execution of the tests and the test evaluation.

(4) Where the structure is made of individual components such as pultruded sections, tests should be carried out on typical elements in accordance with 7.3.2, unless sufficient relevant data are available from the component manufacturer.

(5) The test procedures should be in accordance with the advisory clauses in 7.3.5.3.

(6) In order to evaluate the experimental behaviour, relevant material properties should be obtained from suitable specimens cut from the structure after the test, from locations that have been only lightly stressed. Failing this, representative elements may be used, for example untested units made by the same process and using the same materials. Finally, relevant data provided by the manufacturer may be used when there is sufficient experience of the materials and manufacturing process. Where standard components are used, the third option is likely to give the most realistic values.

(7) It may also be appropriate to test individual connections or individual components in accordance with 7.3.3 and 7.3.2.

7.3.5.3 Load tests of completed structures or parts of structures

P(1) Testing may be carried out to assess the structure as built and to decide whether or not it meets the requirements of the original design.

P(2) The loading shall accurately model that on the structure, without causing unnecessary damage or distress.

(3) Loading a structure to its design ultimate load may impair its subsequent performance in service. While such overload tests may sometimes be justified, for example to check an unproven design, it is generally recommended that the structure be loaded to a level appropriate to the serviceability limit states. If sufficient measurements of deformation are taken, then they can be used to calibrate the original design in predicting the ultimate strength and long-term performance of the structure.

(4) The total test load to be carried should be the greater of:

- the sum of the characteristic dead load and 1.15 times the characteristic imposed load
- 1.125 times the sum of the characteristic dead and imposed loads.

(5) In deciding on a suitable figure, due allowance should be made for finishes, partitions, etc. and for any load sharing that could occur in the finished structure.

(6) Test loads should be applied and removed incrementally while observing all proper safety precautions. The rate of loading and unloading should be appropriate to the intended use of the structure as the rate will influence the response.

(7) The test load should be applied at least twice, with a minimum of one hour between tests and allowing five minutes after a load increment is applied before recording deformation measurements.

(8) For structures in which deformation is critical, consideration should also be given to an additional application of load, which should be left in place for 24 hours.

(9) Residual deflection after the first application and removal of the test load will indicate movement at joints, which will not be recoverable. The response of the structure should be assessed on the second loading, taking the residual deflection as the datum when determining the load-induced deflection. The structure should return to this datum deflection on removal of the second loading. Failing this the loading should be repeated until full recovery is achieved before determining the load- induced deflection.

(10) In order to evaluate the experimental behaviour, relevant material properties should be obtained from suitable specimens cut from lightly stressed parts of the structure, taking due account of the effect of subsequent repairs. Failing this representative samples may be tested, for example units made by the same process and using the same materials. Finally, relevant data provided by the manufacturer may be used when there is sufficient experience of the materials and manufacturing process. Where standard components are used, the third option is likely to give the most realistic values.

(11) Where the structure is formed of standard components it may be appropriate to test similar components, in accordance with 7.3.2, if all parties are satisfied that they are representative of those in the complete structure.

7.4 ADDITIONAL TESTS FOR SPECIAL PURPOSES

P(1) When the loads to be carried by the structure are predominantly dynamic, additional tests may be required, on components, connections or on the complete structure.

P(2) Similarly, additional tests may be required when the structure is subjected to an unusual environment or combination of environmental loads.

P(3) Where long-term effects are poorly understood additional tests, on the complete structure or on elements thereof, may be required as part of the control and maintenance of the completed structure, see 8.9.

P(4) Additional tests may be required when a structure is damaged and subsequently repaired.

(5) The manufacturer of standard components should provide data on the structural performance, under short-term, static loads. Where dynamic loads, such as fatigue, blast or impact are significant, additional tests may be required to supplement the manufacturer's data, using the approaches detailed in 7.3.2 to 7.3.5.

(6) Similarly, standard information on the durability of FRP will generally be based on data obtained under standard ambient conditions. Where there is an unusual combination of environmental loads, for example an aggressive atmosphere combined with elevated temperatures, additional tests on components or laminates may be required, using the approaches outlined in 7.3.1 and 7.3.2.

(7) Testing within *Control and maintenance of the completed structure*, see 8.9, will be required when deterioration of the component could lead to a significant loss of strength or stability or an inability to satisfy the serviceability requirements. This could consist of in-situ non-destructive testing, such as by ultrasonics, or the removal and testing of individual components. The latter should be in accordance with 7.3.2.

(8) When a structure is damaged, for example as a result of accidental overload or impact, tests may be required, either on the damaged structure or after repair. The degree of testing will depend on the severity of the damage and should be agreed by all concerned. Testing should be in accordance with 7.3.2 to 7.3.5 as appropriate.

(9) In addition, tests may be required when the loading or environmental conditions differ significantly from those for which the structure was originally designed.

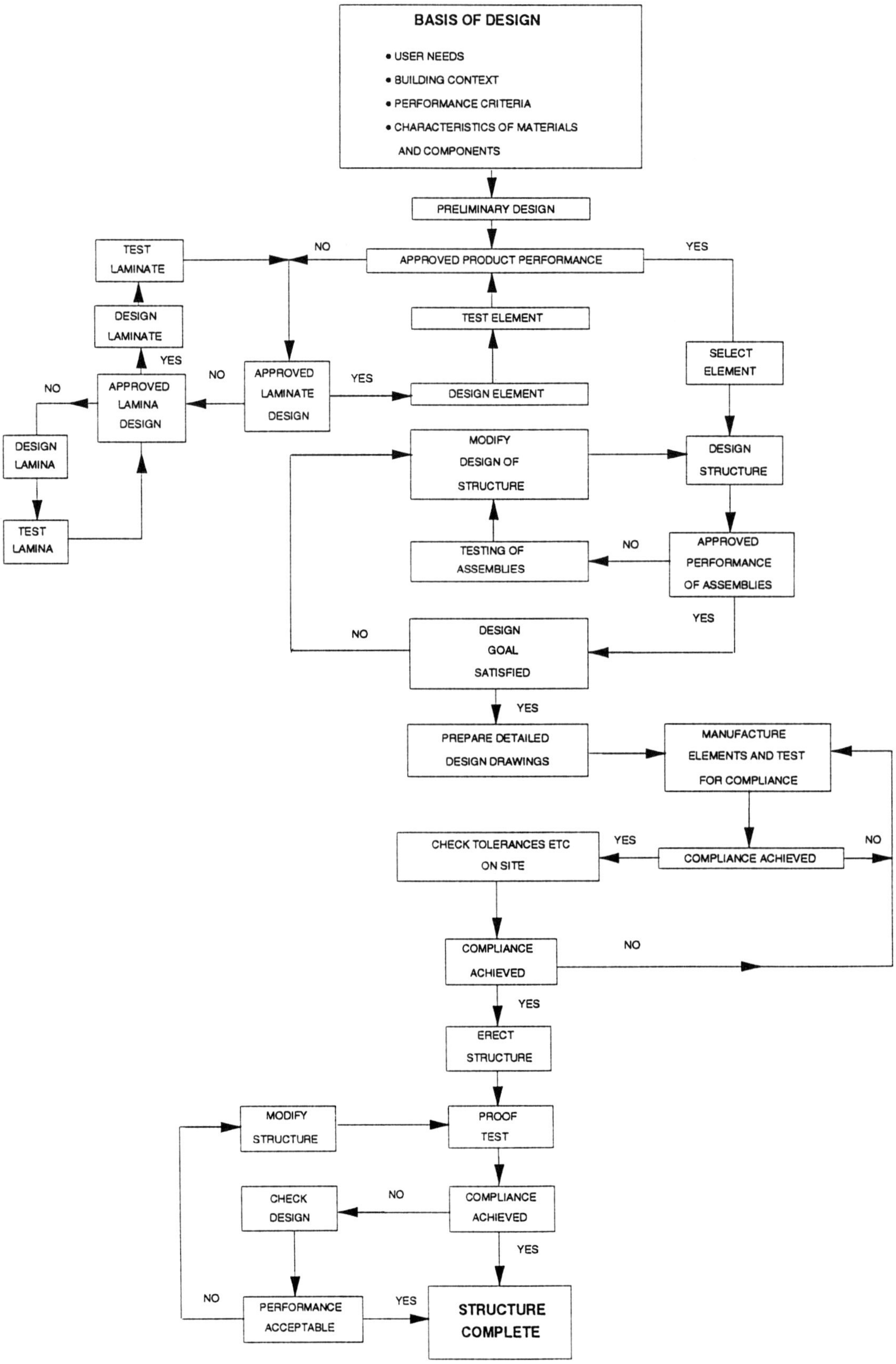

Figure 7.1 Flow diagram of the design and construction process.

CONTENTS

8 QUALITY CONTROL

8 QUALITY CONTROL

8.1 SCOPE AND OBJECTIVES

P(1) This section specifies the minimum control measures necessary for the design and construction of structures made from fibre-reinforced polymeric composites. They comprise essential actions and decisions, as well as checks to be made, in compliance with specifications, standards and the general state-of-the-art, to ensure that all specified requirements are met.

8.2 CLASSIFICATION OF THE CONTROL MEASURES

8.2.1 General

P(1) With regard to the quality control required in 2.1 of this EUROCOMP Design Code, three basic control systems are identified in terms of the parties who may exercise quality control; different objectives are defined for each system:

- internal control
- external control
- compliance control.

8.2.2 Internal control

(1) Internal control is carried out by the designer, the contractor, the fabricator or by the supplier, each within the scope of the specific task in the manufacturing or construction process. It is exercised

- on his/her own "internal" initiative or
- according to "external" rules established by the client or by an independent organisation.

8.2.3 External control

(1) External control, comprising all measures for the client, is carried out by an independent organisation charged with this task by the client or by the relevant authority. External control may consist of:

- the verification of internal control measures (in so far as these are made in accordance with external specifications) or
- additional checking procedures independent from internal control systems.

8.2.4 Compliance control

(1) Compliance control is exercised to verify that a particular service or production function has been carried out in compliance with the specifications previously established.

(2) Compliance control is generally part of the external control.

8.3 VERIFICATION SYSTEMS

P(1) The frequency and intensity of control depend on the consequences of possible mistakes and errors in the various stages of the manufacturing and construction processes. In order to improve the effectiveness of control, different control measures are combined in a verification system.

8.4 CONTROL OF DIFFERENT STAGES

(1) Depending on the purpose and timing of the control, the following stages may be identified:

- control of design
- control of production and construction
- control of completed structure.

8.5 CONTROL OF DESIGN

P(1) Control of design shall conform with the appropriate Euronorm, ISO or National administrative procedures.

8.6 CONTROL OF COMPONENT MANUFACTURE

8.6.1 Objectives

P(1) Control of the component manufacture comprises all the measures necessary to maintain and regulate the quality of the materials and of the workmanship in conformity with specified requirements. It consists of inspection and tests and involves the assessment of test results.

(2) The objects which need to be controlled and the relevant ISO Standards are summarised in Table 8.1.

8.6.2 Elements of component manufacture

P(1) The component manufacture controls include:

- initial tests and checking procedures
- tests and checking in the course of manufacture
- final tests and checks.

(2) Different verification systems may be appropriate for:

- a continuous production; the aim of this system is to achieve a uniform quality of the products in the long-term
- a single product; the aim is mainly to comply with the basic requirements of the project.

(3) For a single product, it may be appropriate to concentrate on precautionary measures, in particular on initial tests and on checks during construction.

8.6.3 Initial tests

P(1) Where necessary, initial tests shall be made before the start of the manufacturing process in order to check that the intended component can be manufactured satisfactorily using the specified materials, equipment and manufacturing methods.

(2) The quality and compatibility of the materials should be shown to be adequate, either by reference to previous experience or by means of prior tests. Only approved materials should be used.

8.6.4 Checks during manufacture

P(1) The properties of the materials and their suitability, the dimensions of the finished components and the equipment used shall be subjected to a regular system of verification during the manufacturing process.

P(2) Important findings shall be filed in written reports, which shall be available to all the parties concerned.

(3) Depending on the degree of reliability required, additional special control measures may be agreed.

Table 8.1 Standards for production and construction control.

Item or process	*Control of materials or production*	*ISO Reference no.*	*Euronorm no.*
Glass reinforcement	Constituent materials	1889, 2078, 2797, 3341, 3375 and 3598	
	Woven mats, etc.	2113, 2559, 3342, 3374, 3616, 4602, 4603 and 4605	
Manufacture of Component	Control of reinforcement	1886, 2558	
	Prepregs, etc.	8604, 8605 and 8606	
	Resins, etc.	60, 176, 584, 1148, 1628, 1675, 2555, 3219, 6186	
	Control of manufacturing process	9001 and 9002	29001 and 29002
Completed Component	Mechanical properties of composite	178, 3268, 3597, 4585 and 4899	61, 63
	Surface properties	179, 180, 868, 2039 and 6609	59
	Density	1183	
	Long-term behaviour (creep)	899 and 6602	

8.7 CONTROL OF COMPONENT DELIVERY

P(1) When components and materials are received at the site, their compliance with the terms of the original order shall be checked.

P(2) Important findings shall be filed in written reports which shall be available to all the parties concerned.

(3) The delivery note should certify that the FRP components were manufactured, marked and treated in accordance with the order.

(4) In addition to (3) above, the delivery note should contain the following information:

- date of manufacture of the unit
- method of storage since manufacture
- date of delivery of the unit
- identification markings and, if required, the reference number of each component in accordance with the customer's order.

(5) Depending on the degree of reliability required, additional special control measures may be agreed. Details of tests that may be required are given in 7.3.2 of this EUROCOMP Design Code.

(6) The delivery notes for any other construction materials, such as adhesives for the bonding of joints, should contain the following information:

- identification of each item, along with its source
- the batch number and date of manufacture if appropriate
- any special storage instructions.

(7) Where components are assembled together either at the place of manufacture or at some place other than the final location, a written record of the assembly, as given in 8.8 of this EUROCOMP Design Code, should accompany the delivery note for the components.

8.8 CONTROL OF ASSEMBLY

P(1) When components are assembled, either at the place of manufacture or in their final location (or elsewhere), the work shall be carried out in accordance with the specification.

P(2) Important findings shall be filed in written reports which shall be available to all the parties concerned.

(3) The units to be joined should be identified clearly in accordance with the specification.

(4) The units to be joined should be checked for twist, bow and flatness, within the tolerance specified.

(5) Preformed or predrilled holes should be correctly located, to within the tolerance specified.

(6) Similarly any cast or formed ends should be as defined, within the tolerance specified.

(7) Any remedial action required as a result of one or more of the items in (4), (5) and (6) being beyond the specified tolerances shall be carried out only with the agreement of all the parties concerned, and

recorded in accordance with P(2).

(8) A written record of the assembly of the components shall be filed which should include the following information:

- identification of the assembly and the components
- confirmation that the work was carried out by competent staff, who had received the necessary training in accordance with the specified method
- the environmental conditions, such as temperature and humidity, in so far as they influence the curing of any adhesive used
- the names of staff involved and, where appropriate, the responsible engineer
- any deviation from the specified method, as indicated by (7).

(9) When assembly is carried out at the place of manufacture of the units, or at some place other than the final location, the written record in (8) shall accompany the assembled components and form part of the control of delivery, as given in 8.7.

(10) Depending on the degree of reliability required, additional control measures may be agreed. This may be necessary for highly stressed connections, when tests may be required to determine their adequacy, as given in 7.3.3, or the adequacy of complete assemblies, as given in 7.3.4.

8.9 CONTROL AND MAINTENANCE OF THE COMPLETED STRUCTURE

(1) A planned control programme should specify the control measures and inspections to be carried out in service where long-term compliance with the basic requirements of the project is not adequately ensured.

(2) All the information required for the structure's use in service and its maintenance should be made available to the person who assumes responsibility for the complete structure.

Part 2 EUROCOMP Handbook

CONTENTS
EUROCOMP HANDBOOK

CONTENTS

1 BASIC INFORMATION

1 BASIC INFORMATION

1.1 FOREWORD

The aim of this EUROCOMP Handbook is to amplify the information contained in the EUROCOMP Design Code, so that users of the code can understand the decisions that have been taken during the drafting process. In particular it is intended to cover areas in which there is insufficient experience of the use of polymeric composite materials, or insufficient experimental data, to formulate precise design clauses. Here only guidance can be given.

The EUROCOMP Handbook is not intended to be a text book and hence topics are only covered in sufficient depth for the designer to understand the materials, their properties and their behaviour in a structure. It is assumed that the user of the EUROCOMP Design Code is familiar with the process of design using conventional construction materials such as concrete and steel. Thus the EUROCOMP Background Document attempts to highlight those areas in which the properties of polymeric composites differ from those of conventional materials.

The EUROCOMP Background Document should be read in parallel with the EUROCOMP Design Code. With the exception of Chapter 1, which is discussed below, each section of the EUROCOMP Design Code has an equivalent section in the EUROCOMP Background Document, using the same system of numbering. For clarity, where it is felt that the EUROCOMP Design Code is self explanatory, the section heading is shown in the EUROCOMP Background Document with the brief statement "No additional information".

Chapter 1 in the EUROCOMP Design Code follows closely the introductory chapters of the Eurocodes for steel and for concrete. It consists largely of statements of the Principles of design, which do not require amplification. Hence Chapter 1 of the EUROCOMP Background Document is largely devoted to an introduction to polymeric fibre composite materials, chiefly for those unfamilar with their properties and the manufacturing techniques involved.

1.2 SPECIAL TERMS USED IN THE EUROCOMP HANDBOOK

A number of special terms are listed in 1.4.2 of the EUROCOMP Design Code. Further terms used in the EUROCOMP Handbook are defined in the relevant sections.

1.3 MATERIALS

1.3.1 General

The EUROCOMP Design Code is mainly concerned with glass fibre reinforced thermosetting polymer composites. Properties of the basic materials are given in Chapter 3 and of typical composites in Chapter 4. This section of the EUROCOMP Handbook will therefore deal primarily with those composites. However, it is desirable that the composite be compared with some of the alternative options that are available.

The most important fibres in order of market volume are glass, carbon and aramid. Hence although glass fibre reinforcement composites will dominate this section, other composites using carbon and aramid will not be excluded. Other fibres, such as polyethylene, are used as the reinforcement for composites, but their use is currently of insufficient volume to justify inclusion.

Where the discussion is specific to these materials, the abbreviations carbon FRP and aramid FRP are used.

This introductory section, unlike that in the EUROCOMP Design Code, is not restricted to the relatively high structural performance materials in which fibres with high elastic modulus, high strength and high aspect ratio are combined with a compatible polymeric matrix. It also includes those composites which utilise the fibre primarily as a crack stopping device, such as moulding compounds. The processes in which they are utilised (hot press moulding) is outside the scope of the code itself, but they fulfil a useful role in their ability to be produced in complex shapes at high production speed. Their properties are not necessarily outstanding but they may be adequate for their purpose and should not be ignored.

The fibre type, matrix type, volume fraction and fibre arrangement are all variables which may be selected to obtain the optimum performance requirements. These combine into a range of options which is literally infinite.

The volume fraction of fibres is dependent on two main factors, the fibre geometry and the process being used to produce the composite.

The arrangement of the fibres is totally variable but there are three main categories. When all the fibres lie in one direction the construction is termed uni-directional. When a proportion of the fibres lie at 90° to the remainder it is termed bi-directional. This is achieved either by the use of woven or non-woven fabric or by the use of discrete layers of fibres each uni-directional but successively laid at 90°. In addition, reinforcement fabrics are now available with fibres at ±45° as well as 0° and 90°. Finally when the fibres are randomly distributed and are in-plane the arrangement is termed random.

1.3.2 Matrix system

1.3.2.1 Introduction

The family of polymeric composites splits into those which use thermosetting resins and those which use thermoplastic resins. The EUROCOMP programme is only concerned with the thermosets.

The matrix system consists of a number of components:

- thermoseting resin
- catalysts
- accelerator (depending on the process)
- fillers (optional in some processes)
- pigments (colouring for the product)
- release agents (aids removal from the mould)
- inhibitors (help control the cure process)
- low shrink or low profile additives (improve surface finish and thickness dimensions)
- monomer (to adjust resin mix viscosity)
- fire retardant additives
- various sundry items.

1.3.2.2 Function of the matrix

The matrix material has several duties to perform: it transfers load between the fibres, it protects the notch sensitive fibres from abrasion and it forms a protective barrier between the fibres and the environment. As a protective barrier it prevents attack from moisture, chemicals and oxidation. It also plays the dominant role in providing shear, transverse tensile and compression properties to the composite. The behaviour of the composite under the effects of temperature is also governed by the matrix performance.

1.3.2.3 Processability of a resin system

The processability of a resin system determines to what extent it can be used in a variety of processes. Within this field are such factors as viscosity, shelf life, cure regime, etc. The processability and the thermo-mechanical performance of a resin system are the two factors which most effectively characterise resin systems used in composites. The thermoset resins most commonly used in composites are: unsaturated polyester, urethane methacrylate, vinyl ester, epoxy and phenolic. The typical range of properties of these resins are given in Table 1.1. More specific information is given in Chapter 3 of the EUROCOMP Design Code.

Table 1.1 Indicative properties of resin castings.

Property	*Units*	*Range of values*
Specific gravity		1.08-1.25
Tensile strength	N/mm^2	50-85
Tensile modulus	kN/mm^2	2.2-3.7
Flexural strength	N/mm^2	80-130
Flexural modulus	kN/mm^2	2.9-4.2
Hardness Barcol 939/1		40-45
Elongation at break	%	1.2-6.5
Water absorption by weight	%	1-2
Glass transition temperature	°C	50-160

1.3.3 Reinforcements

1.3.3.1 Rovings, mats, woven fabrics and non-crimp multi-axial (NCMA)

Reinforcement materials are available in a variety of formats which have been developed with two primary aims, which are not necessarily compatible. These are to achieve the physical properties required by a suitable arrangement of the fibres and at the same time to achieve an acceptable method of manufacture. The forms which are available are:

Uni-directional, in which all (or a very high proportion) of the fibres lie in one direction. This is supplied in the form of continuous strands or bobbins for either internal or external unwinding. It is known as 'roving' if glass fibre or 'tow' if carbon fibre.

Chopped fibre mats; the strands are chopped into short lengths (5-50 mm) and bound together into a loose fabric with a more or less random fibre

orientation. Continuous filament mats are of a similar form of construction but, as the name implies, the fibres are continuous.
Woven fabric; using conventional cloth weaving techniques, glass yarns are woven together to form a fabric, with different properties along and across the fabric. Woven rovings are of a similar form.

Non-crimp multi-axial: the glass rovings are laid down into layers or plies of variable weight and orientation. They are then stitched together using a light thread.

1.3.3.2 Alternative fibre reinforcements

Although the EUROCOMP Design Code is restricted to glass fibre as the reinforcement, it is necessary to consider the alternative reinforcement fibres in order to appreciate the characteristics, benefits and disadvantages of glass fibre as a reinforcement option.

The reinforcement types which will be discussed are glass fibre, carbon fibre and aramid fibre. Each of these is not a single fibre type but more a family of fibres with a range of properties. Typical properties of a range of reinforcement fibres are given in Table 1.2 The strength and strain figures

Table 1.2 Initial fibre properties before processing.

Family	*Fibre*	*Property*			
		Tensile modulus (kN/mm^2)	*Ultimate strength (N/mm^2)*	*Strain (%)*	*Specific gravity*
Glass					
	E	72	3450	4.8	2.54
	A	70	3030	4.3	2.50
	S2	88	4600	5.7	2.47
	R	86	4400		2.55
	ECR, C	69	3030	4.4	2.49
Carbon					
Toray	T300	230	3530	1.5	1.77
	T800	294	5490	1.9	1.81
	T1000	294	7060	2.4	1.80
	M40J	377	4400		1.77
Aramid					
Kevlar	49	125	2750	2.4	1.44
	29	83	2750	4.0	1.44
Teijin	Technora	70	3000	4.4	1.39

quoted for fibres should be taken as the maximum upper limit available. The fibre performance is often degraded in processing at the composite manufacturing stage. The fibre modulus, however, is not degraded in processing which makes the modulus of the composite highly predictable.

1.3.3.3 Glass fibre

As a broad generalisation glass fibres can be categorised into two sets: those with a modulus around 70 kN/mm^2 and with low to medium strength (i.e. E, A, C and ECR), and those with a modulus around 85 kN/mm^2 and with higher strength (i.e. R - S2 - glass) (reference 1.1).

Although E-glass fibres are reasonably strong, they are very susceptible to damage, both physical and environmental, which reduces their strength. Care must be taken in both the manufacture of the fibre and the production of the composite to minimise this reduction in strength, particularly if the composite is strength critical.

S2 - and R-glass fibres are stronger and stiffer than E-glass fibres and they are therefore used in the more demanding applications where their extra cost can be justified.

The specific gravity of glass fibre is about 2.5, which is higher than the specific gravity of other reinforcing fibres but lower than that of metals used in construction (aluminium has a specific gravity of about 2.8 and steel 7.8).

Glass fibres are the most commonly used reinforcing fibres. They have good properties, very good processing characteristics and they are inexpensive.

The processing characteristics required of glass fibres include: choppability, low static build up, conformance to complex shape, etc. and resin compatibility requirements such as fast wet out, good fibre/matrix adhesion, etc. Thus glass fibres are available with characteristics which are designed to make them suitable for whichever of the composite manufacturing processes is to be used.

E-glass fibres devitrify at about 675°C and R-glass fibres at about 780°C. They are therefore reasonably resistant to the effects of temperature and have an excess of performance available when used with polymeric matrices. These degrade at much lower temperatures and therefore limit the temperature performance of the composite material.

Glass fibres are insulators of both electricity and heat and thus their composites exhibit very good electrical and thermal insulation properties. They are transparent to radio frequency radiation and hence have found extensive use in radar antenna applications.

E-glass fibres with a modulus of about 70 kN/mm^2 can only produce composites with modest moduli when compared with steel. As an absolute limit, assuming unidirectional fibres and the highest feasible fibre volume fraction of say 75%, the stiffest E-glass fibre composite has a modulus of 50 kN/mm^2. But a more achievable maximum for practical purposes is 45 kN/mm^2. At right angles to this, in the transverse direction, the modulus approaches that of the resin itself at about 4 kN/mm^2. An E-glass laminate made from random glass fibres with a more modest fibre volume fraction of say 20% would have a modulus of only 5 kN/mm^2. Hence random E-glass fibre laminates are relatively flexible.

The use of S2 - or R-glass improves the composite modulus to about 60 kN/mm^2 for uni-directional and 25 kN/mm^2 for non-crimped bi-directional constructions. However, they are both more expensive than E-glass and are only available in a fairly limited range of material types and resin compatibilities. Probably the most important virtue of S2 - and R-glass is that the strength is considerably higher than that of E-glass.

E-glass is highly resistant to most chemicals but it is attacked by both mild acids and mild alkalis. The extensive use of glass fibre reinforcement in chemical plants is reliant upon the corrosion resistance of the polymer matrix and its ability to ensure that the glass fibres do not become exposed to the environment (see 1.3.14, stress corrosion).

Other glass fibres, notably ECR-glass show an improvement over E-glass in a corrosive environment which may gain access to it. Bare E-glass in distilled water retains about 65% of its short term ultimate tensile strength after 100 days. This compares with R-glass, which has a strength retention under the same conditions of 75% after 100 days (reference 1.2).

1.3.3.4 Carbon fibre

The term carbon fibre covers a whole family of materials which encompass a large range of strengths and stiffnesses. Generally the use of carbon fibre allows the production of composites with very high stiffness and high strength.

The elastic modulus, strength, strain and density of a range of carbon fibres commercially available is shown in Table 1.2. The most universally used carbon fibres have a tensile modulus of about 230 kN/mm^2, tensile strength of 3200-3500 N/mm^2 and strain to failure of 1.5%. They are typified by T300 fibre from Toray. Their cost/performance ratio justifies their use in many commercially driven applications.

Carbon fibres have also been developed with extremely high performance but at a cost which only justifies their use on rare occasions. T1000 fibre, available from Toray, has a tensile strength of around 7000 N/mm^2 and modulus of 294 kN/mm^2 (reference 1.3).

1.3.3.5 Aramid fibres

Aromatic ether amide or "aramid" fibres are organic, man made fibres which are available in various forms for use in composites. They are characterised by having reasonably high tensile strength, a medium modulus and a very low density. Their composites fit well into a gap in the range of stress/strain curves left by the family of carbon fibres at one extreme and glass fibres at the other.

There are three major commercial suppliers: Dupont produce Kevlar aramid in several versions; Akzo produce under the trade name Twaron, while the Teijin company have an aramid which they market under the trade name of Technora (reference 1.4).

Aramid fibres are insulators of both electricity and heat. They are resistant to organic solvents, fuels and lubricants. As unprotected fibres they are highly tenacious and do not behave in a brittle manner, as do both carbon and glass fibres.

There are two distinct types of aramid fibre, those in which the elastic modulus is about the same as for glass fibre, typically 60-70 kN/mm^2, and those with a modulus of about twice this level.

Kevlar 29 falls into the first category and Kevlar 49 into the higher modulus category. The higher modulus material is the more usual for composites. But the lower modulus aramids do have applications in composites when high strain to failure or high work to failure is required. Both types have a tensile stress/strain curve which is essentially linear to failure. In compression the curve is non-linear, with a lower failure stress as a result of a different mode of failure.

The specific gravity of aramids is in the range 1.39-1.44. Therefore the main advantage of aramids is their strength/weight and stiffness/weight ratios. Hence they can have an advantage over many carbon and glass fibres if either specific strength or specific stiffness is the requirement. It should be noted however that some aramids have relatively low compressive strength, perhaps one-quarter of the tensile strength, which makes their use in either compressive or flexural situations problematic.

Aramids, such as Kevlar 49, are chemically quite stable and have high resistance to neutral chemicals. But they are susceptible to attack particularly by strong acids and also by bases. However, Technora aramid fibre has extremely high strength retention in both acids and alkalis.

1.3.4 Compatibility of resin and fibres

It is obviously important that the reinforcement and the resin are compatible with one another. Consequently the reinforcement fibre is produced with a surface coating which makes it specifically compatible with a resin type or types.

It is primarily important that there should be good adhesion between the two to ensure that stresses are transferred into, and thus carried by, the fibre. On the other hand it can be detrimental if adhesion is too good. This can make the composite brittle as the excessive adhesion between the fibre and the resin does not allow the growth of cracks along the fibre in an impact situation. Hence a crack can only propagate through the fibre and thus through the composite.

There are other ways in which they must be compatible. For instance the resin must be able to penetrate the fibre bundle and then be able to wet the surface of the individual filaments. At the same time the resin must not have a viscosity which is too low as it could simply drain from the reinforcement before cure has taken place. The cure temperature of the resin must be less than the temperature which would degrade the performance of the reinforcement fibre.

To ensure that the resins and fibres are compatible the materials suppliers recommend particular products for specific use.

1.3.5 Mechanical characteristics of composites

1.3.5.1 General

The properties of a composite vary with position and the angle of the reinforcement in relation to the direction under consideration. In a fibre direction, they are generally completely elastic up to failure and exhibit no yield point or region of plasticity (although there may be a region of pseudo-plasticity). Thus in spite of the relatively low elastic modulus they tend to have low strain to failure when compared with metals which have a long inelastic response beyond the yield point. The work done to failure, represented by the area under the stress/strain curve, is thus relatively small when compared to that of many metals. This need not be a problem but is a factor which must be given due consideration, particularly in the design of details such as joints. The designer needs to pay special attention to the fact that local yielding can not be relied upon to dissipate excessive stress. This may be particularly important when considering the affects of changes in ambient conditions, such as temperature and moisture, and in the design of bolted connections (see Chapter 5).

The properties of the composite are affected by fibre and matrix type, the relative quantity of each in the composite and the directionality of the fibres. A composite which has all the fibres aligned in one direction, i.e. is uni-directional, is stiff and strong in that direction, but in the transverse direction it will have low modulus and low strength.

A composite with equal amounts of fibre in the longitudinal and transverse directions has equal strength and stiffness in the two directions. However, neither would be as high as in the uni-directional case. In theory, if that same amount of fibre was randomly laid in-plane (isotropically), then the resulting composite would have equal strength and stiffness in all in-plane directions but less than in the bi-directional case (in the axes of the fibres).

The directionality of the reinforcement has a significant effect on the amount of reinforcement which can be packed into a composite.

The properties of composites are also affected by the curing system and the post-curing procedure.

1.3.5.2 Volume fraction of fibre

The fibre volume fraction depends heavily on the method of manufacture. A uni-directional composite may have a fibre volume fraction as high as 75%. However, this can only be achieved if all the fibres are highly aligned and closely packed. A more typical fibre volume fraction for uni-directional composites is 65%. If the fibre configuration is changed to put fibres in other directions, then the maximum fibre packing is reduced further. A typical fibre volume fraction for bi-directional reinforcement (woven fibre) is 40% and a typical volume fraction for random in-plane reinforcement (chopped strand mat) is 20%.

The effect on tensile strength and stiffness when the fibre arrangement is changed is a reduction from unity for uni-directional to half for bi-directional and to three-eights for random (in-plane) reinforcement (reference 1.5).

When the effect of volume fraction is also taken into account the result is shown in Table 1.3.

Table 1.3 The effect of arrangement and volume fraction of fibre.

	Reinforcement directionality factor	*Typical volume fraction* V_f	*Product*
Uni-directional	1	65%	0.65
Bi-directional non-crimp	0.6	50%	0.3
Bi-directional woven	0.5	40%	0.2
Random in-plane	0.375	20%	0.075

Thus the mechanical properties of uni-directional laminates are very different from those of random laminates even if the same fibre type and resin type are used in each. A factor of 0.65/0.075 or about eight-fold can be expected.

1.3.5.3 Typical properties

Typical mechanical properties of a variety of glass fibre reinforced composites are given in Chapter 4 of the EUROCOMP Design Code.

A uni-directional glass fibre composite, with 65% fibres, will have a tensile strength of about 700 N/mm^2 and a tensile modulus of about 40 kN/mm^2.

A uni-directional aramid fibre composite with a similar volume of fibres will have a tensile strength of about 1400 N/mm$^{2.}$ However, its compressive yield strength is about one-sixth of this at 230 N/mm^2. This also affects the flexural performance, giving a value of about 300 N/mm^2.

Uni-directional composites produced from Toray T300 or similar carbon fibre reinforcement have typical properties as follows:

Longitudinal tensile modulus	125-135 (kN/mm^2)
Longitudinal tensile strength	1700-1800 (N/mm^2)

The specific modulus of carbon fibre composites (i.e. the ratio of elastic modulus to density) is their most important characteristic. A uni-directional composite of Torayca M46 has a modulus of 255 kN/mm^2 which compares with about 210 kN/mm^2 for steel . The specific density of composites of the two fibres is about 1.6, giving specific moduli as shown in Table 1.4 together with those for uni-directional E-glass and steel.

Table 1.4 Specific moduli of composites and of steel.

Material	*Specific modulus*
M46 uni-directional composite	160
M46 bi-directional composite	70
E-glass uni-directional composite	23
Steel	27

1.3.6 Fatigue

The fatigue behaviour of glass FRP is discussed in Chapter 4. The modulus is relatively low and hence the composite is required to work at high strain which can approach the strain to failure of the matrix. This can allow a fatigue process to occur, resulting in a reduced life. Glass FRP's are therefore generally more sensitive to fatigue than those composites with fibres of higher modulus.

Bi-directional E-glass/epoxy has a fatigue performance for zero-tension cyclic loading of monotonic strength of 419 N/mm^2 and a slope of the curve of 42.5 N/mm^2 per decade of cycles (reference 1.7).

Uni-directional carbon FRP composites are relatively insensitive to tension fatigue when loaded in the fibre direction. Therefore they have excellent fatigue performance when compared with metals and other composites even at very high stress levels (reference 1.8).

Even quasi-isotropic composition carbon FRPs, where the matrix performance has a more dominant role, can show a two - to four - fold improvement in fatigue resistance over steel and aluminium. It should be noted however that when stressed in the direction transverse to the fibres or in compression the fatigue life is substantially reduced.

Uni-directional Kevlar 49 composites in tension/tension fatigue have very good performance which is superior to S-glass and E-glass composites. However, in flexural fatigue they have poor fatigue strength at low cycles, probably as a result of the poor static flexural performance of Kevlar 49. But at a high level of cycles this is less apparent.

1.3.7 Creep

Thermosetting polymers are viscoelastic at ambient temperature and unreinforced are susceptible to creep . Fibre reinforcements are generally elastic and creep resistant but some types may show the phenomena of stress rupture if exposed to high stresses in adverse environments. Composites reinforced with unidirectional stiff fibres and stressed in the fibre direction can be highly creep resistant as the fibre attracts a high proportion of the load in line with its stiffness relative to the polymer matrix. The behaviour of composites under static loading is a function of many material variables: polymer type and degree of cure; fibre type, orientation, straightness and volume fraction; the stability of the fibre polymer matrix interface; and the moisture content of the polymer. The main external variables determining creep are stress and temperature.

Bi-directional woven construction laminates do not perform as well as unidirectional materials. The presence of crimp in the fibres gives rise to a local stress concentration in the polymer which may cause cracking and allows the fibre to straighten out under stress. Non-woven continuous fibre reinforcements are preferred in critical applications. The least resistant constructions to creep are the random discontinuous short fibre reinforcements, as they contain a multitude of fibre end stress concentrations and higher shear stresses developed in the polymer to couple between fibres.

Creep behaviour of composites normally follows a primary, secondary and tertiary phase as shown in Figure 1.1. Whether each successive region is reached depends on the factors above, loading level and time. For low strains the initial viscoelastic behaviour is usually recoverable on removal of stress. The first and second stages are associated with a delayed equalisation of internal stresses between fibre and matrix with shake down of any local peak stresses. The tertiary stage leading to rupture results from progressive local failure mechanisms and continuous readjustment of internal stresses and is associated with matrix and laminar cracking, interface instability, individual fibre slippage and breakage.

Resistance to creep is dependent on the alignment of fibres to match the external loading and minimise stresses in the matrix. When subjected to tensile stresses carbon composites resist long term creep very well. In off axis situations creep rates will be higher and in compression the contribution by the matrix to local fibre stability is critical and lower allowable stresses are required.

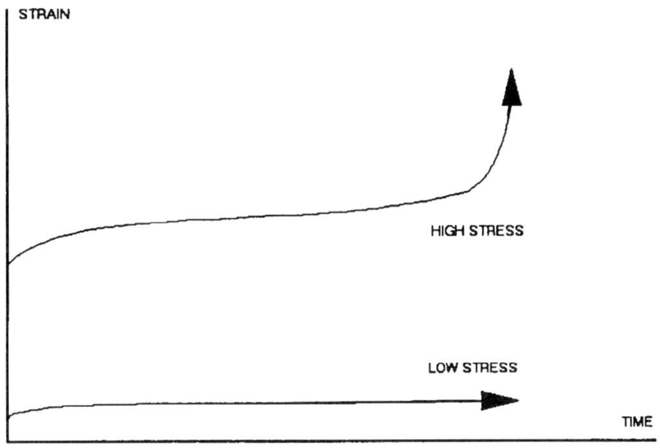

Figure 1.1 Creep behaviour of composites at high and low loads.

Although aramid fibres have high inherent tensile strength, particularly in uni-directional construction, in composites they tend to have creep rates very much higher than similar glass or carbon composites (reference 1.9).

1.3.8 Impact

Glass fibre composites have very good resistance to impact due to their high strain to failure, when compared with other fibres.

Aramid FRPs have excellent impact resistance, particularly to ballistic impact. On a weight basis they are superior to glass fibre composites, which themselves offer good ballistic impact resistance, but they are more expensive.

Carbon FRPs have a strain to failure which is typically about 1.5%. They also have totally elastic behavour up to failure and therefore their ability to absorb impact energy without damage is limited. To improve the impact performance, fibres with higher strain to failure, such as glass or aramid, may be incorporated.

1.3.9 Effects of temperature

1.3.9.1 General

At elevated temperatures all resins soften and reach a stage at which the polymer passes from a glass-like state to a rubbery state. This is the glass transition temperature (Tg) and this less 10-20°C is generally taken as indicating the upper limit to the usable temperature range for all normal applications. The range of values of Tg, which is very dependent on the type of resin, is given in Table 1.1. The reinforcement, which has a lower coefficient of thermal conductivity than the resin, tends to increase the effective Tg of the whole composite to a level above that of the basic resin.

The most widely used method for measuring the Tg is DSC (differential scanning calometry). However, composites are often highly filled materials, which complicates the determination of the Tg by the DSC method. Another useful and perhaps more informative tool in assessing the thermal performance is DMTA (dynamic mechanic thermal analyses), where the changes in E modulus are analysed as a function of temperature. This enables, for example, the study of the changes in mechanical behaviour around and beyond the Tg, which may differ significantly for different fibre/resin combinations, from a drastic drop to a moderate decrease. As shown by the upper curve of Figure 1.2, a decrease in storage modulus E' is observed around the Tg of the material. The loss tangent curve ($\tan \delta = E''/E'$, where E'' = loss modulus) gives information on the polymer itself, e.g. of the crosslinking.

Low temperatures tend to make polymers less flexible and therefore there may be a tendency towards damage by fatigue but generally both the strength and the stiffness of composites are unimpaired by the effect of low temperature.

1.3.9.2 Thermal expansion coefficients

The thermal expansion coefficients of thermoset resins are typically in the range 50×10^{-6} to 110×10^{-6}/°C. This is much higher than those of the reinforcement fibres. That of E-glass fibre for example is about 5×10^{-6}/°C, whereas carbon fibres and aramid fibres can have small negative coefficients of expansion, which can result in a near zero or even negative thermal expansion coefficient of the composite.

The thermal expansion of the composite depends not only on the type of reinforcement and the type of matrix but also on the geometry of reinforcement, its volume fraction and of course the amount and type of any filler present. Because it is a function of the fibre orientation, the thermal expansion will depend on the direction considered. Thermal expansion coefficients for some E-glass composites are shown in Table 1.5.

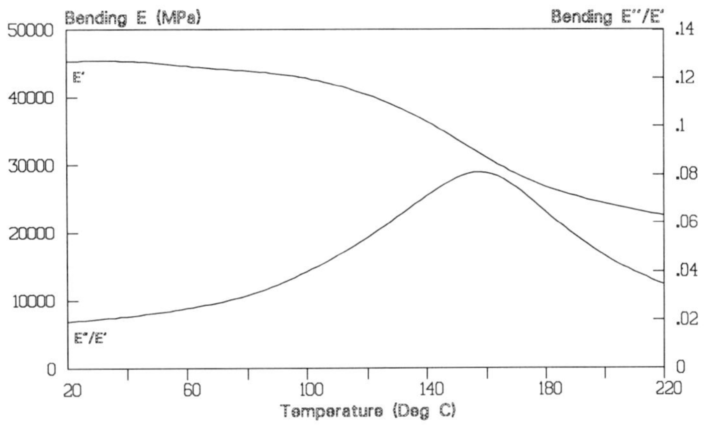

Figure 1.2 The effects of temperature on the dynamic mechanical behaviour of one polyester/glass fibre composite (isophthalic polyester Neste S 560 Z, E-glass, manufactured by pultrusion). The DSC method gives the Tg of the pure S 560 Z resin casting as 135°C. Owing to certain additives, the Tg of the FRP material in question could not be determined with DSC.

Table 1.5 Thermal expansion coefficients and volume fractions for some E-glass polyester composites.

	Uni-Dir. 0°	*Uni-Dir. 90°*	*Bi-Dir. fibre*	*Mat + roving pultrusion*
V_f	65%	65%	65%	37%
Thermal expanstion Coefficient 10^{-6}/°C	8.6	14.1	9.8	11

1.3.10 Fire properties

FRP composites usually contain a large proportion of filler and/or glass fibre, neither of which supports combustion. The resin and additives are the components in the composite which can burn. (The exception to this is phenolic resin which tends to char).

The fire performance of composites covers a wide range from highly flammable to non-burning and is governed by:

(a) The matrix type

Polyester and vinyl ester resin without additives burn readily.

Urethane methacrylate though not intrinsically better than polyester or vinyl ester in fire, does have better smoke and toxicity performance. It also allows the use of a higher filler loading which can improve fire performance.

Epoxy resins have similar fire performance to polyester when viewed against more resistant polymers such as phenolic resin or modified methacrylate resins with aluminium trihydrate.

Phenolic resin is the most fire resistant of these polymers and can be considered non-combustible (reference 1.10). However, it suffers badly under thermal shock, when it is prone to cracking.

(b) Fire retardant additives

These improve surface spread of flame, ignitability, etc. but are generally toxic chemicals and therefore contribute adversely to the toxicity rating and smoke generation.

(c) Quantity and type of filler

The more inert filler there is in the composite, the less combustible the matrix material and hence the fire performance is improved. Also some fillers generate water when subjected to fire and this reduces the level of heat generated.

(d) Reinforcement

Again, the more incombustible reinforcement there is in the composite the less material there is to burn. Hence the reinforcement arrangements which allow a high fibre volume fraction, such as uni-directional and bi-directional, give better fire performance than random reinforcement. Also long continuous fibres are less prone to mechanical damage and therefore maintain the integrity of the structure longer than do chopped fibres.

The most common fire performance criteria are:

- surface spread of flame
- fire penetration
- ease of ignition
- fuel contribution
- limiting oxygen index, etc.

The particular risk assessment determines the most suitable set of criteria.

FRP composites can be formulated to give a wide range of fire performance requirements, by the use of a high volume fraction of fibre and/or suitable filler loading and/or fire retardant additives. Typical specifications which are achievable include:

- surface spread of flame to BS476 Part 7 Class 1
- fire propagation to BS476 Part 6 Class 0
- and fire penetration and spread of flame to BS476 Part 3 Class AA.

These specifications are demanding for polyester or vinylester resin systems. They are more easily achieved with urethane methacrylate which allows a higher level of filler loading. Phenolic resin composites can achieve this performance without fire retardant additives but they are difficult to process.

1.3.11 Chemical resistance

Glass FRP has been used extensively in chemical plant structures for many years because of their inherent resistance to corrosive environments, whether acidic or alkaline. The most suitable resin system is selected on the advice of the resin manufacturer to suit the particular circumstances.

1.3.12 Moisture effects

All polymers are susceptible to the absorbtion of moisture. In general this results in a reduction in mechanical properties and glass transition temperature.

The glass transition temperature can be reduced to 75% of its dry value by the effect of a moisture content of 4% and its flexural strength can be reduced to 50% of its dry value by the effect of a 1.5% moisture content.

The greater the amount of moisture, the greater is loss of property. However, resin systems can be selected which have excellent resistance to the effects of moisture. The designer must take the effects of moisture into account and if the composite is to be subjected to a wet environment then "wet properties" must be used in the design process.

1.3.13 Resistance to weathering

The general term "weathering" covers the effects of temperature, moisture, sunlight, wind, dust, acid rain, etc. on the properties of composites. There are several mechanisms involved, most notably leaching of chemical constituents from the resin, sunlight attack causing embrittlement, and erosion of the resin as a result its degradation and the effect of wind and airborne particles. Thus it is the fact that the environment is allowed access to the fibres that gives rise to most of the adverse structural consequences of weathering.

The weathering process is effectively countered by the use of resins applicable to the environment, ultra-violet stabilisers in the matrix and if necessary chemical resistant fibres (C-glass, ECR-glass or polyester) in the surface of the composite.

1.3.14 Stress corrosion

Glass fibre composites are susceptible to a phenomenon known as stress corrosion where a laminate under stress in an acidic or basic environment can fail catastrophically at very low stress if the environment can gain access to the fibres by diffusion or, more likely, a crack (reference 1.11).

Under normal circumstances the aggressive medium is isolated from the reinforcement fibre by the matrix resin. However, if for some reason there is a crack present in the laminate, then the medium can gain access to the fibre, which is attacked and loses its strength. The fibre has now lost its ability to act as a crack stopper. If the composite is under stress, then the fibre fails and the crack can proceed, thus allowing the process to repeat until the laminate fails.

It can be argued that the environment gains access to the glass fibre, either by diffusion or via surface cracks, but work has shown that polyester resin has low permeability to hydrochloric acid (reference 1.12). Hence the mechanism requires a crack, a significant stress and an environment which is aggressive to the reinforcement. Unless all three are present stress corrosion will not take place or will be substantially reduced.

The resistance of the glass fibre to the aggressive environment is very important. It has been shown that bare E-glass is rapidly attacked by various acids including modest concentrations of either sulphuric acid, or hydrochloric acid (reference 1.13). The behaviour of the composite can be improved by the use of a chemically resistant grade of glass such as ECR-glass. This shows a much prolonged retention of strength in mild acids in comparison with E-glass. Alternatively a thermoplastic liner can be used. To avoid stress corrosion it is more important that the resin is tough, to reduce cracking, rather than chemically resistant (reference 1.14).

Generalisations are dangerous and particularly so in this field. Therefore reference should be made to relevant data or if these are not available then accelerated stress corrosion trials should be carried out.

1.3.15 Blistering

Blistering is a phenomenon which can occur in gel coated, polyester, vinylester and epoxy laminates reinforced with glass fibre after lengthy contact with water. Boat hulls and swimming pools are therefore particularly prone to be affected.

The blistering process is thought to be osmotic. Water molecules, very slowly, permeate the resin, and any voids in the laminate pick up moisture.

If the resin system is hydrolytically unstable, i.e. able to decompose by reaction with water, then salts can be formed in the water. The gel coat acts as a semi-permeable membrane and an osmotic cell is created where pressure builds up which can be sufficient to create blisters.

Much work has been done in this field and it has been concluded that there are several aspects which are important, such as raw material selection, production techniques and working environment. Current recommendations to minimise the likelihood of blistering are:

- use good water resistant resins such as isophthalic polyester and very good water resistant resins for the gel coat
- use in the laminate surface a reinforcement which has a non-hydrolysable binder such as "powder bound" chopped strand mat rather than "emulsion bound" chopped strand mat.
- use a surface tissue in addition to the gel coat. Ensure thoroughly impregnated and wetted glass fibre and the correct resin/glass ratio (references 1.15, 1.16 and 1.17).
- working conditions in the workshop and on site should be clean, dust free and dry.

1.4 PROCESSES

1.4.1 Introduction

A basic understanding of manufacturing processes is particularly pertinent to the designer using composite components. This is because of the profound effect that the process has on not only detail design but also the choice of reinforcement and properties of the resulting composite.

As the EUROCOMP Design Code is concerned with the structural use of reinforced plastic, suitable manufacturing processes will be those that lead to high volumes of products with a low level of variability.

The aim of this section is to provide a reference to those processes which are within the scope of the code, i.e. which satisfy the above criteria, so that they may be understood by a non-specialist in composites and enable him/her to use the code.

It is not intended or necessary to be fully exhaustive on the subject, but it is important that users of the code have ready access to a description and a critique of the processes in order that they may assess the viability of a particular process for a specific application.

1.4.2 Scope

In order to minimise the number of variables within the programme it was decided to restrict the processes to the minimum feasible number commensurate with producing a valid code. Hence three of the primary processes have been specifically denoted as within the scope of the code.

They are hand lay-up, pultrusion and resin transfer moulding (RTM). These have been identified as a sensible, minimum range of processes which are necessary to produce components for construction applications.

These processes are described in detail and characterised to ensure as far as possible that the designer is familiar with them.

Other primary processes are introduced so that the designer is acquainted with them. The range of processes described is not complete but is considered adequate for a general understanding.

The processes which are of primary importance are listed in Table 1.6.

1.4.3 Lay-Up

1.4.3.1 Summary

This was one of the first processes used to produce FRP. There are now many more sophisticated processes but this remains a fundamental option with several advantages which make it attractive in the right circumstances. It is a simple process requiring little capital expenditure. It is particulary relevant for large and/or complex shapes. However, it is not suitable for making products in large numbers as it is labour intensive. Nor is it suitable if two finished surfaces are required.

As pressure is not applied to the laminate during the curing process, high fibre content cannot be achieved and therefore only modest but in many cases adequate properties are possible. The use of woven (or knitted) rovings provides enhanced properties compared with chopped strand mat. Hand lay-up is, for example, the process used to produce the hull components for glass FRP minesweepers, demonstrating the scale of the production that can be achieved.

Table 1.6 Processes for forming composites.

Type	*Process*	*Specific to code*
Open Lay-up	Hand lay-up	Yes
	Spray lay-up	No
Intermediate	Moulding	No
	RTM/resin injection	Yes
	Autoclave/vacuum bag	No
Compression moulding	Hot press moulding	No
Continuous	Pultrusion	Yes
Winding processes	Filament winding	No
Other process which are of secondary importance:		
	Centrifugal casting	No
	Pressure bag	No
	Plymol	No
	Pullwinding	No
	Pullforming	No

1.4.3.2 Process description

The process uses an open mould which is generally itself FRP previously produced from a timber pattern.

The mould is prepared for use by cleaning and then by the application of release agents to ensure subsequent release of the component. A "gel coat" is then applied to the surface of the mould by brushing or spraying. This consists of a similar resin to that used for the main laminate but it is more thixotropic (hence the name) and often more chemically resistant. This provides the protective outer barrier and the aesthetic surface of the laminate. The gel coat is allowed to cure to a tacky consistency, when it is ready for the application of the main laminate. The latter consists of layers of chopped strand mat (CSM), or woven roving or a combination of both. Laminating resin is brushed on to the gel coat, after which the CSM is laid down. The CSM is impregnated with further resin by stippling with a brush and then rolled with a "split washer" roller to consolidate the laminate and to remove entrapped air. The laminate conforms to the mould surface by the adhesion of the resin. Sufficient layers of reinforcement are applied until the desired thickness is achieved. The laminate is then left to cure.

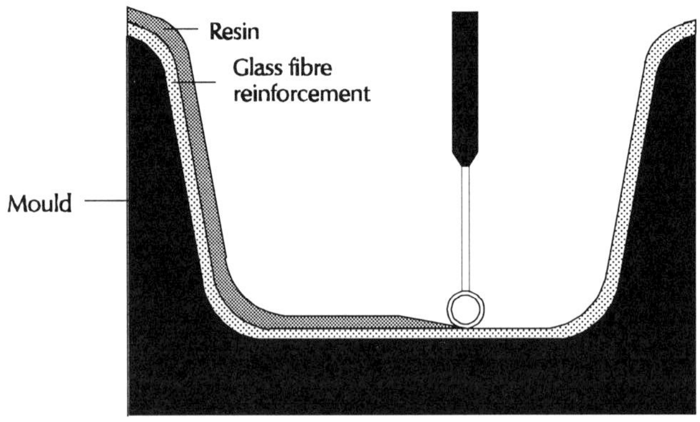

Figure 1.3 Open mould for lay-up.

A typical moulding produced by hand lay-up is perhaps 3 m^2, but components range from as small as 0.5 m^2 or less, to 300 m^2 for marine vessels and chemical plant.

The process is labour intensive, a typical production rate for the application of the laminate being about 3 kg/hour. However, it is very versatile and as the capital required for moulds is minimal, short production runs can be justified. The usual viable range of production quantity is from 1 to 500 components.

Lay-up is a cold moulding process and therefore cure is achieved by the use of a catalyst (e.g. a peroxide) and also an accelerator (typically cobalt napthanate). This may be assisted by the application of heat, generally by the use of an oven. When the laminate is sufficiently cured, it may be released from the mould and excess material at the edge trimmed back. Often the released part is post-cured at elevated temperature to reach the temperature performance or mechanical properties required for the application.

Cure of the laminate takes place at atmospheric pressure, which helps to make the process simple and inexpensive. However, a high level of fibre content is not possible and hence high mechanical performance properties are not achievable. This does not mean they are unacceptable; it simply results in thicker laminates than those from processes which can achieve high fibre content. The extra material cost is balanced against low capital cost, etc.

In summary, the process lends itself to those applications which require low production quantities, perhaps a complex shape and maybe larger size than could be produced by any other method. It is also suitable for those applications which do not require particularly high intrinsic properties.

1.4.4 Pultrusion

1.4.4.1 Summary

Pultrusion is a continuous process for the production of constant section profiles in composite materials. The fibre reinforcement is first impregnated with resin mix. It is pulled through a heated steel die which causes the resin to react and cure. As it is continuous, it is very efficient in both capital and labour employed. The process achieves high fibre content, and thus high mechanical properties, along the length of the pultruded section, but not in the transverse direction. Hollow sections can be readily produced but all sections have to be straight. It is particularly suitable for large production runs of beam and column sections.

1.4.4.2 Process description

The resin is applied to the reinforcement either by the use of a dip tank or by injection. The former approach is the most common. In this the reinforcement is pulled through a tank containing the resin mix. The tank is replenished as necessary, to ensure that the process is continuous. The injection method is a cleaner option. In this the reinforcement is pulled through a cavity fixed to the front of the die. The resin is injected under pressure into the cavity and, thus, it impregnates the reinforcement.

The temperature of the die is such that the resin system is caused to react and cure within the die.

The profile is pulled by either reciprocating pullers or a caterpillar haul-off and it is then automatically cut to length with a saw.

In essence, pultrusion is an apparently simple process. However, the production of complex and consistently high quality profiles requires a high degree of technical resource and experience. For example, the precise positioning of the reinforcement materials, in front of the die entrance, is of paramount importance. It is achieved by a carefully engineered "in-feed" system. This guides, locates, folds and tensions the various layers of reinforcing mat and uni-directional fibre. It also wipes excess resin from the reinforcement.

The shape of the die determines the profile shape. But the quantities of reinforcement and resin mix are critical to ensure that the die is filled properly. Resin systems must be highly reactive to cure in the available time. Machine speed, die temperature and resin reactivity are parameters which interact and must be balanced.

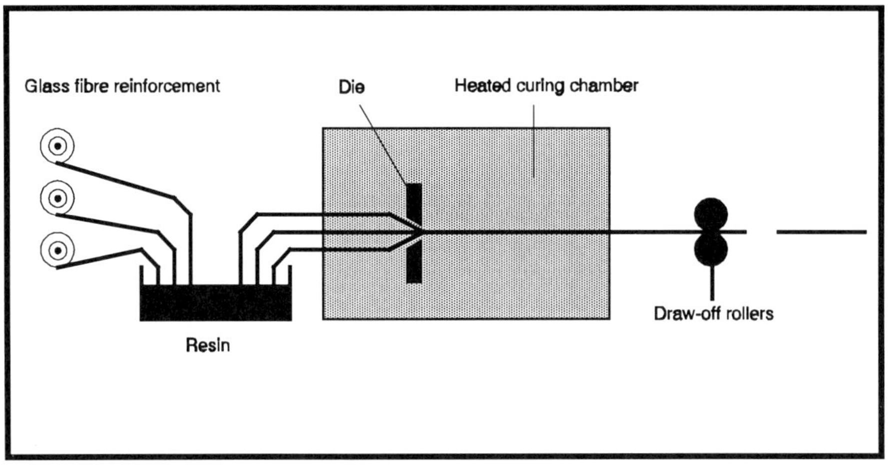

Figure 1.4 The pultrusion process.

The reinforcement used is generally a combination of uni-directional roving and random mat (continuous filament mat). This combination provides a reasonably optimised set of properties in the longitudinal and the transverse directions respectively. Other reinforcement types may be used, including woven roving, chopped strand mat (non-soluble binder), needled mat and multi-axial non-woven.

Isophthalic polyester is the most commonly used resin but vinylester is important in corrosion resistance applications. Urethane methacrylate has very good fire, smoke and toxicity performance. Epoxy is used in the higher performance applications and almost always with carbon fibre. Phenolic resins are being developed for pultrusion, to take advantage of their excellent fire performance.

1.4.4.3 Preferred characteristics of pultruded profiles

The shapes which may be produced can be as simple as a rod or as complex as a thin walled multi-cavity hollow box section. Within limitations the complexity of the profile is simply a matter of having the right die produced, but some characteristics of the component design are critical and are described below.

1.4.4.4 Size

Pultruded profiles can be as small as 3 mm diameter or up to 1 m wide and 250 mm deep and are produced at a rate of between 180 m and 12 m per hour, depending on size and complexity.

1.4.4.5 Quantity

Owing to the capital investment, pultrusion is generally not a viable option if the quantity of profile is less than about 5000 m, but above this level it is an extremely efficient process. Some manufacturers keep stocks of standard sections, enabling relatively small quantities to be obtained at reasonable cost.

1.4.4.6 Corner radii

Mat and fabric reinforcements tend to follow the line of least resistance and will move away from the die surface at corner details, particulary if the wall thickness increases at the corner. This creates a resin rich area on the outside of the radius which is vulnerable to cracking. This can be alleviated by an increase in fibre content or filler content but the reinforcement conforms to the shape more easily and stress concentrations are reduced if the corner radius is as large as possible and if the wall thickness is constant throughout the radius. The preferred minimum corner radius for internal radii is 1.5 mm and in order to maintain a uniform wall thickness throughout the corner, the external radius should equal the internal radius plus wall thickness.

1.4.4.7 Wall thickness

The process speed is, amongst others factors, a function of how quickly the resin can be heated.

Thick section parts require a longer time for the heat to penetrate and start the reaction than do thin sections. Therefore, they run more slowly. Once the curing process starts, the exothermic reaction can produce temperatures which are too high, resulting in in-built stresses and possible cracking. It is, therefore, very desirable with pultruded profiles to minimise the wall thickness.

Hence when thick sections are unavoidable, a less reactive temperature/cure schedule is employed to ensure control of the process. Therefore, thick sections run more slowly than thin sections. A practical range for wall thickness is:

Minimum thickness = 2 mm

Maximum thickness = 20 mm.

However, in extreme cases, a range of 1 - 50 mm may be achievable. Bar of 75 mm diameter has been produced.

1.4.4.8 Draught angle

Components produced from any of the moulding processes generally require a draught angle to allow their release from the mould. This is not the case with pultruded profiles. The die can be produced "square" if required. However, compensation for shrinkage, which tends to "pull in" the profile is required. Hence dies generally are produced with an allowance of about 1° per 90° for reduction in angle due to shrinkage. The quality of the die manufacture can have a significant effect on the dimensional tolerance achieved. Over specification of tolerance can lead to unnecessary costs.

1.4.5 Resin transfer moulding (RTM)

1.4.5.1 Summary

RTM is a closed mould process in which the component is formed between a matched pair of moulds. The moulds are often themselves made from FRP and are therefore relatively low cost. The process typically has a cycle time of 15 to 30 minutes.

It is neither a high volume process nor a hand process and thus it is referred to as an intermediate process.

1.4.5.2 Process description

The moulds are cleaned and prepared by the application of a suitable release agent. Previously prepared reinforcement, usually continuous strand mat but other reinforcement types may be used, is placed in the lower mould and the two mould halves are brought together and clamped. The reinforcement must fit the contours and perimeter of the mould accurately. Hence pre-formed mat is often the preferred option.

The resin is machine-mixed immediately prior to it being injected under pressure, typically 5 bar, into the cavity between the two mould halves through a port usually in the middle of one of the mould surfaces. A "pinch-off" device is necessary at the perimeter of the mould to allow air to pass through but restrict the flow of resin. This creates a back pressure which ensures that the mould cavity is fully filled. Pumping continues until resin is seen to penetrate the pinch-off at the furthest extremity of the mould. The entry port is sealed to maintain pressure and the resin left to gel and cure.

The mould surfaces are kept in the correct location by either clamping devices or a low pressure press. Injection pressures are not high. Nevertheless, flimsy moulds will distort.

Resin systems are generally, but not essentially, cold cure. They are mixed and then injected into the mould generally within seconds but injection may take 5-10 minutes for large complex shapes. They can be extremely reactive and consequently have very short gel and cure times. Low viscosity/high reactivity versions of polyester, phenolic resin and epoxy resin are the most commonly used.

High temperature release agents are necessary owing to the high exothermic reactions inherent in the process. The use of FRP and low cost tooling methods do not lend themselves to mechanical part ejection devices. Consequently release tends to be a hand process and labour intensive.

Continuous filament mat (CFM) is the most commonly used reinforcement, but chopped strand mat (CSM) with thermo-deformable binders allows the production of pre-formed reinforcement shapes and is becoming more common.

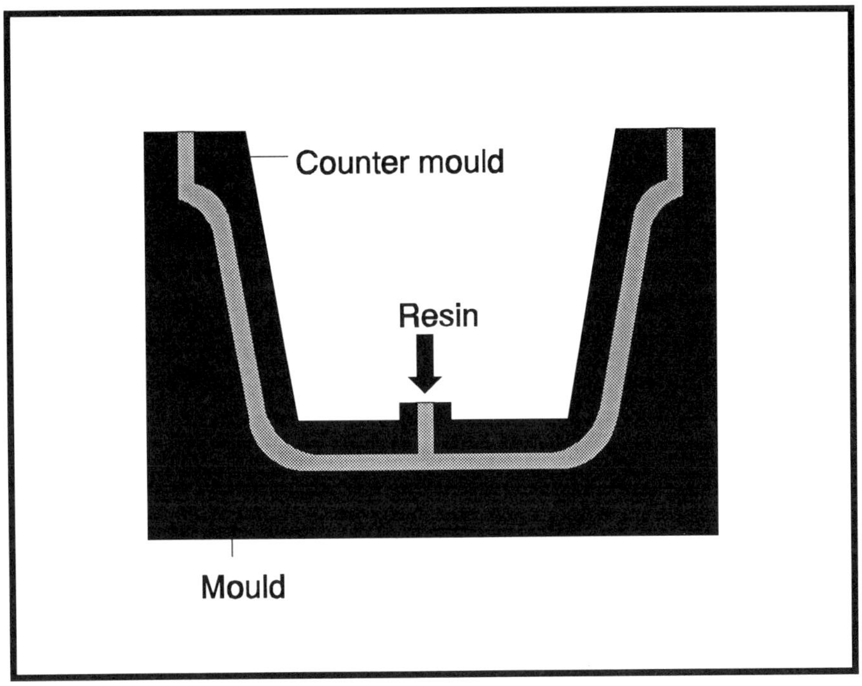

Figure 1.5 Resin transfer moulding.

Mouldings up to about 10 m^2 have been made by RTM but a more typical size is about 0.5 m^2, produced at a rate of typically three per hour. This rate and the relatively low investment required in tooling and the mixing/injection machine result in a minimum viable production quantity of about 2000 components. However, if the equipment exists then the process becomes viable at production quantities of several hundred.

1.4.6 Spray lay-up

1.4.6.1 Summary

The spray-up machine consists of a gun for spraying catalysed/ accelerated resin and a unit which chops continuous glass fibre. They are simultaneously sprayed on to an open mould coated with a releasing agent. The laminate is rolled and then allowed to cure, after which the moulding is released and trimmed to size. The properties and applications are similar to hand lay-up but the use of the spray machine produces a much faster lamination time.

1.4.6.2 Process description

Continuous glass fibre (roving) is chopped in the chopper unit of the spray gun by the action of a rotating blade against a polyurethane roller. The resulting chopped strands are blown from the gun by the exhaust from the air motor which drives the chopper. Concurrently resin is sprayed from the gun and the streams of resin and chopped glass fibre meet approximately at the mould surface.

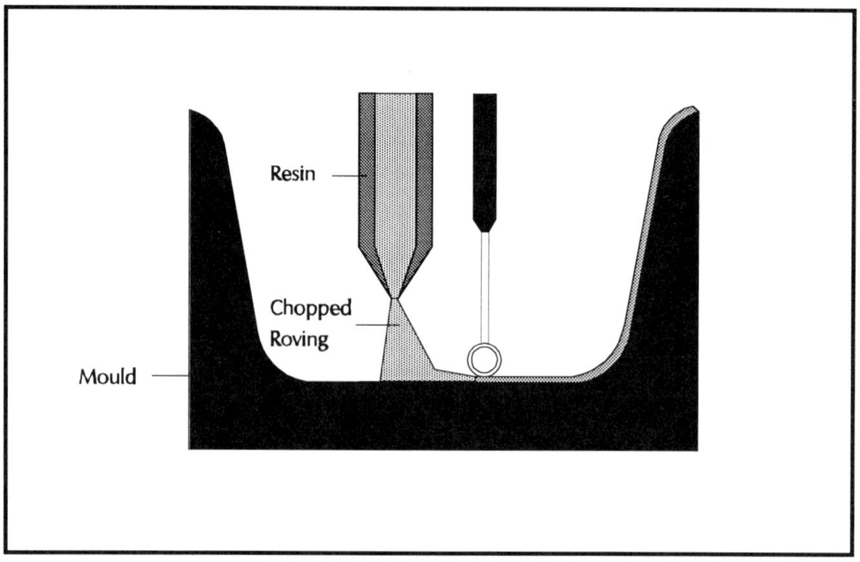

Figure 1.6 Spray-up moulding.

There are two systems of spray-up in common use: twin-pot and catalyst injection. The twin-pot machine uses one pressure pot for catalysed resin and another for accelerated resin. They are simultaneously fed to the spray gun and individually sprayed each through its own nozzle. The two streams are adjusted to meet at about 300 mm in front of the mould surface where they mix. As the two streams only meet outside the spray head premature gelation cannot occur. The twin pot machine is relatively simple to operate but has the disadvantage that the pressure pots require frequent replenishment which is time consuming and particularly inconvenient for large mouldings.

The catalyst injection method overcomes this disadvantage by having one supply of resin which is pre-accelerated. The catalyst is injected into the resin at the gun continuously while the gun is operated. The resin is pumped to the gun and this allows the process to be continuous.

Whichever system is used the reaction process only starts at or beyond the gun. This implies that highly reactive, fast curing resin systems may be used with little fear of premature gelation. This then allows rapid demould and cycle times.

It is possible to automate the spray-up process with, for instance, a robot, but the hand operated method is most common.

The ratio of glass fibre to resin is adjusted by running each separately for a fixed time period (typically 15 sec.) and weighing the quantities of each. The gun is adjusted to give the desired ratio, which is typically 2.5:1 resin : glass (by weight).

Spray guns can deposit glass fibre at the rate of 2-5 kg/min, but they are generally operated only in short bursts, giving an overall material throughput of about 20 kg of laminate per hour. Unlike hand lay-up the method cannot be used with woven (or knitted) fabrics.

After the spray pass, the operator uses a split washer roller to consolidate the laminate. After about 1 to 2 minutes the strands break down into their individual filaments and a second pass with the roller conforms the laminate to the mould surface and completes air removal and consolidation.

The typical size of moulding produced by spray is perhaps 10 m^2 but it is suitable for any size larger than about 2 m^2. Moulds which are much smaller than that are inefficient and difficult to control. Generally spray machines are operated a rate of about 20 kg of laminate per hour. This translates, for a 3 mm thick laminate, to a surface area of about 5 m^2.

1.4.6.3 Materials used

Special glass fibre roving grades are made which are particularly suitable for spray-up. They have good choppability together with low static and fast wet through.

Usually only orthophthalic and isophthalic polyester resins are used in spray-up. They are normally highly reactive to take full advantage of the process characteristics.

1.4.7 Cold press moulding

1.4.7.1 Summary

This is a technique in which low pressures and low temperatures are used. Hence light duty and low cost tooling can be used. Production rates and capital expenditure are moderate. The size of component is restricted to the size of press available; a typical component has an area of perhaps 0.5 m^2. The process lends itself to production runs of typically 500 off, but is economic in the range 100 to 5000 components off.

1.4.7.2 Process description

The cycle starts with mould preparation. The mould is cleaned and a release agent is applied. The reinforcement is then placed in the lower mould. Several layers may be applied, depending on the thickness of the mould cavity to be filled. The reinforcement is allowed to overlap the pinch-off area of the mould.

As this is a cold moulding process there is no heat to promote the cure and consequently the resin mix requires both catalyst and accelerator. This must be mixed immediately prior to each shot (a batch would immediately start to cure). Once thoroughly mixed the resin is poured on to the middle of the reinforcement.

The press is then closed and a pressure of about 5 bar is applied. As the mould halves close, the resin and reinforcement are compressed. The resin is pushed towards the perimeter of the mould, the pinch-off area. Because the reinforcement overlaps the pinch-off it is compressed more than within the mould. This restricts the flow of resin but allows the passage of air. Hence as the resin flows to the pinch-off a back pressure is created which ensures that the resin flows to all parts of the cavity.

The reactivity of the resin mix is high, so the cure process commences very rapidly and is completed in about 15 minutes typically, but can be as rapid as 5 minutes if the component characteristics are favourable. As the resin cures, an exotherm is generated which heats the mould. Thus the amount of accelerator is progressively reduced after the first two or three mouldings, until steady state conditions are reached. A compromise is obtained to achieve the fastest cycle time without either product or mould degradation due to excessive temperature.

After cure is complete the press may be opened and the component removed. There is a small amount of "flash" around the periphery of the component which is then machined off.

1.4.7.3 Materials used

The reinforcement material must allow the flow of resin, not only through its thickness but in-plane and over relatively large distances. Also the reinforcement must be capable of conformity to complex mould shapes and variations in thickness.

These factors dictate the type of reinforcement which can be used. Continuous filament mat and needle mat (both glass fibre) are most common, but woven roving and multi-axial (glass, carbon, etc.) can also be used.

The most popular resin system is polyester, owing to its versatility and suitability to the process. However, for more demanding applications epoxy or phenolic resins are also used.

1.4.8 Autoclave/vacuum bag moulding

1.4.8.1 Summary

Layers of fibre pre-impregnated with resin (prepreg) are applied to a mould and rolled. A rubber bag is placed over the lay-up and the air is removed by means of a vacuum pump. Generally the mould is then placed in an autoclave, which applies further pressure and temperature to cure the resin, or simply in an oven.

The process is slow and laborious but produces excellent properties. The size of component which can be produced is governed by the size of autoclave or oven available. A typical size of moulding is 2 m^2 but components up to 100 m^2 have been produced. The cycle time is in the range 1 to perhaps 20 hours but is typically 12 hours. Hence the process does not lend itself to mass production and is usually restricted to runs of less than 500 and, more typically, 50 off.

1.4.8.2 Process description

The mould is cleaned and a release agent or release film is applied. The backing film is removed from the prepreg surface and successive layers of prepreg are placed on the mould and rolled to conform to the mould surface. The layers are applied in a predetermined order and angle to achieve the laminate properties required.

A porous release fabric is applied to the lay-up to allow the passage of air and excess resin. This is followed by bleeder/breather layers which allow any air to be drawn from the laminate when the vacuum is applied. A vacuum bag is then placed over the laminate, sealed down and the vacuum is applied. The curing cycle now starts either in an oven or in an autoclave which applies both heat and pressure (typically 180°C and 10-20 bar). The action of the pressure from the autoclave or vacuum consolidates the laminate and the individual layers become fused.

Some prepregs have the precise quantity of resin present which is required. More often there is an excess of resin to ensure good consolidation.

The process was originally only carried out by hand but sophisticated equipment has now been developed for machine placement of the prepreg. As an alternative to prepregs, hand lay-up systems are also used.

Computer controlled machines allow the precise placement of tapes of prepreg and eliminate problems associated with hand operations.

1.4.8.3 Materials

Carbon, glass and aramid fibre prepregs may be used. Alternatively dry fibre, in any format, may be used.

Almost exclusively epoxy resin is used. The process produces excellent properties and hence the use of a high performance resin system is justified.

1.4.9 Hot press moulding

1.4.9.1 Summary

Hot press moulding is a mass production method for the manufacture of composite components. A hydraulic press is used capable of applying a pressure of about 100 tonnes/m^2 of platen area. A matched metal tool is located between the platens, the cavity between the tool halves having been machined to produce the shape required. The tool is heated to about 130-170°C either from the heated platens or directly (by cartridge heaters, etc.).

The polymer mix (thermoset) and reinforcement are placed in the tool, which is then closed. After 2-3 minutes cure is completed, the tool is opened and the component removed.

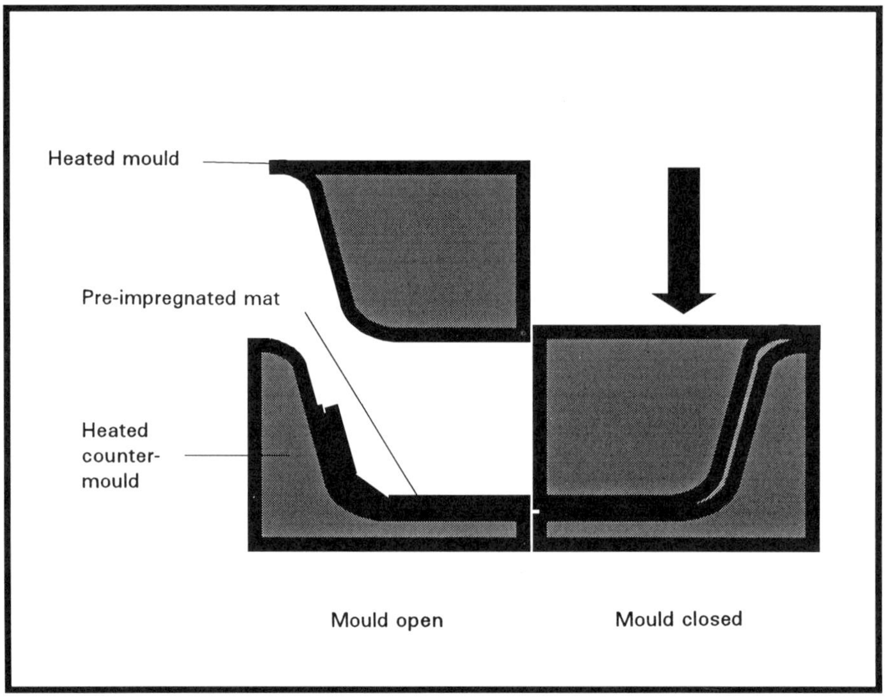

Figure 1.7 Hot press moulding.

The size of mouldings which can be compression moulded is dependent on the size of press available but 0.5 m^2 is typical, with a maximum of about 3 m^2.

Owing to the high capital investment in the press and tools, compression moulding is only suitable for high volume requirements, typically 10000 components and greater.

1.4.9.2 Process description

Hot press moulding can be carried out either using dry reinforcement to which a resin mix is added when in the tool or by the use of a previously produced compound of resin and reinforcement.

If dry reinforcement is being used, then it is cut to a size which is generally slightly larger than the area to be moulded. This ensures that there is reinforcement in the periphery of the moulding. Sufficient layers are prepared to give the required fibre volume fraction in the finished component. The total weight is checked. The resin mix which has been previously prepared in a batch contains fillers, release agents, a catalyst, etc. but not an accelerator. The heat of the tool provides the acceleration to the process.

The tool, which is pre-heated to the required processing temperature, is cleaned and a release agent is applied. The reinforcement is placed in the lower half of the tool and the resin mix is poured on to the fibre. The pouring technique must avoid air entrapment. Hence it is preferable not to pour with a circular motion. Generally only one-third to one-half of the reinforcement is wetted at this stage.

If DMC is being used, then a sufficient amount is weighed, or if SMC is being used, it is cut to the predetermined pattern required and check weighed. The compound is placed in the lower half of the tool.

Once all the material is in-place the press is closed.

With the "wet resin" technique, the closure of the die and the consequent hydraulic pressure forces the resin through the reinforcement, pushing a front of air to, and through, the pinch-off area. When SMC or DMC are used the closure of the press compresses and heats it. Its viscosity reduces rapidly and it is able to flow to the pinch-off zone of the die.

The pinch off zone in the perimeter of the die cavity is large enough to allow air to pass through (0.05-0.25 mm) but restricts the flow of the much more viscous resin. Hence a pressure is generated which encourages the resin to flow to any remaining voids in the die cavity. Thus the whole cavity is filled with only a very small "flash" of resin passing through the pinch-off. As this is very thin and without reinforcement, it is easily removed from the moulding after release from the die (deflashed).

The applied pressure at closure of the die is maintained while cure takes place. The volume of resin in the die changes as a result of the chemistry of cure and thermal expansion or contraction effects. The applied pressure therefore maintains the shape of the component during the transition from liquid to solid.

The cure time is dependent on the type of resin, the level of curing agent used and the thickness of the component. Thick sections take a long time to heat through, plus they can generate excessive exothermic temperature. Hence die temperature is generally lower for thick parts, which are therefore slower to mould. A typical cure time is two minutes, after which the press is opened and the component released (generally with the aid of a mechanical ejector mechanism within the tool).

1.4.9.3 Materials

As indicated earlier, the materials may be either in the form of compounds or resins and fibres separately. Sheet moulding compound (SMC), dough moulding compound (DMC) and bulk moulding compound (BMC) use polyester resin plus filler, catalyst, pigment, low profile additive, etc.

Epoxy prepreg also may be compression moulded. This may have glass or carbon fibre as the reinforcement in uni-directional or bi-directional geometry.

Fibres may be in the form of continuous strand mat or chopped strand mat with insoluble binder. Preforms may be used which are produced

(a) by spraying chopped glass fibre and a binder on to a former of the required shape, or

(b) by thermally deforming a CSM which has a thermoplastic binder.

Either polyester or epoxy resins are used for wet moulding using very similar mixes to those in compounds.

1.4.10 Filament winding

1.4.10.1 Summary

In the filament winding process a mandrel is rotated upon which are wound reinforcing fibres which have been impregnated with resin. The fibre is controlled by an applicator which traverses the length of the mandrel as it rotates, thus building up the laminate thickness. A surface of revolution is produced, the type (cylinder, cone, etc.) depending on the mandrel shape. Variations include extra axes of movement to allow the winding of domed ends, spheres, conical shapes, etc.

A typical filament wound tube is 300 mm diameter and 2 m long, but cylinders as large as 12 m diameter and others 38 m long have been produced.

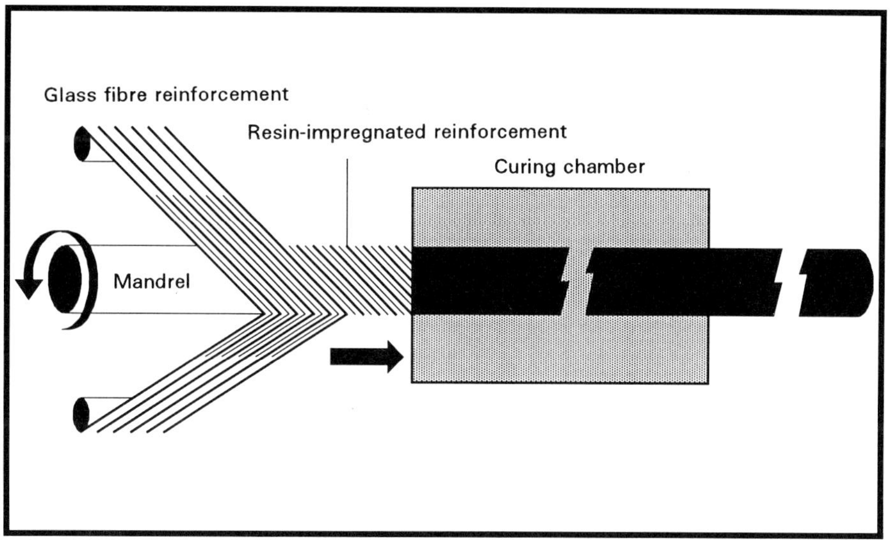

Figure 1.8 Filament winding.

The process is highly controlled and produces composites with excellent properties. However, it is time consuming, the cycle time being generally in the range 1 hour to as much as 24 hours but more typically perhaps 4 hours.

1.4.10.2 Process description

Continuous reinforcement is placed and oriented on the mandrel surface by the position of the pay-out eye. In the most simple filament winding machine the pay-out eye is fixed to the carriage and therefore can only traverse the length of the mandrel. Hence the angle at which it is placed is dependent on the rotational speed of the mandrel and the traverse rate of the carriage. The angle of the reinforcement fibre can be any helical angle from almost axial to hoop.

The use of extra axes of movement (rotational or translational) allow the fibre to be placed on more complex shapes.

The resin is generally applied by an impregnation bath, which is mounted on the carriage. The reinforcement passes between bars in the resin bath which spreads the fibre and allows the resin to penetrate the fibre bundle. Alternatively pre-impregnated fibre may be used.

The fibres conform to the mould surface by virtue of the winding tension. However, on a frictionless surface the fibres will take the shortest route between two points on the surface (this is the geodesic path). Therefore it is necessary either to wind a geodesic path or to generate sufficient frictional force to keep the fibres in the chosen path (by using a tacky resin or tacky prepreg).

A "cold cure" may be used if a non-prepreg system is used, but this creates a pot life problem in the resin bath. Normally heat is applied to the mandrel while it is still on the machine. Alternatively the mandrel is removed and placed in an oven, thus allowing further winding to take place.

After cure the component is removed from the mandrel either by collapsing the mandrel or by a mechanical puller device, typically hydraulic to generate sufficient force. Complex shapes may not allow the mandrel to be withdrawn. This problem may be overcome by the use of a collapsible mandrel or by producing the mandrel from a meltable material such as low melting point alloys or wax.

1.4.10.3 Materials

Glass fibre roving, carbon fibre and aramid as rovings or yarns, or woven tapes of these fibres may all be used. Glass fibre and thermoplastic veils may be used on the mould surface or as a finishing layer to produce resin rich corrosion barriers.

Most commonly, epoxy resin is used because the process lends itself to demanding applications which justify the expense. However, any of the thermosets can be, and are, used commensurate with the application requirements.

1.4.11 Secondary processes

There are a multitude of other processes, variations on a particular process and combinations of processes to suit specific component requirements. A selection of the more important ones are briefly described.

1.4.11.1 Centrifugal casting

A spinning mandrel has reinforcement fibre and resin sprayed on to its inner surface. Centrifugal force presses the laminate on to the mould surface so that it consolidates. The mandrel and laminate is then moved into an oven to allow cure to take place.

The process is used to make pipes and tubes typically from 200-300 mm diameter upwards and 10 m long. It is far faster than filament winding but uses random reinforcement, and the fibre content is low and hence properties are relatively low. The major application is piping, in which case the inherent tendency for the process to produce a resin rich inner surface is advantageous.

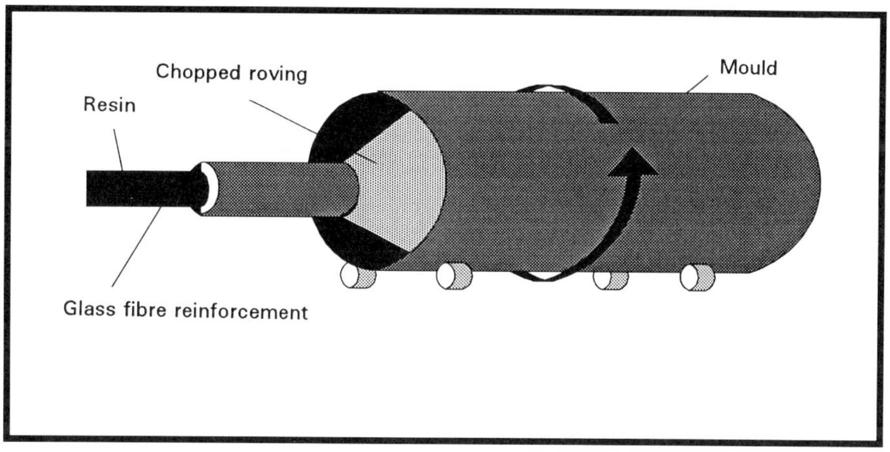

Figure 1.9 Centrifugal moulding.

1.4.11.2 Pressure bag

An outer mould usually made of FRP is prepared as for hand lay-up, but dry reinforcement is laid in place on the mould. Resin which has been catalysed and accelerated is poured on to the reinforcement, after which a rubber bag is placed in the mould and inflated. This pushes the resin through the reinforcement and consolidates the laminate. After cure the bag is deflated and the moulding released. This is particularly suitable for components which have undercuts.

1.4.11.3 Plymol

This is a variation of the pressure bag process but is used for the production of straight and tapered cylinders. The size range is from about 75 mm diameter up to 300 mm in lengths of up to 13 m.

A thermoplastic film is used to form the mould for the outer surface of the cylinder. The reinforcement fibre, in the form of CSM, woven roving or similar is cut to size and laid on to the sheet. A rubber bag in the form of a cylinder with closed ends is placed on the reinforcement. Resin (catalysed and accelerated) is poured on to the reinforcement and the longitudinal edges of the sheet are brought together, forming a tube, and clamped.

The resin cures in about one hour, after which the mould is disassembled and the moulding removed.

1.4.11.4 Pullwinding

This is a variation of the pultrusion process for the manufacture of hollow tubes. The die used is similar to that used in conventional pultrusion but the mandrel which forms the inner surface of the tube is considerably longer in pullwinding. The conventional in-feed system used in pultrusion is replaced with a carousel which rotates about the axis of the mandrel. The carousel carries several packages of uni-directional reinforcement which are wound on to the die mandrel and pulled into the die. The resin is applied by injection into a die cavity.

1.4.11.5 Pullforming

This is a marriage of two processes: pultrusion and compression moulding. A pultrusion in-feed system is employed for management of the reinforcement and to apply the resin mix but the conventional pultrusion die is replaced with one which will open and close. This allows variations in the cross-section and linearity of the component which are not possible with pultrusion.

1.5 COMPARISON WITH CONVENTIONAL STRUCTURAL ENGINEERING MATERIALS

1.5.1 General

In comparison with steel and concrete, polymeric composites provide the structural engineer with a much wider range of mechanical properties to optimise the design. The designer must avoid the temptation to simply change the construction material without considering the overall design. Similarly, composites are not suitable for every application and should only be used when they show a benefit over conventional materials. The following is provided as initial guidance when approaching the design of FRP structures for the construction industry.

1.5.2 Stiffness and strength

Glass has a stiffness comparable with unreinforced concrete but tensile strengths in excess of steel. It follows that designs will tend to be controlled by stiffness requirements both with respect to deflection serviceability and buckling. Thus choice of section for stiffness and stability often tends to control the design, with in many cases a reserve of strength being provided in order to accommodate uncertainties in strength prediction, and assurance that a serviceability limit state is reached before an ultimate limit state. More detailed comparisons are given in Chapter 4.

1.5.3 Costs

When considering FRP in relation to conventional structural materials it is essential that one takes into account not only the basic material cost but also the costs associated with construction and the whole life costs, including any long term maintenance. Material prices are generally quoted per unit weight. This can be extremely misleading in comparison with say steel as the composite will have a significantly lower density. In addition it may be significantly stronger, though what proportion of this strength can be utilised in any given situation will depend on type of structure and its loading. Usually the net result is that the price differential will be considerably lower than one would expect based on the material price alone. For example, Meier et al (reference 1.18) describe a particular application in which 6.2 kg of carbon FRP was used in place of 175 kg of structural steel. Though the price per kg of the carbon FRP was 50 times that of the steel, the total cost of the FRP used on the project was only 1.8 times as great. This differential was largely offset by significantly reduced construction costs as a result of the lighter weight of the material.

1.5.4 Connections

The design approaches used for the connection of components made of FRP are covered in detail in Chapter 5 of the EUROCOMP Design Code, with workmanship aspects being covered in Chapter 6. The most usual methods of connection are mechanical, using bolts or similar, or adhesive bonding. There are significant differences between the behaviour of FRP connections and the equivalent in metals.

With bolted connections in metals, high peak stresses where the bolt bears against the edge of the hole are dissipated by local yielding. Simple assumptions may thus be made about load sharing between a group of bolts. Similarly relatively large tolerances may be used for the bolt holes, allowing easier assembly of the connection. The same does not apply to FRP, which generally has a linear elastic response up to failure, and hence the peak stresses may lead to local crushing of the material.

With steel, the use of friction grip bolts is common, most of the load being transferred by frictional forces between the components and only a small proportion in bearing through the bolts. Composites generally have a smooth surface and highly torqued bolts will lead to local crushing below the washer. Hence bolt joints rely on nearly all the load being transferred by the bolts themselves in bearing.

One final point to be remembered is that if the fibre orientation is largely uni-directional, and hence the transverse properties are relatively low, forming a hole through the material can significantly reduce its strength.

With adhesive bonding, the behaviour of the connection is very dependent on the standard of workmanship, in particular the preparation of the surface and the curing of the adhesive. A major difference between bonding FRP and metal is that the surface layer, which will have a significant influence on the behaviour of the connection, may have a lower strength than the rest of the component. For example, the fibre reinforcement may be concentrated in the body of the component while the surface may be largely unreinforced. Similarly, through-thickness strengths will be lower than in-plane strengths, and thus bonded joints will be prone to failure by peeling, except when the loading is solely axial.

REFERENCES

1.1 Smith C S, *Design of Marine Structures in Composite Materials,* Elsevier Applied Science, London, 1990.

1.2 Vetrotex Data sheet, High performance fibres: R-glass, 1987.

1.3 Tuhru Hiramatsu, Tomitake Higuchi and Junichi Matsui, Torayca T1000 ultra high strength fibre and its composites properties. *In Proc 8th Int Conf SAMPE* La Baule, 1987.

1.4 Imuro H and Yoshida N, Differences between Technora and PPTA aramid. *In 25th Int Man-made Fibres Conf,* Austria, 1986.

1.5 Krenchel H, *Fibre Orientation,* Akademisk Forlag, Copenhagen, 1964.

1.6 Duthie A C, The preliminary design of composite structures, BPF RPG Congress, Nottingham, 1986.

1.7 Mandell J F and Meier U, Fatigue crack propagation in E-glass/epoxy composites, ASTM STP 569, 1975.

1.8 Walton J M and Yeung Y C T, Flexible tension members from composite materials. *In Proc 6th Arctic Eng Symp,* Vol III, American Soc of Mech Eng, 1987.

1.9 Anderson J J, *Creep Properties of Composite Rods*, British Ropes internal report, 1987.

1.10 Forsdyke K L, Phenolic resin composites for fire and high temperature applications, Int Conf FRC 84 PRI, Liverpool, 1984.

1.11 Jones F R, Stress corrosion, Int Conf FRC 88 PRI, Liverpool, 1988.

1.12 Caddock B D, Evans K E and Hull D, The role of diffusion in the micromechanisms of stress corrosion cracking of E-glass/polyester composites, In Proc IMechE 2nd Int Conf Fibre Reinforced Composites, Liverpool, 1986.

1.13 Cochram D and Scrimshaw G F, Acid corrosion of FRP, 35th Conf RP/CI SPI, Washington, 1980.

1.14 Hogg P J, Price J N and Hull D, Stress corrosion of FRP, PRI Int Conf. FRC, Liverpool, 1984.

1.15 Birley A W, Dawkins J V and Strauss H E, Blistering in glass reinforced polyester laminates, BPF Congress, Brighton, 1984.

1.16 Crump S, A study of blister formation in gel coated laminates, 41st Conf. RP/CI SPI, Washington, 1986.

1.17 Norwood L S, Edgell D W and Hankin A G, Blister performance of FRP systems in aqueous environments, BPF Congress, Brighton, 1980.

1.18 Meier U et al, Strengthening of structures with Carbon FRP laminates; research and applications in Switzerland *In Advanced Composite Materials in Bridges and Structures* (Ed K W Neale and P Labossiere), Canadian Society for Civil Engineering, Montreal, Canada, 1992.

ACKNOWLEDGEMENT

The figures in this chapter showing various processing techniques are based on diagrams in literature published by Vetrotex.

CONTENTS

2 BASIS OF DESIGN

2 BASIS OF DESIGN

This Chapter draws on references 2.1 to 2.3 and 2.6 throughout.

2.0 NOTATION - SECTIONS 2.1 TO 2.4 (See also sections 1.6 and 1.7 of EUROCOMP Design Code)

No additional information.

2.1 FUNDAMENTAL REQUIREMENTS AND WARNING OF FAILURE

2.1.1 Fundamental requirements

No additional information.

2.1.2 Warning of failure

No additional information.

2.2 DEFINITIONS AND CLASSIFICATIONS

2.2.1 Limit states and design situations

2.2.1.1 Limit states

Limit state design requires that design against collapse, loss of stability (e.g. due to compression buckling of columns or plates, torsional buckling of columns, lateral torsional buckling of beams and sway buckling of frames), excessive deformation and matters affecting the safety of the structure and people be considered separately from design against local damage, minor deformation, etc.

Loss of stability from increased second-order effects (i.e. changes in the effects of actions or in the forces in members resulting from deformation of the components, members or of the structure as a whole) due to creep may be dealt with by using ultimate limit state loads and the long-term structural stiffness, e.g. with a creep factor applied to the modulus of elasticity.

2.2.1.2 Design situations

No additional information.

2.2.2 Actions

2.2.2.1 Definitions and principal classifications

No additional information.

2.2.2.2 Characteristic values of actions

No additional information.

2.2.2.3 Representative values of variable actions

ψ : a modifying factor applied to the variable actions.

This factor is applied just to the variable actions (Q_K). It takes account of the reduced probability of loads in combination ($\psi < 1$), rarity or frequency of occurrence ($\psi <> 1$), fatigue and dynamic effects ($\psi > 1$).

Although it is a modifying factor for (variable) actions, it is the main means of regulating such actions to allow for changes in material properties resulting from load effects, e.g. cyclic loads, fatigue or reversing loads, and impact loads.

Because it is applicable only to variable actions and not to permanent actions, it is not used to take account of loss of strength due to sustained loads: this is already taken account of in the material properties used for design. Thus, for sustained variable actions, $\psi = 1$.

The quasi-permanent value of a variable action, $\psi_2 Q_k$ ($\psi_2 \leq 1$), is simply that part of a variable action which is effectively permanent when considering long-term effects.

2.2.2.4 Design values of actions

γ_G : partial safety factors for permanent actions.

The value of γ_G varies according to whether upper or lower design values will produce the most unfavourable result and whether a normal or accidental design situation is being considered.

γ_Q : partial safety factors for variable actions.

Similar considerations as for γ_G apply, the minimum value being zero for loading in one sense.

For both, γ_G and γ_Q, the value of the partial safety factor applies solely to the particular action being considered. Each is made up of two subsidiary factors:

$\gamma_{G,1}$, $\gamma_{Q,1}$, the load variation factor, takes account of the possibility of unfavourable deviation of the various loads from the values considered in deriving the characteristic loads;

$\gamma_{G,2}$, $\gamma_{Q,2}$, the structural performance factor, takes account of possible inaccurate assessment of the combined effects of loading, unforeseen stress redistribution within the structure, variations in the dimensional accuracy achieved in the structure insofar as they affect its response, and the importance of the limit state being considered.

It should be noted that the load combination and sensitivity component is now taken account of by ψ.

Although, in general, the values of γ_G and γ_Q are common to all the Eurocodes, they are not wholly independent of the structural material as $\gamma_{G,2}$, $\gamma_{Q,2}$, the structural performance factor, is intended to cater among other things for uncertainties in the assessment of structural response to actions.

For FRP composite design, such uncertainties in the assessment of the structural response to actions can vary more than for other structural materials. Nevertheless, it is desirable for the same or a similar set of partial safety factors for permanent actions and variable actions to be used in the EUROCOMP Design Code as in the Eurocodes generally. Therefore, in the EUROCOMP Design Code, the supplementary $\gamma_{G,2}$ and $\gamma_{Q,2}$ factors are, wherever possible, incorporated into the design expressions for conditions that must be satisfied for compliance with the EUROCOMP Design Code by means of the coefficient of model uncertainty (see 2.2.2.5).

2.2.2.5 Design values of the effects of actions

The coefficient of model uncertainty is variously denoted as C, C_x or K in the text or, exceptionally, for approximate formulae, is incorporated in the numeric coefficients.

As its name suggests, it allows an additional partial safety coefficient to be introduced into a particular design expression without having to modify the existing set of γ_f or γ_m partial safety coefficients. The purpose is to cater for:

- uncertainty in the analysis or assessment of structural or material behaviour
- uncertainty in the assessment of the response to actions;

  or, where more than one mode of failure is possible, to ensure that the most benign mode, e.g. one giving adequate warning of failure, occurs before, say, a brittle failure.

2.2.3 Material properties

2.2.3.1 Characteristic values

Characteristic material properties for fibre reinforced composites cannot be so rigorously defined as for most other structural materials. Standard data are either not generally available or cannot be relied upon except for preliminary design. Because of its non-homogeneity and anisotropy, up to 21 elastic constants may, in theory, be required for a full analysis of a complex laminate. In practice though, due to symmetry, these reduce to:

Young's moduli	E_{11}, E_{22} (= E_{33})
Shear moduli	G_{12} (= G_{13})
Poisson's ratio	ν_{12} (= ν_{13}), ν_{23}

Both stiffness and strength vary with the composition, with temperature and other environmental conditions, with production processes and with time. (Compare this with steel, for which standard grades are produced under stringent quality control conditions with a guaranteed minimum yield strength. Constitutive properties vary little, whatever the composition and environmental conditions).

With fibre reinforced composites, therefore, material design data has normally to be obtained from the testing of specimens which, for maximum reliability, should represent as closely as possible the conditions of the final product. Details of testing procedures are given in Chapter 7 of the EUROCOMP Design Code.

For FRP composite materials, characteristic values are required for at least the following elastic constants or material properties:

For each ply of a laminate:

- E_1, E_2, σ_{11}, σ_{22}, (tension and compression)
- ν_{12}, G_{12}, τ_{12}

and for the laminate as a whole:

- the interlaminar shear strength
- the through-thickness tensile strength
- local damage criteria (e.g. first-ply failure, surface crazing, or matrix cracking).

For each panel of a pultruded section:

- E_l, E_t, σ_l, σ_t, (tension and compression)
- ν_{lt}, G_{lt}, τ_{lt}

These are required not only for the particular constituent materials (formulation and proportions) but for the particular production process, and for the particular temperature and environmental conditions of use of the final product.

It may not be practical to obtain the full range of characteristic properties necessary to carry out a rigorous design of a particular structure, simply because of the large number of tests required to obtain a significant result. Accordingly, certain properties may be derived from others, using one of the alternative methods given in Chapter 4, provided that appropriate partial material safety factors are used, reflecting the reliability of the derived property. The method used must refer back to a characteristic property or to characteristic properties that can be verified for quality control purposes (see Chapter 7). This is why short-term values are used for the characteristic material properties.

2.2.3.2 Design values

No additional information.

2.2.4 Geometrical properties

No additional information.

2.2.5 Load arrangements and load cases

No additional information.

2.3 DESIGN REQUIREMENTS

2.3.1 General

No additional information.

2.3.2 Ultimate limit states

2.3.2.1 Verification conditions

No additional information.

2.3.2.2 Combinations of actions

No additional information.

2.3.2.3 Design values of permanent actions

No additional information.

2.3.3 Partial safety factors for ultimate limit states

2.3.3.1 Partial safety factors for actions on building structures

This section draws on references 2.4 and 2.5. Otherwise no additional information.

2.3.3.2 Partial safety factors for materials

This section draws on references 2.4 and 2.5.

The partial material safety factor, γ_m, allows for uncertainties in the assumed properties of the material in the final structure. Those uncertainties given in the EUROCOMP Design Code are, where possible, a combination of those given in Eurocodes and those currently used in the design of glass FRP composites. In the EUROCOMP Design Code, γ_m is made up, generally, of three subsidiary factors: the first two of which, in principle, are common to all Eurocodes.

$\gamma_{m,1}$ takes account of unfavourable deviations of the material properties from the specified characteristic values and of possible differences between the material property in the structure or component and that derived from the test specimens.

$\gamma_{m,1}$ also takes account of the level of uncertainty in the derivation of the laminate or panel characteristics.

For the properties of a multi-ply laminate, there are three alternative methods of deriving the laminate characteristics, with reducing levels of uncertainty:

1. Properties of constituent materials (i.e. fibre and matrix) are derived from test specimen data
   Properties of individual plies are derived from theory
   Properties of the laminate are derived from theory

2. Properties of individual plies are derived from test specimen data
   Properties of the laminate are derived from theory

3. Properties of the laminate are derived from test specimen data.

The value of $\gamma_{m,1}$ depends on which alternative method has been used to derive the laminate characteristic properties.

$\gamma_{m,2}$ takes account of uncertainties in the material properties due to the nature of the constituents (e.g. type and form of fibre reinforcement), of possible local weaknesses arising from the manufacturing process and of unfavourable geometric deviations resulting from manufacturing tolerances.

$\gamma_{m,3}$ takes account of changes in the material properties over time and as a result of environmental effects. (It should be noted that $\gamma_{m,3}$ is not a true partial safety factor in this application, since it does not represent uncertainties, but is being used as a modification factor.)

2.3.4 Serviceability limit states

No additional information.

2.4 DURABILITY

(See Chapter 3 and 4.13.)

No additional information.

2.5 ANALYSIS

2.5.1 General provisions

2.5.1.0 Notation (see also 1.6 and 1.7 of EUROCOMP Design Code)

No additional information.

2.5.1.1 General

No additional information.

2.5.1.2 Load cases and combinations

No additional information.

2.5.1.3 Imperfections

No additional information.

2.5.1.4 Second order effects

No additional information.

2.5.1.5 Design by testing

(See Chapter 7 of EUROCOMP Design Code).

No additional information.

2.5.2 Idealisation of the structure

2.5.2.0 Notation (see also 1.6 and 1.7 of EUROCOMP Design Code)

No additional information.

2.5.2.1 Structural models for overall analysis

(No additional information).

2.5.2.2 Geometrical data

(See Chapter 6.)

2.5.3 Calculation methods

2.5.3.0 Notation (see also 1.6 and 1.7 of EUROCOMP Design Code)

No additional information.

2.5.3.1 Basic considerations

No additional information.

2.5.3.2 Types of structural analysis

(a) *Ultimate limit states*

No additional information.

(b) *Serviceability analysis*

No additional information.

2.5.3.3 Simplification

No additional information.

2.5.4 Effects of anisotropy

The performance of laminates mainly depends on:

- content, orientation and length of the fibres
- the properties of the material used for the matrix
- the quality of the bond between the fibres and the matrix material
- method of manufacture.

Reinforcements can be uni-directional, cross ply, angle ply or random in their arrangement. In any one direction, the mechanical properties of the composites will be proportional to the volume fraction of the fibres oriented in that direction and the components due to other fibre directions.

The fibres and the resin material interact and redistribute loads. The effectiveness of such transference, however, depends largely upon the quality of the bond between them.

While composite materials owe their unique balance of properties to the combination of matrix and fibres, it is the fibre system that is primarily responsible for strength and stiffness. However, the fibre dominates the field in terms of volume, properties and design versatility.

A single lamina with uni-directional reinforcement has high mechanical properties along its longitudinal axis, and low-to-moderate properties along its transverse axis. This is the primary difference from a structural analysis and design standpoint between composites and metals.

Metals are normally homogeneous and isotropic in nature and their reaction to an applied load can be defined by knowing two of the three basic elastic constants (the modulus of elasticity E, the shear modulus G and Poisson's ratio ν). A basic uni-directional lamina on the other hand is orthotropic in nature, having three mutually perpendicular planes of symmetry of elastic properties. Four independent elastic constants (E_1, E_2, G_{12} and ν_{12}) characterise the basic uni-directional lamina.

For many composite applications the laminate is not orthotropic, but anisotropic, though this can exist in many particular forms, one of which occurs when an orthotropic laminate is loaded in a direction which does not coincide with one of the principal axes, or when the lay-up is symmetric but not balanced about a principal reference axis. A $[0^\circ/\pm45^\circ/90^\circ]_s$ laminate is balanced and quasi-isotropic. As a general rule, all laminates should be symmetrically laid up about their mid-plane to avoid stretching bending coupling.

After the selection of the fibre/resin material, the design process concentrates on the lamination rationale: the number of plies required, the angle of orientation and the stacking sequence for a particular section. There exists no universal lamination geometry that can satisfy all possible loadings. Lamination geometry should be based on the anticipated stress state, i.e. magnitude, direction and combined biaxial and shear stresses, and the strength and/or stiffness requirements to realize the potential structural efficiency of the composite. Stiffness requirements may be based on laminate flexural and in-plane mechanical properties. The freedom of design choices offered by both the material properties and the laminate configuration require a systematic approach to determine the lamination geometry.

2.5.5 Determination of the effects of time dependent deformation of the composite

This section draws on reference 2.7.

2.5.5.0 *Notation* (see also 1.6 and 4.7 of EUROCOMP Design Code)

No additional information.

2.5.5.1 *General* (see also 4.7 of EUROCOMP Design Code)

No additional information.

REFERENCES

2.1 ENV 1991-1: Eurocode number 1: Basis of design and actions on structures part 1: Basis of Design, European Committee for Standardisation, 1991.

2.2 ENV 1992-1-1: Eurocode number 2: Design of concrete structure part 1: General rules and rules for buildings, European Committee for Standardisation, 1992.

2.3 ISO 2394: 1986: General principles on reliability of structures, International Standards Organisation, Geneva, 1986

2.4 BS 4994: 1987: Design and construction of vessels and tanks in reinforced plastics. British Standards Institution, London, 1987.

2.5 Head P R and Templeman R B, Application of limit state design principles to composite structural systems. *In Polymers and polymer composites in construction* (Ed L C Hollaway), Thomas Telford. London. 1990.

2.6 Quinn J A, *Design Data - Fibreglass Composites,* Fibreglass Ltd, 1981.

2.7 Hollaway L, *Polymer Composites for Civil and Structural Engineering.* Chapman and Hall, London, 1993.

CONTENTS

3 MATERIALS

3 MATERIALS

3.1 REINFORCEMENT

3.1.1 Fibres

The properties of fibres are the key to the structural performance and application potential of composites. Fibres are either made by drawing from a melt, as in the case of glass and thermoplastic fibres, or produced as precursors by spinning or extrusion for chemical or thermal conversion, as in the case of carbon fibres. Resulting fibre properties depend on the chemical nature of the backbone of the fibre and alignment of the molecular structure. The ability to control and eliminate flaws in fibre manufacturing processes is paramount to obtaining high strength reinforcing materials.

The EUROCOMP Design Code is restricted to the use of E-, C- and ECR-glass fibres, although these are only a few of the types of reinforcing materials available. The chemical compositions of these fibres are shown in Table 3.1. In addition there are other glass fibre compositions: R - and S-glass, which have similar properties and are noteworthy on account of their having 40% more strength and 20% more stiffness than E-glass. They also perform better at elevated temperatures but are less common on account of cost and are used mostly in high performance applications. In addition to the amorphous range of glasses there are various grades of carbon fibres, which have stiffness up to four times that of glass, and lightweight aramid fibres, which have very high tensile strength but poor compression properties. Most fibres are similar in size, diameter 6-15 microns, and can be produced in a range of planar reinforcements for manufacturing composites. There is now a tendency for some suppliers to move towards fibres of 20-30 microns. These have the advantages that they cost less, are easier to impregnate, and have fewer health and safety problems. Strength properties are lower than for the smaller diameter fibres.

The properties of a range of fibres are discussed in reference 3.1 which, together with reference 3.2, is an excellent source of collated information on basic materials and property data for reinforced plastics.

3.1.1.1 E-glass

E-glass has a low alkali composition which produces the excellent electrical insulation properties and is the most commonly used glass fibre in FRP owing to its good mechanical and physical properties, versatility, ready availability and economic price. Of the common glasses it shows much less degradation with time to attack by water.

3.1.1.2 C-glass

C-glass is a medium - high alkaline glass. It is used in most critical chemical environments when E-glass is not acceptable. In many cases C-glass is only used for the surface layers of the laminate.

Table 3.1 Indicative glass fibre compositions (wt.%).

Components	*E-glass*	*C-glass*	*ECR-glass*
Silicon oxide	54	65	54
Aluminium oxide	15	4	15
Calcium oxide	17	13	21
Magnesium oxide	5	3	5
Sodium oxide	<1	8	<1
Potassium oxide	<<1	2	<1
Boron oxide	8	5	0
Barium oxide	-	1	0

3.1.1.3 ECR-glass

ECR-glass is an E-type glass having enhanced chemical resistance especially in acid environments. Owing to its low alkali content, its electrical properties are better than those of C-glass.

3.1.2 Rovings

Fibre glass rovings are made directly from the melt spinning of glass. Conventional rovings are made by collecting together the individual fibres from each of the bushings at the melt and winding them together to form a strand of fibres. In this process binders and coupling agents are usually applied to suit the intended end use and to limit degradation of fibre properties when the fibre is handled in the additional processing stages. Various weights of rovings are available made up from a multiplicity of fibre strands and which are defined by a tex number. This is the weight in grams of a kilometre of fibre. Rovings are packaged on drums and are used directly in filament winding and pultrusion processing and form the fibre stock for the weaving of cloths. A typical pultrusion machine might use 2400 and 4800 tex fibres.

3.1.3 Mats

Mats are the cheapest form of planar reinforcements but are more expensive than uni-directional rovings. They are made by chopping rovings into short lengths and assembling them into a handleable mat with a binder which may be insoluble or soluble in water. The latter should be avoided for applications where water immersion is foreseen as this can be the cause of osmosis effects and lead to loss of composite integrity. Thermoplastic binders are also used to allow mats to be formed into shaped preforms for resin transfer moulding processing. The short fibres in mats are oriented in plane at random and thus mats are non-aligned materials.

3.1.3.1 Chopped strand mat (CSM)

Chopped strand mat consists of a random array of short uniform lengths of fibre strands, in a 25-50 mm range, bound together by a binder for ease of handling. A range of weights, giving different layer thickness, are available in widths of up to 2 m. While these materials are easy to use, the open and variable nature of the reinforcement often produces resin rich areas and heterogeneous microstructures. These materials because of their non-aligned nature are only capable of producing low volume fraction composites of up to 25%, depending on method of manufacture, and their use is limited in structural applications.

3.1.3.2 Continuous filament mat (CFM)

This is made from continuous swirls of fibre strands and has some advantage over chopped mats in that there is continuity of fibres, which are interlocked albeit still in a random array. The overall nature is still similar to chopped fibre mats, continous filament mats but are more open and springy and their structural use is again limited. They are however used in pultrusions in combination with rovings and fabrics.

3.1.4 Woven roving

Many rovings (WRs) are woven into heavy weave fabrics for applications that require a rapid build up of thickness over large areas. Woven rovings are bi-directional reinforcements having a warp (unlimited length in roll direction) and weft (finite width of roll) construction of continuous untwisted tows of fibres. Woven rovings of different areal weights are determined by the warp and weft filament tow density used in forming the cloth, but these will normally be balanced in a plain weave construction. There are many forms in which woven roving may be supplied. These are defined by the number of warp yarns (ends) in the length and the number of filling yarns (picks) in the width directions. There are also a number of weave patterns, such as plain, basket, twill and satin, analogous to those of woven fabrics. The ratio of picks to ends determines the level of anisotropy in the composite. The weave pattern controls the handling and forming characteristics of the fabric. Special unbalanced cloths can be produced for particular applications and their selection and specification will be determined by the requirements of the component being constructed. Woven rovings are usually heavier and thicker than yarn based fabric materials and are used extensively in thick section ship constructions and for smaller boats. Woven rovings having a coarse texture (400-600 g/m^2)

often cannot be laid together without leading to very resin rich interfaces with voids and poor adhesion, resulting in low values of interlaminar shear strength. In such instances improvements can be made by the inclusion of chopped strand mat between the layers of the woven rovings. Fibre volume fractions can achieve about 40% but are dependent on construction and method of composite manufacture.

3.1.5 Fabrics

A fabric is a lighter weight material constructed of interlaced yarns, fibres or filaments, usually with a planar structure. Typical glass-fibre fabrics are manufactured by interlacing warp (lengthwise) yarns and fill (crosswise) yarns on conventional weaving looms. Such fabrics are woven into a variety of styles which permit quite exact control over thickness, weight and strength. The principal factors which define a given style are fabric count, warp yarn, fill yarn and weaves.

The fabric count refers to the number of warp yarns (ends) per centimetre and number of filling yarns (picks) per centimetre. For example, a fabric count of 8.5 x 4.5 means that there are 8.5 ends per centimetre running in the warp (lengthwise) direction and 4.5 picks per centimetre running in the fill (transverse) direction. Fabric count plus the properties of the warp and fill yarns used to weave fabrics are the principal factors which determine fabric strength.

The weave of a fabric refers to how warp yarns and fill yarns are interlaced. Weave determines the appearance and some of the handling and functional characteristics of a fabric. Selection of a pattern style usually follows the need for ease of handling and ability to drape around double and compound curvatures. Woven fabrics are characterised by a degree of crimp, set by the weaving pattern style. Among the popular weave patterns are plain, twill, satin and uni-irectional. Woven fabrics are normally much finer and lighter per ply than woven rovings.

Plain weave (linen weave)

In this the weft is threaded successively above and below each warp thread, and then inversely in the following pass. It is the oldest and most common textile weave and is the firmest, most stable construction, providing porosity and minimum slippage. Strength is uniform in both directions if identical amounts of fibres are used. See Figure 3.1.

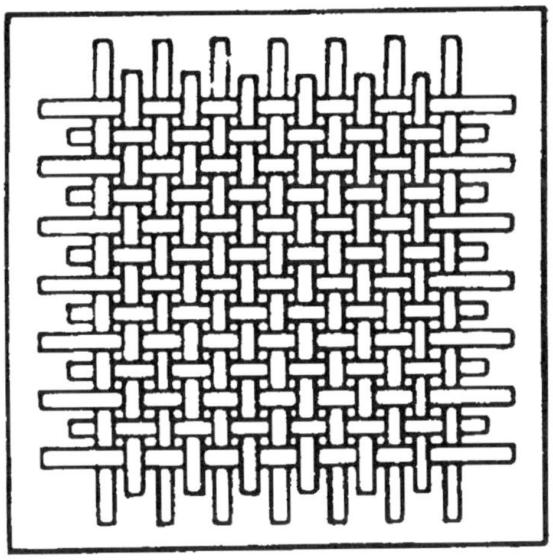

Figure 3.1 Plain wave (linen weave).

Twill weaves

Here the warp and weft threads are crossed in a programmed sequence and frequency in order to obtain the diagonal lines characteristic of this type of weave. Figures 3.2 and 3.3 illustrate examples of 3 on 1 and 2 on 2 twill weaves respectively. Twill weaves drape better than plain weave.

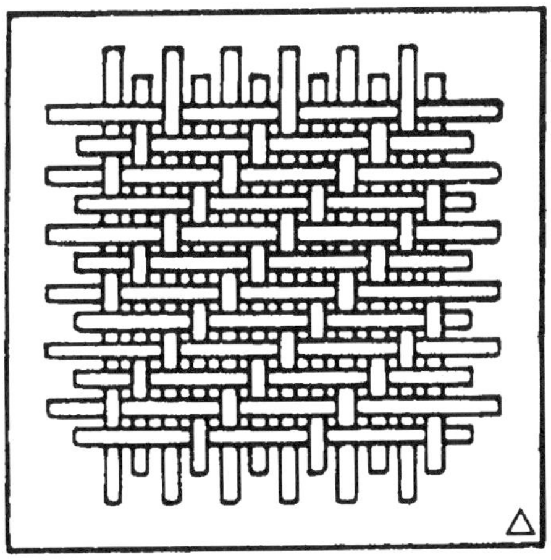

Figure 3.2 Twill weave (3 on 1).

Figure 3.3 Twill weave (2 on 2).

Satin weaves

In this, one warp end is woven over several successive fill yarns, then under one fill yarn. A configuration having one warp end passing over four and under one fill yarn is called a five-harness satin weave. The higher the satin number, the higher the count of warp and weft threads. The satin weave is more pliable than the plain weave. It conforms readily to compound curves and can be woven to a very high density. Satin weaves are less open than other weaves and have a high strength in both directions. See Figure 3.4.

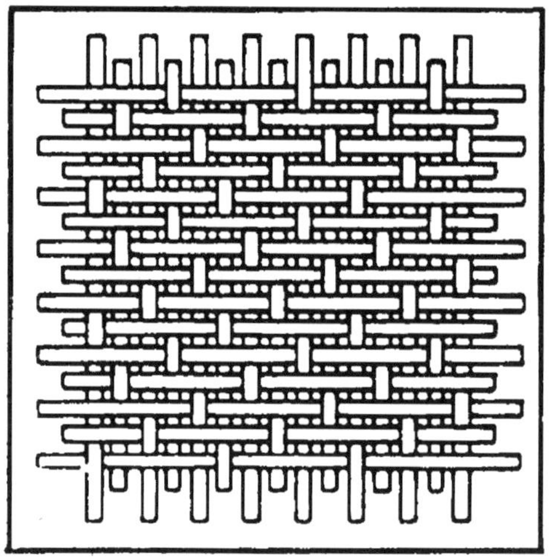

Figure 3.4 Satin weave.

Uni-directional weave

This involves weaving a great number of larger yarns in one direction with fewer and generally smaller yarns in the other direction. Such weaving can be adapted to any of the basic textile weaves to produce a fabric of maximum strength in one direction, usually along the roll length. The threads are parallel and simply held together. See Figure 3.5.

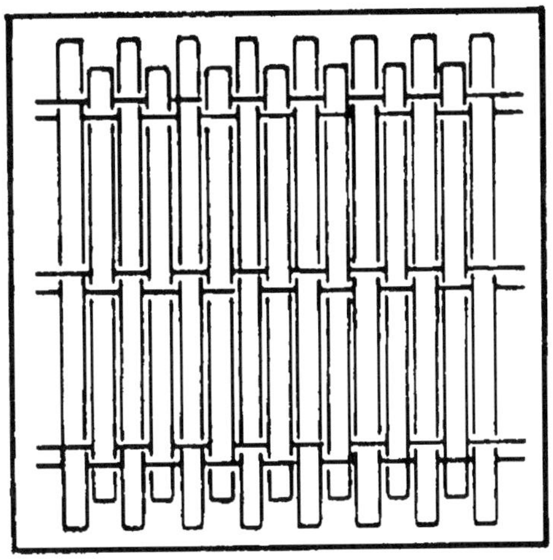

Figure 3.5 Uni-directional weave.

3.1.6 Non-crimp fabric

Crimp is the undulating deviation of fibres as a result of the weaving process and has an influence on the properties of the composite. Significant crimp will prevent layers from bedding together and leads to resin rich areas in the section and low volume fractions of reinforcement. Crimped fibres can also introduce higher local shear stresses and lower local mechanical properties. Non-crimped fabrics, now commonly referred to as multiaxial fabrics, are planar textiles produced by assembling fibres, yarns, and rovings in arrays with little or no crimp. Uni-directional weaves approach this, as the lighter weft yarns do not significantly cause bending of the thicker warp strands. For bi-directional non-crimp fabrics two crossed uni-directional arrays are held together either with very light stitching through the thickness or sometimes with a random spot binder between the plies. These materials are more expensive than woven reinforcements and can be more difficult to handle, but they also allow more freedom in putting together a desired directional lay-up, for example if 45° plies are required for improved in plane shear strength.

3.1.7 Prepregs

Prepregs were developed for producing the optimum in composite properties and are the combination of reinforcing fibres with intermediate partially cured resins. They may be any of the variety of reinforcements discussed above coated uniformly with a controlled amount of polymer. Many resin types can be used but for the more high performance applications to which these materials are aimed, because of extra processing cost, epoxy and phenolic systems have been generally developed more. In a prepreg the polymer is in a B-staged condition, which is an intermediate partially crosslinked resin condition allowing ease of handling with a degree of tack for assembling into a laminated composite pack.

Prepregs offer high mechanical properties with control of high fibre volume fraction, but require pressure facilities (autoclaves, vacuum tables) to consolidate and elevated temperatures to finalise cure and achieve their high properties. They also have a limited shelf life and need to be stored in freezers to extend their usable life.

3.1.8 Three dimensional integrated fibre preforms

Integrated reinforcement preforms can be made by advanced automated weaving processes allowing tailoring of continuous fibres into three dimensional arrays to suit a specific design requirement. The weaving process allows a flat preform to be produced which can be opened out into a three dimensional reinforcing form by inserting a suitable core shaped to produce the required structural section. Weaving variables allow uni-directional and cross-plied fibres to be woven simultaneously and interlinked to form a unified preform such that on opening the preform specific fibre configurations can be achieved in each part of the component. Mixed fibre preforms are possible. The preforms are then infiltrated with resins and cured to form rigid structural components such as box, I beam or sandwich sections. The preform can be designed to give uni-directional fibres in the flanges and $\pm$ 45° fibres in the webs. These preforms may subsequently be used as part of a continuous processing fabrication route, e.g. in pultrusions, or infiltrated as discrete shapes by resin transfer moulding methods. Integrated fibre preforms are most cost effective where large volumes of similar shaped composite components are required and have the advantage of improving the strength where fibres change direction through the continuity of the fibres and weave interlocking.

3.2 RESINS

3.2.1 General

No additional information.

3.2.2 Polyester resins

Unsaturated polyester resins are one type of a group of thermosetting polymer materials that have been widely used for marine and chemical plant applications. Their merits stem from user convenience, economic cost and flexibility of formulation, allowing slow or rapid conversion from low viscosity fluids into rigid solids when initiated and cured (crosslinking) either at room temperature or at an elevated temperature. In a typical general purpose room temperature curing polyester, styrene is used as a crosslinking monomer through a free radical mechanism activated by a peroxide catalyst and cobalt salt accelerator. This polymerisation produces a three dimensional array of polymer chains linking the initial unsaturated groups of the polyester polymer.

The rate of conversion and state of final crosslinking (degree of cure) is critical to the development of the expected mechanical properties and stability of the composite to environmental factors. Although room temperature curing systems give good properties, it is good practice where appropriate to cure above ambient temperature, as this achieves a higher degree of cure and gives generally more environmental stability.

There are many different types of polyester resins formulated for specific processing routes and applications. General purpose orthophthalic polyesters are based on a combination of maleic and phthalic anhydride monomers with a glycol (commonly propylene glycol) to form an unsaturated polymer, with reactive double bonds, and which allows three dimensional crosslinks to be made with styrene. Adjustment to the backbone of the unsaturated polyester with other starting monomers can produce different types of polyester having chain flexibility and resilience, heat resistance, low shrinkage during cure, weather resistance particularly to UV, chemical inertness, and fire retardancy. Unfortunately, however many, of the desirable properties cannot be induced without lowering others, particularly mechanical properties, and for any specific application some compromise is inevitable. The reader is referred to references 3.3, 3.4 and 3.5 for more details of the types of polyesters, their chemistry and their curing mechanisms.

Fire retardant polyesters contain a blend of chlorine, bromine, phosphorus or antimony based chemical groups and offer reduced flammability but with somewhat reduced mechanical properties and wet durability. The smoke emissions are reduced but the fumes may still be unhealthy. Additives may be particulate or the fire retardant groups embedded into the polymer backbone by using substituted acids. HET acid polyesters (hexachloromethylene-tetra-hydrophthalic acid) are an example of the latter.

Polyesters can be blended with particulate fillers to confer a range of features; increased viscosity, use of fire retardants, improved hardness and abrasion, opaqueness, ultra-violet stabilisation, colour and cost reduction. Mechanical properties are influenced by the incorporation of fillers.

Products

The main use of polyester resins is in combination with materials such as glass fibre to create reinforced plastics combining the best properties of both.

A wide range of different polyester resins can be produced by varying the proportion of raw materials used or altering production parameters. Polyester resins are generally classified according to the main raw material used in their manufacture.

General polyesters are usually orthophthalic acid based resins, which are used in applications based on open moulds and hand and spray lay-up moulding. Resins with a low level of styrene emission are ideal for these applications.

The raw materials used in producing speciality polyester resins differ. Self-extinguishing grades, used in applications requiring high standards of fire safety, usually include halogen compounds. Other grades offer high-performance mechanical properties, heat and chemical resistance, or elasticity.

Speciality polyesters also include gel coats and top coats. These are pigmented polyester resins and are used for coating the surfaces of glass-reinforced laminated mouldings.

Properties and use

Polyesters are almost always used in combination with a reinforcing component. The most widely used reinforcement is glass fibre, but aramid and carbon fibres are also used to some extent. Reinforced polyesters combine light weight with high strength and good resistance to weathering and chemical attack.

The popularity of polyester in reinforced plastics is based on its ease of use, low cost and the ability to produce good composite properties after room temperature curing. It has good structural properties, which make it possible to produce large, complex mouldings, such as boat hulls, in one piece. Many application processes are relatively cheap, which means that even small series of mouldings can be manufactured cost-effectively.

Reinforced polyesters have been widely used in the boat building industry since the early sixties. Most yachts and power boats are now made of polyester resins and glass fibre. Reinforced polyester plastics are also widely used in pipe and container structures, owing to their good chemical and corrosion resistance. Polyesters have increased their role in vehicle and transportation equipment production, where many parts previously manufactured from metal have been replaced with components produced from reinforced plastics.

In the construction industry, polyesters have become widely used in prefabricated panels, floor coverings and moisture barriers. One significant area of use is in sanitary equipment, such as bathtubs, jacuzzis and shower basins.

Other polyester products include flag-poles, doors, waste bins, ski rack enclosures, ski poles, fish hatchery basins, water slides and storage vessels. Many reinforced plastic products are protected by an outer gel coat layer which can be built in during composite construction most easily in hand and spray lay-up processing.

3.2.3 Vinyl ester resins

Vinyl esters are thermosetting resins that consist of a polymer backbone with an acrylate or methacrylate termination. The backbone component of vinyl ester resins can be derived from epoxide, polyester or urethane but those based on epoxide resins have most commercial significance. Bisphenol A epoxy formed vinyl esters were designed for chemical resistance and commonly formulated for viscosity for use in filament winding of chemical containers. Typically styrene is used as a reactive dilutent to modify viscosity. Phenolic novolac epoxies are used to produce vinyl esters with higher temperature capability and good solvent resistance, particularly in corrosive environments, and their FRP composites have demonstrated initial economy and better life cycle costs compared with metals.

Vinyl ester resins are similar to unsaturated polyester resins in that they are cured by a free radical initiated polymerisation. However, they differ from the polyesters in that the unsaturation is at the ends of the molecule rather than along the polymer chain. Unlike polyesters, vinyl esters show a greater resistance to hydrolysis as well as lower peak exotherm temperatures and less shrinkage upon cure. Cured vinyl ester resins exhibit excellent resistance to acids, bases and solvents. They also show improved strain to failure, toughness and glass transition temperatures over polyesters. They can be used in filament winding, pultrusion, resin injection, vacuum moulding and conventional hand lay-up.

3.2.4 Modified acrylic resins (Modar)

Methacrylate or MODAR resins are thermosetting materials which can be cured by thermal and chemical techniques. These materials are differentiated from other systems in their ability to be cured quickly within a few minutes, without affecting their ultimate mechanical performance. Normally in polyester systems high rates of cure are undesirable as they are accompanied by high exotherms, causing cracking and brittleness.

Like polyesters they have a high mould shrinkage property.

Experience of structural use in FRP applications is much more limited than with the other resins discussed.

3.2.5 Phenolic resins

Phenolic resins are the oldest synthetic polymers. They were synthesised in the 19th century and their commercial value identified as early as 1905, with Bakelite. Traditional engineering uses for these thermosetting materials include moulding materials, grinding wheels, foundry binders and friction material binders. All these applications took advantage of the material's excellent heat and dimensional stability and adhesive qualities.

Phenolics are now growing into the FRP composite areas of hand lay-up, open and closed mould techniques, pultrusion and filament winding. The driving force behind this recent growth is the concern over flammability and smoke generation of materials used in transportation and in certain types of construction. Use of phenolics is limited to those fabrication processes and sizes of structure which allow high temperature curing.

Chemistry of phenolics

Phenolic resins are the reaction product of one or more of the phenols with one or more of the aldehydes or ketones. Commercial phenolics are produced chiefly from phenol and formaldehyde as resoles. Traditional phenolics are supplied as laminating solutions of partly reacted resins in an alcohol, such as isopropyl alcohol, for the manufacture of prepreg reinforcements.

Phenolic resins have been developed to allow their use as laminating resins in a water solution. After impregnation, cure is achieved by the addition of an acid catalyst, resulting in an exothermic reaction, and with emission of water vapour, which tends to cause an inherent high void content in composites. This problem, combined with a degree of chemical affinity for moisture, can lead to a level of water absorption up to three times greater than with an isophthalic polyester resin.

Resorcinol-modified phenolic resin requires a two component system, of which one component consists of a resin containing the resorcinol and the second a similar resin containing the paraformaldehyde. When mixed, in roughly equal quantities, polymerization starts. An enhanced reactivity of the resorcinol-containing resin for a low temperature cure occurs in the paresence of paraformaldehyde.

Novolacs or two-step phenolic resins require crosslinking agents to cure. The resulting thermosetting resin is an insoluble and infusible material. Novolacs are used in moulding compounds.

Novolac formation occurs under acidic conditions at 100°C with the excess of phenol (about one mole phenol to 0.7 to 0.85 mole formaldehyde is reacted). Novolac curing is accomplished by addition of crosslinking agents, usually HEXA (hexamethylenetramine) or HMTA whose reaction products form urea and formaldehyde when heated (110°C).

3.2.6 Epoxy resins

Epoxy resins offer the best mechanical properties for FRP for a number of reasons. They are well known as adhesives, adhere well to a wide variety of materials, fillers and reinforcing fibres, can be formulated in a wide range of resins for particular processing routes, and can be cured with a variety of curing agents to give a broad range of properties after cure. Most common curing mechanisms do not release volatiles or water and hence the shrinkage after cure is lower than for polyesters, vinyl esters, methacrylates and phenolic resins. Cured epoxy systems are not only resistant to moisture and chemicals but also provide excellent electrical insulation properties. Temperature stability is also high when cured above the service temperature.

Many common epoxies are based on the diglycidyl ether of bisphenol A (DEGBA) which is a short chain aromatic molecule terminating on each end with the epoxy group. All epoxy resins have these terminating end groups which can bond chemically with each other and with other molecules to form a large three dimensional network during the cure process. Most commercial epoxies are a blend of materials, of which mono and difunctional ethers are formulated as reactive diluents to control processing viscosity. Crosslinking is promoted by a variety of curing agents, of which amines and anhydrides are the more common, and which in some cases require an accelerator to be added to increase the rate of reaction. Nearly all epoxies require an elevated temperature curing cycle to optimise and stabilise mechanical properties and this can be limiting to the production of large structural items without expenditure for heat curing facilities.

The chemistry of this subject is complex and with the variety of epoxy resins and curing agents the reader is referred to references 3.3, 3.4 and 3.5 and should consult with materials suppliers for more detail.

The cost of epoxies is higher than polyesters but they are very versatile and widely used in filament winding, pultrusion, resin injection and in the manufacture of prepregs.

3.3 CORES

3.3.1 General

Sandwich construction is commonly used with composites to increase structural efficiency, with the FRP forming the outer skins and bonded to a variety of core materials.

Materials available as cores are in three forms:

- basically lightweight (e.g. balsa-wood)
- lightweight because foamed
- lightweight because honeycombed.

Materials for cores include:

Metals - steel, aluminium, titanium as hexagonal honeycomb (maximum shear strength, expensive).

Wood - chipboard, plywood, strawboard, woodwool/ cement.

Plastics - rigid polyurethane foam, expanded polystyrene, honeycomb, crosslinked PVC foam (one of the most important core materials), polymethacrylimides, other plastics foams (excellent thermal insulation, weaker and more expensive than paper).

Others
- low density porous masses of glass fibre or mineral wool with resin.
- porous cement or concrete.
- sand/resin.
- paper or cardboard honeycomb (moderate insulation, good strength at minimum cost).
- syntactic foam (microballoons in a polymer matrix to provide high compressive properties).

Comparison of various core materials are given in Tables 3.2, 3.3, 3.4 and 3.5.

3.3.2 Foam

3.3.2.1 General

Foams have been extensively used in the manufacturing of FRP structures. They can simply be used as non-structural support formers in producing a desired shape, e.g. using low density foams to construct hat stiffening of panels, or more efficiently in high density forms with thinner skins and sandwich construction. Also in the former case very low density foams may be used unless the method of construction of the FRP requires significant pressure. In the latter case structural foams with good compression and shear rigidity are required, thus demanding higher density materials. Foams may be made from many plastic materials (ABS, cellulose acetate, epoxies, phenolics, -polycarbonate, polypropylene, polyurethanes, polyimides and crosslinked PVC) in various densities, and are essentially produced by using a foaming agent as part of the system and injecting the required amount by weight into a closed mould. Good control of mixing and distribution of the initial liquid is required to produce consistency in mechanical properties which are a function of both the foam type and density. Graded density foams can be made as well as integral sandwich foams where the surfaces are solid but the core cellular.

Further information on the structure and properties of foams are given in references 3.7 and 3.8 and from suppliers information, in references 3.6 and 3.8.

Table 3.2 Qualitative comparison of core materials.

Form	*Material*	*Advantages*	*Limitations*
Solid	Balsa-wood	Robust, easy to shape	Variable, affected by moisture
Foamed	PVC Polymethacrylimides Polymethoenylimide Polyurethane Melamine	Temperature limitation Controlled density Can be foamed in situ Service temperature to 200°C	Poor fire performance
Honeycomb (1) Aluminium	Pure Al strip sometimes epoxy coated	High strength, low weight	Can corrode next to carbon fibre reinforced plastic
(2) Nomex[R]	Aramid paper strip dipped in phenolic resin	Good strength weight ratio, good fatigue resistance	More difficult to shape-expensive
(3) Titanium	Ti alloy strip	High temperature resistance	Cost and weight

Nomex[R] Trade name of Hexcel's synthetic (nylon) paper.

Table 3.3 Properties of sandwich core materials.

Core material/ trade name	*Classification*		*Cell size (in)*	*Density (kg/m³)*	*Lengthwise shear strength (N/mm²)*	*Lengthwise shear modulus (N/mm²)*	*Compressive strength (N/mm²)*	*Temperature resistance (°C)*	*Notes*
End-grain balsa	Wood		-	100-250	0.3-20.0	2.2-8	1.2-3.6		
Rohacell[R])	Plastics	)PM1*	-	30-90	-	-		175	
Klegecel[R])		)PVC	-	30-50	0.5/1.0	-		80	
Phenolic)	Foams	)Phenolic	-	40-180	01/0.5	-	0.2/0.9	250	Good fire resistance
Honeycomb	Al sheets		-	30-133	10-5.5	-	1.0-11	-	High mechanical strength
Honeycomb	Nomex[R]		1/8	27-150	0.6-3.6	3.7-17	0.9-12.5		Good fatigue strength and moisture resistance
Honeycomb	Nomex[R]		3/16	90	2.7	80	10		
Honeycomb	Nomex[R]		1/2	128	-	-	7		
Foam	Syntectic			800-1200	-	-	30-100		Very high mechanical strength

* Polymethoenylimide.

Table 3.4 Properties of PVC foam cores (Divinycell H Sandwich Core Material, reference 3.6).

Quality	*Units*	*H 30*	*H 45*	*H 60*	*H 80*	*H 100*	*H 130*	*H 160*	*H 200*	*H 250*
Density ASTM D 1622	kg/m³	36	46	60	80	100	130	160	200	250
Compressive strength* ASTM D 1621	N/mm² @ +20°C	0.3	0.6	0.8	1.2	1.7	2.5	3.4	4.4	5.8
Compressive modulus* ASTM D 1621 procedure B	N/mm2 @ +20°C	20	40	60	85	125	175	230	310	400
Tensile strength* ASTM D 1623	N/mm² @ +20°C	0.9	1.3	1.6	2.2	3.1	4.2	5.1	6.4	8.8
Tensile modulus* ASTM D 1623	N/mm² @ +20°C	28	42	56	80	105	140	170	230	300
Tensile strength** ISO 1926	N/mm² @ +20°C	0.8	1.2	1.5	2.0	2.4	3.0	3.9	4.8	6.4
Shear strength** ASTM C 273	N/mm² @ +20°C	0.35	0.5	0.7	1.0	1.4	2.0	2.6	3.3	4.5
Shear modulus** ASTM C 273	N/mm² @ +20°C	13	18	22	31	40	52	66	85	108

* perpendicular to the plane. ** parallel to the plane

Table 3.5 Properties of rigid PMI foam cores (Rohacell S grade with flame retardant additions) reference 3.8).

Properties	*Units*	*51S*	*71S*	*110S*	*200S*	*Standard DIN*
Gross density	kg/m³	52	75	110	205	53 420
Compressive strength	N/mm²	0.75	1.53	2.97	7.66	53 421
Compressive modulus	N/mm²	45	73	130	388	53 421
Compression	%	9	8	5	11	53 421
Tensile strength	N/mm²	1.12	1.97	3.27	8.48	53 455
Tensile modulus	N/mm²	52	91	157	352	53 457
Elongation at break	%	3.8	4.1	3.6	4.1	53 455
Shear strength	N/mm²	0.70	1.27	2.28	5.47	53 294
Shear modulus	N/mm²	22	34	58	138	53 294
Dimensional stability under heat	°C	190	190	190	200	53 424

3.3.2.3 Structural foams

The term "structural foams" designates components possessing full density skins and cellular cores, similar to structural sandwich constructions, or to bones, whose surfaces are solid but whose cores are cellular. For structural purposes, they have favourable strength and stiffness-to-weigh ratios, because of their sandwich type configuration. Frequently, they can provide the necessary structural performance at a reduced cost of materials.

Several processes are employed for the manufacture of structural foams. In the high-pressure process, the first step is to fill a mould solidly with the resin under pressure. While it is still soft, the mould is expanded or a core retracted, which provides an interior space to be foam-filled by expansion and foaming inward of the still-soft resin or by injecting foam into the interior space. In either case, the result is a dense skin surrounding a cellular core. In the low-pressure process, a mould is partially filled and the molten resin expands to fill the mould, forming a skin upon contact with the walls of the mould.

3.3.2.2 Reinforced foams

The fairly dense varieties of thermoplastic and thermosetting foams may be reinforced, usually with short glass fibres, but other fibres and metal, and other reinforcements such as carbon black, may be employed. The reinforcing agents are generally introduced into the basic ingredients and are blown along with them, to form part of and to reinforce the walls of the cells. When this is done, it is not unusual to obtain increased mechanical properties, especially in thermosets, of 400 to 500% with glass fibre contents up to 50% by weight. With liquid foams, such as some of the thermosets, reinforcing fibres may be introduced into the mould itself before or after foaming, but increases in mechanical properties are likely to be much lower than those above.

3.3.2.4 Others

In the processing of other foams, such as syntactic foams, instead of employing a blowing agent to form bubbles in the mass, preformed bubbles of glass, ceramics or plastic are embedded in a matrix of unblown resin. In multi foams such preformed bubbles are combined with a foamed resin to provide both kinds of cells. Reduction in weight is an obvious objective. However, this may be accompanied by other properties. A mixture of microspheres and resin can be formulated into a mouldable mass that can be shaped, or pressed into cavities and moulds, much as can moulding sand and clay. Properties of the finished hardened or cured mass can be tailored by suitable formulation. Synthetic "wood", for example, is provided by a mixture of polyester resin and small hollow glass spheres. Epoxy foams are commonly syntactic.

3.3.3 Honeycombs

3.3.3.1 General

Honeycombs were first made of paper by the Chinese some 2000 years ago. The product only found its use in the aerospace industry around 1940. Honeycombs are formed from virtually any thin-sheet material connected together in a manner that resembles the honeycomb made by bees hence the name (see Figure 3.6). The structural performance is particularly high in direct compression and shear owing to the directionality of the shaped form.

Honeycomb products are manufactured from impregnated resin sheet material (paper, glass fabric, etc.) or metal foil by bonding at staggered intervals and then expanding into the required shape, most commonly a hexagon (see Figure 3.7). The size of the unit structure and wall thickness, which will determine the density, is critical in resisting direct compression loads.

Metallic honeycombs

Metallic honeycombs are commonly made of aluminium, corrosion resistant steel, titanium and nickel based alloys.

Non-metallic honeycombs

The most common non-metallic honeycombs are made of Nomex, fibreglass and kraft paper.

3.3.3.2 Aluminium

The most commonly used grade is 5052 alloy, which contains 97% aluminium and 2.5% chromium with a corrosion resistant coating applied. The foil used in manufacturing of the honeycomb is strain hardened with a fully hard H191 temper which conforms with the US specification MIL-A-81596.

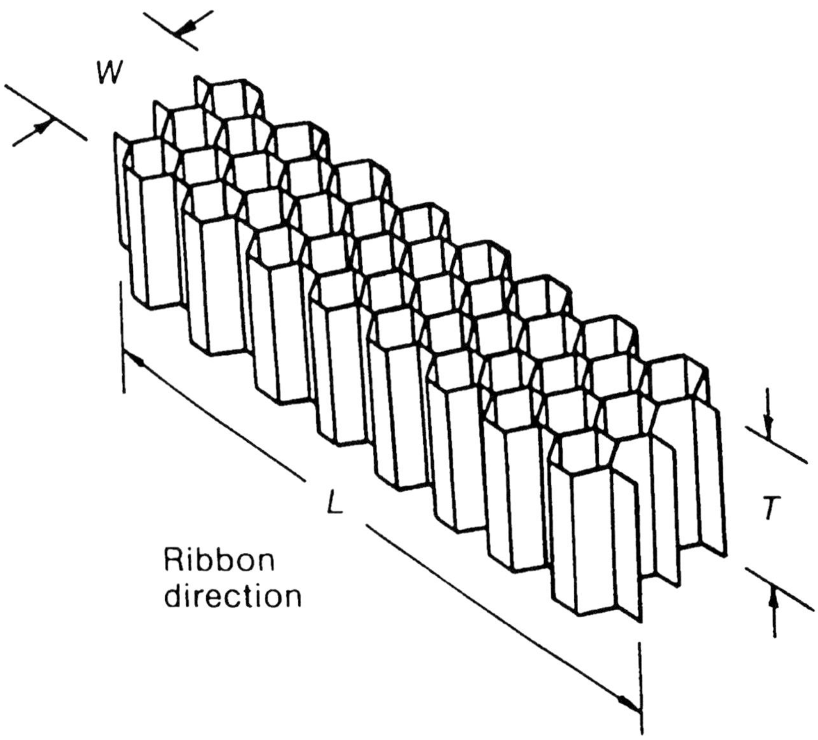

Figure 3.6 A typical honeycomb.

3.3.3.3 *Aramid*

Manufactured from Nomex high temperature resistant fibrous aramid sheet. A coating of usually phenolic resin is applied after the sheets of aramid have been bonded together and expanded into honeycomb form. It is the combination of aramid and phenolic resin which gives this type of honeycomb its superior strength, toughness and chemical resistance.

Aramid honeycombs are normally formed into hexagonal shaped cells for optimum mechanical properties. It is also possible to over-expand the material so that the cells become rectangular in shape. Over-expanded honeycomb has improved drapability for use in production of curved sandwich structures.

3.3.3.4 *Others*

Other honeycombs made of fibreglass, kraft paper, graphite and Kevlar are also available for specific applications.

3.3.4 Solids

Wood is the most appropriate material for a solid sandwich core and has been used extensively for many applications, such as doors and partitions. For lightweight constructions flat or end grain balsa is still used, although this is being superseded by synthetic materials. Bonded microspheres of inorganic materials, glass, ceramics, etc. to form syntactic foams are useful core materials.

3.4 GEL COATS

3.4.1 Base resin

Gel coats are normally applied first to the mould surface of a laminate, particularly for hand lay-up processing, to produce a decorative, glossy protective surface finish. As there are many faults that can arise with gel coats (reference 3.5), the formulation and application of the base resin is very important in determining the integrity and performance of the gel coat. Only the very best resin formulations are used as a base for gel coats and generally isophthalic acid is preferred over orthophthalic acid. The linear unsaturated polyster resin is produced in combination with a glycol, of which propylene glycols and ethylene glycol are most widely used in formulations, though neopentyl glycol can be used to obtain better water absorption resistance.

3.4.2 Pigments

Pigments are used to confer colour and to modify the effects of absorbed radiations. Choice of pigments influences the viscosity, the dispersion behaviour and sometimes the cure characteristics of the final gel coat. Some colours are now hard to produce because of the difficulty of having the right pigment and because of recent legislation that has prohibited using heavy metal pigments. This has put a challenge to gel coat producers and their pigment suppliers to find suitable new pigments. Because surface chemistry plays a vital role in the pigment dispersion this is very difficult.

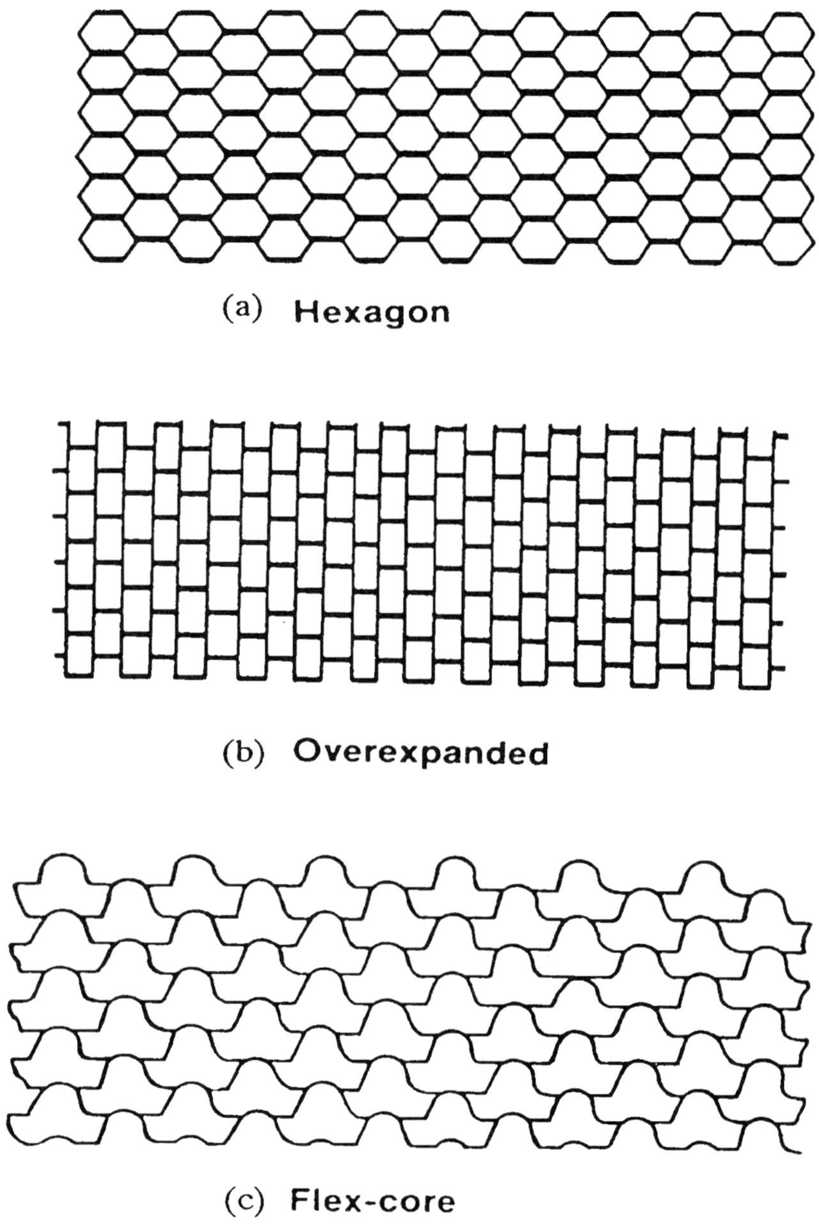

Figure 3.7 Common types of Honeycomb.

3.4.3 Additives

To enhance the handling properties of the gel coat there is a wide range of additives available. Most important are thixotropic agents, usually fumed silica. Bentonites are also used because of lower price. Other important additives are different kinds of surfactant. Among these are the air release, wetting and levelling additives and additives preventing separation of pigments or thixotropic agents. The additives have to be carefully chosen so they match the pigment in use.

3.4.4 Processing

The critical step for gel coat production is the dispersion of the pigments to the base resin to give the right colour and colour solidity. The latter is important to prevent show through of the fibre reinforcement and to obtain consistency of finish. Bead mills and roller mills are used for the dispersion of pigments and additives.

3.4.5 Gel coat properties

Since no reinforcement is used with the gel coat, the thixotropy has to be sufficient to prevent sagging on vertical mould surfaces. Base viscosity also has to be adjusted to the application method. Gel coats available today can be sprayed, but many are still required to be applied by brush. For spraying, an increased styrene content is used, which besides reducing viscosity compensates for evaporation during the spraying process.

Gel time for gel coats is always short, much shorter than for any ordinary polyester. This is due to two important factors. Short cycle times for the moulds are always desired. On top of that it is not desirable to put additives preventing styrene evaporation into gel coats. These would adversely affect the adhesion of subsequent laminates; therefore the gel coat has to be quite fast reacting to prevent excess styrene loss through evaporation. To adjust this gel time, proper care has to be taken by the gel coat manufacturer to ensure that other properties of the gel coat, such as colourfastness, are not compromised.

Application of coatings to FRP exposed top surfaces, i.e. those not in contact with a mould, may be required for similar reasons to using gel coats. These flow coats mainly differ in their method of application and consolidation. Flow coats should be applied if possible when the exposed surface is still at the "green" stage in order to obtain the highest adhesion if similar materials are being used to the main resin system. Adhesion to time-delayed exposed FRP surfaces can be low and considerable additional effort may be required in surface preparation to attain the required bond strengths.

Failure to use high quality materials and to obtain compatibility with the underlying material may lead to loss of adhesion and blistering through osmosis when subject to a saline environment.

3.5 SURFACE VEILS

Surface veils are lightweight thin surfacing materials usually constructed from swirl mats either in glass or from a synthetic fibre. Their main use is to act as a decorative surface, shielding from view the texture of the underlying structural reinforcement. But they also may be used in conjunction with gel coats for added support. C-glass surface veils are used for high chemical and environmental resistance. In pultrusion they are used to provide a smooth finish and protect fibres from being exposed at the surface.

3.6 ADDITIVES

3.6.1 Fillers

Fillers are mostly added to polyester resins help to reduce costs, but their diluent effect will also reduce any peak exotherm generated during cure and lead to lower shrinkage. In some instances where smooth surface finishes are required this may be beneficial. Clays and carbonates are common fillers with typical additions of up to 30% by weight. The effect of using fillers needs to be well tested on mechanical and weathering properties, as fillers can increase viscosity, reduce wetting of the base resin to fibres and cause voids in the composite. Environmental exposures of laminates containing fillers have generally indicated less favourable results than those without. Fillers may also interfere with the curing mechanism of resins. Fillers can also modify some properties beneficially, e.g. hardness.

3.6.2 Pigments

Dyes and pigments can be added to resins to obtain a desirable colour finish and since in most cases they are added in small amounts, less than 1% by weight, they will not affect material properties unduly. However, they may affect the rate of cure and this need to be checked in advance of composite manufacture. Some pigments also have merit in absorbing ultra-violet radiation, which will assist in improving weathering properties.

3.6.3 Flame retardants

Improved fire retardancy can be conferred to most resins by the incorporation of other materials. These materials may be specifically formulated for chemical incorporation into the resin during cure or exist as particulates in physical contact with the resin. Most organic phosphate esters and halogenated hydrocarbons, as well as antimony trioxide, are physically combined with the resin, and some halogen compounds are combined chemically. High concentrations of aluminium tri-hydrate (ATH) act as both filler and an effective low smoke/toxicity additive by releasing water of hydration at the flame temperature.

Typically polyester resins may be improved considerably by adding ATH or halogen additives. Vinyl esters can be treated with halogens. However, in both cases smoke emission still constitutes some problems because of the presence of styrene. Better relative performance may be achieved with Modar resins, as they allow greater filling ratios with ATH than polyesters, owing to the initial lower original viscosity and there is lower smoke emission due to the absence of styrene.

The addition of such materials can lower the strength, and affect the colour and weathering properties and may prevent certain types of processing being adopted as viscosity increases with filler content.

3.6.4 Low profile additives

Shrinkage is important in processing, as in most mouldings the final shape is produced, and while some resins such as epoxies, vinyl esters and phenolics have low mould shrinkage, others including polyesters and methacrylate resins have high shrinkage. Shrinkage control can be obtained with the addition of low profile additives, most of which are thermoplastic. Some are added by predissolving in styrene (e.g. acrylics, polystyrene, polyvinyl acetate) and others as particulates (e.g. polyethylene). These additives impart a smooth surface with dimensional stability and have found wide use in hot moulding (SMC, DMC). Some resin transfer moulding processes are now using low profile additives for producing more accurate parts.

REFERENCES

3.1 Hancox N L and Mayer T M, *Design Data for Reinforced Plastics - A Guide for Engineers and Designers,* Chapman and Hall, London, 1994.

3.2 Johnson A F, Engineering design properties of GRP BPF & NPL. Publication No. 215, London, 1978.

3.3 (eds) Grayson M and Eckroth D *Encyclopedia of Chemical Technology*, 3rd Edition, Vol. 18, John Wiley & Sons, New York, 1982.

3.4 (eds) Waterman N A and Ashby M F *Elsevier Materials Selector,* Vol. 3, Elsevier Applied Science, Oxford, 1991.

3.5 (ed) Lubin G *Handbook of Composites,* Von Nostrand Reinhold, New York, 1982.

3.6 Divinylcell Technical Manual - High performance core material for sandwich construction H-Grade, Divinycell International Ltd, Gloucester.

3.7 Gibson L J and Ashby M F, *Structure and Properties of Cellular Solids* Pergamon Press, Oxford, 1988.

3.8 Rohacell (PMI) Data Sheet, Rohm Technical Products, 1990.

3.9 Balakrishna T and Wang T G, Comparison of mechanical properties of some foams and honeycombs, *Journal of Materials Science*, 25, 5157-5162, 1990.

CONTENTS

4 SECTION AND MEMBER DESIGN

4 SECTION AND MEMBER DESIGN

4.1 ULTIMATE LIMIT STATE

It is a fundamental assumption of material behaviour that for short term loading the material is linear elastic to ultimate failure. Under long term loading the material is linear visco-elastic with recovery on removal of load, providing the strain does not cause permanent material deterioration.

4.2 SERVICEABILITY LIMIT STATE

No additional information.

4.3 MEMBERS IN TENSION

4.3.1 Scope and definitions

Tension members are those structural elements which are subject to direct axial stress without significant bending. Examples are hangers, struts, ties and braces in frames. They are designed to have adequate resistance to the applied loads, taking into account any reductions in the cross-sectional area and stress concentrations due to discontinuities.

As glass fibre composites have a relatively low Young's modulus, axial strain can be significant. Hence it may be neccesary to apply a serviceability limit state to this criterion to ensure that if for instance the tension member is used as a brace, it adequately prevents excessive sway in the structure. In any event there is a limiting allowable strain to ensure that the resin does not craze.

4.3.2 Design procedures

Consider as an example a tension member manufactured from uni directional E-glass fibres and polyester resin by the pultrusion process. The design criteria are listed below:

Length	5 m
Diameter	To be determined
Load, N_k	50 kN
Elongation	20 mm max

Characteristic material properties (see EUROCOMP Design Code, Table 4.13).

Longitudinal modulus,	$E_{x,t,k}$	=	41 kN/mm^2
Ultimate tensile strength,	$f_{x,t,k}$	=	690 N/mm^2

Limit states (see EUROCOMP Design Code 2.2).
The following limit states will be considered:

Ultimate (ULS) : Tensile failure

Serviceability (SLS) : Deflection

Assume $\gamma_{m1} = 2.25$
$\gamma_{m2} = 1.1$
$\gamma_{m3} = 1.2$

Hence $\gamma_m = 2.97$, say 3

Partial coefficients			ULS	SLS
Partial load coefficient	γ_f		1.5	1.0
Partial material coefficient	γ_m	strength	3.0	-
		stiffness	-	1.0

Design (see EUROCOMP Design Code 4.3.2)

1. Tensile failure

Design material strength (ULS), $f_{x,t,d} = f_{x,t,k}/\gamma_m$
$= 690/3.0$
$= 230\ \text{N/mm}^2$

Design load (ULS), $N_{Sd} = \gamma_f N_k$
$= 1.5 \times 50$
$= 75\ \text{kN}$

Required cross-sectional area of rod is given by:

$A \geq 75000/230$
$\geq 326\ \text{mm}^2$

Required rod diameter, $D = 20.4\ \text{mm}$

2. Deflection

Design tensile modulus (SLS), $E_{x,t,d} = E_{x,t,k}/1.0$
$= 41$

Design load (SLS), $N_{Sd} = \gamma_f N_k$
$= 50\ \text{kN}$

Deflection, d $= N_{Sd} l/(E_{x,t,d} A)$
$= 50000 \times 5000/(41000 \times 326)$
$= 18.7\ \text{mm}$

Note the SLS for deflection, 20 mm, has not been exceeded.

Therefore required rod diameter to meet deflection SLS is 22.6 mm.

Conclusion

To achieve the load ULS and the SLS, a rod diameter of 20.4 mm is required.

4.4 MEMBERS IN COMPRESSION

4.4.1 Scope and definitions

This section applies only to those compression members for which the load is applied coincident with the centroid of the member. If the load application is eccentric, then bending stresses as well as axial stresses occur. The effect of combined bending and axial compression is considered elsewhere.

The effects of global (Euler) and local buckling dominate the design of compression members in any material. This is particularly so when designing FRP compression members. The inherent relatively low elastic modulus of FRP requires a full consideration of all possible modes of buckling.

Global buckling of a member is considered with the assumption that the material is isotropic and that the relevant elastic modulus is that in the weak axis of the member.

Local buckling of the individual elements of a section, covered in 4.7 of the EUROCOMP Design Code, requires a knowledge of both the longitudinal and the transverse flexural stiffness of the element.

4.4.2 Design procedures

Determine the maximum axial design compressive force that can be safely applied to the glass FRP section shown in Figure 4.1. It is manufactured from E-glass random mat plus roving and polyester resin, by the pultrusion process.

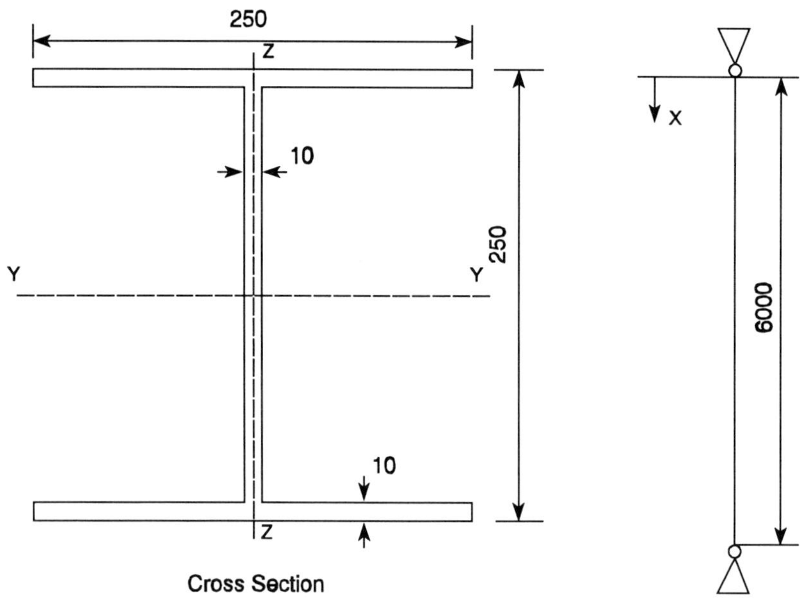

Figure 4.1 Example of compression member.
Assume that the column is pin-ended.

Characteristic material properties (see EUROCOMP Design Code Data, Table 4.12).

Longitudinal compressive modulus,	$E_{x,c,k}$	=	17.2 kN/mm^2
Transverse compressive modulus,	$E_{y,c,k}$	=	5.5 kN/mm^2
Ultimate compressive strength, Long,	$f_{x,c,k}$	=	207 N/mm^2
Longitudinal shear modulus,	$G_{xy,k}$	=	2.9 kN/mm^2
Poisson's ratio (x, y direction),	$\upsilon_{xy,k}$	=	0.33
Poisson's ratio (y, x direction),	$\upsilon_{yx,k}$	=	0.11
	γ_{m1}	=	1.5
	γ_{m2}	=	1.1
	γ_{m3}	=	1.2
Hence	γ_m	=	1.98, say 2

Section properties

Second moment of area (Y - Y axis),	I_{YY}	=	8.22 x 10^7 mm^4
Second moment of area (Z - Z axis),	I_{ZZ}	=	2.61 x 10^7 mm^4
Area,	A	=	7300 mm^2

Limit states (see EUROCOMP Design Code 2.2).

The following limit states will be considered:

Ultimate (ULS) : Compressive failure
Local buckling (short column failure mode)
Lateral bending (long column failure mode)

Assume partial coefficients as follows:

Partial load coefficient	γ_f		2.5
Partial material coefficient	γ_m	strength	2.0
		stiffness	1.0

Design

1. Compressive failure

Design compressive strength (ULS), $f_{x,c,d}$ $= f_{x,c,k}/\gamma_m$
$= 207/2.0$
$= 103.5\ \text{N/mm}^2$

2. Local buckling

(a) Web (EUROCOMP Design Code 4.7.4.4)

The local buckling stress for a long plate with "pinned" longitudinal edges is:

$$\sigma_{x,cr,b} = 2\ \pi^2\{(D_xD_y)^{1/2} + H_0\}/tb^2$$ (EUROCOMP Design Code, equation (4.9))

where:

maximum inside width of plate element, $b =$ 230 mm

web thickness, t = 10 mm

$D_x = E_xt^3/12(1 - \upsilon_{xy}\upsilon_{yx})$ (EUROCOMP Design Code, Table 4.1)

design longitudinal modulus, $E_{x,d}$ $= E_{x,k}/\gamma_m$
$= 17.2/1.0$
$= 17.2\ \text{kN/mm}^2$

Therefore:

$D_x = 17.2 \times 10^3 \times 10^3/12(1 - (0.33 \times 0.11))$
$= 1490\ \text{Nm}$

$D_y = E_{y,d}t^3/12(1 - \nu_{xy}\nu_{yx})$

Design transverse modulus, $E_{y,d}$ = $E_{y,k}/\gamma_m$
= 5.5/1.0
= 5.5 kN/mm^2

Therefore:

D_x = $5.5 \times 10^3 \times 10^3 / 12(1 - (0.33 \times 0.11))$
= 476 Nm

D_0 = $D_{xy} + 2D'_{xy}$

where:

D_{xy} = $\nu_{yx}E_{x,d}t^3/12(1 - \nu_{xy}\nu_{yx})$

= $0.11 \times 17.2 \times 10^3 \times 10^3/12(1-(0.33 \times 0.11))$

= 163 Nm

Stiffness constant, D'_{xy} = $G_{xy,d}t^3/12$

Design shear modulus, $G_{xy,d}$ = $G_{xy,k}/\gamma_m$
= 2.9/1.0
= 2.9 kN/mm^2

Therefore:

D'_{xy} = $2.9 \times 10^3 \times 10^3/12$
= 241 Nm

and

H_0 = $148.4 \times 10^3 + (2 \times 216.7 \times 10^3)$
= 645 Nm

Finally

$\sigma_{x,cr,b}$ = $2\,\pi^2\{(1490 \times 475)^{1/2} + 645\}/(10 \times 230^2) \times 1000$
= 55.4 N/mm^2

(b) Flange

The local buckling stress for a long plate with one "free" and one "pinned" longitudinal edge is:

$\sigma_{x,cr,b} = \pi^2\{(D_x(b/a)^2)+(12D'_{xy}/\pi^2)\}/tb^2$ (EUROCOMP Design Code Equation (4.10))

where:

a = half wave length of the buckle = the length of the plate = 6000 mm

b = width of outstanding flange = 120 mm

t = thickness of flange = 10 mm

$\sigma_{x,cr,b} = \pi^2\{(1490(120/6000)^2)+(12 \times 241/\pi^2)\}/10 \times 120^2 \times 1000$

= 20.1 N/mm^2

3. Long column (Euler) buckling

$\sigma_{x,cr,b} = N_{c.Rd}$

where:

$N_{c.Rd} = k\pi^2 E_{x,d}\, I/L^2$

column has "pin" ends; thus k = 1.0

Therefore

$N_{c.Rd} = \pi^2 \times 17.2 \times 10^3 \times 2.61 \times 10^7/6000^2 = 123000$ N

$\sigma_{x,cr,b} = 123000/7300$
$= 16.86$ N/mm^2

The lowest value of stress from (1), (2) and (3) gives the critical failure stress: i.e. (3).

The column design load,

$N_{sd} = \sigma_{x,cr,b}A$
$= 16.86 \times 7300$
$= 123$ kN

Characteristic (allowable load)

$= N_{sd}/\gamma_f$
$= 123.6/2.5 = 49.2$ kN

4.5 MEMBERS IN FLEXURE

4.5.1 General

The deflection of a beam is a function of (span/depth)^3. Therefore there is a significant benefit to be gained from an increase in the depth of a beam. This allows materials of relatively low modulus to be used without significant penalty.

Take a simple example. If the modulus of steel and glass FRP are taken as 200 and 20 N/mm^2 respectively, then the increased depth required of a glass FRP beam to have the same bending deflection as a steel beam is:

$(E_{steel}/E_{GRP})^{1/3} = (200/20)^{1/3} = 2.1$

Therefore the glass FRP beam would be 2.1 times the depth of the steel beam for the same flexural rigidity.

Hence materials with a low modulus can be used as beams if the application can tolerate an increase in beam depth.

However, beams with a low span/depth ratio have deflection due to shear which can be of the same order of magnitude as deflection due to bending. This is the case with glass FRP composites because of their typically greater depth and also because their shear modulus is relatively low. Hence deflection due to shear can be appreciable. Sandwich laminates tend to be constructed with low density, low shear modulus core materials and are a further example of where shear deflection is significant.

Figure 4.2 illustrates the effect the inclusion of shear deflection has on the allowable applied load. As an example a simply supported beam with Uniformly Distributed Load has been taken. The beam used is a 12" x 6" (300 x 150 mm) I beam (see reference 4.1). The two curves represent the allowable load plotted against span, when shear deflection is ignored and when it is included.

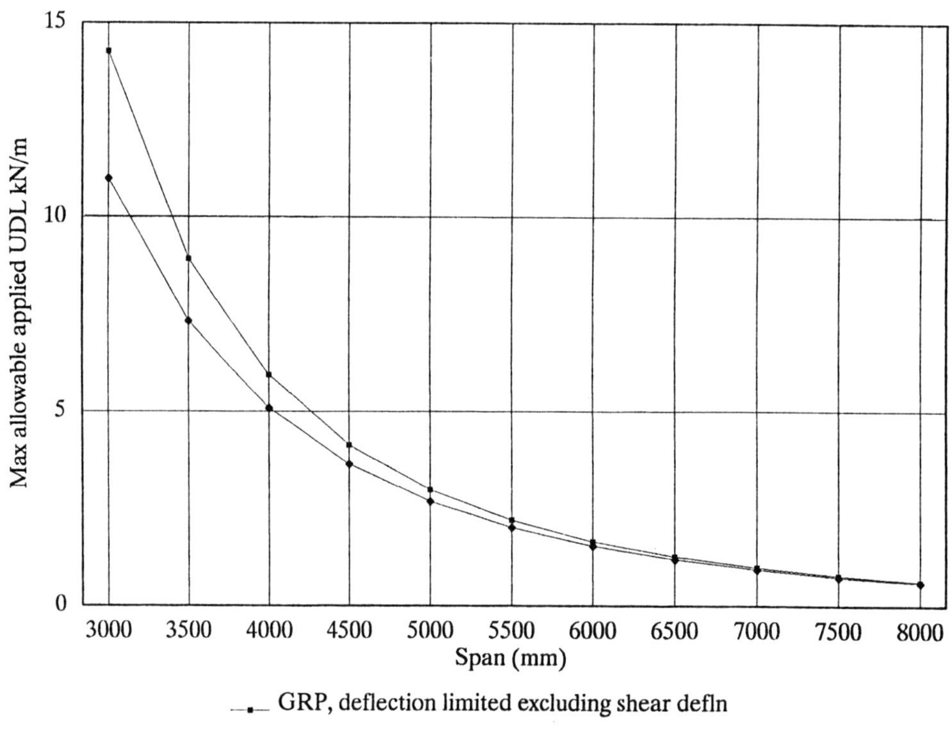

Figure 4.2 Effect of inclusion of shear deflection term on allowable UDL.

It can be seen that at short spans the simple model which excludes shear deflection would incorrectly allow a significantly higher load than the more precise model. When the span exceeds in this case about 7 m (which is a span/depth ratio of 23), the difference between the two models is sufficiently small to be ignored for many applications.

As a general rule of thumb, glass FRP beams with a span/depth ratio greater than 25 have negligible shear deflection in comparison to bending deflection. But for those with a smaller span/depth ratio, the deflection due to shear should be determined.

It should be noted that the elastic modulus in bending is not necessarily the same as that in tension or compression. It may be determined analytically or by physical test. If it is determined by test it should be ensured that the test span is sufficiently large that deflection due to shear is negligible. The shear rigidity of a section may be defined as the product of shear modulus and the cross-section of the shear area.

A convenient method for the determination of both shear rigidity and flexural rigidity is described in reference 4.2. In this test load/deflection curves are determined for a number of different spans. The data are then plotted as a straight line such that EI and GA are determined from the gradient and the intercept.

4.5.2 Design procedures

Determine the maximum uniformly distributed design force that can be safely applied to the glass FRP section shown below. It is manufactured from E-glass random mat plus roving and polyester resin, by the pultrusion process. This example also includes strength in bending, 4.5.3.

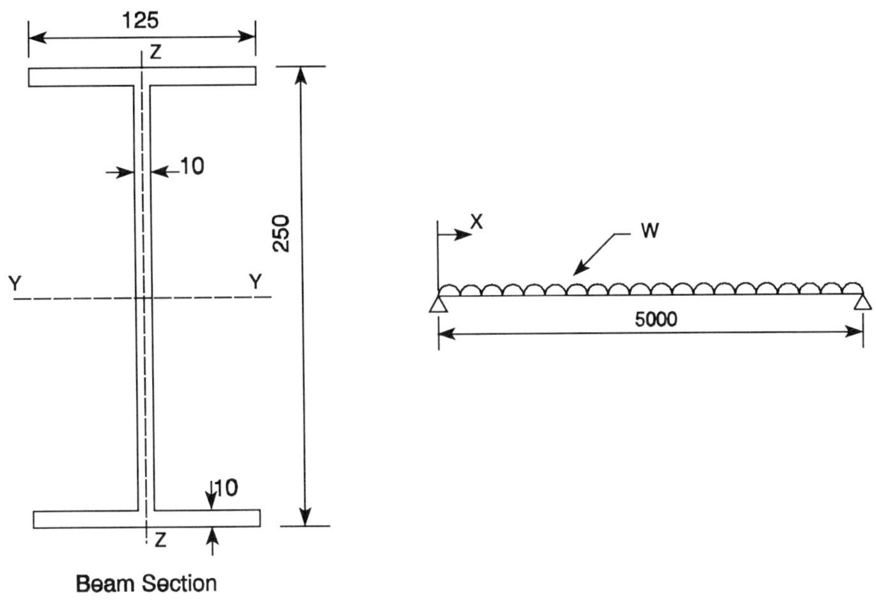

Figure 4.3 Example of uniformly loaded beam.

Assume that the beam is simply supported at its ends, is laterally supported and the deflection limit is set at span/200.

Characteristic material properties (see EUROCOMP Design Code Table 4.12).

Longitudinal tensile modulus,	$E_{x,t,k}$	=	17.2 kN/mm^2
Longitudinal compressive modulus,	$E_{x,c,k}$	=	17.2 kN/mm^2
Transverse tensile modulus,	$E_{y,t,k}$	=	5.5 kN/mm^2
Transverse compressive modulus,	$E_{y,c,k}$	=	5.5 kN/mm^2
Ultimate tensile strength, long.,	$f_{x,t,k}$	=	207 N/mm^2
Ultimate compressive strength, long.,	$f_{x,c,k}$	=	207 N/mm^2
Ultimate tensile strength, trans.,	$f_{y,t,k}$	=	48 N/mm^2
Ultimate compressive strength, trans.,	$f_{y,c,k}$	=	103 N/mm^2
Longitudinal shear modulus,	$G_{xy,k}$	=	2.9 kN/mm^2
Longitudinal shear strength,	$\tau_{xy,k}$	=	31 N/mm^2
Poisson's ratio (x, y direction),	$\nu_{xy,k}$	=	0.33
Poisson's ratio (y, x direction),	$\nu_{yx,k}$	=	0.11

Section Properties

Second moment of area (Y - Y axis),	I_{YY}	=	4.62 x 10^7 mm^4
Second moment of area (Z - Z axis),	I_{ZZ}	=	3.27 x 10^6 mm^4
Section modulus (Y - Y axis),	W_{YY}	=	3.69 x 10^5 mm^3
Section modulus (Z - Z axis),	W_{ZZ}	=	5.24 x 10^4 mm^3
Effective polar moment of area,	J_{eff}	=	1.6 x 10^5 mm^4
Area,	A	=	4800 mm^2
Area of web,	A_w	=	2300 mm^2

The following limit states will be considered:

Serviceability (SLS) :
- Deformation : due to bending
- Deformation : due to shear

Ultimate (ULS) :
- Flexural strength
- Shear strength
- Bearing strength
- Web buckling due to flexure
- Web buckling due to shear
- Web buckling due to flexure and shear
- Web crippling
- Compression flange buckling
- Lateral torsional buckling

Partial coefficients			ULS	SLS
Partial load coefficient	γ_f		2.5	1.0
Partial material coefficient	γ_m	strength	2.0	-
		stiffness	1.1	1.1

Design

1. Deformation (SLS)

Design material longitudinal modulus, $E_{x,d} = E_{x,k}/\gamma_m$
$= 17.2 \times 10^3/1.0$
$= 17.2 \times 10^3$ N/mm^2

Design material shear modulus, $G_{xy,d} = G_{xy,k}/\gamma_m$
$= 2.9 \times 10^3/1.0$
$= 2.9 \times 10^3$ N/mm^2

Maximum allowable deflection, d = span/200
= 5000/200
= 25 mm

Deflection due to bending and shear

$$d = (5WL^3/384E_{x,d}I_{YY})+(0.125WL/A_wG_{xy,d})$$

Therefore maximum allowable load

$$W = d/((5L^3/384E_{x,d}I_{YY})+(0.125L/A_wG_{xy,d}))$$

$$= 25/((5 \times 5000^3/384 \times 17.2 \times 10^3 \times 4.62 \times 10^7)+ (0.125 \times 5000/2300 \times 2.9 \times 10^3)) = 1082 \text{ N}$$

2. Flexural strength

Design material strength (tensile and compressive), $f_{x,d} = f_{x,k}/\gamma_m$
$= 207/2.0$
$= 103.5$ N/mm^2

Design load, $N_{Sd} = 8xf_{x,d}W_{YY}/L$
$= 8 \times 103.5 \times 3.69 \times 10^5/5000$
$= 61.1 \times 10^3$ N

3. Shear strength

Design material shear strength (ULS),
$\tau_{xy,d} = \tau_{xy,k}/\gamma_m$
$= 31/2.0$
$= 15.5$ N/mm^2

Design load, $N_{Sd} = 2V_{xy,d}$

where:

design shear load, $V_{xy,d} = \tau_{xy,d}A_w$
$= 15.5 \times 2300$
$= 35650$ N

Therefore

design load, $N_{Sd} = 2 \times 35650$
$= 7.13 \times 10^4$ N

4. Web buckling due to flexure (ULS)

The critical buckling stress for the web in pure bending,

$\sigma_{x,cr,b} = (k\pi^2 D_x)/(d_w{}^2 t_w)$

where:

k = 20 (simple support along flange/web junction)

d_w	=	depth of web =	230 mm
t_w	=	thickness of web =	10 mm

D_x = $E_{x,d}t^3/12(1 - \nu_{xy}\nu_{yx})$
= $17200 \times 10^3/12(1 - 0.33 \times 0.1)$
= 1490 Nm

$$\sigma_{x,cr,b} = \frac{20\pi^2\ 1490 \times 1000}{230^2 \times 10} = 556\ \mathrm{N/mm^2}$$

(EUROCOMP Design Code Equation (4.26))

5. Web buckling due to shear (ULS)

The critical buckling stress for the web in pure shear,

$\tau_{x,cr,b} = (k\pi^2 E_{x,d}t_w{}^2)/(12(1 - \nu_{xy}{}^2)d_w{}^2)$
(EUROCOMP Design Code, Equation (4.28))

where:

k = 5.35 (simple support along both edges of the web)

$\tau_{x,cr,b}$ = $(5.35 \times \pi^2 \times 5.5 \times 10^3 \times 10^2)/(12(1 - 0.33^2)230^2)$
= 51.4 $\mathrm{N/mm^2}$

6. Local buckling of compression flange

The local buckling stress for a long plate with one "free" and one "pinned" longitudinal edge is:

$\sigma_{x,cr,b} = \pi^2\{(D_x(b/a)^2)+(12D'_{xy}/\pi^2)\}/t_f b^2$
(EUROCOMP Design Code, Equation (4.38))

where:

a = half wave length of the buckle = the length of the beam = 5000 mm b = width of outstanding flange = 57.5 mm

stiffness constant, D'_{xy} = $G_{xy,d}t^3$ / 12

design shear modulus, $G_{xy,d}$ = $G_{xy,k}/\gamma_m$
= 2.9/1.0
= 29 kN/mm^2

Therefore:

D'_{xy} = $29 \times 10^3 \times 10^3/12$
= 241.7×10^3

$\sigma_{x,cr,b}$ = $\pi^2\{(1490.0 \times 10^3(57.5/5000)^2)+(12 \times 241.7 \times 10^3/\pi^2)\}/(10 \times 57.5^2)$
= 87.8 N/mm^2

From the above calculations the serviceability limit state of combined flexural and shear deflection limit the allowable load.

7. Calculate the minimum length of bearing, n, for the two end supports to resist web crippling (transverse strength and local buckling).

7.1 Strength

Minimum length of bearing, S_s = $R_u/(t_w f_{y,c,d})$
(EUROCOMP Design Code, Equation (4.32))

where:

R_u = reaction at support = $N_{sd}\gamma_f/2$
= 9800 x 2.5/2
= 12250 N

$f_{y,c,d}$ = $f_{y,c,k}/\gamma_m$
= 103/2.0
= 51.5 N/mm^2

S_s = 12250/(10 x 51.5)
= 24 mm

7.2 Buckling

Minimum length of bearing, n = $12 \times d_w^2 R_u/(\pi^2 E_{y,c,d} t_w^3)$
= $12 \times 230^2 \times 12250/(\pi^2 \times 5.0 \times 10^3 \times 10^3)$ = 158 mm

Therefore buckling dominates, use n = 158 mm (minimum),

8. Calculate the maximum distance between lateral bracing, l_b, to resist lateral torsional buckling.

The critical buckling moment is:

$$M_{b,cr} = C_1 P_{ey} \left[k \frac{I_w}{I_{zz}} + \frac{GJ}{P_{ey}} \right]^{1/2}$$

where:

buckling coefficient C_1 = 1.13

buckling coefficient k = 1.0

Euler column load for buckling in the weak direction

$P_{e,y} = \pi^2 E_{z,b,d} I_{zz}/kL^2$

L = unbraced length

Now critical buckling flexural stress ≥ design material strength

$f_{x,d}$ = 103.5 N/mm^2

Therefore solving for L_b produces a maximum unbraced length of 1350 mm. Hence, a 5000 mm long beam requires 4 braces at 5000/4 = 1250 mm centres plus one at each support.

The limiting deflections of span/150 and span/200 are values typically in current use.

The limiting deflections given of span/250 and span/500 are derived from ISO 4356 and should ensure generally satisfactory performance for buildings such as dwellings, offices, public buildings and factories. Care should be taken to ensure that there are no special circumstances which would render them inappropriate for the structure considered. Further information on problems resulting from deflections and limiting values can be obtained from ISO 4356.

4.5.3 Strength in bending

See 4.5.2 above.

4.6 MEMBERS IN SHEAR

For an example of the design method see 4.5.2 above. Otherwise no additional information.

4.7 STABILITY

For an example of the design method see 4.5.2 above. Otherwise no additional information.

4.8 COMBINATION MEMBERS

No additional information.

4.9 PLATES

4.9.0 NOTATION

No additional information.

4.9.1 General

Laminated plates form one of the basic structural elements for the use of composite materials. In this document laminated plates are defined as Plates Type I, II or III, depending on the degree of directional homogeneity of their material properties.

Analysis of isotropic thin plates, Plates Type I, is an established field in which in-plane loading and bending are analysed separately. The former is described by in-plane stress theory and the latter is examined using plate theory. This separation is possible because the two loadings are uncoupled; when they occur together the resultant is achieved by superposition.

The laminated plates discussed in this code are symmetric laminates, there is no extension/bending coupling, hence $B_{ij}=0$, and classical thin plate theory can be used. Membrane action is not considered.

The classical assumptions of thin plate theory are:

(i) the thickness of the plate is much smaller than the in-plane dimensions (minimum dimension is greater than 50 times the plate thickness),
(ii) the strains in the deformed plate are small compared to unity,
(iii) normals to the undeformed plate surface remain normal to the deformed surface,
(iv) vertical deflection does not vary through the thickness,
(v) stress normal to the plate surface is negligible.

On the basis of assumptions (iii), (iv) and (v) the displacement field in the plate can be written as a set of partial differential equations from which the stress and strain relationships can be derived.

The differential equations for the force and moment resultants, in terms of stress and strain, are then solved to obtain the maximum bending moments, twisting moments and deflections.

Where these equations cannot be solved explicitly, energy methods, such as that proposed by Galerkin and described by Whitney (reference 4.3), are employed to obtain approximate solutions.

Design tables for three different material definitions with various loading and edge support conditions are provided in this document, Tables 4.1 to 4.38. Advanced numerical analysis techniques can be used for more general conditions.

Information on the buckling behaviour of plates has been prepared by Turvey and Marshall (reference 4.4).

4.9.2 Plates Type I (isotropic plates)

Plates Type I are defined as plates whose material properties do not change at any point in the plate. Hence, for Plates Type I $D_{11} = D_{22}$ and $D_{16} = D_{26} = 0$.

Closed form solutions and design tables are available in standard text and design books (e.g. Roark & Young (reference 4.4 and 4.5) and Timoshenko (reference 4.6) for standard support conditions and various types of loading.

4.9.3 Plates Type II

A laminate in which the stacking sequence below the midplane is a mirror image of the sequence above the midplane is called a symmetric laminate. For symmetric laminates the coupling terms of its stiffness matrix, B_{ij}, vanish.

When there is no coupling between the bending and twisting coefficients of a symmetrically stacked laminate, i.e. $D_{16} = D_{26} = 0$, the laminate is referred to as orthotropic.

When bending/twisting coupling is undesirable it can be eliminated, or in most cases significantly reduced, by judicious arrangement of the laminae within the laminate. Bending/twisting coupling can only be completely eliminated by using uni-directional or cross-ply construction.

Uni-directional construction implies that all layers have the same orientation. A cross-ply construction which gives no bending/twisting coupling would consist of 0° and 90° plies only.

(In most instances these restrictions will be neither practical nor economical. However, if a symmetric laminate were constructed with many plies, and the plies with the same angular orientation were not grouped together, the magnitude of the D_{16} and D_{26} terms would reduce with respect to the remaining terms of the D matrix.)

The governing differential equation for a plate in the bending is given by:

$$D_{11}\frac{\partial^4 w}{\partial x^4} + 2(D_{12} + 2D_{66})\frac{\partial^4 w}{\partial x^2 \partial y^2} + D_{22}\frac{\partial^4 w}{\partial y^4} = q \tag{4.1}$$

4.9.3.1 Simply-supported

The following boundary conditions are applicable to a plate with side dimensions a and b, simply-supported around its four edges.

$$w(0, y) = w(a, y) = M_x(0, y) = M_x(a, y) = 0 \tag{4.2}$$

$$w(x, 0) = w(x, b) = M_y(x, 0) = M_y(x, b) = 0 \tag{4.3}$$

The moment boundary conditions take the form

$$\text{at } x = 0, a \qquad D_{11}\frac{\partial^2 w}{\partial x^2} + D_{12}\frac{\partial^2 w}{\partial y^2} = 0 \tag{4.4}$$

$$\text{at } y = 0, b \qquad D_{12}\frac{\partial^2 w}{\partial x^2} + D_{22}\frac{\partial^2 w}{\partial y^2} = 0 \tag{4.5}$$

For the cases where the transverse load can be expanded in the double Fourier sine series, the load q(x, y) is given by :

$$q(x, y) = \sum_{m=1}^{\infty}\sum_{n=1}^{\infty} q_{mn} \sin\frac{m\pi x}{a} \sin\frac{n\pi y}{b} \tag{4.6}$$

where the Fourier coefficients are determined from the relationship,

$$q_{mn} = \frac{4}{ab}\int_0^b\int_0^a q(x, y)\sin\frac{m\pi x}{a}\sin\frac{n\pi y}{b}\,dx\,dy \tag{4.7}$$

A solution to Equation (4.1) that satisfies the boundary conditions of Equations (4.2) through (4.5) is of the form

$$w = \sum_{m-1}^{\infty}\sum_{n=1}^{\infty} A_{mn} \sin\frac{m\pi x}{a} \sin\frac{n\pi y}{b} \tag{4.8}$$

By substituting Equation (4.8) into Equation (4.1) we obtain the infinite series solution

$$w = \frac{a^4}{\pi^4}\sum_{m=1}^{\infty}\sum_{n=1}^{\infty}\frac{q_{mn}}{D_{mn}} \sin\frac{m\pi x}{a} \sin\frac{n\pi y}{b} \tag{4.9}$$

where

$$D_{mn} = D_{11}m^4 + 2(D_{12} + 2D_{66})(\frac{mna}{b})^2 + D_{22}(\frac{na}{b})^4$$

Substituion of Equation (4.9) into the laminate constitutive equations gives the expressions for the moment resultants, Equations (4.10):

$$M_x = \frac{a^2}{\pi^2}\sum_{m=1}^{\infty}\sum_{n=1}^{\infty}\frac{q_{mn}}{D_{mn}}(m^2D_{11} + n^2\frac{a^2}{b^2}D_{12}) \sin\frac{m\pi x}{a} \sin\frac{n\pi y}{b}$$

$$M_y = \frac{a^2}{\pi^2}\sum_{m=1}^{\infty}\sum_{n=1}^{\infty}\frac{q_{mn}}{D_{mn}}(m^2D_{12} + n^2\frac{a^2}{b^2}D_{22}) \sin\frac{m\pi x}{a} \sin\frac{n\pi y}{b} \qquad (4.10)$$

$$M_{xy} = -2\frac{a^3/b}{\pi^2} D_{66} \sum_{m=1}^{\infty}\sum_{n=1}^{\infty}\frac{mnq_{mn}}{D_{mn}} \cos\frac{m\pi x}{a} \cos\frac{n\pi y}{b}$$

(a) Uniformly distributed loading

For the case of uniform loading, the integral in Equation (4.7) yields

$$q_{mn} = \frac{16q_0}{\pi^2 mn} \text{ (m and n odd)} \qquad (4.11)$$

$$q_{mn} = 0 \text{ (m or n even)}$$

Substitution of Equation (4.11) into Equations (4.10) give the moment resultants under this loading configuration.

(b) Triangular load

For the case of a linearly varying triangular load applied over the complete plate area, with a maximum intensity of q, the integral in Equation (4.7) yields,

$$q = \frac{q_0 x}{a}$$

(c) Patch/point load

For the case of a patch/point load, the integral in Equation (4.7) yields

$$q_{mn} = \frac{16P}{\pi^2 mn\, cd} \sin\frac{m\pi\zeta}{a} \sin\frac{m\pi\xi}{b} \sin\frac{m\pi c}{2a} \sin\frac{n\pi d}{2b} \qquad (4.12)$$

where c and d are dimensions of the rectangle parallel to the x and y directions respectively and ζ and ξ are the x and y coordinates respectively of the centre of the rectangle.

4.9.3.2 Clamped plates under uniform loading

Exact solutions for the response for these edge conditions and under general loading conditions are not available. Equations have been derived by Whitney, [4.3] using energy methods.

(a) Uniformly distributed loading

For the case of uniformly distributed loading, the loading integral is given in 4.9.3.1 (a).

Using polynomial shape functions, the deflection of the plate may be expressed as:

$$w = \frac{6.125q_0(x^2 - ax)^2(y^2 - by)^2}{7D_{11}b^4 + 4(D_{12} + 2D_{66})a^2b^2 + 7D_{22}a^4} \qquad (4.13)$$

The deflection at the centre of the plate is given by :-

$$w_{max} = 0.00342\frac{q_0a^4}{D_{11} + 0.571(D_{12} + 2D_{66})(a/b)^2 + D_{22}(a/b)^4} \qquad (4.14)$$

The moments can be calculated using the following equations:-

$$M_x = -(D_{11}\frac{\partial^2 w}{\partial x^2} + D_{12}\frac{\partial^2 w}{\partial y^2})$$

$$M_y = -(D_{12}\frac{\partial^2 w}{\partial x^2} + D_{22}\frac{\partial^2 w}{\partial y^2})$$

$$M_{xy} = -2D_{66}\frac{\partial^2 w}{\partial x \partial y}$$

where D_{ij} are the respective terms of the bending submatrix of the plate stiffness matrix (see 4.10 *Laminate Design*).

The maximum moment occurs at the middle of each edge and is given by

$$M_{max} = -0.513q_oa^2 \qquad (4.15)$$

(b) Triangular loading

The loading for a linearly varying triangular load distribution is given in 4.9.3.1 (b) The tables in the EUROCOMP Design Code were produced by solving Equation (4.1) using the Galerkin method (reference 4.3) and a total of ten harmonics.

(c) Patch/point loading

The loading for a patch/point load is given in 4.9.3.1 (c). The tables in the EUROCOMP Design Code were produced by solving Equation (4.1) using the Galerkin method, (reference 4.3).

4.9.4 Plates Type III

Plates Type III are defined as symmetrically laminated plates whose material properties are different in all directions with respect to the axis of the plate. This class of laminates includes plates in which bending-twisting coupling (non-zero D_{16} and D_{26} terms) exists.

The general equation for plates Type III under transverse loading is,

$$D_{11}\frac{\delta^4 w}{\delta x^4} + 4D_{16}\frac{\delta^4 w}{\delta_x^3 \delta y} + 2(D_{12} + 2D_{66})\frac{\delta^4 w}{\delta^2 x \delta y^2} + 4D_{26}\frac{\delta^4 w}{\delta x \delta y^3} + D_{22}\frac{\delta^4 w}{\delta y^4} = q \qquad (4.16)$$

4.9.4.1 Uniformly distributed loading

The loading for a uniformly distributed load is given in Equation (4.11). The tables in the EUROCOMP Design Code have been produced by solving Equation (4.16) using the Galerkin method (reference 4.3) and appropriate boundary conditions.

4.9.5 Shells

An axisymmetrically loaded cylindrical shell shall be defined as a cylinder subjected to a uniform internal or external pressure. (The notation is consistent with the notation used for laminate design.)

The general solution for lateral displacement, w(x), of a composite cylindrical shell (length L, radius R, wall thickness h), under axially symmetric loads can be expressed as:

$$w(x) = \frac{M_0}{2\alpha^2 D_{11}} e^{-\alpha x}(\sin \alpha x - \cos \alpha x) - \frac{Q_0}{2\alpha^3 D_{11}} e^{-\alpha x}\cos \alpha x$$

$$+ \frac{M_L}{2\alpha^2 D_{11}} e^{-\alpha(L-x)}[\sin \alpha(L - x) - \cos \alpha(L - x)]$$

$$+ \frac{Q_L}{2\alpha^3 D_{11}} e^{-\alpha(L-x)}\cos \alpha(L - x) + \frac{1}{4\alpha^4 D_{11}} p(x)$$

where

$$\alpha^4 = \frac{3(1 - \nu_{\theta x}\nu_{x\theta})}{R^2 h^2}\frac{D_{22}}{D_{11}}$$

and

$$\frac{dw}{dx} = \frac{M_0}{\alpha D_{11}} e^{-\alpha x} \cos \alpha x + \frac{Q_0}{2\alpha^2 D_{11}} e^{-\alpha x}(\sin \alpha x + \cos \alpha x)$$

$$- \frac{M_L}{\alpha D_{11}} e^{-\alpha(L-x)} \cos \alpha(L - x)$$

$$+ \frac{Q_L}{2\alpha^2 D_{11}} e^{-\alpha(L-x)}[\sin \alpha(L - x) + \cos \alpha(L - x)]$$

$$+ \frac{1}{4\alpha^4 D_{11}} \frac{dp(x)}{dx}$$

The terms D_{11} and D_{22} should be calculated from the individual lamina properties and similarly $\nu_{x\theta}$ and $\nu_{\theta x}$ should be taken as $-d_{12}/d_{11}$ and $-d_{12}/d_{22}$ respectively.

The constants M_0, M_L, Q_0 and Q_L are constants of integration determined from the boundary conditions. (Either $w_{0,L}$ are prescribed or $Q_{0,L}$ are zero and either $(dw/dx)_{0,L}$ are prescribed or $M_{0,L}$ are zero.)

Then using

$$M_x = -D_{11}\frac{d^2w}{dx^2} = M_0 e^{-\alpha x}(\sin \alpha x + \cos \alpha x)$$

$$+ \frac{Q_0}{\alpha} e^{-\alpha x} \sin \alpha x$$

$$+ M_L e^{-\alpha(L-x)}[\sin \alpha(L - x) + \cos \alpha(L - x)]$$

$$- \frac{Q_L}{\alpha} e^{-\alpha(L-x)} \sin \alpha(L-x) - \frac{1}{4\alpha^4} \frac{d^2p(x)}{dx^2}$$

and

$$M_\theta = \nu_{\theta x} M_x$$

and,

$$Q_x = -D_{11} \frac{d^3w}{dx^2} = -2M_0 \alpha e^{-\alpha x} \sin \alpha x + Q_0 e^{-\alpha x}(\cos \alpha x - \sin \alpha x)$$

$$+ 2M_L \alpha e^{-\alpha(L-x)} \sin \alpha(L - x)$$

$$- Q_L e^{-\alpha(L - x)}[-\cos\alpha(L - x) + \sin\alpha(L - x)]$$

$$- \frac{1}{4\alpha^4} \frac{d^3p(x)}{dx^3}$$

the force and moment stress resultants are calculated.

The above force and moment stress resultants can then be used to determine the stresses at locations h_k above and below the mid-surface of the cylinder thickness. The stress equations are:

$$\sigma_x = \frac{M_x h_k}{h^3/12}, \quad \sigma_\theta = \frac{M_\theta h_k}{h^3/12}$$

$$\sigma_{xz} = \frac{3Q_x}{2h}\left[1 + \left(\frac{h_k}{h/2}\right)^2\right]$$

These stresses and equivalent elastic properties, calculated by classical lamination theory, can then be used in the following equations to calculate the strains in the shell:

$$\varepsilon_x = \frac{1}{E_x}(\sigma_x - \nu_{\theta x}\sigma_x)$$

$$\varepsilon_\theta = \frac{1}{E_\theta}(\sigma_\theta - \nu_{\theta x}\sigma_x)$$

$$\varepsilon_{x\theta} = \frac{1}{2G_{x\theta}}\sigma_{x\theta}$$

(Note: E_x, E_θ and $G_{x\theta}$ are equivalent global properties (see 4.9.2 if the shell comprises several laminae, or 4.9.1 for a single ply shell). For the shell reference system the 'θ' direction is analogous to the '2' direction for uni-directional laminae.)

If the Hart-Smith failure envelope is being used, the failure criterion can be evaluated at this stage.

In the event that the Tsai-Wu technique will be implemented the respective lamina stresses should be calculated from the above strains, as explained in 4.10.4.

4.9.6 EUROCOMP plate tables: worked examples

Eight illustrative worked examples, not necessarily using glass fibres, are given below and are compared with results using Roark (reference 4.5) and finite element analysis in Table 4.39.

Example 1

Plate dimensions $a = b = 1.0$ m

Ply lay-up 4 plies, assumed isotropic

Ply thickness $t = 2.5$ mm

Elastic ply properties $E_1 = E_2 = 38$ kN/mm^2
$\nu = 0.3$

Uniformly distributed load (UDL) $q_o = 1.0$ kN/m^2

Edge conditions Simply supported on all sides - S

STEP 1 Using Section 4.9 of EUROCOMP Design Code determine D_{ij},

$D_{11} = D_{22} = 3.48$ kNm
$D_{16} = D_{26} = 0.0$

Therefore, plate is PLATE TYPE I

STEP 2 Calculate 'D' ratios,

$D_{11}/D_{22} = 1.0$

STEP 3 Identify appropriate EUROCOMP Design Code coefficient table,

Plate Type 1
$a/b = 1.0$
Uniformly distributed loading
Simply supported on all sides

Therefore, TABLE 4.1

STEP 4 Determine deflection and moment coefficients, using interpolation if necessary, and use Equations (4.46)-(4.48) of the EUROCOMP Design Code to calculate deflections and moments,

$\alpha = 0.00406$
$\beta_x = 0.04792$
$\beta_y = 0.04792$

Deflection $w = \alpha\, q_o\, b\, a^3 / D_{22}$
Moments $M_x = \beta_x\, q_o\, a^2$
$M_y = \beta_y\, q_o\, a^2$

giving $w = 1.17 \times 10^{-3}$ m
$M_x = 0.048$ kNm
$M_y = 0.048$ kNm

Example 2

Plate dimensions	a	=	b	= 1.0 m
Ply lay-up	4 plies, assumed Isotropic			
Ply thickness	t	=	2.5 mm	
Elastic ply properties	E_1	=	E_2	= 38 kN/mm^2
	ν	=	0.3	
Triangular load	TL	at y = 0	q_o	= 0.0 kN/m^2
		at y = 1.0m	q_o	= 1.0 kN/m^2
Edge conditions	Clamped supports on all sides - C			

STEP 1 Using Section 4.9 of EUROCOMP Design Code determine D_{ij},

D_{11} = D_{22} = 3.48 kNm

D_{16} = D_{26} = 0.0

Therefore plate is PLATE TYPE I

STEP 2 Calculate 'D' ratios,

D_{11}/D_{22} = 1.0

STEP 3 Identify appropriate EUROCOMP Design Code coefficient table,

Plate Type I

a/b = 1.0

Triangular loading

Clamped supports on all sides

Therefore, TABLE 4.2

STEP 4 Determine deflection and moment coefficients, using interpolation if necessary, and use Equations (4.46)-(4.48) of the EUROCOMP Design Code to calculate deflections and moments

α = 0.000693

β_x = 0.0224

β_y = 0.0294

giving

w = 1.99 x 10^{-4} m

M_x = 0.022 kNm

M_y = 0.029 kNm

Example 3

Plate dimensions	a	=	b	= 1.0 m
Ply lay-up	4 plies, assumed isotropic			
Ply thickness	t	=	2.5 mm	
Elastic ply properties	E_1	=	E_2	= 38 kN/mm^2
	ν	=	0.3	
Patch load	PL	over central area a/10 x b/10		
	q_o	=	1.0 kN/m^2	
Edge conditions	Clamped supports on all sides - C			

STEP 1 Using Section 4.9 of EUROCOMP Design Code determine D_{ij},

D_{11} = D_{22} = 3.48 kNm

D_{16} = D_{26} = 0.0

Therefore plate is PLATE TYPE I

STEP 2 Calculate 'D' ratios,

D_{11}/D_{22} = 1.0

STEP 3 Identify appropriate EUROCOMP Design Code coefficient Table,

Plate Type I

a/b = 1.0

Patch loading

Clamped supports on all sides

Therefore, TABLE 4.2

STEP 4 Determine deflection and moment coefficients, using interpolation if necessary, and use Equations (4.46)-(4.48) of the EUROCOMP Design Code to calculate deflections and moments

α = 0.000053

β_x = 0.00194

β_y = 0.00194

giving w = 1.52×10^{-5} m

M_x = 1.94×10^{-3} kNm

M_y = 1.94×10^{-3} kNm

Example 4

Plate dimensions	a	=	1.0 m
	b	=	2.0 m
Ply lay-up	$[0°]_4$		(orthotropic)
Ply thickness	t	=	2.5 mm
Elastic ply properties	E_1	=	50 kN/mm^2 E_2 = 30kN/mm^2
	ν_{12}	=	0.3 ν_{21} = 0.18
	G_{12}	=	15.31 N/mm^2 (assume $G_{13} = G_{23} = G_{12}$)

Uniformly distributed load (UDL) q_o = 1.0 kN/m^2

Edge conditions Simply supported on all sides - S

STEP 1 Using Section 4.9 of EUROCOMP Design Code determine D_{ij},

D_{11} = 4.405 kNm
D_{22} = 2.643 kNm
D_{12} = 0.793 kNm
D_{66} = 1.276 kNm
D_{16} = D_{26} = 0.0

Therefore plate is PLATE TYPE II

STEP 2 Calculate 'D' ratios,

D_{11}/D_{22} = 1.67
D_{12}/D_{11} = 0.18
D_{66}/D_{22} = 0.48

STEP 3 Identify appropriate EUROCOMP Design Code coefficient table,

Plate Type II
a/b = 0.5
Uniformly distributed loading
Simply supported on all sides

Therefore, TABLE 4.4

STEP 4 Determine deflection and moment coefficients, using interpolation if necessary, and use Equations (4.46)-(4.48) of the EUROCOMP Design Code to calculate deflections and moments

α = 0.0039
β_x = 0.1039
β_y = 0.0309

giving w = 2.9 x 10^{-3} m
M_x = 0.104 kNm
M_y = 0.03 kNm

Example 5

Plate dimensions	a	=	1.0 m
	b	=	2.0 m
Ply lay-up	$[0°]_4$		(orthotropic)
Ply thickness	t	=	2.5 mm
Elastic ply properties	E_1	=	50 kN/mm^2 E_2 = 30kN/mm^2
	ν_{12}	=	0.3 ν_{21}= 0.18
	G_{12}	=	15.31 N/mm^2 (assume G_{13} = G_{23} = G_{12})
Triangular load	TL	at y = 0	q_o = 0.0 kN/m^2
		at y = 2.0 m	q_o = 2.0 kN/m^2
Edge conditions	Clamped supports on all sides - C		

STEP 1 Using Section 4.9 of EUROCOMP Design Code determine D_{ij},

D_{11} = 4.405 kNm
D_{22} = 2.643 kNm
D_{12} = 0.793 kNm
D_{66} = 1.276 kNm
D_{16} = D_{26} = 0.0

Therefore plate is PLATE TYPE II

STEP 2 Calculate 'D' ratios,

D_{11}/D_{22} = 1.67
D_{12}/D_{11} = 0.18
D_{66}/D_{22} = 0.48

STEP 3 Identify appropriate EUROCOMP Design Code coefficient table,

Plate Type II
a/b = 0.5
Triangular Loading
Clamped supports on all sides

Therefore, TABLE 4.24

STEP 4 Determine deflection and moment coefficients, using interpolation if necessary, and use Equations (4.46)-(4.48) of the EUROCOMP Design Code to calculate deflections and moments

α = 0.00061
β_x = 0.07236
β_y = 0.01887

giving
w = 9.3 x 10^{-4} m
M_x = 0.144 kNm
M_y = 0.04 kNm

Example 6

Plate dimensions	a	=	1.0 m
	b	=	2.0 m
Ply lay-up	$[0°]_4$		(orthotropic)
Ply thickness	t	=	2.5 mm
Elastic ply properties	E_1	=	50 kN/mm² E_2 = 30kN/mm²
	ν_{12}	=	0.3 ν_{21} = 0.18
	G_{12}	=	15.31 N/mm² (assume $G_{13} = G_{23} = G_{12}$)
Patch load	PL	over central area a/10 x b/10	
	q_o	=	1.0 kN/m²
Edge conditions	Simply supported on all sides - S		

STEP 1 Using Section 4.9 of EUROCOMP Design Code determine D_{ij},

D_{11} = 4.405 kNm
D_{22} = 2.643 kNm
D_{12} = 0.793 kNm
D_{66} = 1.276 kNm
D_{16} = D_{26} = 0.0

Therefore plate is PLATE TYPE II

STEP 2 Calculate 'D' ratios,

D_{11}/D_{22} = 1.67
D_{12}/D_{11} = 0.18
D_{66}/D_{22} = 0.48

STEP 3 Identify appropriate EUROCOMP Design Code coefficient table,

Plate Type II
a/b = 0.5
Patch loading
Simply supported on all sides

Therefore, TABLE 4.8

STEP 4 Determine deflection and moment coefficients, using interpolation if necessary, and use Equations (4.46)-(4.48) of the EUROCOMP Design Code to calculate deflections and moments

α = 0.000126
β_x = 0.006143
β_y = 0.00353

giving

w = 9.52×10^{-5} m
M_x = 6.10×10^{-3} kNm
M_y = 3.50×10^{-3} kNm

Example 7

Plate dimensions	a	=	1.0 m
	b	=	2.0 m
Ply lay-up	$[0°]_4$		(orthotropic)
Ply thickness	t	=	2.5 mm
Elastic ply properties	E_1	=	50 kN/mm^2 E_2= 30kN/mm^2
	ν_{12}	=	0.3 ν_{21} = 0.18
	G_{12}	=	15.31 N/mm^2 (assume $G_{13} = G_{23} = G_{12}$)
Patch load	PL	over central area a/10 x b/10	
	q_o	=	1.0 kN/m^2
Edge conditions	Clamped supports on all sides - S		

STEP 1 Using Section 4.9 of EUROCOMP Design Code determine D_{ij},

D_{11} = 4.405 kNm
D_{22} = 2.643 kNm
D_{12} = 0.793 kNm
D_{66} = 1.276 kNm
D_{16} = D_{26} = 0.0

Therefore plate is PLATE TYPE II

STEP 2 Calculate 'D' ratios,

D_{11}/D_{22} = 1.67
D_{12}/D_{11} = 0.18
D_{66}/D_{22} = 0.48

STEP 3 Identify appropriate EUROCOMP Design Code coefficient table,

Plate Type II
a/b = 0.5
Patch loading
Clamped supports on all sides

Therefore, TABLE 4.20

STEP 4 Determine deflection and moment coefficients, using interpolation if necessary, and use Equations (4.46)-(4.48) of the EUROCOMP Design Code to calculate deflections and moments

α = 0.000045
β_x = 0.005609
β_y = 0.001502

giving w = 3.39×10^{-5} m
M_x = 5.60×10^{-3} kNm
M_y = 1.50×10^{-3} kNm

Example 8

Plate dimensions	a	=	3.0 m
	b	=	4.0 m
Ply lay-up	[90°/+45°/−45°/0°/90°/0°/−45°/45°/90°]		

Ply thickness	t (0°, 90°) = 1.0 mm t (+45°,−45°) = 4.0 mm		
Elastic ply properties	E_1	=	50 kN/mm² E_2 =30 kN/mm²
	ν_{12}	=	0.3 ν_{21} = 0.18
	G_{12}	=	15.31 N/mm² (assume $G_{13} = G_{23} = G_{12}$)

Uniformly distributed load (UDL) q_o = 1.0 kN/m²

Edge conditions Simply supported on all sides - S

STEP 1 Using Section 4.9 of EUROCOMP Design Code determine D_{ij},

D_{11} = 29.93 kNm
D_{22} = 34.11 kNm
D_{12} = 7.95 kNm
D_{66} = 12.43 kNm
D_{16} = D_{26} = 2.74 kNm

Therefore plate is PLATE TYPE III

STEP 2 Calculate 'D' ratios,

D_{11}/D_{22} = 0.88
D_{12}/D_{11} = 0.27
D_{66}/D_{22} = 0.36
D_{16}/D_{11} = 0.06 (= D_{26}/D_{11})

STEP 3 Identify appropriate EUROCOMP Design Code coefficient tables,

Plate Type III
a/b = 0.75
Uniformly distributed loading
Simply supported on all sides,

Therefore interpolate using Table 4.5 (D_{16}/D_{11} = 0.0) and Table 4.29 (D_{16}/D_{11} = 0.35) and then between the two sets of results

STEP 4 Determine deflection and moment coefficients, using interpolation if necessary, and use Equations (4.46)-(4.48) of the EUROCOMP Design Code to calculate deflections and moments

α = 0.00617
β_x = 0.0605
β_y = 0.0544

giving
w = 0.0194 m
M_x = 0.55 kNm
M_y = 0.49 kNm

Table 4.1 Plate design coefficients.
Support conditions: simple. Loading: various loads. Materials: plate type I.

D_{11}/D_{22} = 1.0 a/b = 0.25/0.5/0.75/1.0

a/b		0.25		0.5		0.75		1.0	
Poisson's ratio		0.15	0.3	0.15	0.3	0.15	0.3	0.15	0.3
Uniform load	α/0.01	0.321	0.321	0.506	0.506	0.497	0.497	0.406	0.406
	β_x/0.01	12.3	12.3	9.91	10.1	6.67	7.15	4.23	4.79
	β_y/0.01	2.03	3.87	3.19	4.64	4.10	5.03	4.23	4.79
Patch Load	α/0.01	0.0158	0.0158	0.0160	0.0160	0.0142	0.0142	0.0114	0.0114
	β_x/0.01	0.939	0.996	0.540	0.584	0.357	0.395	0.250	0.283
	β_y/0.01	0.509	0.642	0.370	0.444	0.298	0.346	0.250	0.283
Triangular load	α/0.01	0.208	0.208	0.272	0.272	0.254	0.254	0.205	0.205
	β_x/0.01	8.08	8.21	5.36	5.60	3.43	3.73	2.14	2.45
	β_y/0.01	3.12	3.97	2.880	3.52	2.69	3.10	2.44	2.69

Table 4.2 Plate design coefficients.
Support conditions: clamped. Loading: various loads. Materials: plate type I.

D_{11}/D_{22} = 1.0 a/b = 0.25/0.5/0.75/1.0

a/b		0.25		0.5		0.75		1.0	
Poission's Ratio		0.15	0.13	0.15	0.3	0.15	0.3	0.15	0.3
Uniform load	α/0.01	0.402	0.0497	0.125	0.119	0.14	0.146	0.126	0.126
	β_x/0.01	11.1	-12.0	-9.62	-9.62	-6.61	-5.36	-4.58	-4.60
	β_y/0.01	-11.4	-5.81	-3.54	-3.54	-2.11	-5.47	-4.58	-4.60
Patch load	α/0.01	0.0068	0.0061	0.0068	0.0067	0.0066	0.0065	0.0053	0.0053
	β_x/0.013	-0.729	-0.673	-0.406	-0.393	0.332	-0.310	-0.201	0.194
	β_y/0.013	0.307	0.359	-0.169	0.297	0.291	0.255	-0.201	0.194
Triangular load	α/0.01	0.106	0.0487	0.0772	0.0773	0.0887	0.0849	0.0693	0.0693
	β_x/0.01	-12.8	-5.95	-4.51	-4.52	-3.28	-3.36	-2.21	-2.24
	β_y/0.01	-20.8	-6.95	-2.30	-2.26	-1.13	-3.34	-2.99	-2.90

Table 4.3 Plate design coefficients.
Support conditions: simple. Loading: uniform. Materials: plate type II.

$D_{16}/D_{11} = D_{26}/D_{11} = 0.0$ a/b = 0.25

D_{66}/D_{22}	D_{11}/D_{22}	0.5		2.0		4.0		10.0	
	D_{12}/D_{11}	0.15	0.3	0.15	0.3	0.15	0.3	0.15	0.3
0.1	$\alpha/0.01$	0.665	0.659	0.163	0.163	0.081	0.081	0.032	0.032
	$\beta_x/0.01$	12.7	12.7	12.5	12.5	12.5	12.5	12.5	12.6
	$\beta_y/0.01$	2.06	4.04	1.88	3.79	1.90	3.78	1.91	3.78
0.3	$\alpha/0.01$	0.63	0.622	0.163	0.162	0.081	0.081	0.032	0.032
	$\beta_x/0.01$	12.1	12.0	12.5	12.5	12.5	12.5	12.5	12.5
	$\beta_y/0.01$	2.32	4.12	1.91	3.80	1.90	3.78	1.90	3.78
0.6	$\alpha/0.01$	0.572	0.565	0.161	0.160	0.081	0.081	0.0326	0.0325
	$\beta_x/0.01$	11.0	10.9	12.4	12.3	12.5	12.5	12.5	12.5
	$\beta_y/0.01$	2.33	3.95	1.94	3.79	1.91	3.78	1.907	3.77
1.0	$\alpha/0.01$	0.505	0.499	0.158	0.156	0.080	0.080	0.0325	0.032
	$\beta_x/0.01$	9.68	9.63	12.1	12.0	12.4	12.4	12.5	12.5
	$\beta_y/0.01$	2.19	3.60	1.94	3.73	1.91	3.76	1.90	3.77

Table 4.4 Plate design coefficients.
Support conditions: simple. Loading: uniform. Materials: plate type II.

$D_{16}/D_{11} = D_{26}/D_{11} = 0.0$ a/b = 0.5

D_{66}/D_{22}	D_{11}/D_{22}	0.5		2.0		4.0		10.0	
	D_{12}/D11	0.15	0.3	0.15	0.3	0.15	0.3	0.15	0.3
0.1	$\alpha/0.01$	1.14	1.08	0.331	0.313	0.166	0.158	0.06571	0.0631
	$\beta_x/0.01$	11.2	11.0	12.9	12.5	12.9	12.5	12.6	12.4
	$\beta_y/0.01$	6.06	7.311	2.68	4.416	2.15	3.95	1.93	3.73
0.3	$\alpha/0.01$	0.890	0.854	0.307	0.291	0.161	0.152	0.651	0.0624
	$\beta_x/0.01$	8.69	8.58	12.0	11.6	12.5	12.1	12.5	12.2
	$\beta_y/0.01$	4.71	5.73	2.58	4.17	2.14	3.85	1.934	3.69
0.6	$\alpha/0.01$	0.666	0.646	0.276	0.263	0.152	0.145	0.0640	0.0613
	$\beta_x/0.01$	6.43	6.42	10.8	10.5	11.9	11.5	12.4	12.0
	$\beta_y/0.01$	3.53	4.33	2.40	3.82	2.09	3.70	1.92	3.64
1.0	$\alpha/0.01$	0.498	0.487	0.243	0.233	0.142	0.136	0.0626	0.0597
	$\beta_x/0.01$	4.75	4.78	9.51	9.32	11.1	10.8	12.1	11.8
	$\beta_y/0.01$	2.64	3.26	2.17	3.42	2.00	3.49	1.90	3.56

Table 4.5 Plate design coefficients.
Support conditions: simple. Loading: uniform. Materials: plate type II.

$D_{16}/D_{11} = D_{26}/D_{11} = 0.0$ a/b = 0.75

D_{66}/D_{22}	D_{11}/D_{22}	0.5		2.0		4.0		10.0	
	D_{12}/D_{11}	0.15	0.3	0.15	0.3	0.15	0.3	0.15	0.3
0.1	$\alpha/0.01$	1.08	1.0108	0.419	0.375	0.228	0.202	0.0946	0.0838
	$\beta_x/0.01$	7.35	7.32	11.4	10.8	12.3	11.5	12.6	11.8
	$\beta_y/0.01$	8.43	8.75	4.13	5.11	2.94	4.16	2.18	3.57
0.3	$\alpha/0.01$	0.773	0.733	0.362	0.329	0.210	0.188	0.091	0.0814
	$\beta_x/0.01$	5.13	5.226	9.84	9.49	11.3	10.7	12.2	11.4
	$\beta_y/0.01$	5.87	6.232	3.55	4.46	2.72	3.87	2.12	3.47
0.6	$\alpha/0.01$	0.538	0.518	0.301	0.277	0.188	0.170	0.087	0.0779
	$\beta_x/0.01$	3.50	3.62	8.11	7.95	10.1	9.71	11.6	10.9
	$\beta_y/0.01$	4.00	4.31	2.93	3.74	2.44	3.50	2.04	3.33
1.0	$\alpha/0.01$	0.382	0.372	0.245	0.229	0.164	0.150	0.082	0.0736
	$\beta_x/0.01$	2.44	2.55	6.55	6.52	8.86	8.59	10.9	10.4
	$\beta_y/0.01$	2.79	3.04	2.38	3.08	2.15	3.11	1.93	3.15

Table 4.6 Plate design coefficients.
Support conditions: simple. Loading: uniform. Materials: plate type II.

$D_{16}/D_{11} = D_{26}/D_{11} = 0.0$ a/b = 1.0

D_{66}/D_{22}	D_{11}/D_{22}	0.5		2.0		4.0		10.0	
	D_{12}/D_{11}	0.15	0.3	0.15	0.3	0.15	0.3	0.15	0.3
0.1	$\alpha/0.01$	0.795	0.740	00.4079	0.353	0.245	0.207	0.110	0.0917
	$\beta_x/0.01$	4.13	4.35	8.79	8.52	10.6	9.93	11.8	10.7
	$\beta_y/0.01$	8.03	7.94	4.82	5.14	3.48	4.09	2.41	3.301
0.3	$\alpha/0.01$	0.569	0.540	0.339	0.300	0.218	0.187	0.105	0.0877
	$\beta_x/0.01$	2.90	3.12	7.24	7.18	9.39	8.96	11.1	10.2
	$\beta_y/0.01$	5.65	5.71	3.96	4.33	3.090	3.69	2.28	3.15
0.6	$\alpha/0.01$	0.398	0.384	0.270	0.245	0.187	0.164	0.0973	0.0822
	$\beta_x/0.01$	1.98	2.17	5.71	5.80	8.01	7.80	10.3	9.64
	$\beta_y/0.01$	3.88	3.98	3.12	3.49	2.63	3.21	2.11	2.95
1.0	$\alpha/0.01$	0.284	0.276	0.212	0.196	0.158	0.141	0.0887	0.0760
	$\beta_x/0.01$	1.39	1.53	4.43	4.59	6.69	6.64	9.40	8.89
	$\beta_y/0.01$	2.71	2.81	2.42	2.76	2.20	2.74	1.92	2.72

Table 4.7 Plate design coefficients.
Support conditions: simple. Loading: patch. Materials: plate type II.

$D_{16}/D_{11} = D_{26}/D_{11} = 0.0$ a/b = 0.25

D_{66}/D_{22}	D_{11}/D_{22}	0.5		2.0		4.0		10.0	
	D_{12}/D_{11}	0.15	0.3	0.15	0.3	0.15	0.3	0.15	0.3
0.1	$\alpha/0.01$	0.0322	0.0311	0.0111	0.0103	0.00632	0.00575	0.00289	0.00255
	$\beta_x/0.01$	0.958	0.973	1.29	1.31	1.47	1.48	1.69	1.68
	$\beta_y/0.01$	0.794	0.889	0.546	0.665	0.457	0.587	0.371	0.517
0.3	$\alpha/0.01$	0.0272	0.0265	0.0101	0.00955	0.00592	0.00546	0.00278	0.00248
	$\beta_x/0.01$	0.819	0.836	1.18	1.213	1.38	1.40	1.63	1.633
	$\beta_y/0.01$	0.634	0.726	0.484	0.603	0.421	0.552	0.355	0.501
0.6	$\alpha/0.01$	0.0226	0.0221	0.00906	0.00864	0.00546	0.00509	0.00265	0.00238
	$\beta_x/0.01$	0.690	0.707	1.07	1.09	1.28	1.30	1.55	1.56
	$\beta_y/0.01$	0.496	0.581	0.417	0.534	0.378	0.509	0.334	0.481
1.0	$\alpha/0.01$	0.0187	0.0184	0.00805	0.00775	0.00498	0.00469	0.00249	0.00227
	$\beta_x/0.01$	0.583	0.597	0.959	0.989	1.17	1.20	1.46	1.49
	$\beta_y/0.01$	0.392	0.467	0.356	0.468	0.336	0.464	0.311	0.457

Table 4.8 Plate design coefficients.
Support conditions: simple. Loading: patch. Materials: plate type II.

$D_{16}/D_{11} = D_{26}/D_{11} = 0.0$ a/b = 0.5

D_{66}/D_{22}	D_{11}/D_{22}	0.5		2.0		4.0		10.0	
	D_{12}/D_{11}	0.15	0.3	0.15	0.3	0.15	0.3	0.15	0.3
0.1	$\alpha/0.01$	0.0339	0.0322	0.119	0.0110	0.00685	0.00622	0.00322	0.00282
	$\beta_x/0.01$	0.557	0.568	0.789	0.813	0.922	0.952	1.11	0.150
	$\beta_y/0.01$	0.566	0.616	0.393	0.459	0.334	0.406	0.275	0.356
0.3	$\alpha/0.01$	0.0265	0.0255	0.0107	0.0100	0.00641	0.00587	0.00310	0.00274
	$\beta_x/0.01$	0.452	0.466	0.718	0.745	0.864	0.899	1.07	1.11
	$\beta_y/0.01$	0.458	0.504	0.351	0.417	0.309	0.382	0.263	0.344
0.6	$\alpha/0.01$	0.0201	0.019	0.00947	0.00894	0.00587	0.00543	0.00294	0.00262
	$\beta_x/0.01$	0.358	0.373	0.637	0.666	0.793	0.832	1.017	1.06
	$\beta_y/0.01$	0.362	0.402	0.305	0.368	0.278	0.351	0.247	0.329
1.0	$\alpha/0.01$	0.0152	0.0149	0.00818	0.00779	0.00530	0.00495	0.00276	0.00249
	$\beta_x/0.01$	0.286	0.299	0.557	0.587	0.720	0.760	0.955	1.00
	$\beta_y/0.01$	0.287	0.321	0.262	0.321	0.248	0.318	0.229	0.311

Table 4.9 Plate design coefficients.
Support conditions: simple. Loading: patch. Materials: plate type II.

$D_{16}/D_{11} = D_{26}/D_{11} = 0.0$ a/b = 0.75

D_{66}/D_{22}	D_{11}/D_{22}	0.5		2.0		4.0		10.0	
	D_{12}/D_{11}	0.15	0.3	0.15	0.3	0.15	0.3	0.15	0.3
0.1	$\alpha/0.01$	0.0297	0.0277	0.0121	0.0109	0.00708	0.00627	0.00333	0.00287
	$\beta_x/0.01$	0.365	0.376	0.574	0.590	0.680	0.705	0.837	0.875
	$\beta_y/0.01$	0.473	0.496	0.317	0.359	0.265	0.314	0.217	0.272
0.3	$\alpha/0.01$	0.0216	0.0206	0.0105	0.00965	0.00651	0.00582	0.00319	0.00277
	$\beta_x/0.01$	0.284	0.298	0.509	0.531	0.631	0.661	0.804	0.846
	$\beta_y/0.01$	0.371	0.395	0.282	0.323	0.245	0.293	0.208	0.263
0.6	$\alpha/0.01$	0.0154	0.0149	0.00887	0.00821	0.00582	0.00526	0.00301	0.00263
	$\beta_x/0.01$	0.219	0.232	0.439	0.464	0.571	0.605	0.759	0.805
	$\beta_y/0.01$	0.287	0.309	0.244	0.282	0.221	0.268	0.195	0.250
1.0	$\alpha/0.01$	0.0113	0.0110	0.00731	0.00686	0.00510	0.00467	0.00279	0.00247
	$\beta_x/0.01$	0.172	0.184	0.374	0.400	0.509	0.545	0.708	0.758
	$\beta_y/0.01$	0.226	0.245	0.207	0.243	0.196	0.241	0.181	0.236

Table 4.10 Plate design coefficients.
Support conditions: simple. Loading: patch. Materials: plate type II.

$D_{16}/D_{11} = D_{26}/D_{11} = 0.0$ a/b = 1.0

D_{66}/D_{22}	D_{11}/D_{22}	0.5		2.0		4.0		10.0	
	D_{12}/D_{11}	0.15	0.3	0.15	0.3	0.15	0.3	0.15	0.3
0.1	$\alpha/0.01$	0.0218	0.0204	0.0111	0.00980	0.00693	0.00591	0.00335	0.00278
	$\beta_x/0.01$	0.243	0.257	0.433	0.449	0.534	0.555	0.673	0.706
	$\beta_y/0.01$	0.392	0.403	0.275	0.298	0.227	0.257	0.182	0.220
0.3	$\alpha/0.01$	0.0160	0.0152	0.00942	0.00842	0.00621	0.00539	0.00318	0.00266
	$\beta_x/0.01$	0.191	0.2057	0.377	0.399	0.489	0.514	0.642	0.679
	$\beta_y/0.01$	0.306	0.320	0.241	0.265	0.208	0.239	0.173	0.212
0.6	$\alpha/0.01$	0.0115	0.0111	0.00763	0.00697	0.00539	0.00476	0.00295	0.00250
	$\beta_x/0.01$	0.149	0.161	0.320	0.344	0.435	0.466	0.602	0.644
	$\beta_y/0.01$	0.237	0.250	0.205	0.229	0.186	0.216	0.162	0.201
1.0	$\alpha/0.01$	0.00842	0.00822	0.006111	0.00569	0.00458	0.00412	0.00269	0.00232
	$\beta_x/0.01$	0.117	0.127	0.269	0.294	0.382	0.415	0.556	0.603
	$\beta_y/0.01$	0.186	0.198	0.173	0.196	0.163	0.193	0.150	0.188

Table 4.11 Plate design coefficients.
Support conditions: simple. Loading: triangular. Materials: plate type II.

$D_{16}/D_{11} = D_{26}/D_{11} = 0.0$ a/b = 0.25

D_{66}/D_{22}	D_{11}/D_{22}	0.5		2.0		4.0		10.0	
	D_{12}/D_{11}	0.15	0.3	0.15	0.3	0.15	0.3	0.15	0.3
0.1	$\alpha/0.01$	0.450	0.436	0.122	0.118	0.0612	0.0593	0.0242	0.0237
	$\beta_x/0.01$	8.78	8.64	9.62	9.39	9.61	9.41	9.51	9.39
	$\beta_y/0.01$	5.41	5.96	3.29	4.09	2.60	3.51	2.02	3.0650
0.3	$\alpha/0.01$	0.386	0.377	0.116	0.113	0.060	0.0581	0.0241	0.0235
	$\beta_x/0.01$	7.46	7.39	9.15	8.96	9.41	9.22	9.47	9.33
	$\beta_y/0.01$	4.12	4.65	2.87	3.66	2.38	3.29	1.94	2.97
0.6	$\alpha/0.01$	0.323	0.317	0.109	0.106	0.058	0.0564	0.0239	0.0232
	$\beta_x/0.01$	6.18	6.15	8.54	8.40	9.11	8.937	9.382	9.23
	$\beta_y/0.01$	3.06	3.540	2.42	3.18	2.13	3.0167	1.83	2.85
1.0	$\alpha/0.01$	0.268	0.264	0.101	0.0988	0.0558	0.0542	0.0235	0.0229
	$\beta_x/0.01$	5.08	5.08	7.87	7.77	8.72	8.57	9.24	9.09
	$\beta_y/0.01$	2.30	2.70	2.028	2.73	1.88	2.72	1.72	2.71

Table 4.12 Plate design coefficients.
Support conditions: simple. Loading: triangular. Materials: plate type II.

$D_{16}/D_{11} = D_{26}/D_{11} = 0.0$ a/b = 0.5

D_{66}/D_{22}	D_{11}/D_{22}	0.5		2.0		4.0		10.0	
	D_{12}/D_{11}	0.15	0.3	0.15	0.3	0.15	0.3	0.15	0.3
0.1	$\alpha/0.01$	0.603	0.571	0.192	0.178	0.100	0.092	0.0407	0.0377
	$\beta_x/0.01$	5.95	5.87	7.77	7.48	8.15	7.77	8.23	7.84
	$\beta_y/0.01$	5.17	5.52	3.05	3.64	2.35	3.06	1.77	2.62
0.3	$\alpha/0.01$	0.464	0.444	0.174	0.163	0.0952	0.0884	0.039	0.0370
	$\beta_x/0.01$	4.52	4.51	6.99	6.79	7.69	7.38	8.05	7.68
	$\beta_y/0.01$	3.87	4.20	2.62	3.21	2.14	2.84	1.70	2.53
0.6	$\alpha/0.01$	0.344	0.333	0.153	0.144	0.0884	0.0826	0.0386	0.0359
	$\beta_x/0.01$	3.30	3.34	6.10	5.97	7.10	6.86	7.79	7.46
	$\beta_y/0.01$	2.81	3.08	2.18	2.73	1.89	2.57	1.60	2.41
1.0	$\alpha/0.01$	0.256	0.250	0.132	0.126	0.0809	0.0761	0.0371	0.0346
	$\beta_x/0.01$	2.42	2.46	5.22	5.16	6.45	6.28	7.47	7.18
	$\beta_y/0.01$	2.05	2.27	1.79	2.29	1.64	2.29	1.49	2.28

Table 4.13 Plate design coefficients.
Support conditions: simple. Loading: triangular. Materials: plate type II.

$D_{16}/D_{11} = D_{26}/D_{11} = 0.0$ a/b = 0.75

D_{66}/D_{22}	D_{11}/D_{22}	0.5		2.0		4.0		10.0	
	D_{12}/D_{11}	0.15	0.3	0.15	0.3	0.15	0.3	0.15	0.3
0.1	$\alpha/0.01$	0.54	0.511	0.218	0.195	0.121	0.107	0.0517	0.0451
	$\beta_x/0.01$	3.68	3.71	6.08	5.86	6.83	6.45	7.30	6.77
	$\beta_y/0.01$	4.87	4.96	2.95	3.27	2.26	2.70	1.66	2.23
0.3	$\alpha/0.01$	0.392	0.372	0.188	0.170	0.111	0.099	0.0497	0.0436
	$\beta_x/0.01$	2.59	2.66	5.20	5.08	6.23	5.95	7.00	6.54
	$\beta_y/0.01$	3.56	3.68	2.50	2.83	2.03	2.46	1.577	2.14
0.6	$\alpha/0.01$	0.273	0.263	0.156	0.143	0.0991	0.0893	0.0471	0.0416
	$\beta_x/0.01$	1.78	1.86	4.27	4.24	5.50	5.32	6.61	6.22
	$\beta_y/0.01$	2.54	2.65	2.04	2.34	1.76	2.18	1.46	2.0155
1.0	$\alpha/0.01$	0.195	0.189	0.126	0.118	0.0864	0.0789	0.0440	0.0392
	$\beta_x/0.01$	1.24	1.31	3.43	3.46	4.76	4.67	6.14	5.84
	$\beta_y/0.01$	1.83	1.92	1.63	1.91	1.49	1.89	1.33	1.87

Table 4.14 Plate design coefficients.
Support conditions: simple. Loading: triangular. Materials: plate type II.

$D_{16}/D_{11} = D_{26}/D_{11} = 0.0$ a/b = 1.0

D_{66}/D_{22}	D_{11}/D_{22}	0.5		2.0		4.0		10.0	
	D_{12}/D_{11}	0.15	0.3	0.15	0.3	0.15	0.3	0.15	0.3
0.1	$\alpha/0.01$	0.398	0.371	0.207	0.179	0.126	0.106	0.0581	0.0478
	$\beta_x/0.01$	2.05	2.18	4.49	4.42	5.55	5.28	6.42	5.91
	$\beta_y/0.01$	4.26	4.21	2.81	2.91	2.15	2.38	1.55	1.91
0.3	$\alpha/0.01$	0.286	0.271	0.172	0.153	0.112	0.0965	0.0550	0.0457
	$\beta_x/0.01$	1.45	1.58	3.71	3.74	4.91	4.76	6.06	5.63
	$\beta_y/0.01$	3.12	3.14	2.34	2.48	1.91	2.14	1.45	1.82
0.6	$\alpha/0.01$	0.201	0.193	0.137	0.124	0.0965	0.0845	0.0508	0.0428
	$\beta_x/0.01$	1.00	1.11	2.93	3.02	4.19	4.14	5.58	5.25
	$\beta_y/0.01$	2.23	2.27	1.87	2.02	1.62	1.87	1.33	1.69
1.0	$\alpha/0.01$	0.143	0.140	0.108	0.100	0.0812	0.0725	0.0462	0.0395
	$\beta_x/0.01$	0.709	0.794	2.28	2.40	3.49	3.52	5.04	4.83
	$\beta_y/0.01$	1.61	1.66	1.47	1.62	1.35	1.59	1.19	1.55

Table 4.15 Plate design coefficients.
Support conditions: clamped. Loading: uniform. Materials: plate type II.

$D_{16}/D_{11} = D_{26}/D_{11} = 0.0$ a/b = 0.25

D_{66}/D_{22}	D_{11}/D_{22}	0.5		2.0		4.0		10.0	
	D_{12}/D_{11}	0.15	0.3	0.15	0.3	0.15	0.3	0.15	0.3
0.1	$\alpha/0.01$	0.194	0.193	0.0539	0.0548	0.0295	0.029	0.0134	0.0129
	$\beta_x/0.01$	-12.2	-12.2	-13.6	-13.7	-14.9	-14.9	-16.9	-16.3
	$\beta_y/0.01$	-4.92	-4.41	-4.82	-4.13	-3.46	-4.49	-2.54	-4.91
0.3	$\alpha/0.01$	0.186	0.186	0.0528	0.0531	0.0290	0.028	0.013	0.0127
	$\beta_x/0.01$	-11.7	-11.6	-13.3	-13.3	-14.6	-14.6	-16.7	-16.1
	$\beta_y/0.01$	-3.93	-3.54	-4.39	-4.01	-3.26	-4.38	-2.51	-4.84
0.6	$\alpha/0.01$	0.179	0.195	0.0513	0.0512	0.0282	0.028	0.0129	0.0125
	$\beta_x/0.01$	-11.0	-11.2	-12.9	-12.9	-14.3	-14.1	-16.4	-15.8
	$\beta_y/0.01$	-1.69	15.1	-3.94	-3.88	-3.01	-4.25	-2.46	-4.75
1.0	$\alpha/0.01$	0.156	0.160	0.0496	0.0494	0.0274	0.0271	0.0126	0.0122
	$\beta_x/0.01$	-11.2	-10.7	-12.5	-12.4	-13.9	-13.7	-16.0	-15.4
	$\beta_y/0.01$	-12.2	-7.62	-3.53	-3.74	-2.75	-4.12	-2.40	-4.64

Table 4.16 Plate design coefficients.
Support conditions: clamped. Loading: uniform. Materials: plate type II.

$D_{16}/D_{11} = D_{26}/D_{11} = 0.0$ a/b = 0.5

D_{66}/D_{22}	D_{11}/D_{22}	0.5		2.0		4.0		10.0	
	D_{12}/D_{11}	0.15	0.3	0.15	0.3	0.15	0.3	0.15	0.3
0.1	$\alpha/0.01$	0.297	0.291	0.0868	0.0854	0.046	0.0454	0.0189	0.00961
	$\beta_x/0.01$	-9.47	-9.35	-11.3	-10.4	-11.1	-13.7	-11.3	-17.3
	$\beta_y/0.01$	-5.99	-5.93	-2.52	-3.17	-1.73	-4.12	-3.19	-5.19
0.3	$\alpha/0.01$	0.271	0.266	0.084	0.0830	0.0453	0.045	0.019	-0.0247
	$\beta_x/0.01$	-8.69	-8.56	-10.9	-10.4	-7.63	-10.3	-11.5	-10.4
	$\beta_y/0.01$	-5.53	-5.48	-2.42	-3.15	-1.87	-3.11	-2.37	-18.4
0.6	$\alpha/0.01$	0.239	0.236	0.0810	0.0796	0.044	0.0436	0.0188	0.0233
	$\beta_x/0.01$	-7.73	-7.65	-10.1	-10.0	-11.63	-10.8	-11.5	-18.7
	$\beta_y/0.01$	-4.99	-4.94	-2.27	-3.04	-1.75	-3.27	-1.93	-5.62
1.0	$\alpha/0.01$	0.207	0.205	0.0769	0.0756	0.0427	0.0423	0.0185	0.0195
	$\beta_x/0.01$	-6.81	-6.72	-9.92	-9.71	-11.0	-10.8	-11.1	-10.7
	$\beta_y/0.01$	-4.43	-4.38	-2.13	-2.93	-1.67	-3.25	-1.70	-3.23

Table 4.17 Plate design coefficients.
Support conditions: clamped. Loading: uniform. Materials: plate type II.

$D_{16}/D_{11} = D_{26}/D_{11} = 0.0$ a/b = 0.75

D_{66}/D_{22}	D_{11}/D_{22}	0.5		2.0		4.0		10.0	
	D_{12}/D_{11}	0.15	0.3	0.15	0.3	0.15	0.3	0.15	0.3
0.1	$\alpha/0.01$	0.292	0.286	0.106	0.102	0.0593	0.056	0.0253	0.025
	$\beta_x/0.01$	-6.41	-6.21	-9.10	-8.81	-9.76	-9.55	-10.7	-10.6
	$\beta_y/0.01$	-6.93	-6.92	-2.86	-2.86	-1.98	-2.88	-1.61	-3.20
0.3	$\alpha/0.01$	0.266	0.260	0.101	0.0982	0.0564	0.0561	0.0254	0.0245
	$\beta_x/0.01$	5.31	5.43	-8.66	-8.39	-9.73	9.46	-10.3	7.96
	$\beta_y/0.01$	-7.06	-2.97	-2.73	-2.80	-1.65	-2.78	-1.57	-2.09
0.6	$\alpha/0.01$	0.263	0.273	0.0983	0.0912	0.0544	0.0517	0.0249	0.0241
	$\beta_x/0.01$	11.4	14.1	-7.63	-7.52	-9.33	-9.12	-10.2	-10.0
	$\beta_y/0.01$	-10.5	-13.6	-3.548	-2.13	-1.56	-2.71	-1.54	-3.02
1.0	$\alpha/0.01$	-0.142	-0.0903	0.0886	0.08631	0.0527	0.0538	0.0229	0.0238
	$\beta_x/0.01$	-62.2	-47.4	-7.11	-6.92	-8.29	11.6	-10.6	11.0
	$\beta_y/0.01$	-31.0	18.4	-2.86	-2.73	-1.81	-2.96	-1.56	-2.89

Table 4.18 Plate design coefficients.
Support conditions: clamped. Loading: uniform. Materials: plate type II.

$D_{16}/D_{11} = D_{26}/D_{11} = 0.0$ a/b = 1.0

D_{66}/D_{22}	D_{11}/D_{22}	0.5		2.0		4.0		10.0	
	D_{12}/D_{11}	0.15	0.3	0.15	0.3	0.15	0.3	0.15	0.3
0.1	$\alpha/0.01$	0.213	0.207	0.107	0.101	0.0650	0.0614	0.0300	0.283
	$\beta_x/0.01$	-3.67	-3.58	-6.96	-6.53	-8.29	-7.74	-9.13	10.8
	$\beta_y/0.01$	-6.85	-6.72	-3.73	-4.12	-2.40	-2.34	-1.38	-2.33
0.3	$\alpha/0.01$	0.186	0.182	0.100	0.0956	0.0623	0.058	0.0295	0.0275
	$\beta_x/0.01$	-3.26	-3.19	-6.50	-6.39	-7.98	-7.43	-8.85	-8.47
	$\beta_y/0.01$	-6.08	-5.98	-3.47	-3.35	-2.31	-2.25	-1.34	-2.54
0.6	$\alpha/0.01$	0.157	0.154	0.0913	0.0874	0.0587	0.0556	0.0286	0.0260
	$\beta_x/0.01$	-2.79	-2.81	-5.97	-5.79	-7.55	-7.02	7.96	-7.70
	$\beta_y/0.01$	-5.26	-5.10	-3.19	-3.09	-2.19	-2.16	-1.19	-2.32
1.0	$\alpha/0.01$	0.131	0.129	0.0816	0.0785	0.0546	0.0518	0.0276	0.0262
	$\beta_x/0.01$	-2.39	-2.35	-5.41	-5.23	-7.06	-6.91	-8.74	9.87
	$\beta_y/0.01$	-4.44	-4.39	-2.89	-2.81	-2.05	-2.08	-1.32	1.12

Table 4.19 Plate design coefficients.
Support conditions: clamped. Loading: patch. Materials: plate type II.

$D_{16}/D_{11} = D_{26}/D_{11} = 0.0$ a/b = 0.25

D_{66}/D_{22}	D_{11}/D_{22}	0.5		2.0		4.0		10.0	
	D_{12}/D_{11}	0.15	0.3	0.15	0.3	0.15	0.3	0.15	0.3
0.1	$\alpha/0.01$	0.00867	0.00850	0.00298	0.00279	0.00166	0.00153	0.00075	-0.00069
	$\beta_x/0.015$	-0.591	-0.584	-0.797	-0.757	-0.880	-0.828	-0.993	-0.932
	$\beta_y/0.015$	-0.314	-0.344	-0.119	-0.227	-0.132	-0.248	-0.149	-0.279
0.3	$\alpha/0.01$	0.00832	0.00817	0.00291	0.00278	0.00164	0.00154	0.000751	0.000698
	$\beta_x/0.015$	-0.569	-0.561	-0.779	-0.752	-0.869	-0.829	-0.983	-0.930
	$\beta_y/0.015$	-0.305	-0.357	-0.117	-0.225	-0.130	-0.249	-0.147	-0.279
0.6	$\alpha/0.01$	0.00770	0.00680	0.0027	0.00271	0.00159	0.00152	0.0007384	0.000695
	$\beta_x/0.015$	-0.532	-0.476	-0.751	-0.733	-0.849	-0.821	-0.968	-0.925
	$\beta_y/0.015$	-0.373	-1.16	-0.112	-0.220	-0.127	-0.246	-0.145	-0.277
1.0	$\alpha/0.01$	0.00754	0.00734	0.00264	0.00259	0.00154	0.00148	0.000721	0.000686
	$\beta_x/0.015$	-0.512	-0.505	-0.715	-0.704	-0.821	-0.801	-0.949	-0.915
	$\beta_y/0.015$	0.09298	-0.152	-0.107	-0.211	-0.123	-0.240	-0.142	-0.274

Table 4.20 Plate design coefficients.
Support conditions: clamped. Loading: patch. Materials: plate type II.

$D_{16}/D_{11} = D_{26}/D_{11} = 0.0$ a/b = 0.5

D_{66}/D_{22}	D_{11}/D_{22}	0.5		2.0		4.0		10.0	
	D_{12}/D_{11}	0.15	0.3	0.15	0.3	0.15	0.3	0.15	0.3
0.1	$\alpha/0.01$	0.0111	0.0109	0.00341	0.00331	0.00187	0.00173	0.000944	0.00133
	$\beta_x/0.015$	-0.391	-0.375	-0.476	-0.481	-0.532	-0.571	-0.676	0.919
	$\beta_y/0.015$	-0.257	-0.262	-0.136	-0.145	-0.110	-0.202	-0.101	-0.213
0.3	$\alpha/0.01$	0.0109	0.0100	0.00333	0.00325	0.00185	0.00176	0.000914	0.00295
	$\beta_x/0.015$	-0.348	-0.351	-0.468	-0.462	-0.563	-0.523	-0.647	1.588
	$\beta_y/0.015$	-0.245	-0.245	-0.130	-0.138	-0.103	-0.157	-0.0971	0.787
0.6	$\alpha/0.01$	0.00906	0.00894	0.0033	0.00315	0.00182	0.00175	0.000896	0.000637
	$\beta_x/0.015$	-0.322	-0.304	0.796	-0.468	-0.506	-0.496	-0.630	0.516
	$\beta_y/0.015$	-0.224	-0.222	-0.118	-0.140	-0.101	-0.149	-0.0945	-0.161
1.0	$\alpha/0.01$	0.00790	0.00781	0.00307	0.00301	0.00178	0.00171	0.000879	0.000795
	$\beta_x/0.015$	-0.282	-0.281	-0.428	-0.413	-0.499	-0.477	-0.617	-0.574
	$\beta_y/0.015$	-0.202	-0.202	-0.115	-0.124	-0.097	-0.145	-0.0926	-0.172

Table 4.21 Plate design coefficients.
Support conditions: clamped. Loading: patch. Materials: plate type II.

$D_{16}/D_{11} = D_{26}/D_{11} = 0.0$ a/b = 0.75

D_{66}/D_{22}	D_{11}/D_{22}	0.5		2.0		4.0		10.0	
	D_{12}/D_{11}	0.15	0.3	0.15	0.3	0.15	0.3	0.15	0.3
0.1	$\alpha/0.01$	0.00930	0.00916	0.00377	0.00384	0.00169	0.00189	0.000814	0.000757
	$\beta_x/0.015$	-0.0538	-.00039	0.175	0.409	-0.417	0.109	-0.227	-0.122
	$\beta_y/0.015$	-0.319	-0.262	-0.164	-0.202	-0.061	-0.113	-0.118	-0.111
0.3	$\alpha/0.01$	0.00766	0.00605	0.00353	0.00341	0.00218	0.00154	0.000977	0.000706
	$\beta_x/0.015$	-0.0931	-0.0931	0.131	0.160	0.428	-0.494	0.406	-0.684
	$\beta_y/0.015$	-0.173	0.298	-0.165	-0.159	-0.136	-0.116	-0.0523	-1.338
0.6	$\alpha/0.01$	0.00295	0.00126	0.00319	0.00323	0.00197	0.00209	0.000834	0.000765
	$\beta_x/0.015$	-0.673	-0.912	0.0282	0.206	0.178	0.646	-0.0545	-0.131
	$\beta_y/0.015$	0.302	0.642	-0.125	-0.109	-0.110	-0.100	-0.066	-0.132
1.0	$\alpha/0.01$	0.0253	0.0211	0.00291	0.00279	0.00171	0.00167	0.00113	0.000812
	$\beta_x/0.015$	3.97	2.93	0.0491	0.00995	-0.0540	0.210	1.24	0.174
	$\beta_y/0.015$	1.65	-1.46	-0.113	-0.122	-0.0596	-0.080	-0.111	-0.126

Table 4.22 Plate design coefficients.
Support conditions: clamped. Loading: patch. Materials: plate type II.

$D_{16}/D_{11} = D_{26}/D_{11} = 0.0$ a/b = 1.0

D_{66}/D_{22}	D_{11}/D_{22}	0.5		2.0		4.0		10.0	
	D_{12}/D_{11}	0.15	0.3	0.15	0.3	0.15	0.3	0.15	0.3
0.1	$\alpha/0.01$	0.00784	0.00766	0.00394	0.00371	0.00239	0.00217	0.00111	0.00104
	$\beta_x/0.015$	-0.137	-0.141	-0.284	-0.323	-0.330	-0.287	-0.431	0.450
	$\beta_y/0.015$	-0.267	-0.270	-0.152	-0.179	-0.100	-0.210	-0.064	-0.0832
0.3	$\alpha/0.01$	0.00687	0.00674	0.00369	0.00352	0.00230	0.00216	0.00111	0.00101
	$\beta_x/0.015$	-0.156	-0.129	-0.260	0.248	-0.318	-0.360	0.359	0.354
	$\beta_y/0.015$	-0.229	-0.251	-0.137	-0.139	-0.0965	-0.108	-0.0539	-0.079
0.6	$\alpha/0.01$	0.00585	0.00574	0.00337	0.00323	0.00217	0.00206	0.00108	0.00108
	$\beta_x/0.015$	-0.108	-0.127	-0.242	0.231	-0.304	-0.316	0.496	1.17
	$\beta_y/0.015$	-0.234	-0.178	-0.127	-0.127	-0.0921	-0.094	-0.0484	0.292
1.0	$\alpha/0.01$	0.00489	0.00482	0.00302	0.00291	0.00203	0.00193	0.00104	0.000986
	$\beta_x/0.015$	-0.101	0.103	-0.222	0.213	-0.286	0.289	-0.352	0.424
	$\beta_y/0.015$	-0.173	-0.178	-0.117	-0.117	-0.0869	-0.0877	-0.0559	-0.0660

Table 4.23 Plate design coefficients.
Support conditions: clamped. Loading: triangular. Materials: plate type II.

$D_{16}/D_{11} = D_{26}/D_{11} = 0.0$ a/b = 0.25

D_{66}/D_{22}	D_{11}/D_{22}	0.5		2.0		4.0		10.0	
	D_{12}/D_{11}	0.15	0.3	0.15	0.3	0.15	0.3	0.15	0.3
0.1	$\alpha/0.01$	0.0494	0.043	0.0244	0.0193	0.0101	0.0088	0.0040	0.0036
	$\beta_x/0.01$	-3.08	-2.72	-6.12	-4.92	-5.25	-4.71	-5.35	-5.13
	$\beta_y/0.02$	-1.20	1.04	-4.26	-2.84	-2.05	-1.53	-1.07	-0.794
0.3	$\alpha/0.01$	0.0399	0.0282	0.0225	0.0188	0.0099	0.0088	0.0039	0.0036
	$\beta_x/0.01$	-2.48	-1.77	-5.65	-4.76	-5.44	-4.65	-5.28	-5.07
	$\beta_y/0.02$	0.674	0.846	-3.744	-2.68	-1.94	-1.50	-1.03	-0.78
0.6	$\alpha/0.01$	0.0110	-0.184	0.0208	0.0182	0.0096	0.0088	0.0039	0.0036
	$\beta_x/0.01$	-0.623	11.3	-5.206	-4.59	-4.98	-4.58	-5.20	-4.99
	$\beta_y/0.02$	6.48	50.2	-3.23	-2.49	-1.80	-1.44	-0.987	-0.77
1.0	$\alpha/0.01$	0.129	0.0827	0.0193	0.0177	0.0094	0.0087	0.0038	0.0036
	$\beta_x/0.01$	-9.06	-4.88	-4.83	-4.40	-4.83	-4.50	-5.10	-4.92
	$\beta_y/0.02$	-21.2	-10.4	-2.79	-2.29	-1.66	-1.38	-0.932	-0.758

Table 4.24 Plate design coefficients.
Support conditions: clamped. Loading: triangular. Materials: plate type II.

$D_{16}/D_{11} = D_{26}/D_{11} = 0.0$ a/b = 0.5

D_{66}/D_{22}	D_{11}/D_{22}	0.5		2.0		4.0		10.0	
	D_{12}/D_{11}	0.15	0.3	0.15	0.3	0.15	0.3	0.15	0.3
0.1	$\alpha/0.01$	0.150	0.147	0.0469	0.0459	0.0255	0.0277	-0.0055	0.0041
	$\beta_x/0.01$	-4.36	-4.28	-5.36	-5.13	-5.69	7.05	-7.76	-30.1
	$\beta_y/0.02$	-4.27	-4.16	-1.56	-1.35	-0.725	1.46	-7.45	6.22
0.3	$\alpha/0.01$	20.1	0.132	0.0452	0.0444	0.0252	0.0269	0.0009	0.0032
	$\beta_x/0.01$	-3.94	-3.91	-5.26	-5.07	-5.44	6.06	-2.80	-12.3
	$\beta_y/0.02$	-3.77	-3.68	-1.44	-1.20	-1.57	-0.86	-4.73	-10.4
0.6	$\alpha/0.01$	0.117	0.115	0.0428	0.0424	0.0247	0.0270	0.0043	0.0098
	$\beta_x/0.01$	-3.50	-3.45	-11.5	-4.91	-5.53	6.21	-2.52	4.51
	$\beta_y/0.02$	-3.21	-3.13	-1.62	-1.01	-0.428	1.04	-3.25	0.506
1.0	$\alpha/0.01$	0.100	0.0990	0.0406	0.0402	0.0242	0.0289	0.0060	0.0092
	$\beta_x/0.01$	-3.05	-2.98	-4.79	-4.69	-5.52	7.30	-3.41	-4.86
	$\beta_y/0.02$	-2.64	-2.56	-1.02	-0.738	-0.666	2.41	-2.45	-0.601

Table 4.25 Plate design coefficients.
Support conditions: clamped. Loading: triangular. Materials: plate type II.

$D_{16}/D_{11} = D_{26}/D_{11} = 0.0$ a/b = 0.75

D_{66}/D_{22}	D_{11}/D_{22}	0.5		2.0		4.0		10.0	
	D_{12}/D_{11}	0.15	0.3	0.15	0.3	0.15	0.3	0.15	0.3
0.1	$\alpha/0.01$	0.150	0.146	0.0567	0.054	0.0316	0.0308	0.0141	0.0138
	$\beta_x/0.01$	-2.88	-2.83	-4.54	-4.34	-5.12	-4.81	-5.66	-5.40
	$\beta_y/0.02$	-4.86	-4.79	-2.13	-2.06	-1.38	1.49	0.919	1.58
0.3	$\alpha/0.01$	0.128	0.148	0.053	0.0519	0.0306	0.0294	0.0136	0.0126
	$\beta_x/0.01$	-2.77	4.40	-4.29	-4.11	-4.92	-4.64	-5.77	-3.22
	$\beta_y/0.02$	-4.47	0.940	-2.01	-1.97	1.24	1.41	0.38	2.69
0.6	$\alpha/0.01$	0.130	0.134	0.0509	0.0486	0.0295	0.0282	0.0133	0.0134
	$\beta_x/0.01$	8.17	6.21	-3.77	-3.75	-4.72	-4.52	-5.52	-5.00
	$\beta_y/0.02$	-5.55	-7.02	-2.08	1.77	1.18	1.88	0.852	1.36
1.0	$\alpha/0.01$	-0.114	-0.0501	0.0509	0.0480	0.0283	0.0267	0.0131	0.0135
	$\beta_x/0.01$	-48.9	-24.9	6.33	5.57	-4.15	-3.68	-5.29	5.81
	$\beta_y/0.02$	-18.1	10.6	2.83	3.38	1.44	2.15	1.19	0.82

Table 4.26 Plate design coefficients.
Support conditions: clamped. Loading: triangular. Materials: plate type II.

$D_{16}/D_{11} = D_{26}/D_{11} = 0.0$ a/b = 1.0

D_{66}/D_{22}	D_{11}/D_{22}	0.5		2.0		4.0		10.0	
	D_{12}/D_{11}	0.15	0.3	0.15	0.3	0.15	0.3	0.15	0.3
0.1	$\alpha/0.01$	0.107	0.104	0.0552	0.0522	0.0338	0.031	0.0159	0.0148
	$\beta_x/0.01$	-1.74	-1.72	2.75	-3.3731	-4.13	-3.86	-4.96	-3.79
	$\beta_y/0.02$	-4.05	-4.00	-2.55	-2.5548	-1.64	1.37	-0.880	1.27
0.3	$\alpha/0.01$	0.0942	0.0922	0.0514	0.0489	0.0323	0.0304	0.0156	0.0146
	$\beta_x/0.01$	-1.58	-1.56	-3.21	-3.1157	-3.97	-3.77	-4.81	-4.12
	$\beta_y/0.02$	-3.64	-3.50	-2.27	-2.1649	-1.56	-1.44	-0.855	1.39
0.6	$\alpha/0.01$	0.0798	0.0780	0.0467	0.0447	0.0304	0.0287	0.0151	0.0118
	$\beta_x/0.01$	-1.39	-1.33	-2.95	-2.8560	-3.71	-3.57	-4.53	-6.12
	$\beta_y/0.02$	-3.00	-3.36	-2.07	-1.9830	-1.46	-1.38	-0.78	-4.43
1.0	$\alpha/0.01$	0.066	0.0652	0.0417	0.0401	0.0282	0.0267	0.0145	0.0139
	$\beta_x/0.01$	-1.16	-1.14	-2.67	-2.5941	-3.55	-3.40	-4.46	4.22
	$\beta_y/0.02$	-2.82	-2.74	-1.85	-1.7877	-1.35	-1.26	-0.76	0.824

Table 4.27 Plate design coefficients.
Support conditions: simple. Loading: uniform. Materials: plate type III.

$D_{16}/D_{11} = D_{26}/D_{11} = 0.35$ a/b = 0.25

D_{66}/D_{22}	D_{11}/D_{22}	0.5		2.0		4.0		10.0	
	D_{12}/D_{11}	0.15	0.3	0.15	0.3	0.15	0.3	0.15	0.3
0.1	$\alpha/0.01$	0.670	0.668	0.104	0.207	-0.143	0.082	0.012	0.208
	$\beta_x/0.01$	12.8	12.7	6.14	22.1	-34.3	13.2	2.49	117.0
	$\beta_y/0.01$	1.51	3.64	-1.84	9.33	-4.54	3.49	1.19	15.9
0.3	$\alpha/0.01$	0.644	0.637	0.338	0.158	0.093	0.057	0.010	0.053
	$\beta_x/0.01$	12.3	12.2	47.0	11.2	15.2	5.80	1.50	25.3
	$\beta_y/0.01$	2.17	4.50	21.4	2.87	1.71	1.03	1.33	4.81
0.6	$\alpha/0.01$	0.586	0.578	0.161	0.161	0.057	0.099	-0.005	0.039
	$\beta_x/0.01$	11.2	11.1	12.1	12.2	6.66	19.3	-6.31	16.6
	$\beta_y/0.01$	2.32	3.98	1.58	3.55	-0.054	5.81	1.67	3.81
1.0	$\alpha/0.01$	0.515	0.509	0.160	0.159	0.143	0.078	0.061	0.034
	$\beta_x/0.01$	9.86	9.80	12.2	12.1	32.1	11.1	26.6	13.8
	$\beta_y/0.01$	2.20	3.64	1.79	3.65	6.43	3.27	0.96	3.48

Table 4.28 Plate design coefficients.
Support conditions: simple. Loading: uniform. Materials: plate type III.

$D_{16}/D_{11} = D_{26}/D_{11} = 0.35$ a/b =0.5

D_{66}/D_{22}	D_{11}/D_{22}	0.5		2.0		4.0		10.0	
	D_{12}/D_{11}	0.15	0.3	0.15	0.3	0.15	0.3	0.15	0.3
0.1	$\alpha/0.01$	1.31	1.23	0.514	0.748	0.270	0.17	0.107	0.109
	$\beta_x/0.01$	13.0	12.5	41.3	25.8	27.1	10.9	27.3	25.3
	$\beta_y/0.01$	6.71	8.15	4.26	33.4	4.82	2.43	1.59	6.53
0.3	$\alpha/0.01$	0.966	0.921	2.68	0.321	0.560	1.23	0.248	0.113
	$\beta_x/0.01$	9.41	9.22	366.7	10.2	101.9	-1028.4	22.6	26.3
	$\beta_y/0.01$	5.00	6.11	109.6	2.20	21.7	-161.0	5.11	6.79
0.6	$\alpha/0.01$	0.698	0.675	0.310	0.293	-1.23	0.231	0.109	0.137
	$\beta_x/0.01$	6.67	6.64	12.0	11.4	-1059.0	17.1	21.2	44.3
	$\beta_y/0.01$	3.63	4.46	2.02	3.69	27.0	8.25	3.50	9.51
1.0	$\alpha/0.01$	0.512	0.500	0.268	0.255	0.138	0.119	0.111	0.007
	$\beta_x/0.01$	4.82	4.85	10.3	10.0	43.4	3.68	24.0	-7.62
	$\beta_y/0.01$	2.68	3.31	1.97	3.41	-9.18	0.433	3.14	-0.899

Table 4.29 Plate design coefficients.
Support conditions: simple. Loading: uniform. Materials: plate type III.

$D_{16}/D_{11} = D_{26}/D_{11} = 0.35$ a/b = 0.75

D_{66}/D_{22}	D_{11}/D_{22}	0.5		2.0		4.0		10.0	
	D_{12}/D_{11}	0.15	0.3	0.15	0.3	0.15	0.3	0.15	0.3
0.1	$\alpha/0.01$	1.33	1.20	-0.991	-0.154	0.241	11.89	0.258	0.16
	$\beta_x/0.01$	9.20	8.79	50.8	-192.8	-14.8	2808.6	161.6	19.1
	$\beta_y/0.01$	10.8	10.7	71.6	-104.6	4.29	207.7	1.42	4.93
0.3	$\alpha/0.01$	0.85	0.799	0.694	0.462	0.006	0.265	0.280	0.145
	$\beta_x/0.01$	5.55	5.60	30.9	4.19	-49.7	64.4	14.3	15.9
	$\beta_y/0.01$	6.47	6.79	8.66	3.41	6.48	24.3	2.04	4.44
0.6	$\alpha/0.01$	0.564	0.541	0.399	0.350	0.265	0.333	0.187	0.126
	$\beta_x/0.01$	3.59	3.70	10.4	9.85	53.3	32.7	20.8	10.5
	$\beta_y/0.01$	4.15	4.47	2.85	4.08	16.5	9.63	2.59	3.73
1.0	$\alpha/0.01$	0.392	0.381	0.29	0.266	0.265	0.270	0.149	0.087
	$\beta_x/0.01$	2.45	2.56	7.71	7.43	9.92	22.3	14.7	-26.8
	$\beta_y/0.01$	2.83	3.08	2.35	3.21	0.545	10.0	2.51	1.36

Table 4.30 Plate design coefficients.
Support conditions: simple. Loading: uniform. Materials: plate type III.

$D_{16}/D_{11} = D_{26}/D_{11} = 0.35$ a/b = 1.0

D_{66}/D_{22}	D_{11}/D_{22}	0.5		2.0		4.0		10.0	
	D_{12}/D_{11}	0.15	0.3	0.15	0.3	0.15	0.3	0.15	0.3
0.1	$\alpha/0.01$	0.948	0.859	0.008	0.722	-0.581	0.030	-1.21	0.30
	$\beta_x/0.01$	4.72	4.90	-29.7	-29.3	131.8	32.1	-184.8	44.4
	$\beta_y/0.01$	9.87	9.43	-8.32	-4.63	-13.8	11.1	5.80	8.93
0.3	$\alpha/0.01$	0.620	0.583	0.529	0.718	0.158	-0.017	2.64	0.232
	$\beta_x/0.01$	3.02	3.25	-18.3	27.0	5.33	-26.8	329.7	37.8
	$\beta_y/0.01$	6.17	6.61	-22.2	16.2	5.83	10.6	8.54	7.60
0.6	$\alpha/0.01$	0.415	0.399	0.425	0.343	-0.017	-0.462	0.504	0.199
	$\beta_x/0.01$	2.00	2.19	8.82	8.19	-35.1	-93.1	61.7	14.5
	$\beta_y/0.01$	4.02	4.11	3.54	4.40	19.2	-32.9	3.49	3.17
1.0	$\alpha/0.01$	0.29	0.282	0.267	0.238	0.251	0.286	0.248	0.103
	$\beta_x/0.01$	1.38	1.53	5.49	5.43	65.5	18.8	34.6	100.4
	$\beta_y/0.01$	2.75	2.85	2.63	3.03	29.1	5.44	2.35	9.47

Table 4.31 Plate design coefficients.
Support conditions: simple. Loading: uniform. Materials: plate type III.

$D_{16}/D_{11} = D_{26}/D_{11} = 0.65$ a/b = 0.25

D_{66}/D_{22}	D_{11}/D_{22}	0.5		2.0		4.0		10.0	
	D_{12}/D_{11}	0.15	0.3	0.15	0.3	0.15	0.3	0.15	0.3
0.1	$\alpha/0.01$	5.01	-1.42	0.379	1.00	0.163	0.190	0.063	0.068
	$\beta_x/0.01$	-86.4	-60.6	32.7	94.6	25.0	33.3	23.9	27.9
	$\beta_y/0.01$	198.8	-72.2	6.52	36.0	4.19	9.70	3.52	7.6
0.3	$\alpha/0.01$	0.662	0.662	-0.854	0.167	0.177	0.261	0.064	0.071
	$\beta_x/0.01$	12.8	12.7	-74.6	41.4	29.6	48.0	24.8	29.9
	$\beta_y/0.01$	1.36	3.42	-25.2	5.36	4.37	14.0	3.43	8.03
0.6	$\alpha/0.01$	0.620	0.613	0.156	0.197	0.261	-0.11	0.066	0.078
	$\beta_x/0.01$	11.9	11.8	1.77	9.02	46.3	-21.5	26.5	33.8
	$\beta_y/0.01$	2.11	3.92	2.17	4.26	6.83	-7.62	3.35	8.90
1.0	$\alpha/0.01$	0.54	0.534	0.493	0.097	-0.002	0.061	0.070	0.100
	$\beta_x/0.01$	10.3	10.2	66.1	5.52	4.12	-0.085	28.6	45.4
	$\beta_y/0.01$	2.13	3.66	-4.68	1.08	-1.73	0.998	3.55	11.7

Table 4.32 Plate design coefficients.
Support conditions: simple. Loading: uniform. Materials: plate type III.

$D_{16}/D_{11} = D_{26}/D_{11} = 0.65$ a/b = 0.5

D_{66}/D_{22}	D_{11}/D_{22}	0.5		2.0		4.0		10.0	
	D_{12}/D_{11}	0.15	0.3	0.15	0.3	0.15	0.3	0.15	0.3
0.1	$\alpha/0.01$	4.89	1.24	0.653	0.470	0.592	0.429	0.605	0.250
	$\beta_x/0.01$	-71.8	-18.8	-2.62	-85.2	22.4	-30.2	62.1	17.0
	$\beta_y/0.01$	-36.8	-38.2	5.48	-24.6	1.18	-3.48	-8.12	0.894
0.3	$\alpha/0.01$	1.33	1.21	0.164	0.354	0.455	0.322	0.433	0.218
	$\beta_x/0.01$	13.2	12.4	25.5	-11.9	5.57	56.0	41.8	6.52
	$\beta_y/0.01$	6.39	8.06	-5.60	1.90	1.57	9.26	-4.99	-0.480
0.6	$\alpha/0.01$	0.800	0.76	0.199	-0.366	0.322	0.296	0.308	0.181
	$\beta_x/0.01$	7.51	7.40	-20.2	-102.9	54.1	-7.67	25.0	-37.0
	$\beta_y/0.01$	3.96	4.91	5.44	-3.18	-0.623	1.14	-2.76	-7.90
1.0	$\alpha/0.01$	0.549	0.534	0.185	0.510	0.252	0.203	0.226	0.158
	$\beta_x/0.01$	5.02	5.04	47.4	25.6	4.06	5.44	6.86	65.9
	$\beta_y/0.01$	2.73	3.40	-15.0	7.17	2.64	3.12	-1.34	11.8

Table 4.33 Plate design coefficients.
Support conditions: simple. Loading: uniform. Matrials: plate type III.

$D_{16}/D_{11} = D_{26}/D_{11} = 0.65$ a/b = 0.75

D_{66}/D_{22}	D_{11}/D_{22}	0.5		2.0		4.0		10.0	
	D_{12}/D_{11}	0.15	0.3	0.15	0.3	0.15	0.3	0.15	0.3
0.1	$\alpha/0.01$	0.321	1.77	0.813	0.401	-0.539	-1.45	-0.072	-0.166
	$\beta_x/0.01$	-86.8	21.2	28.0	11.2	-45.9	-27.7	-14.3	-57.4
	$\beta_y/0.01$	-122.6	25.4	14.5	11.2	-3.17	-4.69	4.01	-6.87
0.3	$\alpha/0.01$	1.31	1.12	0.306	2.36	-0.365	2.08	-0.082	0.031
	$\beta_x/0.01$	9.46	8.09	4.87	331.5	57.0	85.7	-15.2	76.5
	$\beta_y/0.01$	12.2	10.6	11.2	-39.7	12.2	27.0	4.11	25.6
0.6	$\alpha/0.01$	0.648	0.615	-0.067	0.346	2.08	0.490	-0.110	-0.157
	$\beta_x/0.01$	3.86	3.97	58.0	-5.75	80.1	15.1	-26.6	-23.8
	$\beta_y/0.01$	4.67	4.97	-7.14	5.43	14.8	10.8	3.76	7.14
1.0	$\alpha/0.01$	0.418	0.405	2.28	0.229	0.354	0.298	1.44	-0.124
	$\beta_x/0.01$	2.47	2.59	-42.4	-2.33	-4.23	39.5	938.6	16.1
	$\beta_y/0.01$	2.92	3.18	49.3	-20.1	14.3	3.72	88.9	0.784

Table 4.34 Plate design coefficients.
Support conditions: simple. Loading: uniform. Materials: plate type III.

$D_{16}/D_{11} = D_{26}/D_{11} = 0.65$ a/b = 1.0

D_{66}/D_{22}	D_{11}/D_{22}	0.5		2.0		4.0		10.0	
	D_{12}/D_{11}	0.15	0.3	0.15	0.3	0.15	0.3	0.15	0.3
0.1	$\alpha/0.01$	2.23	0.974	0.807	0.327	-0.279	-0.357	-0.032	-0.040
	$\beta_x/0.01$	9.91	-25.2	-3.85	-4.20	40.3	-17.8	-4.74	-9.22
	$\beta_y/0.01$	20.7	-20.4	11.7	3.82	0.990	-5.17	4.81	0.749
0.3	$\alpha/0.01$	0.862	0.769	0.236	-0.804	-0.270	-0.797	0.023	-0.048
	$\beta_x/0.01$	3.40	3.82	-6.80	-98.0	-15.5	-31.7	-2.73	-6.86
	$\beta_y/0.01$	8.83	8.34	4.04	-38.5	-1.75	-11.0	0.298	1.38
0.6	$\alpha/0.01$	0.469	0.446	0.220	0.026	-0.797	1.11	-0.044	-0.066
	$\beta_x/0.01$	2.03	2.25	12.2	6.90	-29.0	27.8	6.21	-0.608
	$\beta_y/0.01$	4.48	4.52	9.54	9.65	-6.85	14.6	3.47	2.65
1.0	$\alpha/0.01$	0.308	0.298	1.65	0.533	0.580	1.43	-0.046	-0.059
	$\beta_x/0.01$	1.36	1.52	-51.3	6.30	6.42	219.0	-6.73	-21.4
	$\beta_y/0.01$	2.85	2.95	-40.5	23.1	6.64	58.9	2.93	-0.057

Table 4.35 Plate design coefficients.
Support conditions: simple. Loading: uniform. Materials: plate type III.

D_{16}/D_{11} = D_{26}/D_{11} = 1.0 a/b = 0.25

D_{66}/D_{22}	D_{11}/D_{22}	0.5		2.0		4.0		10.0	
	D_{12}/D_{11}	0.15	0.3	0.15	0.3	0.15	0.3	0.15	0.3
0.1	$\alpha/0.01$	0.063	1.40	0.319	0.302	0.167	0.158	0.069	0.065
	$\beta_x/0.01$	18.6	49.3	26.8	24.1	30.1	27.8	31.8	29.5
	$\beta_y/0.01$	-41.4	25.3	1.23	5.59	0.250	5.27	-0.334	5.03
0.3	$\alpha/0.01$	1.34	1.18	0.296	0.281	0.161	0.153	0.068	0.064
	$\beta_x/0.01$	24.7	27.6	22.6	18.3	28.3	26.1	31.09	28.8
	$\beta_y/0.01$	7.35	11.0	1.60	5.20	0.389	5.08	-0.268	4.96
0.6	$\alpha/0.01$	0.989	0.675	0.266	0.249	0.153	0.145	0.066	0.063
	$\beta_x/0.01$	20.0	15.8	-17.5	26.6	25.7	23.4	29.9	27.9
	$\beta_y/0.01$	10.0	6.04	10.0	3.59	0.594	4.82	-0.177	4.85
1.0	$\alpha/0.01$	0.606	0.597	0.258	0.230	0.143	0.137	0.064	0.061
	$\beta_x/0.01$	11.6	11.4	28.6	18.0	21.9	14.1	28.6	26.7
	$\beta_y/0.01$	1.98	3.76	1.96	6.91	1.00	4.56	-0.065	4.72

Table 4.36 Plate design coefficients.
Support conditions: simple. Loading: uniform. Materials: plate type III.

D_{16}/D_{11} = D_{26}/D_{11} = 1.0 a/b = 0.5

D_{66}/D_{22}	D_{11}/D_{22}	0.5		2.0		4.0		10.0	
	D_{12}/D_{11}	0.15	0.3	0.15	0.3	0.15	0.3	0.15	0.3
0.1	$\alpha/0.01$	0.037	-3.78	-1.54	-0.694	-0.292	-0.575	-0.061	-0.105
	$\beta_x/0.01$	-3.77	-127.5	-121.7	74.7	-44.4	-103.7	-19.0	-43.5
	$\beta_y/0.01$	17.8	22.5	-7.82	37.6	2.92	-17.4	7.10	-1.23
0.3	$\alpha/0.01$	-1.90	1.42	-2.10	2.67	-0.373	-0.236	-0.069	-0.139
	$\beta_x/0.01$	28.2	2.18	22.2	54.5	-49.4	30.3	-21.9	-66.7
	$\beta_y/0.01$	-148.3	37.0	29.0	6.10	4.87	21.2	7.13	-5.51
0.6	$\alpha/0.01$	0.780	1.63	1.03	0.643	-0.236	-0.029	-0.084	-2.47
	$\beta_x/0.01$	29.8	17.9	25.4	14.8	36.0	54.0	-29.0	-1947.2
	$\beta_y/0.01$	-25.9	14.8	3.66	8.21	18.8	7.59	7.14	-373.2
1.0	$\alpha/0.01$	0.660	0.634	0.688	0.321	-0.544	7.36	-0.128	-0.045
	$\beta_x/0.01$	5.65	5.62	161.5	37.9	-7.98	330.1	-54.6	30.9
	$\beta_y/0.01$	2.98	3.77	-65.3	2.06	10.4	-18.9	6.58	15.4

Table 4.37 Plate design coefficients.
Support conditions: simple. Loading: uniform. Materials: plate type III.

$D_{16}/D_{11} = D_{26}/D_{11} = 1.0$ a/b = 0.75

D_{66}/D_{22}	D_{11}/D_{22}	0.5		2.0		4.0		10.0	
	D_{12}/D_{11}	0.15	0.3	0.15	0.3	0.15	0.3	0.15	0.3
0.1	$\alpha/0.01$	0.438	-0.106	-0.260	-0.401	-0.087	-0.070	-0.003	-0.013
	$\beta_x/0.01$	-9.72	-8.70	-3.62	-5.24	-3.61	-0.332	-6.21	-3.26
	$\beta_y/0.01$	-14.6	-27.0	-0.367	-1.61	6.91	1.72	8.78	4.51
0.3	$\alpha/0.01$	0.759	1.42	-0.474	-0.975	0.055	-0.114	-0.006	-0.015
	$\beta_x/0.01$	-89.0	13.5	-4.56	15.3	-0.707	-1.97	-5.14	-2.69
	$\beta_y/0.01$	152.2	-6.16	-0.931	0.209	-7.33	2.57	7.83	4.24
0.6	$\alpha/0.01$	0.218	1.07	-4.19	1.58	-0.114	-0.050	-0.010	-0.018
	$\beta_x/0.01$	-67.3	8.84	-71.9	20.4	-2.60	39.6	-3.98	-1.77
	$\beta_y/0.01$	43.6	13.2	-21.9	12.4	3.59	11.4	6.81	3.85
1.0	$\alpha/0.01$	0.490	0.469	0.439	0.229	-0.189	-0.589	-0.014	-0.022
	$\beta_x/0.01$	2.50	2.64	-0.251	-8.25	30.2	-62.4	-2.84	0.34
	$\beta_y/0.01$	3.24	3.50	3.72	0.358	5.77	2.38	5.84	2.96

Table 4.38 Plate design coefficients.
Support conditions: simple. Loading: uniform. Materials: plate type III.

$D_{16}/D_{11} = D_{26}/D_{11} = 1.0$ a/b = 1.0

D_{66}/D_{22}	D_{11}/D_{22}	0.5		2.0		4.0		10.0	
	D_{12}/D_{11}	0.15	0.3	0.15	0.3	0.15	0.3	0.15	0.3
0.1	$\alpha/0.01$	0.267	-0.032	-0.162	-0.24	-0.155	0.255	-0.131	0.003
	$\beta_x/0.01$	11.7	14.2	-6.30	-5.58	3.19	36.6	-35.2	12.0
	$\beta_y/0.01$	2.79	-1.14	0.477	-2.02	0.212	-3.94	-10.39	0.694
0.3	$\alpha/0.01$	1.64	0.438	-0.281	-0.49	-9.80	-0.098	0.057	-0.001
	$\beta_x/0.01$	23.6	14.4	-6.23	-10.8	-249.1	-82.6	26.4	11.7
	$\beta_y/0.01$	3.87	24.0	-1.80	-5.46	144.8	20.6	2.93	0.112
0.6	$\alpha/0.01$	3.35	0.660	-1.38	1.95	-0.098	-0.043	0.012	-0.006
	$\beta_x/0.01$	-29.4	5.71	-32.2	54.8	-79.3	-6.88	12.8	12.6
	$\beta_y/0.01$	373.1	9.02	-14.1	29.0	42.6	2.27	-0.550	-0.77
1.0	$\alpha/0.01$	0.352	0.339	0.350	0.164	-0.055	-0.055	0.000	-0.012
	$\beta_x/0.01$	1.28	1.47	16.6	20.3	-7.09	-0.009	11.04	17.47
	$\beta_y/0.01$	3.15	3.22	5.40	6.55	3.37	1.80	-2.01	-2.76

Table 4.39 Plate design coefficients.
Plate tables summary ANSYS vs. Roark vs. EUROCOMP

Example number	Plate type	Load type	Edge	EUROCOMP (13 harmonics)			'Roark'			ANSYS			Comments on comparative results and values of correction factors in tables
				Defl (mm)	M_x (kNm)	M_y (kNm)	Defl (mm)	M_x (kNm)	M_y (kNm)	Defl (mm)	M_x (kNm)	M_y (kNm)	
1	I	UDL	S	1.17E-3	0.048	0.048	1.16E-3	0.048	0.048	1.17E-3	0.048	0.048	O.K.
2	I	TL	C	1.99E-4	0.026	0.029	1.95E-4	0.27	0.03	1.99E-4	0.026	0.029	O.K.
3	I	PL	C	1.52E-5	1.9E-3	1.9E-3	1.6E-5	2.3E-3	2.3E-3	1.56E-5	2.5E-3	2.5E-3	Factor of 1.3 applied to moment coefficients
4	II	UDL	S	2.9e-3	0.104	0.03	-	-	-	2.51e-3	0.109	0.03	O.K.
5	II	TL	C	9.30e-4	0.144	0.04	-	-	-	7.14e-4	0.106	0.07	Factor of 2.0 applied to M_y coefficient
6	II	PL	S	9.5e-5	6.1e-3	3.5e-3	-	-	-	8.41e-5	6.7e-3	3.7e-3	O.K.
7	II	PL	C	3.4e-5	5.6e-3	1.5e-3	-	-	-	3.54e-5	5.1e-2	2.9e-3	Factor of 1.5 applied to moment coefficients
8	III	UDL	S	1.94e-2	0.55	0.49	-	-	-	1.7E-2	0.5	0.29	O.K.

Notes:

1. The results from the EUROCOMP tables are compared to those obtained using ANSYS and the tables in Roark's Formulae for Stress and Strain.
2. The figures in the table above have been obtained by interpolation. The values in the tables are accurate for the D_i/D_j ratios quoted. The correlation is not necessarily linear, particularly for the case of clamped supports, and hence compensatory factors have been included in some of the tables. (The values in the above table do not include these factors.)
3. All coefficients are additionally factored by 0.01 for ease of presentation. (The cumulative factor for each coefficient is identified in the tables.)

4.10 LAMINATE DESIGN

4.10.0 Notation

The notation used in the EUROCOMP Handbook is consistent with that employed in the EUROCOMP Design Code, although several additional terms are referred to whilst explaining the theory behind the code principles.

A_{ij}	Extensional or membrane terms of laminate stiffness matrix
B_{ij}	Coupling terms of laminate stiffness matrix
D_{ij}	Bending terms of laminate stiffness matrix
E_1	Longitudinal Young's modulus of the lamina
E_{1f}	Longitudinal Young's modulus of the fibres
E_{1m}	Longitudinal Young's modulus of the matrix
E_2	Transverse Young's modulus of the lamina
E_{2f}	Transverse Young's modulus of the fibres
E_{2m}	Transverse Young's modulus of the matrix
G_{12m}	In-plane shear modulus of the matrix
G_{13m}	Out-of-plane shear modulus of matrix (in the 1-3 plane)
G_{23m}	Out-of-plane shear modulus of matrix (in the 2-3 plane)
M_x, M_y, M_{xy}	Moment stress resultants
N_x, N_y, N_{xy}	Force stress resultants
Q_{ij}	Laminate reduced stiffness terms
$\bar{Q}_{ij}$	Laminate transformed reduced stiffness terms
a_{ij}	Extensional or membrane terms of laminate compliance matrix
b_{ij}	Coupling terms of laminate compliance matrix
d_{ij}	Bending terms of laminate compliance matrix
h_k	Distance from middle surface to upper surface of kth layer
h	Total laminate thickness
k_{2f}	Bulk modulus of the fibres
k_{2m}	Bulk modulus of the matrix
k_2	Bulk modulus of the composite
n	Number of laminae in a laminate
V_f	Volume fraction of the fibres
V_m	Volume fraction of the matrix
V_v	Volume fraction of voids
ε_{1t}	Longitudinal failure strain of lamina in tension
ε_{1c}	Longitudinal failure strain of lamina in compression
ε_{2t}	Transverse failure strain of lamina in tension
ε_{2c}	Transverse failure strain of lamina in compression
ε_{ft}	Failure strain of fibres in tension
ε^E_1	Thermal expansional strain in the longitudinal direction
ε^E_2	Thermal expansional strain in the transverse direction
σ_{1t}	Longitudinal tensile strength of the lamina
σ_{1c}	Longitudinal compressive strength of the lamina

σ_{2t}	Transverse tensile strength of the lamina
σ_{2c}	Transverse compressive strength of the lamina
σ_{12s}	Shear strength of the lamina
σ_{ft}	Tensile strength of the fibres
σ_{m}	Tensile/compressive strength of the matrix
σ_{12m}	Shear strength of the matrix
ν_{21}	Minor Poisson's ratio of the lamina
ν_{12f}	Major Poisson's ratio of the fibres
$\chi_{x,y,xy}$	Curvatures

4.10.1 Basic conditions to be satisfied

No additional information.

4.10.2 Lamina stiffness

It is important that engineers have a good understanding of the behaviour of materials when designing a structure made of polymer composite materials. When designing with metals, experienced engineers have a good knowledge of their ductile behaviour which exhibits considerable plastic deformation before failure. As a result of this, in the design of a bolted joint the engineer can safely assume that the applied load is equally shared amongst all bolts. However, this is not true when designing with composites, because they do not have plastic deformation and are very sensitive to stress concentrations at joints and connections.

It is essential to understand the relationship between the fibres and the matrix composing a lamina or a ply and also the factors affecting mechanical behaviour and elastic properties. It is through the understanding of these factors that the engineer is able to specify appropriate types of resin and reinforcement for the design of a polymer composite structure.

4.10.2.1 Unidirectional reinforcement

There are several methods for predicting the elastic properties of unidirectional laminae. Three of the most suitable methods are presented. These are (a) the rule of mixtures method, (b) the self-consistent doubly embedded method and (c) Halpin-Tsai method.

The constitutive equations for the former two techniques are given below, whilst the Halpin-Tsai equations are included in the EUROCOMP Design Code.

One of the most important factors affecting composite properties is the volume percentage of fibres in a composite. Two composites made of the same resin and volume of fibres but different fibre densities will have identical mechanical properties. Also, closely packed fibres decrease fibre spacing, resulting in higher local stress concentrations in the material, and reduce complete penetration of resin into the interstices.

Prior to prediction of the elastic properties, the designer must determine the volume fraction of the fibres and the matrix. Since it is easier to measure weight than volume, the set of Equations (4.17) could be used to determine the volume and the weight fractions of the fibres and the matrix.

$$v_f = \frac{V_f}{V_c} = \frac{w_f\,\rho_m(1 - v_v)}{\rho_f + w_f(\rho_m - \rho_f)}$$

$$v_m = \frac{V_m}{V_c} = \frac{w_m\rho_f(1 - v_v)}{\rho_m + w_m(\rho_f - \rho_m)}$$

$$v_v = \frac{V_v}{V_c} = 1 - \frac{\rho_c w_f}{\rho_f} - \frac{\rho_c}{\rho_m w_m}$$

$$V_c = V_f + V_m + V_v$$

$$v_f + v_m + v_v = 1$$

$$w_f = \frac{W_f}{W_c} = \frac{\rho_f}{\rho_c} v_f \qquad \text{(4.17a to l)}$$

$$w_m = \frac{W_m}{W_c} = \frac{\rho_m}{\rho_c} v_m$$

$$w_c = w_f + w_m$$

$$w_f + w_m = 1$$

$$w_f = \rho_f V_f$$

$$w_m = \rho_m V_m$$

$$w_c = \rho_c V_c$$

(a) *Rule of mixtures method*

This method is based on the simplified assumption of uniform stress or strain and is a conservative method for predicting the elastic properties of a uni-directional composite lamina. The actual properties will always be higher than those predicted by the simple relationship given in Equations (4.18).

$$E_1 = V_f E_{1f} + V_m E_{1m}$$

$$E_2 = \frac{E_{2f} E_{2m}}{E_{2f} (1 - V_f) + E_{2m} V_f}$$

$$G_{12} = G_{13} = \frac{G_{12f} G_{12m}}{G_{12f} (1 - V_f) + G_{12m} V_f}$$

$$G_{23} = \frac{G_{23f} G_{23m}}{G_{23f} (1 - V_f) + G_{23m} V_f}$$

(4.18 a to 4.18 h)

$$\nu_{12} = \nu_{13} = V_f \nu_{12f} + V_m \nu_{12m}$$

$$\nu_{12} = - \frac{\varepsilon_2}{\varepsilon_1}$$

$$\varepsilon_1^E = \frac{\varepsilon_{1f}^E E_{1f} V_f + \varepsilon_{1m}^E E_{1m} (1 - V_f)}{E_{1f} V_f + E_{1m} (1 - V_f)}$$

$$\epsilon_2^E = V_f (\epsilon_{2f}^E + \nu_{12f}\epsilon_{1f}^E) + V_m\epsilon_m^E(1 + \nu_m) - (V_f\nu_{12f} + V_m\nu_m)\, \epsilon_1^E$$

Even though the method is relatively approximate it adequately predicts the longitudinal properties.

This method does not explicitly contain terms associated with the degree of bonding, fibre spacing, fibre shape, packing geometry and other influential factors which can significantly affect the shear moduli and transverse properties of the lamina. However, as this method is conservative, the actual properties will always be higher than those predicted.

(b) *Self-consistent doubly embedded method*

The basic assumption under this method of predicting the elastic properties of uni-directional composite materials is transverse isotropy. This is based on a doubly embedded model, shown in Figure 4.4, consisting of a typical fibre embedded in a micro-region of the continuous matrix phase which in turn is embedded in a homogeneous medium. The homogeneous outer material is taken to be a transversely isotropic material whose properties are identical to the effective properties of the composite material. The set of Equations (4.19) are obtained using a uniform strain field in the longitudinal direction of the fibres.

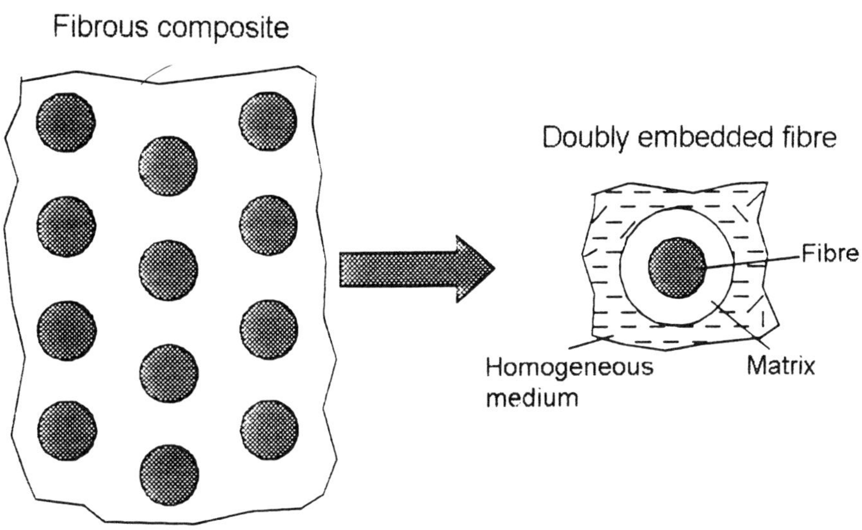

Figure 4.4 Self-consistent doubly embedded model.

The relationships of Equations (4.19) tend to underestimate properties at volume fractions above 0.2-0.3).

$$E_1 = E_{1f} v_f + E_{1m} (1 - v_f) + \frac{4 (\nu_{12m} - \nu_{12f})^2 k_{2f} k_{2m} G_{23m} (1 - v_f) v_f}{(k_{2f} + G_{23m}) k_{2m} + (k_{2f} - k_{2m}) G_{23m} v_f} \tag{4.19a}$$

$$\nu_{12} = \nu_{13} = \nu_{12f} v_f + \nu_{12m} (1 - v_f) + \frac{(\nu_{12m} - \nu_{12f}) (k_{2m} - k_{2f}) G_{23m} (1 - v_f) v_f}{(k_{2f} + G_{23m}) k_{2m} + (k_{2f} - k_{2m}) G_{23m} v_f} \tag{4.19b}$$

$$G_{12} = G_{13} = G_{12m} \frac{(G_{12f} + G_{12m}) + (G_{12f} - G_{12m}) v_f}{(G_{12f} + G_{12m}) - (G_{12f} - G_{12m}) v_f} \tag{4.19c}$$

$$G_{23} = \frac{G_{23m} [k_{2m} (G_{23m} + G_{23f}) + 2G_{23f} G_{23m} + k_{2m} (G_{23f} - G_{23m}) v_f]}{[k_{2m} (G_{23m} + G_{23f}) + 2G_{23f} G_{23m} - (k_{2m} + 2G_{23m})(G_{23f} - G_{23m}) v_f]} \tag{4.19d}$$

$$k_2 = \frac{(k_{2f} + G_{23m}) k_{2m} + (k_{2f} - k_{2m}) G_{23m} v_f}{(k_{2f} + G_{23m}) - (k_{2f} - k_{2m}) v_f} \tag{4.19e}$$

$$E_2 = E_3 = \frac{1}{\dfrac{1}{4\,k_2} + \dfrac{1}{4\,G_{23}} + \dfrac{\nu_{12}^2}{E_1}} \tag{4.19f}$$

$$\nu_{23} = \frac{2\,E_1\,k_2 - E_1\,E_2 - 4\,\nu^2_{12}\,k_2\,E_2}{2\,E_1\,k_2} \tag{4.19g}$$

$$\varepsilon_1{}^E = \frac{\varepsilon_{1f}{}^E\,E_{1f}\,v_f + \varepsilon_{1m}{}^E\,E_{1m}\,(1 - v_f)}{E_{1f}v_f + E_{1m}\,(1 - v_f)} \tag{4.19h}$$

$$\varepsilon_2{}^E = (\varepsilon_{2f}{}^E + \nu_{12f}\,\varepsilon_{1f}{}^E)\,v_f + (\varepsilon_{2m}{}^E + \nu_{12m}\,\varepsilon_{1m}{}^E)\,(1 - v_f)$$

$$- [\nu_{12f}\,v_f + \nu_{12m}\,(1 - v_f)] \left[\frac{\varepsilon_{1f}{}^E\,E_{1f}\,v_f + \varepsilon_{1m}{}^E\,E_{1m}\,(1 - v_f)}{E_{1f}\,v_f + E_{1m}\,(1 - v_f)}\right] \tag{4.19i}$$

$$q_1 = q_{1f}\,v_f + q_{1m}\,(1 - v_f) \tag{4.19j}$$

$$q_2 = q_{2m}\,\frac{(q_{2f} + q_{2m}) + (q_{2f} - q_{2m})\,v_f}{(q_{2f} + q_{2m}) - (q_{2f} - q_{2m})\,v_f} \tag{4.19k}$$

Under the assumption of transverse isotropy, properties perpendicular to the plane of a ply are assumed to be the same as the transverse properties within the plane of the ply. It should be noted that terms related to variations in the cross-sectional shape of the ply fibres and specific fibre packing geometries (e.g. rectangular, square, hexagonal) do not appear in the relationships. This is a result of the geometrical simplifications introduced in Figure 4.5. The simplifications introduced by the embedded cylinder model tend to mask the extent to which neighbouring fibres mutually influence their response characteristics. Consequently, the relationships of Equations (4.19) tend to underestimate properties at high volume fractions of fibres. Nevertheless, the set of Equations (4.19) provides, useful approximations for adequately predicting the elastic properties of uni-directional composite materials.

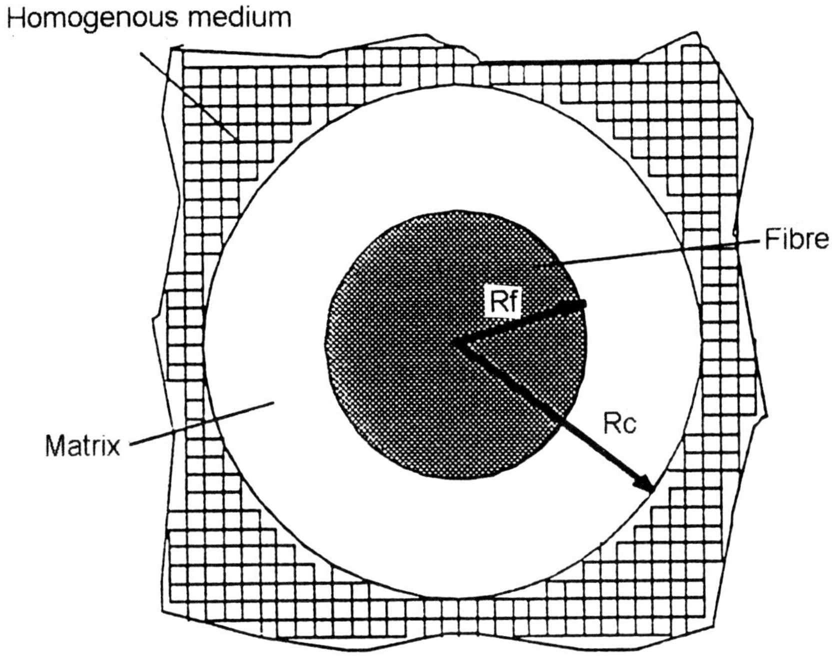

Figure 4.5 Geometry of self-consistent model.

(c) Halpin-Tsai method (a semi-empirical method)

The difficulties associated with predicting accurately the elastic properties of a uni-directional lamina using mathematical closed form solutions, prompted development of a number of semi-empirical relationships. The Halpin-Tsai method (the equations for which are included in the design document) provides the most popular and widely used relationships.

The Halpin-Tsai method relies on the selection of appropriate ξ values. Table 4.5 of the EUROCOMP Design Code gives suitable values for the various directional properties. The Halpin-Tsai method is an empirical one and the ξ have been obtained by comparing the results with experimental data.

4.10.2.2 Other typical reinforcements

Fabric composites have mechanical properties similar to those of laminates made from orthogonal uni-directional layers. However, fibre curvature arising from yarn twist and weave crimp makes fabric reinforcement less efficient than in the case of aligned straight fibres.

The elastic properties of hand lay-up laminates incorporating fabrics/woven roving are susceptible to significant batch to batch variation attributable variations in material properties, cure temperature and quality control, and thus the upper and lower bounds for elastic constants are provided in the code.

In a composite reinforced with randomly distributed fibres only a small percentage of the fibres contribute significantly to strength and stiffness when the material is loaded in any given direction. Consequently the tensile and compressive strengths and extensional moduli are much lower than the corresponding axial properties of a uni-directional composite having the same fibre volume fraction. However, the transverse properties are usually substantially greater than those of a uni-directional composite.

The random nature of the fibrous reinforcement means that this material may be considered isotropic for design purposes.

4.10.3 Laminate stiffness

General considerations

It is recognised that one of the outstanding features of composite materials is their ability to provide a structural component tailor-make to suit individual requirements of loading or stiffness.

A laminate specification notation code is devised to provide both a concise reference and a positive identification of the laminate. In this system of specifying the number of laminate plies and their orientation, angles are listed separated by a slash or comma, with the entire listing enclosed within square brackets. Where there is more than one lamina at any given angle, the number of laminae at that angle is denoted by a numerical subscript (the notation of degree is ignored for simplicity). For example: $[0_3/45_5/90_2]$ indicates 3 plies in the 0° direction, 5 plies in the 45° direction and 2 plies in the 90° direction.

A distinction is made between (+) and (-) angles. However, when there are the same number of (+) and (-) of the same angles, they are usually combined into a (±) type of notation. For example: $[0_3/\pm45_5/90_2]$ indicates 3 plies in the 0° direction, 5 sets of +45/−45 plies and 2 plies in the 90° direction.

If the laminate is symmetrical, the notation slightly differs as: $[0/45/-45/90/0]_s$ which indicates 4 plies in the 0° direction, 2 plies in the +45° direction, 2 plies in the −45° direction and 2 plies in the 90° direction. If the symmetrical laminate has an odd number of plies, the centre ply is overlined to indicate this condition. The subscript s denotes symmetry, indicating a mirror image of the laminate lay-up.

Loading in the longitudinal direction

A composite laminate is usually made of individual layers of lamina orientated at different angles relative to the laminate reference axis. Judicious choice of lamina orientations in a laminate produces stiffness and strength in the required direction of the laminate. Therefore, the properties of a layered laminate are dependent on the the properties of the individual laminae or properties and the configuration of the laminae in the laminate. It is therefore appropriate to consider properties of individual laminae prior to proceeding to determine the behaviour of laminates.

Consider the application of a tensile load in the longitudinal '1' direction, Figure 4.6.

$$\sigma = \begin{bmatrix} \sigma_1 & 0 \\ 0 & 0 \end{bmatrix}$$

$$\varepsilon = \begin{bmatrix} \varepsilon_1 & 0 \\ 0 & -\nu_{12}\varepsilon_1 \end{bmatrix}$$

$$\varepsilon_1 = a_{11}\sigma_1 = \frac{\sigma_1}{E_1}$$

$$\varepsilon_2 = a_{21}\sigma_1 = -\nu_{12}\varepsilon_1 = -\frac{\nu_{12}}{E_1}\sigma_1$$

$$a_{11} = \frac{1}{E_1} \tag{4.20}$$

$$a_{21} = -\frac{\nu_{12}}{E_1} \tag{4.21}$$

where ν_{12} is defined as the negative of the ratio of the strain in the transverse '2' direction to the strain in the longitudinal '1' direction due to a stress in the '1' direction. In other words, from the above

$$\varepsilon_2 = -\nu_{12}\varepsilon_1 \text{ or } \nu_{12} = -\frac{\varepsilon_2}{\varepsilon_1} \tag{4.22}$$

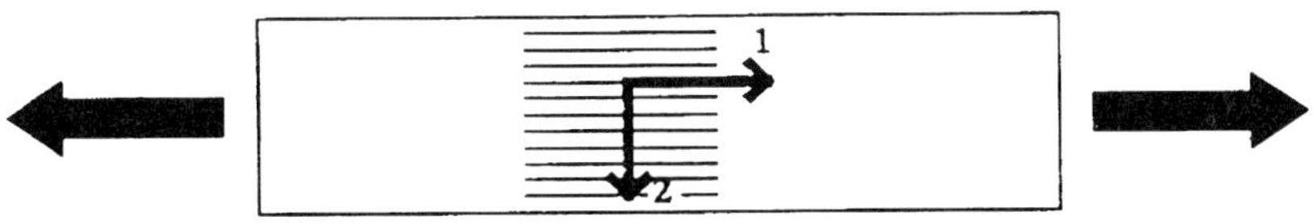

Figure 4.6 Longitudinal tension.

Loading in the transverse direction

Consider a load in the transverse '2' direction, Figure 4.7.

$$\sigma = \begin{bmatrix} 0 & 0 \\ 0 & \sigma_2 \end{bmatrix}$$

$$\varepsilon = \begin{bmatrix} -\nu_{21}\varepsilon_2 & 0 \\ 0 & \varepsilon_2 \end{bmatrix}$$

$$\varepsilon_1 = \mathbf{a}_{12}\sigma_2 = -\nu_{21}\varepsilon_2 = -\frac{\nu_{21}}{E_2}\sigma_2$$

$$\varepsilon_2 = \mathbf{a}_{22}\sigma_2 = \frac{\sigma_2}{E_2}$$

therefore

$$a_{12} = -\frac{\nu_{21}}{E_2} \qquad (4.23)$$

$$a_{22} = \frac{1}{E_2} \qquad (4.22)$$

noting that

$$\frac{\nu_{12}}{E_1} = \frac{\nu_{21}}{E_2} \qquad (4.25)$$

which is a most important relationship, because having physically measured any three of the four quantities in Equation (4.25) the fourth is easily calculated.

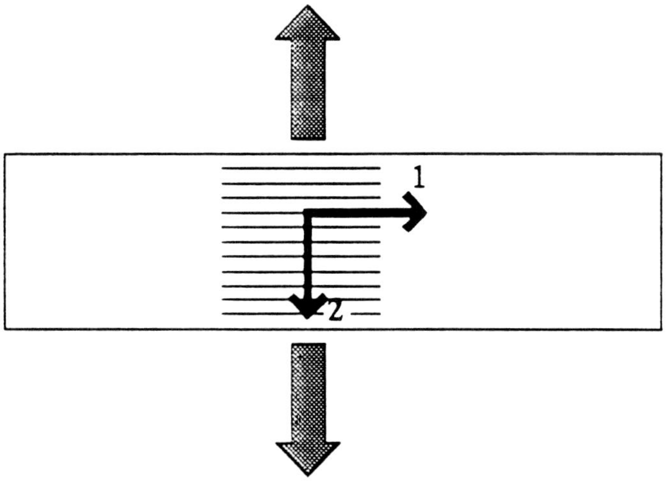

Figure 4.7 Transverse tension.

In-plane shear loading

Consider the application of a shear load, Figure 4.8,

$$\sigma = \begin{bmatrix} 0 & \sigma_{12} \\ \sigma_{21} & 0 \end{bmatrix}$$

$$\varepsilon = \begin{bmatrix} 0 & \varepsilon_{12} \\ \varepsilon_{21} & 0 \end{bmatrix}$$

hence

$$\varepsilon_{12} = \frac{\sigma_{21}}{2G_{21}} = \frac{a_{66}}{2}\sigma_{21}$$

or

$$a_{66} = \frac{1}{G_{21}} = \frac{1}{G_{22}} \tag{4.26}$$

The physical quantities appearing in Equations (4.21 to 4.26) are ε_1, ε_2, G_{12}, ν_{12}, and ν_{21}. However, because of Equation (4.25) the number of physical quantities to be measured for a two-dimensional orthotropic body are reduced from 6 to 4. It is important to note that, if the material were transversely isotropic such that the material were isotropic in the 1 - 2 plane, the physical quantities involved would be $E_1 = E_2$, G_{12} and $\nu_{12} = \nu_{21}$.

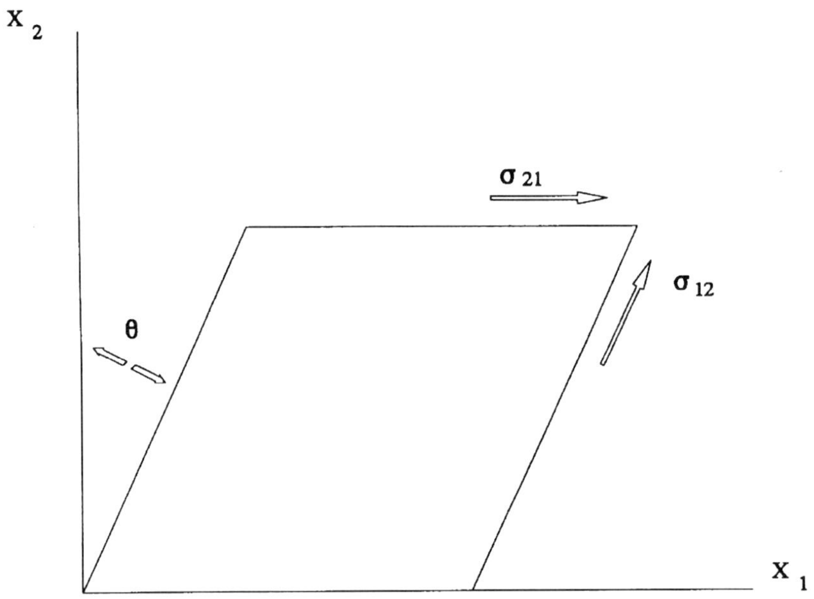

Figure 4.8 Shear test in the plane of x_1 - x_2.

However, because of the isotropic relationship and Equation (4.25) one of these quantities can be calculated:

$$G_{12} = \frac{E}{2(1+\nu)}$$

where $E = E_1 = E_2$ and $\nu = \nu_{12} = \nu_{21}$. The total number of unknown physical quantities becomes two.

Laminae of orthotropic materials

Figure 4.9 shows the positive directions of all stresses. A force equilibrium study can be carried out to relate σ_x, σ_y and σ_{xy} to σ_1, σ_2 and σ_{12}.

$$\sigma_1 = \sigma_x \cos^2\theta + \sigma_y \sin^2\theta + \sigma_{xy}(2\sin\theta \cos\theta) \tag{4.27}$$

$$\sigma_6 = -\sigma_x(\sin\theta \cos\theta) + \sigma_y(\sin\theta \cos\theta) + \sigma_{xy}(\cos^2\theta - \sin^2\theta) \tag{4.28}$$

$$\sigma_2 = \sigma_x \sin^2\theta + \sigma_y \cos^2\theta - \sigma_{xy}(2\sin\theta \cos\theta) \tag{4.29}$$

It can be seen that the Equations (4.27) to (4.29) are directly analogous to Mohr's circle analysis in the determination of the basic

$$\begin{Bmatrix} \sigma_1 \\ \sigma_2 \\ \sigma_6 \end{Bmatrix} = [\,T\,] \begin{Bmatrix} \sigma_x \\ \sigma_y \\ \sigma_{xy} \end{Bmatrix} \tag{4.30}$$

where

$$[\,T\,] = \begin{bmatrix} m^2 & n^2 & 2mn \\ n^2 & m^2 & -2mn \\ mn & mn & (m^2-n^2) \end{bmatrix} \tag{4.31}$$

in which m = cosθ, n = sinθ and θ is defined positive, as shown in Figure 4.10.

Similarly

$$\begin{Bmatrix} \varepsilon_1 \\ \varepsilon_2 \\ \varepsilon_{12} \end{Bmatrix} = [\,T\,] \begin{Bmatrix} \varepsilon_x \\ \varepsilon_y \\ \varepsilon_{xy} \end{Bmatrix} \tag{4.32}$$

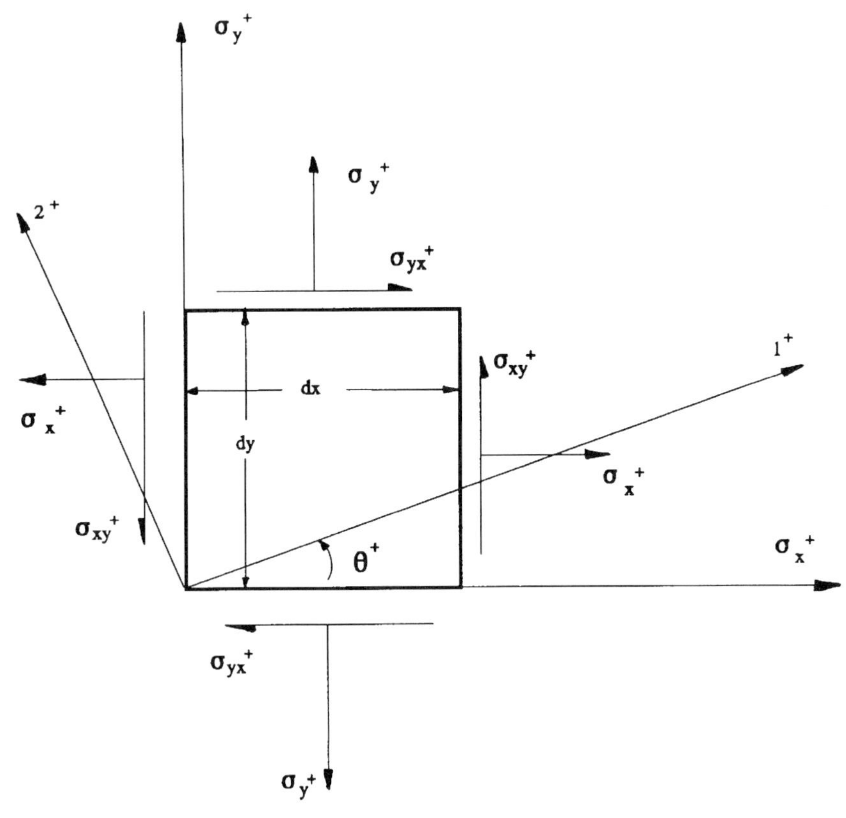

Figure 4.9 Lamina and laminate coordinate system.

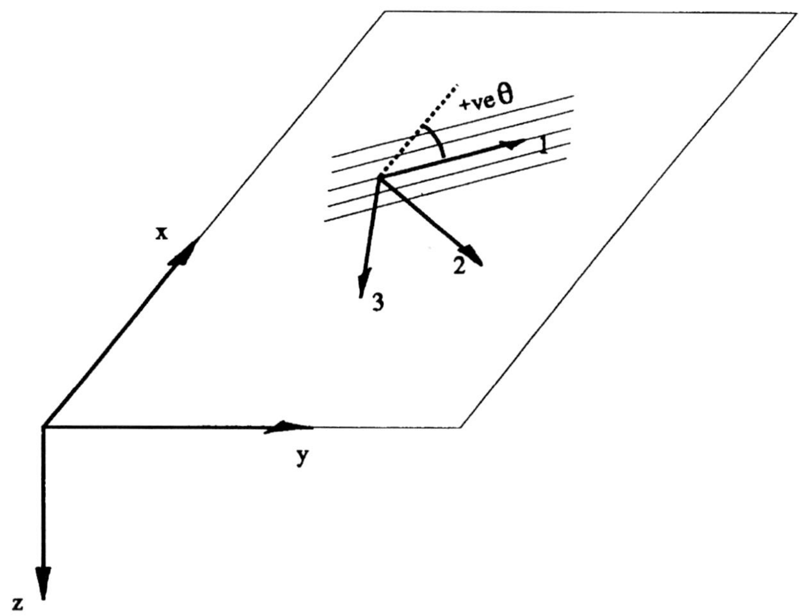

Figure 4.10 Directions of plies in a lamina.

Lamina of composite materials

strength of materials. In matrix form Materials which have two mutually orthogonal planes of elastic symmetry are called orthotropic (a shorthand term for orthogonally anisotropic). Thus

$$\begin{Bmatrix} \sigma_1 \\ \sigma_2 \\ \sigma_{12} \end{Bmatrix} = \begin{bmatrix} Q_{11} & Q_{12} & \\ Q_{21} & Q_{22} & \\ & & 2Q_{66} \end{bmatrix} \begin{Bmatrix} \varepsilon_1 \\ \varepsilon_2 \\ \varepsilon_{12} \end{Bmatrix} \tag{4.33}$$

$$\begin{aligned} Q_{11} &= E_1/(1 - \nu_{12}\nu_{21}) \\ Q_{22} &= E_2/(1 - \nu_{21}\nu_{12}) \\ Q_{12} &= \nu_{21}E_1/(1 - \nu_{12}\nu_{21}) \\ Q_{66} &= G_{12} \end{aligned} \tag{4.34}$$

The resulting inverted matrix is called the **compliance matrix** whose individual terms are

$$\begin{Bmatrix} \varepsilon_1 \\ \varepsilon_2 \\ \varepsilon_{12} \end{Bmatrix} = \begin{bmatrix} \frac{1}{E_1} & -\frac{\nu_{21}}{E_2} & \\ -\frac{\nu_{12}}{E_1} & \frac{1}{E_2} & \\ & & 2\frac{1}{G_{12}} \end{bmatrix} \begin{Bmatrix} \sigma_1 \\ \sigma_2 \\ \sigma_{12} \end{Bmatrix} \tag{4.35}$$

Hence the constitutive relation takes the following form:

$$\begin{Bmatrix} \sigma_1 \\ \sigma_2 \\ \sigma_{12} \end{Bmatrix} = \begin{bmatrix} \dfrac{E_1}{1 - \nu_{12}\nu_{21}} & \dfrac{\nu_{12}E_2}{1 - \nu_{12}\nu_{21}} & \\ \dfrac{\nu_{21}E_1}{1 - \nu_{12}\nu_{21}} & \dfrac{E_2}{1 - \nu_{12}\nu_{21}} & \\ & & 2G_{12} \end{bmatrix} \begin{Bmatrix} \varepsilon_1 \\ \varepsilon_2 \\ \varepsilon_{12} \end{Bmatrix} \qquad (4.36)$$

$$\begin{Bmatrix} \sigma_1 \\ \sigma_2 \\ \sigma_{12} \end{Bmatrix} = [Q] \begin{Bmatrix} \varepsilon_1 \\ \varepsilon_2 \\ \varepsilon_{12} \end{Bmatrix} \qquad (4.37)$$

If the direction of the plies of a lamina is assumed to be inclined at an arbitrary angle θ to the x - y plane, as shown in Figure 4.10, the principles of Equations (4.30) to (4.32) can be utilized to obtain

$$\begin{Bmatrix} \sigma_x \\ \sigma_y \\ \sigma_{xy} \end{Bmatrix} = [\,\overline{Q}\,] \begin{Bmatrix} \varepsilon_x \\ \varepsilon_y \\ \varepsilon_{xy} \end{Bmatrix} \qquad (4.38)$$

where

$$[\,\overline{Q}\,] = [\,T\,]^{-1}\,[\,Q\,T\,]$$

$$[\overline{Q}] = \begin{bmatrix} \overline{Q}_{11} & \overline{Q}_{12} & 2\overline{Q}_{16} \\ \overline{Q}_{12} & \overline{Q}_{22} & 2\overline{Q}_{26} \\ 2\overline{Q}_{16} & 2\overline{Q}_{26} & 2\overline{Q}_{66} \end{bmatrix} \qquad (4.39)$$

and the individual terms of Q_{ij} are:

$$\begin{aligned}
\overline{Q}_{11} &= Q_{11}m^4 + 2(Q_{12} + 2Q_{66})m^2n^2 + Q_{22}n^4 \\
\overline{Q}_{12} &= (Q_{11} + Q_{22} - 4Q_{66})m^2n^2 + Q_{12}(m^4 + n^4) \\
\overline{Q}_{16} &= mn^3Q_{22} + m^3nQ_{11} - mn(m^2 - n^2)(Q_{12} + 2Q_{66}) \\
\overline{Q}_{22} &= Q_{11}n^4 + 2(Q_{12} + 2Q_{66})m^2n^2 + Q_{22}m^4 \\
\overline{Q}_{26} &= m^3nQ_{22} + mn^3Q_{11} + mn(m^2 - n^2)(Q_{12} + 2Q_{66}) \\
\overline{Q}_{66} &= (Q_{11} + Q_{22} - 2Q_{12})m^2n^2 + Q_{66}(m^2 - n^2)^2
\end{aligned} \qquad (4.40)$$

Laminate of composite materials

Any structure of composite materials is formed of numerous layers of lamina bonded together. Figure 4.11 shows a laminate assumed to be comprised of n layers of lamina. A typical layer k is located such that its lower surface is a distance h_k from the mid-plane of the laminate and its upper surface is at a distance h_{k-1}.

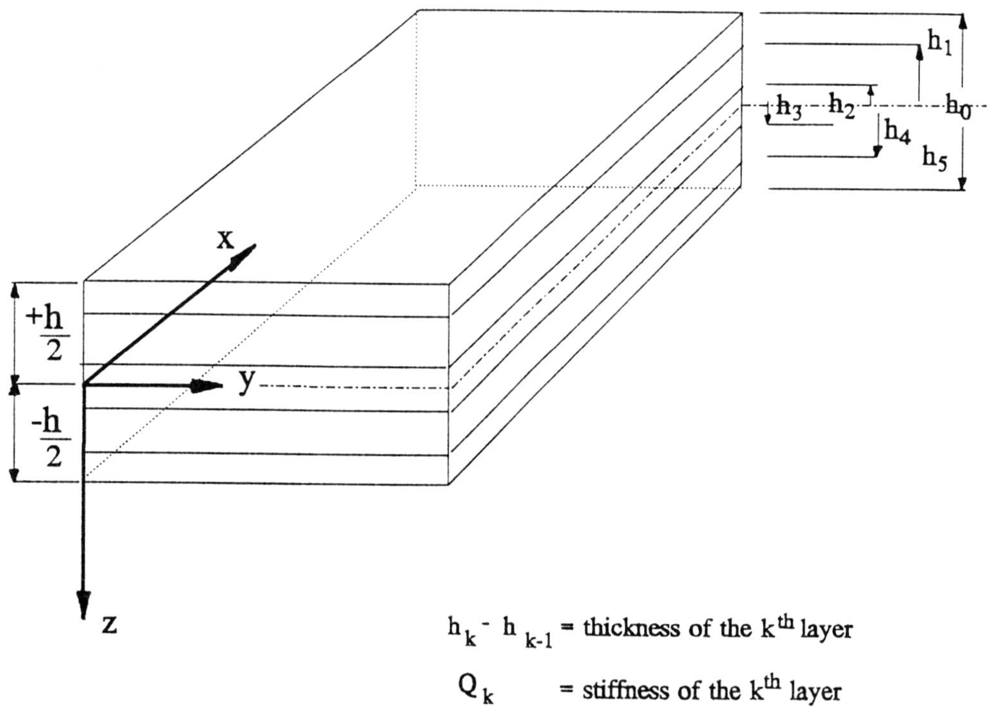

Figure 4.11 Composite laminate.

For a laminated plate the stresses that are integrated across the thickness of the plate or shell are the sum of the stresses across each lamina, as shown in Figure 4.12.

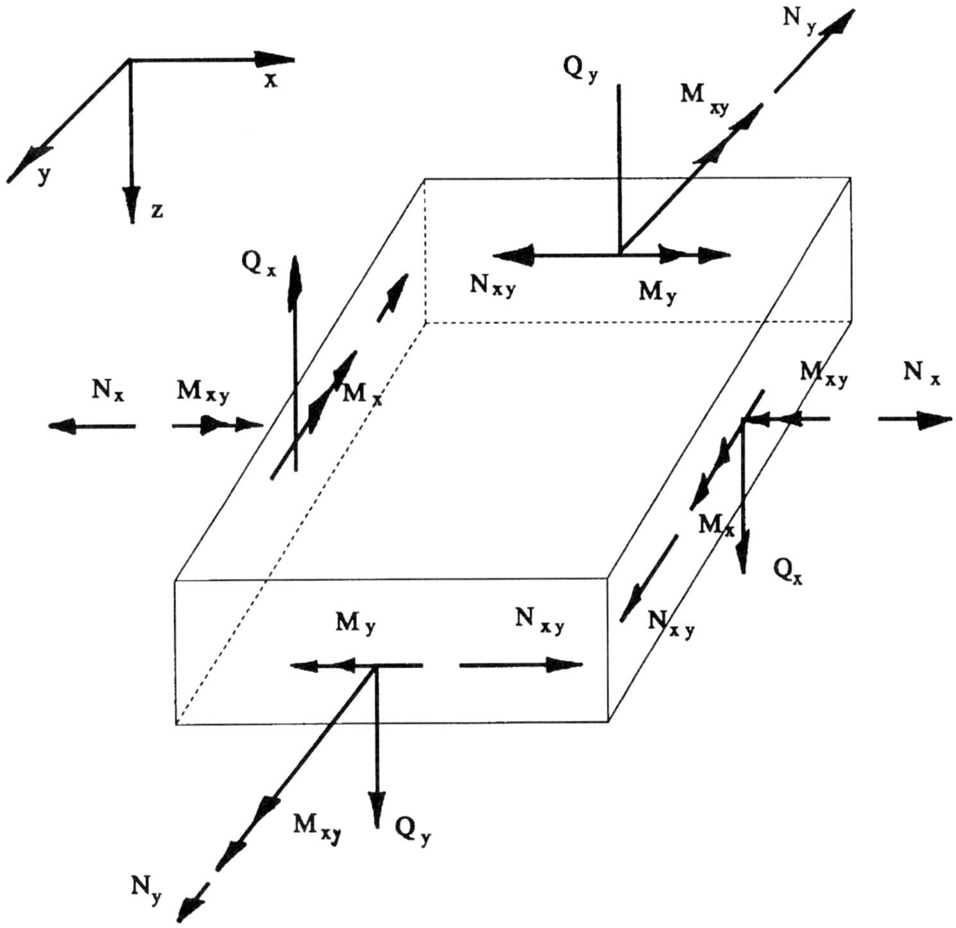

Figure 4.12 Plate moments and shearing forces.

$$\begin{Bmatrix} N_x \\ N_y \\ N_{xy} \end{Bmatrix} = \sum_{k=1}^{n} \left\{ \int_{h_{k-1}}^{h_k} [\,\bar{Q}\,]_k \begin{Bmatrix} \varepsilon_x^o \\ \varepsilon_y^o \\ \varepsilon_{xy}^o \end{Bmatrix} dz + \int_{h_{k-1}}^{h_k} [\,\bar{Q}\,]_k \begin{Bmatrix} \chi_x \\ \chi_y \\ \chi_{xy} \end{Bmatrix} z\, dz \right\} \tag{4.41}$$

where

$$\varepsilon^o{}_x = \frac{\partial u}{\partial x}, \quad \varepsilon^o{}_y = \frac{\partial v}{\partial y}, \quad \varepsilon^o{}_{xy} = \frac{\partial u}{\partial y} + \frac{\partial v}{\partial x}$$

$$\chi_x = -\frac{\partial^2 w^0}{\partial x^2} \quad \chi_y = -\frac{\partial_2 w^0}{\partial y^2} = \chi_{xy} = -2\frac{\partial^2 w^0}{\partial x \partial y}$$

However, since strains $\varepsilon^o_{(...)}$ and curvatures $\chi_{(...)}$ are not functions of z, and within each lamina $[\bar{Q}]_k$ is not a function of z either, Equation (4.41) can be written as

$$\begin{Bmatrix} N_x \\ N_y \\ N_{xy} \end{Bmatrix} = \sum_{k=1}^{n} \left\{ [\bar{Q}]_k \{\varepsilon^o\} \int_{h_{k-1}}^{h_k} dz + [\bar{Q}]_k \{\chi\} \int_{h_{k-1}}^{h_k} z\, dz \right\} \tag{4.42}$$

or in shorthand form

$$\{N\} = [A]\{\varepsilon^o\} + [B]\{\chi\} \tag{4.43}$$

From Equation (4.43) it is seen that the in-plane stress resultants for a laminated plate are not only a function of the mid-surface strains, as in a homogeneous plate, but are also in general a function of the curvature and twisting tautology. Stated in another way, in-plane forces can cause curvatures of the plate, and there is also an interplay between twisting and normal effects.

Similarly

$$\begin{Bmatrix} M_x \\ M_y \\ M_{xy} \end{Bmatrix} = \int_{-\frac{h}{2}}^{\frac{h}{2}} \begin{Bmatrix} \sigma_x \\ \sigma_y \\ \sigma_{xy} \end{Bmatrix} z\, dz = \sum_{k=1}^{n} \int_{h_{k-1}}^{h_k} \begin{Bmatrix} \sigma_x \\ \sigma_y \\ \sigma_{xy} \end{Bmatrix}_k z\, dz$$

$$= \sum_{k=1}^{n} \left\{ \int_{h_{k-1}}^{h_k} [\bar{Q}]_k \{\varepsilon^o\} z\, dz + \int_{h_{k-1}}^{h_k} [\bar{Q}]_k [\chi] z^2\, dz \right\}$$

$$= \left\{ \sum_{k=1}^{n} [\bar{Q}]_k \int_{h_{k-1}}^{h_k} z\, dz \right\} \{\varepsilon^o\} + \left\{ \sum_{k=1}^{n} [\bar{Q}]_k \int_{h_{k-1}}^{h_k} z^2\, dz \right\} \{\chi\} \tag{4.44}$$

$$[M] = [B]\{\varepsilon^o\} + [D]\{\chi\} \tag{4.45}$$

From Equation (4.45) it is seen that for a laminated plate not only are the stress couples a function of curvature, they are also a function of in-plane strains and displacements. Also M_x and M_y are influenced by twisting and M_{xy} is affected by mid-plane normal strains and curvatures.

Equations (4.43) and (4.45) can be written as

$$\begin{bmatrix} N \\ M \end{bmatrix} = \begin{bmatrix} A & B \\ B & D \end{bmatrix} \begin{bmatrix} \varepsilon^o \\ \chi \end{bmatrix} \tag{4.46}$$

It is seen that [A] is the extensional stiffness matrix, [D] is the flexural stiffness matrix and [B] is the bending-stretching coupling matrix.

The result is:

$$\begin{Bmatrix} N_x \\ N_y \\ N_{xy} \\ M_x \\ M_y \\ M_{xy} \end{Bmatrix} = \int_{-\frac{h}{2}}^{\frac{h}{2}} \begin{Bmatrix} \sigma_x \\ \sigma_y \\ \sigma_{xy} \\ z\sigma_x \\ z\sigma_y \\ z\sigma_{xy} \end{Bmatrix} dz =$$

$$\begin{bmatrix} A_{11} & & & & & \\ A_{12} & A_{22} & & & & \\ A_{16} & A_{26} & A_{66} & & & \\ B_{11} & B_{12} & B_{16} & D_{11} & & \\ B_{12} & B_{22} & B_{26} & D_{12} & D_{22} & \\ B_{16} & B_{26} & B_{66} & D_{16} & D_{26} & D_{66} \end{bmatrix} \begin{Bmatrix} \partial u/\partial x \\ \partial v/\partial y \\ \partial u/\partial y + \partial u/\partial x \\ -\partial^2 w/\partial x^2 \\ -\partial^2 w/\partial y^2 \\ -2\,\partial^2 w/\partial x\partial y \end{Bmatrix} \quad (4.47)$$

M_{xy} are the bending and twisting moments per unit length.

The laminate stiffness coefficients occurring in Equation (4.47) are defined

$$A_{ij} = \int_{-\frac{h}{2}}^{\frac{h}{2}} Q_{ij} dz = \sum_{k=1}^{n} (Q_{ij})_k (h_k - h_{k-1}) \qquad i,j = 1,2,6 \quad (4.48)$$

$$B_{ij} = \int_{-\frac{h}{2}}^{\frac{h}{2}} Q_{ij} z\, dz = \frac{1}{2}\sum_{k=1}^{n} (Q_{ij})_k (h_K^2 - h_{K-1}^2) \qquad i,j = 1,2,6 \quad (4.49)$$

$$D_{ij} = \int_{-\frac{h}{2}}^{\frac{h}{2}} Q_{ij} z^2\, dz = \frac{1}{3}\sum_{k=1}^{n} (Q_{ij})_k (h_K^3 - h_{K-1}^3) \qquad i,j = 1,2,6 \quad (4.50)$$

4.10.4 Lamina strength

Two ultimate limit state failure criteria are included in the EUROCOMP Design Code, namely:-

(a) The Hart-Smith failure criterion, (reference 4.7).
(b) The Tsai-Wu failure criterion (reference 4.8).

Whenever possible it is recommended that the Hart-Smith method be employed. It is, however, recognised that in certain instances there may be no experimental strength data available and hence conservative methods for calculating strengths of materials, for use with the Tsai-Wu technique, are included.

When using the Tsai-Wu failure criterion, the laminate strains, obtained by global analysis of the structural system, are used to obtain the respective lamina stresses. The laminate is deemed to have failed at first ply failure, i.e. when the failure criterion is no longer satisfied for any one of its constitutive laminae.

Lamina and laminate strengths are dealt with simultaneously in this document. The strength of a laminate is a function of the strengths of its laminae.

When using the Hart-Smith approach, the laminate failure envelope is constructed by superimposing the failure envelopes of its constitutive laminae or alternatively by using failure data measured at the laminate level to construct a failure envelope in the same manner as for a single ply.

(a) *The Hart-Smith failure criterion*

The Hart-Smith criterion proposes that the strength of laminates be characterised by generalizing the maximum-shear-stress criterion for ductile isotropic metals.

The method is confined to strong stiff fibres embedded in a soft matrix and thus encompassing the fibre/resin composites considered in this code.

This technique has been used extensively in the aerospace industry and it is recommended that it be used in this code.

The theory considers multiple failure modes by superposition, in contrast to the Tsai-Wu failure criterion where there is assumed to be interaction between the various stress components.

In the case of laminate design the stresses normal to the plane of the laminate are typically zero, or negligible, and hence the maximum-shear-stress criterion imposes a limit on the remaining two principal stresses.

The maximum-shear-stress criterion states that yield (or failure) occurs whenever the maximum shear stress, which is necessarily half the difference between the principal stresses, reaches a critical value.

The strain failure envelope for an isotropic material can be adjusted to account for an orthotropic material having multiple Poisson's ratios, the transverse one of which is small for most fibre/polymer composites. The development of the failure envelope, starting from a single measured strength, is illustrated in Figure 4.13. By considering failure in the strain plane it becomes possible to superimpose failures, by different mechanisms, on a common reference plane.

However, as the fibres are supported by a polymer matrix no more stress can be applied to the fibres than the matrix or interfaces will permit and thus other possible failure modes must be accounted for. One of these, for longitudinal compression, is characterised in Figure 4.13. Different cutoffs apply for transverse tension/compression. Therefore by measuring the four characteristic strengths of the lamina, namely longitudinal tension/compression and transverse tension/compression, the complete failure envelope in the strain plane can be constructed.

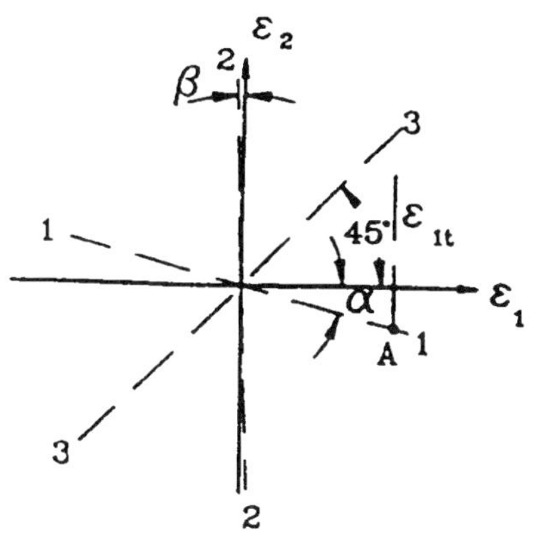

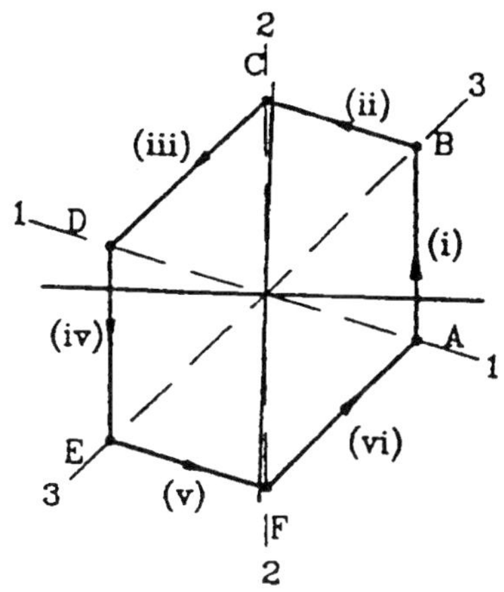

(a) and (b)

(c)

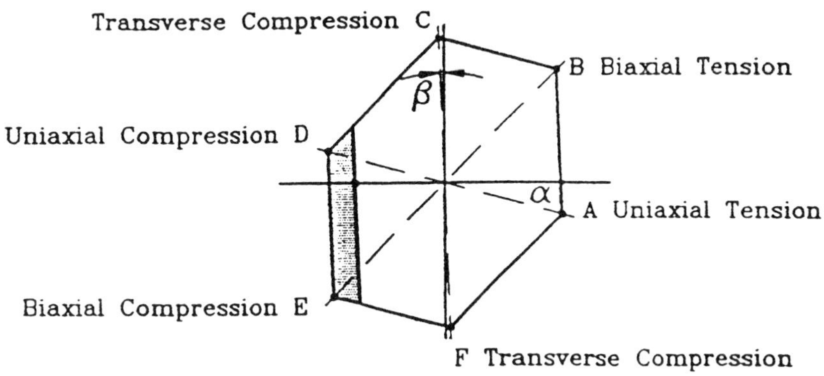

(d)

Figure 4.13 Construction of Hart-Smith failure envelope.

(a) Measure stain to failure in tension $\varepsilon_{1t.}$

(b) Using Poisson's ratios ν_{12} and ν_{21} locate diagonals for the failure envelope and locate first corner 'A'. (Note: $\alpha = \arctan \nu_{12}$ and $\beta = \arctan \nu_{21.}$)

(c) Construct Envelope

- 'B' - line parallel to 2-2 from 'A'
- 'C' - line parallel to 1-1 from 'B'
- 'D' - line parallel to 3-3 from 'C'
- 'E' - line parallel to 2-2 from 'D'
- 'F' - line parallel to 1-1 from 'E'

- Line parallel to 3-3 from 'F' to 'A' closes the failure envelope.

(d) Apply 'cut-offs' as required. If, for example, the strain to failure in compression is less than the strain to failure in tension ($\varepsilon_{1t} < \varepsilon_{1t}$), a 'cut-off' is inclued at $\varepsilon_1 = \varepsilon_{1c}$, as illustrated.

Having established the failure envelope for a single ply (lamina), the failure envelope for a multi-ply laminate can be readily obtained. Figure 4.14 shows how the failure envelope for a four ply laminate, with plies orientated at 0°,90°,−45° and +45° respectively, can be constructed.

Figure 4.14 shows the failure envelope for the 0° ply constructed from data obtained from a longitudinal tensile test. The failure envelopes for the 90°,−45° and +45° piles are given in Figure 4.14 and are obtained by reflecting the 0° envelope in lines drawn at +45°,−22.5° and +22.5° to the reference axes system respectively. Finally, the plies in Figure 4.14 are superimposed to obtain the failure envelope for the laminate.

The shaded area shown in Figure 4.14 corresponds to first ply failure and should be used for design. (The designer should note that for the purposes of illustration the various cut-offs accounting for remaining characteristic strengths have not been considered.)

It is important to note that since the theory is based on **measured** ultimate strains all possible failure modes are accounted for, as the ultimate strains will necessarily be governed by the weakest mode for each kind of in-plane stress.

(b) *The Tsai-Wu failure criterion*

The Tsai-Wu failure criterion considers interaction between the various stress components. The Tsai-Wu failure criterion for plane stress loading conditions is given here in terms of a calculated stress ratio.

A lamina is deemed to have failed when the inequality in Equation (4.51) is no longer satisfied.

$$\begin{aligned} &\frac{1}{\sigma_{1t}} + \frac{1}{\sigma_{1c}}\sigma_1 + \frac{1}{\sigma_{2t}} + \frac{1}{\sigma_{2c}}\sigma_2 \\ &- \frac{1}{\sigma_{1t}\,\sigma_{1c}}\sigma_1^2 - \frac{1}{\sigma_{2t}\,\sigma_{2c}}\sigma_2^2 \\ &+ \frac{1}{\sigma_{12s}^2}\sigma_{12}^2 - \frac{0.5}{\sqrt{\sigma_{1t}\,\sigma_{1c}\,\sigma_{2t}\,\sigma_{2c}}}\sigma_1\,\sigma_2 \leq 1.0 \end{aligned} \tag{4.51}$$

The Tsai-Wu failure criterion requires prior knowledge of the lamina longitudinal tension and compression strengths, the transverse tension and compression strengths and the shear strength in the 1-2 (longitudinal-tranverse) plane of the lamina.

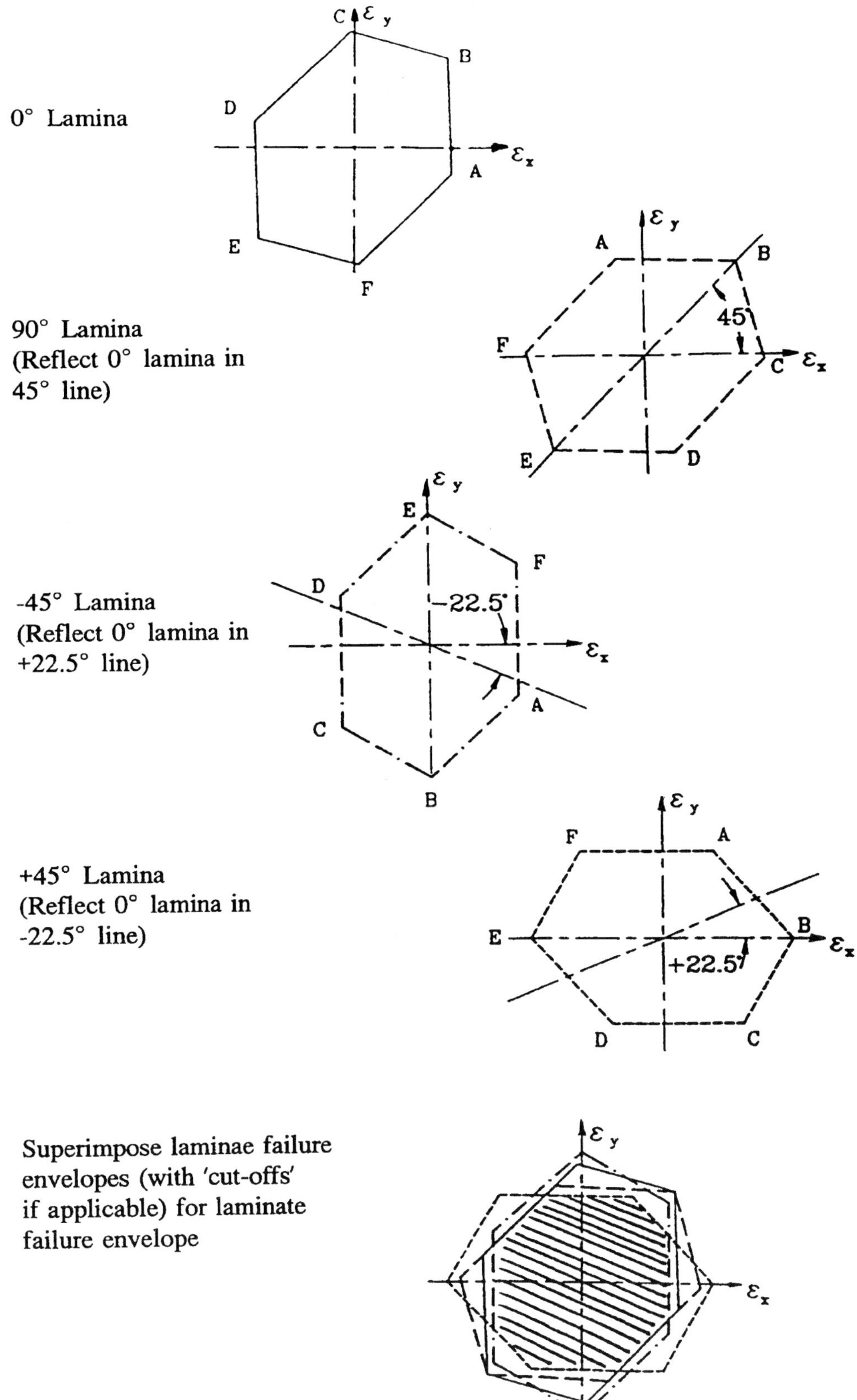

Figure 4.14 Construction of the Hart-Smith failure envelope for a 0°/90°/-45°/+45° Laminate.

For a multi-ply laminate the Tsai-Wu failure criterion should be evaluated for stress components parallel to and perpendicular to the fibre directions, for each lamina of the laminate. The laminate is deemed to have failed if Equation 4.51 is not satisfied for any of its constitutive laminae, i.e. for first ply failure.
The choice of failure criterion for design will be governed by the amount of measured data available to the designer. If measured data are available, the designer should use the Hart-Smith failure criterion.

In the event that no measured data are available the designer can calculate conservative estimates of material strength, using the steps given in the EUROCOMP Design Code and explained in more detail in this section.

The following paragraphs illustrate the complexity of the failure mechanisms associated with fibrous composites and the subsequent difficulty in providing closed-form numerical-expression methods for calculating the strength of such materials.

Hence for the purposes of the EUROCOMP Design Code it is assumed that the strain to failure of the fibres is smaller than that of the matrix and that the matrix can support no load once the fibres reach their failure strain. Therefore Equation (4.58) of the EUROCOMP Design Code is taken as a conservative estimate of the longitudinal strength of the composite.

For the transverse properties the reinforcing effect of the fibres is ignored and the failure strength is taken as that of the matrix.

Note that these calculations are conservative and are included to allow the designer to use the Tsai-Wu failure criterion in the absence of measured data. However, for an economic design it is recommended that the lamina strengths be determined experimentally and that the Hart-Smith failure criterion be employed.

4.10.5 Laminate strength

Laminate strength has been covered in 4.10 *Lamina strength*.

4.10.6 Computer programs for analysis and design

4.10.6.1 General

The unique characteristics of composites cannot be utilized to their full extent without extensive computing. Finite element programs are available for the analysis of structures, but they do not provide efficient tools for the analysis and design of laminates and laminated structural elements such as continuous and notched laminates, bars and plates. A computer program developed specifically for these purposes is a valuable tool for the designer. A lot of time is wasted in mechanical calculations if an appropriate program is not available. The lack of a suitable program may also lead to simplifying assumptions which often result in an inefficient structure or, in the worst case, in a structure which does not meet the requirements.

Computer codes for the analysis and design of laminates and laminated structural elements are discussed in 4.10.6.2 - 4.10.6.5. General purpose finite element (FE) programs are not considered. Summaries on their applicability to composite analyses are available (reference 4.9 - 4.11).

4.10.6.2 Computing needs

Analyses

Important analyses related to the design of composite structures are described below. The need for a computer code for different analyses is identified. Theoretical background for the analyses is not given since the theory is covered in other sections of the EUROCOMP Handbook and in many textbooks (references 4.3, 4.8 and 4.10 - 4.14).

Micromechanical analyses give estimates for the ply engineering constants and hygrothermal expansion coefficients, and for the ply strength in the principal loading conditions. They are used to provide initial ply data for laminate analyses if more reliable test data do not exist. A computer code for micromechanical analyses is useful but it is not necessary since the analyses are rather simple.

Laminate stiffness analysis predicts the constitutive behaviour of a laminate, based on classical lamination theory (CLT). The result is often given in the form of stiffness and compliance matrices. Engineering constants, i.e. the in-plane and flexural moduli, Poisson's ratios and coefficients of mutual influence, are further derived from the elements of the compliance matrix. Analyses are continuously needed in structural design since it is essential to know the constitutive behaviour of laminates forming the structure. The results are also the necessary input data for all other macromechanical analyses. A computer code for the stiffness analysis is a valuable tool on account of the extensive calculations related to the analysis.

Laminate hygrothermal analysis predicts, based on the CLT, the hygrothermal expansion of a laminate, i.e. the thermal expansion corresponding to the unit temperature change and the moisture expansion corresponding to the reference moisture content change. For unsymmetric laminates, two sets of expansion coefficients are typically derived: one for a laminate which is free to curve, another for a laminate with suppressed curvature. The features described above for a laminate stiffness analysis also apply to the hygrothermal analysis.

Laminate load response analysis determines the response of a laminate to a given loading condition, based on the CLT. The results indicate how the laminate and its plies react when subjected to a loading. The analysis is frequently needed in a design process. Since the analysis is also complicated, a computer code that performs the analysis is one of the most important tools of the designer. The tool should have the following features:

- it should calculate the response both to a defined stress state and to a defined strain state since structural requirements typically include both types of loading conditions
- it should allow a stress state both with mechanical forces and moments per unit width and with normalized stresses to be specified. The ability to specify mechanical in-plane forces in a zero-curvature state is essential when unsymmetric laminates are analysed
- it should allow a strain state both with in-plane strains and curvatures and with in-plane and flexural strains to be specified
- it should allow the loading condition to include internal loads, i.e. thermal and moisture loads, since their effect on the laminate response is often significant.

Laminate failure analysis predicts the criticality of a loading condition by comparing actual ply stresses and strains with ply stresses and strains corresponding to the failure of the laminate. Failure criterion functions are used to take into account the combined effect of stress or strain components. The criticality of the loading condition is indicated with a reserve factor which defines how much the load can be increased before failure occurs. The importance of the analysis is evident. The need for a computer code is also obvious since the failure analysis is even more complicated than the load response analysis.

The following specification may be set for the tool performing the analysis:

- it should allow the user to define loading conditions as specified above for the laminate load reponse analysis tool
- it should compute the reserve to the first ply failure (FPF) load and to the last ply failure (LPF) load. The FPF load indicates when the first failure occurs in a ply of the laminate. The LPF load represents the ultimate load-carrying capability of the laminate, i.e. it corresponds to the load that causes the final failure of the last load, carrying ply
- in FPF analyses it should identify the reserve to the failure for each ply of the laminate
- it would be advantageous if the criticality of a loading condition consisting of two parts could be analysed since loads of different origin, e.g. static and dynamic loads, are often applied simultaneously to a structure.

Laminate failure envelope analysis determines the combinations of selected load components that cause the failure of a laminate. Envelopes are constructed from the results of consecutive failure analyses. A computer code for creating failure envelopes is useful since the envelope in graphical form displays clearly the load-carrying capability of the laminate. Such a tool should allow the designer:

- to create both FPF and LPF envelopes
- to create envelopes in stress and strain space for any two load components.

Free-edge analysis is used to solve the three-dimensional stress state close to the free edge of a laminate when the results given by the CLT are not valid. An analytical solution for the stress state exists for the case where a uni-axial strain is applied to the laminate. Numerical analyses, i.e. the FE technique, can be used to solve the stress state in more complicated loading conditions. Three-dimensional failure criteria have been introduced for the prediction of the criticality of stresses. In practice, the free-edge analysis is seldom needed since laminates are often lightly loaded around the free edges. A computer program for the analysis is still important since the analysis is complicated and should be performed as required. This tool should provide:

- an analytical solution for the stress state in uni-axial strain
- a numerical solution for the stress state in more complicated loading conditions
- a failure analysis indicating the criticality of the loading.

Notched laminate analysis predicts the stress state around the hole and evaluates the criticality of these stresses. Stress analyses for notched anisotropic laminates are extensions to the analyses of notched isotropic plates. Specific failure criteria have been developed for failure analyses. The analysis is often needed because requirements set for the assembly, use and maintenance of a structure typically result in notched laminates. Since the analysis is also rather complicated, a computer code should be available at least for the evaluation of stress concentrations around the common circular holes. A possibility to analyse laminates with other shapes of holes is advantageous.

Load response and failure analyses of bars, beams, plates, shells, and mechanical and adhesively bonded joints are common in structural design, as well as *stability and free vibration analyses of bars, plates and shells.* These analyses can always be performed with general purpose FE programs. Nevertheless, computer codes developed specifically for these analyses are useful because FE programs are laborious to use. These tools should have the following features:

- they should take into account, as applicable, the effects of out-of-plane shear deformations, which may be considerable, especially in sandwich structures
- they should consider the local stability of skins in stability analyses of sandwich structures
- they should consider both strength and stability in failure analyses of bars, beams, plates and shells.

Sensitivity analyses predict the performance of laminates and structural elements within a specified tolerance set for a design variable such as ply orientation or fibre fraction. Sensitivity analyses are very important since the designer should always check the effect of manufacturing tolerances on the performance of the structure. Computer codes are extremely useful on account of the extensive calculations related to these analyses. In practice, all the computer codes identified above should provide the possibility to specify a tolerance for each important design variable.

Design

A design task is an inverse problem where solutions meeting specified requirements and targets are sought. In particular the following design tasks are often faced by the designer:

- selection of materials, plies and laminates
- creation of feasible laminates
- optimization of laminates.

Design tools have been developed specifically for these tasks. Tools are typically computer codes, which may be based on analysis algorithms, rule-based inference processes and/or general optimization techniques. Tailored tools are needed for different structural elements because their design principles are dissimilar. Design tools may be useful since many parameters related to a laminate structure complicate the design. However, they are not necessary since each of the tasks can be carried out by solving direct problems, i.e. by creating possible solutions, by running necessary analyses and by studying results.

4.10.6.3 General requirements for a computer program

General requirements for an effective and user-friendly analysis and design program are identified below.

An integrated system, including tools for all important analysis and design tasks, is the best alternative from a designer's point of view as it relieves the designer from having to learn the use of several systems. The system should be designed so that the input data need be fed in only once, i.e. the input data should be available for all analysis and design tools included in the system.

The existence of a database is an important requirement. Without a database all the necessary data need to be typed in each time the system is used. The database should allow the user to store specifications of materials, plies, laminates, structural elements, and loading conditions. In addition, it should allow the user to store results achieved with analysis/design tools. The possibility to create directories and subdirectories for different projects, subsystems and parts should be available.

An efficient and user-friendly user interface is another major requirement. The user interface should allow effective input of data and provide tools for viewing, editing, deleting, searching and sorting the data. Specifying an analysis/design task should be easy and results should be well organized. The possibility to perform comparative analyses should be available, e.g. it should be possible to analyse simultaneously the performance of several laminates or the criticality of several loading conditions. A graphic result viewer should also be available for informative displaying of results. Examples of results for which the graphic viewer is needed are the stress distributions through the thickness of a laminate, failure envelopes, and results of comparative analyses.

Interfaces for exporting and importing data are useful in all computer programs. In particular the following interfaces are considered essential for the system under consideration:

- interfaces for exporting material, ply and laminate data to commonly used FE programs. These are helpful for the designer since FE programs are often needed for global structural analyses and they do not provide efficient tools for producing material, ply and laminate specifications
- interfaces for importing load data from the design database and from the FE programs
- interfaces for importing material and ply data from material databases.

Easy expandability is a key requirement for an analysis and design program since it is impossible to develop software that meets all the needs of a designer. Moreover, enhanced theories are being continuously introduced in the field of composite structures.

Software documentation should include at least a user's guide and a theoretical Handbook. The latter is very important since different approaches are in use in many analyses. Another problem is the lack of established universal notation and conventions.

Finally, *the operating environment* should be compatible with the hardware and software used in design.

4.10.6.4 Programs available

Numerous programs have been developed for the analysis and design of composite laminates and laminated structural elements (reference 4.15). With some exceptions the programs have been developed for personal computers. Well-known, commonly available programs are briefly reviewed below. Individual programs are not evaluated in detail since many of them are being continuously developed. Suppliers or organizations involved in development are identified as a source for further information.

Most of the analysis programs have been developed for laminate analyses. They are based on the classical lamination theory (CLT) and are normally capable of stiffness, load response and failure analyses of laminates. Hygrothermal effects can typically be included in the analyses. Micromechanical analyses are included in some programs. The programs GENLAM/LAMRANK originating from the work of S.W. Tsai are probably the most well-known tools (reference 4.8). Laminate analysis programs have also been developed by the Cranfield Institute of Technology (reference 4.16), Imperial College (reference 4.17), and the University of Delaware (references 4.18 - 4.20).

An example of a program developed for the analyses of structural elements is a simple BASIC program developed by Whitney for bending, buckling and vibration analyses of beams and rectangular plates (reference 4.3). The analysis results are either exact solutions or approximate solutions based on energy principles. Plate analysis programs using the finite element technique have also been introduced (references 4.21 and 4.22). A cylinder analysis program has been introduced for instance by the University of Delaware (reference 4.23).

Programs for the design of laminates and laminated structural elements are much more complicated than the analysis programs. Therefore, only a few design programs have been introduced. Some programs have been developed specifically for the optimization of composite structures. Attempts have also been made to develop expert systems for the design of laminates and laminate structures. Typically these programs are not yet practical tools for a designer.

Several organizations have developed program packages that are capable of performing laminate analyses, structural element analyses and possibly micromechanical analyses. The programs provided by ESDU (Engineering Sciences Data Unit) probably form the most comprehensive set, consisting of almost twenty separate programs (references 4.24 - 4.41). Examples of integrated programs are the Mic-Mac spreadsheet programs (references 4.8)], CEMCAL from the University of Delaware (reference 4.42), ICAN developed by the National Aeronautics and Space Administration (NASA) (reference 4.43), ASCA developed by AdTech Systems Research Inc. (reference 4.44), and LAMDA from the Helsinki University of Technology (reference 4.45). The program CADFIBRE is reported to have a construction and dimensioning module that allows the user to design and dimension structural elements such as springs, shafts, tubes, plates and sandwich panels (reference 4.46).

Examples of recent integrated programs are ESAComp developed for the European Space Agency (ESA) at the Helsinki University of Technology (reference 4.47) and (reference 4.48), DAC from the University of Zaragoza (reference 4.49), and LamTech from the DLR (reference 4.50). All these programs are planned to have versatile analysis capabilities. Design tools are under development at least for the ESAComp system.

The support systems of existing programs are not discussed in detail. However, some typical shortcomings are highlighted in order to provide a check-list for the evaluation of programs:

- though comparative analyses are very useful, many analysis programs allow the user to perform only one analysis at a time, e.g. to analyse one laminate under one loading condition
- the user interfaces of the existing programs are usually not satisfactory. Even the interactive programs are typically very formal,

forcing the user to proceed step by step from the input data to the results. Only a few programs include graphic output for the visualization of data

- some existing programs do not have any database. Thus, the user has to enter all the initial data for every analysis or design task. Even if a database exists, it may be very limited
- external interfaces to other programs needed in the analysis and design of composite structures are not normally provided
- most of the programs do not have a modular structure. This makes it difficult to merge new capabilities into the programs.

4.10.6.5 Program selection

Requirements set for an analysis/design program may differ considerably depending on the application area. Since many of the introduced programs are also continuously under development and new programs are being introduced, it is difficult to recommend any individual program.

The following guidelines may be given for the program selection:

- an analysis program should be available for all composite designers. The needs for the different analyses identified in 4.10.6.2 should be evaluated one by one when the program specification is being prepared. Typically, the possibility to perform laminate stiffness analyses, hygrothermal analyses, load response analyses and failure analyses is a minimum requirement
- design tools may be useful but they are typically not necessary in the design of conventional composite structures
- a serious attempt should be made to find integrated software which includes all the required analysis and design capabilities. The system should further meet all the other requirements identified in 4.10.6.3.

4.11 DESIGN DATA

No additional information.

4.12 CREEP

The information in this section is taken from references 4.51 - 4.62.

4.12.1 Considerations

All polymer materials used in reinforced plastics display some viscoelastic or time-dependent properties. The origins of creep in composites stem from the behaviour of polymers under load together with local stress redistributions between fibre and matrix as a function of time. There is little creep at normal temperatures in the reinforcing fibres. The origin of the creep mechanisms is related to the nature and levels of internal bonding forces between the chains of the polymer, which are influenced by temperature and moisture.

The strength and mechanical rigidity of themosetting polymer resin matrices are dependent on the nature of the polymer chain structure, the density of the crosslinks or length between the chain nodes, and the magnitude of the internal forces holding the polymer chains in position. In general in crosslinked thermosets there are sites where chains are linked by strong chemical bonds and other sites along the chains where rigidity is conferred by lower forces of a physical nature which are temperature determined. In addition some of the internal rigidity is due to physical entanglement of the three dimensional chain structure. The magnitude and dynamic state of these internal physical forces is determined by temperature through the oscillatory motion of the chains. As molecular vibration increases, these local chain attractions can weaken, resulting in a less rigid structure. The "glass transition temperature" point marks the change from a rigid to a more compliant (rubbery) material and is an important property to be considered when selecting materials for creep resistance. Even if the temperature is below this critical point sustained mechanical forces may still overcome the local internal balance of forces, leading to internal rearrangement and deformation.

The magnitude of a creep effect will clearly be dependent on the magnitude of the imposed external loading, the local molecular strains and the energy required for the internal polymer chain movement/ rearrangment. It follows that a basic requirement to minimise creep is to ensure that service temperatures do not approach the glass transition temperature of the polymer and that resin stresses should be minimised. Viscoelastic deformation is a function of time and load level, with the delayed deformation being associated with the kinetics of molecular rearrangement for that temperature. At low loads, although there may be deformation, this may simply result from chain stretching, and on removal of load, full recovery (but delayed with time) can be observed. At higher loads chain scission and permanent chain rearragements will limit the extent of recovery but part of all creep deformations can be recoverable except where significant cracks develop. It should be noted that the uptake of moisture can have a weakening influence on the internal bonding forces, reducing the glass transition temperature and softening the resin. Such effects should be taken into account when selecting matrix materials for FRP composites.

4.12.1.1 Quantification of creep

At low levels of loading creep effects are small and are normally recoverable, although material "shake down" effects should be recognised.

Even with a carefully controlled manufacturing environment there will be regions within the composite of differing compliance and stress variations which will lead to some redistribution of the internal forces.

At higher levels of loading creep may be caused progressively by internal damage, associated with several mechanisms, such as resin cracking at transverse fibres, fibre debonding at interfaces, fibre breakage, delamination between lamina, etc., resulting in increased creep rates, permanent deformation and ultimately failure. There are three distinct stages of deformation with time:

- Stage I, where under the influencing load there is an initial instantaneous deflection and a delayed deformation but with creep rate decreasing with time. This is apparent when stresses and strains are low, less than 0.2%, and results in a steady state of effectively zero creep rate and on removal of load the total deformation is largely recoverable with time

- Stage II, is the subsequent phase, where there is a period of low constant creep rate over a significant time which is associated with small but finite damage growth owing to internal stress rearrangements. A linear rate of creep is usually seen over several decades on a log(time) plot and this stage can still be associated with a significant degree of deformation recovery. Provided that the creep deformation is not excessive, the residual strength of the material is often little different from the original material. This region may be utilised in design if the application is well controlled against environmental effects

- Stage III, the final phase. where there is a progressive increasing creep rate leading on from Stage II which ultimately results in catastrophic failure. At high levels of loading Stage III may follow on immediately from Stage I.

Creep plots may be directly measured by conventional strain or deformation measuring techniques, although problems can arise with instrumentation when small deformations over long time periods are required. Strain and deformation gauges are commonly used and temperature fluctuations must be minimised to obtain reliable data. Deflection or surface strain data are obtained as a function of time for a set of differing loading levels and the data are usually plotted on a logarithmic timescale. The linear representation of the creep behaviour over several decades as shown in Figure 4.15 is often projected but must be viewed with caution. If the data are established for a range of stresses, as is usual, one can construct isochronous stress-strain curves, as shown in Figure 4.16. From these one can see the change in creep modulus with time as a function of stress.

Attempts have been made to predict creep curves from short term data by curve fitting to a polynomial, with moderate success. Difficulties arise when using such equations and transferring the results to different materials, and the physical basis of determining the coefficients has so far proved elusive other than from curve fitting. This is the consequence of a number of factors determining the creep response, the different activation energy levels and external influences, and in practice such constants need to be generated for each type of material, thus limiting universal modelling at present.

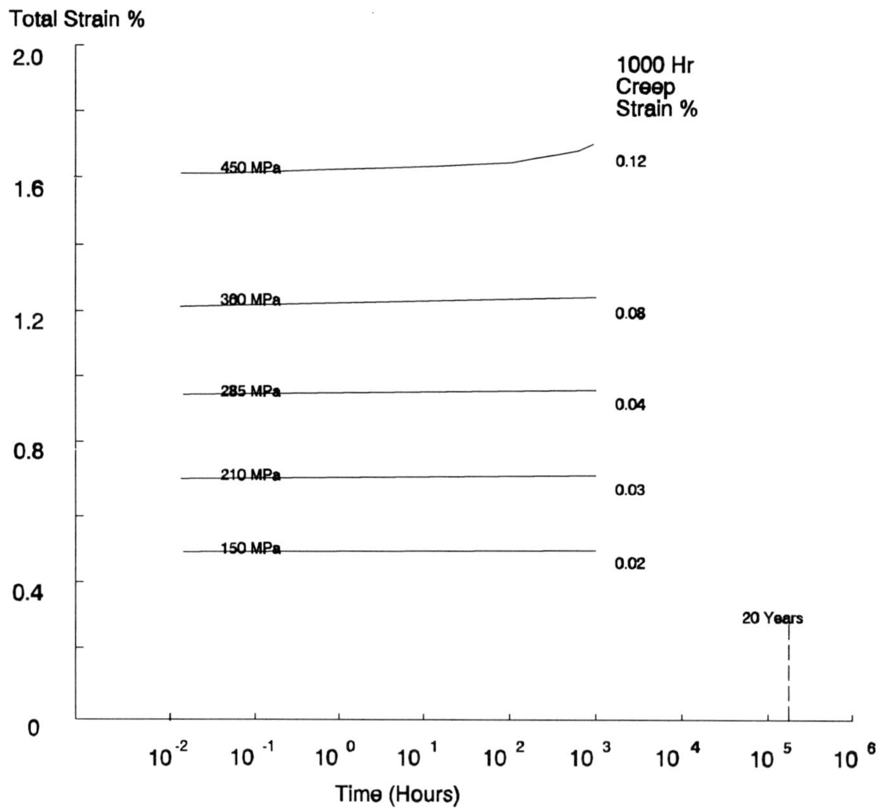

Figure 4.15 Creep curves for UD glass polyester in bending (reference 4.62).

4.12.1.2 Factors affecting magnitude of creep

Mechanisms of creep in FRP materials are related to the progressive changes in the internal balance of forces within the materials resulting from the behaviour of the fibre, adhesion and load transfer at the resin - fibre interface, and from the deformation characteristics of the matrix. Thus any factors which either directly or indirectly cause changes to any of these key areas will affect the creep process.

The polymer matrix is generally the weaker element of any composite material and will determine the overall performance and service use. It has a critical role in transferring load to the fibre and stabilising the fibre from local buckling in compression and shear as well as providing protection against agressive environments. For the reasons discussed above, temperature has a significant influence on composite creep behaviour through the behaviour of the matrix. It is essential to select a polymer with a high reactivity and to fully cure the polymer to obtain a high degree of thermal stability if creep effects are to be controlled. It is also important to select resins with a low moisture uptake and plasticisation behaviour and which have a good temperature margin of at least several tens of degrees between the service operating conditions and the glass transition temperature.

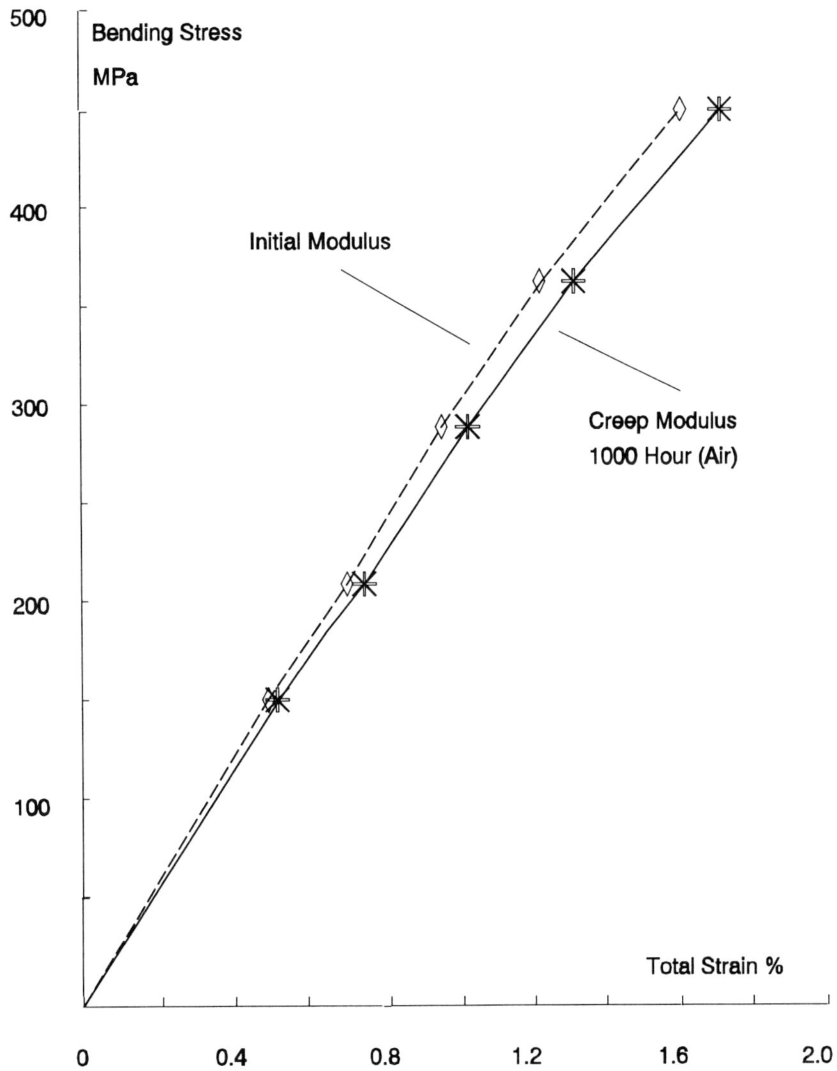

Figure 4.16 Isochronous 1000 hour (air) creep modulus curve for UD-glass polyester in bending at 20°C (reference 4.62).

Adhesion and stability of bonding at the fibre surface is important against resistance to creep, particularly in the presence of moisture, and selection of fibre finishes should ensure compatibility with choice of the resin system to optimise against interface degradation.

The nature of the reinforcement is also important in that higher stiffness fibres attract more load and limit shear deformation in the matrix. Carbon fibre composites show very much reduced creep compared with glass fibres for the same stress. For similar reasons and against a desired loading level, high volume fraction composites have good creep resistance. The highest creep resistance is obtained with unwoven fibres parallel to the external load. In reinforced materials with woven fabrics, the crimp will act as a stress raiser and can cause interlaminar fractures contributing to creep. Of all the potential reinforcements, chopped strand mats and continuous filament mats are the least creep resistant.

Composites are particularly susceptible to creep and relaxation effects in the through-thickness direction for the normal range of planar reinforcements. For loads normal to the plane of the composite, creep will occur, and in many bolted clamped joints the need to design the joint to allow for relaxation of the bearing stress is well recognised.

At low stress levels deformation of material and redistribution of load may occur without evidence of damage, while at high stress levels damage can occur in a short timescale and allow other mechanisms for accelerating degradation, for example resin cracking and moisture penetration.

Examples of creep effects covering environment, fibre type, resin type, effects of temperature, effects of strain and effects of overload are given in a variety of references at the end of this section. Creep effects are also affected by cyclic and intermittent loading. At cyclic frequencies above a few hertz the effects of heating of the material from hysteresis can be a factor in increasing creep.

4.12.1.3 Design method

Of the many studies made on composites to determine their properties and behaviour, reliable data on the long term properties is still an area of concern. This is due mainly to the problem of acquiring consistent data over a long period of time, on materials which are still valid and commercially available, and for which the collection of such data goes against agreed standards for the derivation of engineering data. Long term creep data falls into this category and there are difficulties in extracting meaningful reliable data from the published literature for a document of this type. While there are data on long term behaviour, the problem of arriving at definitive data and making judgements is made difficult by the variety and different combination of materials and often absence of detail in uniquely describing the material in many sources of information. Often such results do not have sufficient data to establish confidence limits and can only be used to indicate trends or comparisons. It is known that certain organisations have acquired long term data for specific applications, e.g. for filament wound pipes and pressure vessel design, but such information is generally commercially protected, giving them the edge in particular areas of design.

The methodology for design of FRP to cater for creep resistance should fundamentally be based on limiting the initial strains in the matrix. This should apply to all primary directions of loading. Where creep and deformation deflections are critical, reinforcements should be continuous, uncrimped, and materials manufactured for high strength and stiffness. In general this implies materials with high volume fractions, in excess of 50%. Axial strain levels should not exceed 0.2% or, if higher, be limited to preventing transverse cracking in the resin. Selection of a resin system to cater for temperature effects should allow a conservative margin of at least 50°C between the service temperature and the glass-resin transition temperature, allowing for the effects of a reduction in the latter from uptake of moisture. Where doubt exists, suitable tests should be conducted to verify the stability of the selected design.

4.12.1.4 Fibre orientation

Fibres should always be aligned with principal loads to limit deflection and to minimise shear distortion in the resin. Fibres should also be continuous to avoid a multiplicity of fibre ends (as in chopped mats), as peak shear stresses are developed here as a consequence of the reinforcement action. Fibres should generally be straight and where fabrics are used the lowest creep effects are associated with low crimp materials, e.g. satin weaves. Creep of fibres is secondary to that of the polymer, but if individual fibres fail there will be a load redistribution, resulting in an incremental deformation. Such events may occur at high loads, or low loads in agressive environments, or in regions of high stress concentration.

4.12.2 Accelerated testing

Accelerated testing has been used to try to predict the creep behaviour of composite materials over long time spans but as yet there is no universal acceptance of accelerated testing methods. The general approach is that of increasing loading levels and/or temperature to cause higher energy activation for the creep processes in air or in specific fluids. By conducting a series of tests at different temperatures an activation energy may be determined and through a kinetic approach of temperature-time superposition plots a master curve can be deduced and predictions made. The approach is valid to a degree but only where the mechanisms are independent and can be isolated against the accelerating factors. Certainly the approach can be useful in indicating sensitivity to temperature and other variables within the range tested. However, as there are many different effects contributing to the overall behaviour, each with different kinetics and not entirely independent, there is always concern over extrapolations even if the experimental data is consistent. Tests of this nature should not be conducted close to significant polymer changes such as second order transitions or where they induce different changes or mechanisms to those expected to occur under normal service. It does need to be emphasised that the prediction of ambient temperature behaviour from tests carried out at elevated temperatures depends for its validity upon (a) identical modes of fluid transport into the composite interior and identical mechanisms of deterioration at ambient temperatures as those operating at higher temperatures and (b) the production of highly reliable data and extrapolation procedures.

Scale, thickness of test specimens and condition of the edges can have a pronounced effect on observed changes under accelerated testing, particularly for moisture uptake and mechanical property changes. All such results should be viewed with caution and many of these will be over pessimistic in predicting longer term properties if scale effects are not recognised.

4.12.3 Rupture

4.12.3.1 General considerations

Rupture may progressively occur in a composite on a micro or macro scale under sustained loading. Inevitably on a micro scale, at and around individual fibres, there will be stress concentrations and a few single fibres may have higher than average stresses such that even at a low level of loading a few single fibres will fail. Such failures are not significant and allow redistribution of internal stresses as part of the inital shakedown of loading. This situation may still occur with time under static loading but at low loads will only contribute on a very small scale to creep deformation and overall can be discounted. Individual fibre failures have been detected by acoustic emmission below 0.05% strain on first loading but are absent on successive loadings until the first level is again exceeded.

At higher loads, at or above resin cracking strains, there may be progressive local damage through individual filament failures, increased relative movement between the polymer and the fibre at fibre ends, and slippage at interfaces. Transverse fibre debonding which results in the loss of stiffness from off-axis fibre laminae may progressively extend into regions of longitudinal fibres, causing interlaminar cracks. Growth of damage is the result of a continued redistribution of internal forces to the point where if there is insufficient equilibrium rapid failure will occur.

4.12.3.2 Stress corrosion

Glass reinforced plastics normally have good corrosion resistance against a variety of chemicals and aggessive environments. This is mainly associated with the inertness of the resin, which can be formulated to confer chemical resistance. The main effects of exposure to fluids or vapour is swelling and debonding as a result of absorption by the matrix. Under stress the resistance of resins and fibres to an environment may be reduced through the interaction of the stress with the internal forces. Indeed stress can accelerate the permeation of fluids into the polymer structure in a similar way to temperature.

The effects of stress environment interactions are severe when the environment weakens the reinforcements, especially if the mechanism itself is accelerated by stress. Under these circumstances surface flaws in the fibre may grow at stresses significantly below the original fracture stress in air, leading to premature fibre rupture under long term loading and lower strains than would be apparent under short term loading. The fibres used in glass FRP manufacture are usually alumina borosilicate E-glass and the problem of static loading is not too severe, high stresses and long exposure times being required for failure. Simple immersion in water or in dilute alkalis is unlikely to cause much degradation but acid environments will. In acid prone environments ECR-glass should be considered. As the polymer confers protection to glass in these cases it is important that stress levels are restricted to prevent resin cracking. In contrast carbon fibres are very resistant to static fatigue and to the effects of environment and, coupled with their higher stiffness, enable the manufacture of more resistant composites. Stress corrosion effects are complex, but manageable, and the choice of fibre and polymer materials for a specific application requires specialist advice. Further information can be obtained from Reference (reference 4.52) which gives details of the mechanisms, case histories, an indication of the time dependent behaviour of several glass FRP materials and guidance on the approaches in design to minimise the effect. The use of barrier coatings to prevent the ingress of aggressive environments is advocated in severe cases using formulated gel coats or thermoplastic liners.

4.13 FATIGUE

The information in this section is based on references 4.63 - 4.72.

4.13.0 Definitions and notation

No further information.

4.13.1 Fundamental requirements

Fatigue design will have one of two objectives:

- to ensure that with an acceptable level of probability, ultimate failure of a structure will not occur before n cycles of a fatigue load
- to ensure that the repair of damage caused by n cycles of a fatigue load will not be required.

These objectives are usually met by limiting either the fatigue stress or fatigue strain below a limit which will give the required fatigue life of n cycles.

However, the fatigue behaviour of FRP composites is difficult to predict owing to the limited test data available and the complex interaction of the particular fatigue loading conditions under consideration.

The conditions on which fatigue behaviour will depend are:

- the resin type and proportion
- the reinforcement type, orientation and proportion
- the nature of the fatigue loading (mean stress and stress amplitude)
- the frequency of the fatigue loading
- the ambient temperature and environment.

Because of this, for applications where fatigue is considered to be a significant problem there is no substitute for carrying out a test programme designed to simulate the in-service fatigue loading conditions. Fatigue tests should be carried out in a similar manner to the established procedures used for testing metals.

The behaviour of steel under fatigue loading exhibits a property known as the fatigue limit. This is a value of the fatigue stress amplitude below which the fatigue life of the element under consideration may be considered infinite. The significance of this property is that for many normal structures in steel where cyclic loadings are low, specific design for fatigue can be ignored.

The limited test data available suggest that, for FRP composites there is no evidence of a fatigue limit at 10^6 cycles (though there is some evidence that one is reached at 10^7 cycles). Thus, all FRP composite structures should consider the effects of fatigue loading and in particular:

- members supporting lifting appliances or rolling loads
- members subject to repeated stress cycles from vibrating machinery
- members subject to wind induced oscillations or stress reversals due to wind loading
- members subject to crowd induced oscillations.

4.13.2 Performance requirements

The fatigue failure of FRP composites is progressive, which is in contrast to that of steel, where ultimate failure occurs rapidly at the end of the fatigue life.

As an initially translucent FRP specimen fails in fatigue, it gradually becomes opaque. When the opacity first occurs, it does so only when the specimen is stressed, but after further stress cycles it becomes progressively permanent and more intense. At final separation of matrix laminates in axial loading, the fracture surface is normal to the loading direction and fibrous in appearance, the resin matrix often being reduced to a white powder. The initial (non-permanent) opacity is caused by debonding of fibres transverse to the load direction, which later cause matrix cracks and permanent opacity in the specimen.

In a fatigue test this internal damage is both cycle dependent and load dependent. Thus, matrix cracking which occurs at 70% UTS in a short term test might be observed at 40% UTS after 10^3 cycles.

Cracks initiated in the resin, on hitting a fibre, turn and run in the direction of the fibre. The matrix degrades, fragments and breaks away. Fibres, otherwise all right in tension, buckle in compression because they are no longer restrained by the supporting matrix.

The progressive nature of fatigue failure in FRP composites requires that the designer should define, with the agreement of the client, the design failure criteria. This will range from the ultimate limit state, when resin/fibre rupture occurs resulting in member collapse, to the serviceability limit state, when fibre debonding or resin cracking occurs resulting in an unsightly appearance or susceptibility to environmental degradation.

Testing has shown that transverse fibre debonding for a number of FRP composites is strain related and independent of the type, proportion and orientation of the glass reinforcement. Thus with a knowledge of the material modulus, the stress to initiate transverse fibre debonding may be calculated for any particular material. Available test data indicate that the strain amplitude to debond varies approximately linearly from 0.3% strain at 10^3 cycles to 0.1% strain at 10^6 cycles.

For ultimate fatigue failure, testing on several FRP composites has shown the fatigue strength to be directly related to the characteristic strength of the material in the direction of the applied loading (short term UTS).

Where a condition of combined fatigue stresses exists the application of uniaxial test data may not be directly relevant. Therefore, a rigorous design assessment will require a suitable test programme. However, as a conservative approach the damage indicator method given in the EUROCOMP Design Code will yield a safe result.

4.13.3 Design methods

The limited test data available on the fatigue loading of FRP composites make it difficult to formulate specific design rules for the infinite combinations of component materials possible. However, if all the available test data are normalised by expressing the fatigue strength for a given fatigue life as a proportion of the characteristic strength (short term ultimate tensile strength), then the relation appears to follow the classic S-N curve form as for metals (except for the absence of a fatigue limit). By fitting a line to the lower bound of the results, scatter, a conservative design rule can be formulated for most FRP composite material combinations.

Clearly it is important to understand the test conditions when formulating design rules based on their results. Some of the testing carried out has been done on FRP composites containing polyester resins, though most testing has been carried out on materials containing epoxide resins. Despite the scatter in the results obtained from the various test programmes, the formulae given in the EUROCOMP Design Code provide a conservative design rule for all FRP composites, subject to the stated provision.

A feature of fatigue tests for metals is that fatigue lives are assumed to be independent of cycle frequency and this is indeed the case for frequencies up to 10 Hz. However, such high frequencies cannot be used for FRP composites since their higher damping properties would cause the specimen to heat up and thus change the failure mechanism. Therefore, tests usually avoid this problem by restricting the frequency to below 10 Hz. Clearly the fatigue cycle frequency needs to be considered in design.

Since the strength properties of FRP composites are more sensitive to temperature change at lower temperatures than metals, the in-service ambient temperature range must be a consideration in the fatigue design.

If the environmental in-service temperature is less than 50°C, fatigue properties may be considered unaffected.

The fatigue strength of a particular FRP composite will also depend on the mean stress present in the component under consideration. For metals, the influence of mean stress on fatigue strength may be predicted by the Goodman Law, which makes use of the UTS, mean stress and fatigue strength with zero mean stress. For FRP composites this law has been found to overestimate fatigue strengths since creep effects are ignored. Modifications to the law have been proposed to account for creep effects which give conservative predictions.

Fatigue tests show that fatigue strength decreases with increasing tensile mean stress for a given fatigue life. Results also show that when compressive strengths are applied there is a small increase in fatigue strengths particularly at low stress amplitudes. This arises because the total time under tensile stress in the load cycle is reduced and it is the tensile stresses which are mainly responsible for damage initiation.

4.14 DESIGN FOR IMPACT

The information in this Section is taken from references 4.58 - 4.63 and reference 4.73 - 4.81.

4.14.1 Fundamental requirements

The impact strength of a component is a measure of its ability to withstand a large force of short duration and is conventionally measured by the energy required to break a standard specimen. The impact strength of FRP composites is mainly provided by the glass fibres which inhibit crack propagation through the composite material. The fibres provide additional energy absorbing mechanisms such as fibre pull out at the fracture face.

The principal aims of impact design are:

- the safety of personnel
- the avoidance of excessive damage to equipment and the structure by ensuring that the structure does not collapse.

The definition of 'excessive' damage will be the subject of discussions between the client and designer unless for a particular case appropriate regulations are available.

4.14.2 Performance criteria

Catastrophic failure of FRP composites is unlikely, owing to their enhanced energy absorbing properties. Therefore, it may be more appropriate to define impact failure in terms of the energy required to produce a certain damage state, rather than ultimate failure. The exact failure criteria adopted should be agreed with the client giving consideration to the consequences of failure. Depending on the application of the component being designed, the consequences of impact failure may range from unacceptable appearance through unacceptable cost of repair/replacement to unacceptable risk to people or equipment.

The impact strength of a particular FRP composite component is not related to a simple fundamental property of the material. The amount of energy absorbed in impact is a complicated function of material composition, component geometry, stress concentrations, impact load magnitude and velocity, ambient temperature and environmental conditions. At present, no simple method is available to predict impact behaviour and therefore all impact design is based on product testing.

In order for test results to be accepted with confidence for design purposes the impact loading used must be of a similar nature to that likely to be experienced in the in-service condition.

Limited testing has been carried out on the influence of weathering on the impact strength of polyester laminates containing both CSM and WR reinforcement. The results suggest that impact strength in temperate climates is unaffected over time, though in tropical climates reductions in strength of 20% and 13% after 4 years were recorded for composites containing CSM and WR reinforcement, respectively.

4.14.3 Design methods

There are two types of impact tests: material tests and product tests. Neither type of test gives meaningful information on the fundamental physical properties of the material, nor do they indicate how a component may behave under a different set of impact conditions. Thus, general rules for structural design cannot be developed from test results.

The most common tests for impact strength are the Izod and Charpy tests. In these tests, a short beam specimen, which may be notched, is struck and broken by a swinging pendulum. The impact strength is defined as the energy needed to break the specimen. Both these tests are arbitrary in their specification. By changing the test conditions quite different results may be obtained and therefore the tests are usually used for quality control and comparison of different materials only.

Drop weight tests are mainly used for quality control of sheet materials. These tests can be applied to FRP laminates and are more appropriate than Izod and Charpy tests since they allow failure criteria other than ultimate failure to be specified.

Because impact strength data from these standard tests cannot be used directly to predict the impact performance of designed components, there is a need for product tests. Certain standard tests exist for specialised component applications such as safety glass and plastics (BS 6206: 1981), but in the general case particular product test programmes will be required to verify design.

Test conditions must be selected to simulate the actual conditions of impact that are likely to be experienced in service, particularly in the case of members subjected to high compression or shear forces.

These conditions shall include not only the component configurations and impact energy but also the nature of the impact load and the environmental conditions to be expected. In general terms consideration should be given to either high velocity light projectiles or low velocity heavy projectiles.

A light projectile moving quickly will transfer little energy to the target as it is likely to perforate it and pass through. Thus damage is likely to be localised. A heavy particle moving slowly is likely to impact a greater amount of energy, resulting in a greater amount of damage which may even affect the stability of the member as a whole.

In general terms FRP composites containing CSM reinforcement have an enhanced impact strength with increasing reinforcement content. For FRP composites containing WR an increase in impact resistance strength with reinforcement content is less marked. FRP composites containing WR generally show twice the impact strength of an equivalent composite containing CSM.

4.15 DESIGN FOR EXPLOSION/ BLAST

The information in this section is taken from references 4.58 and 4.63, 4.73, 4.75, 4.79 - 4.81 and 4.82.

4.15.1 Fundamental requirements

The principal aim of explosion/blast design is the safety of personnel and avoidance of damage to equipment by ensuring that the structure does not collapse. In particular cases, blast-resistance of buildings for example, an additional aim is actually to protect personnel and equipment from injury or blast damage, which would not be the case with an open platform and equipment support structure.

4.15.2 Performance criteria

The loadings to be used for explosion/blast design, where not laid down by national regulatory authorities for health and safety, should be the subject of discussion between the client and designer.

For buildings and elements of building structures, current design practice for explosion and blast is to use the quasi-static approach to dynamic response of the structure adopted in the CIA document "Process Plant Hazard and Control Building Design". This is, in turn, based on ASCE Manual 42 "Design of Structures to Resist Nuclear Weapon Effects".

Blast resistant structures are designed to resist two arbitrary incident over-pressure impulses of 70 kN/m^2 for 20 milliseconds and 20 kN/m^2 for 100 milliseconds. The purpose of the double criteria is to ensure that the structure can resist, on the one hand a 'soft', long duration impulse, and on the other hand a 'hard', short duration impulse. A check is then carried out for adequacy of resistance to an incident over-pressure of 100 kN/m^2 for 30 milliseconds although in this case greater damage is acceptable. These loadings and durations are considered to be equivalent to the explosion of 1 tonne of TNT at a distance of 31.5 m. Principles of design by dynamic methods are given to allow reasonable advantage to be taken of the resilience and energy absorbing properties of the materials or jointing methods of construction in absorbing the energy of the blast. In effect, for each structural member, the methods allows derivation of a static equivalent pressure which, for the particular member's dynamic response characteristics, will be representative of the design over-pressure impulse.

The suction of the blast may be ignored provided structural rebound is taken into account.

Following an explosion, the amount of damage that the structure may sustain or, conversely, the extent of repair that may be necessary should be agreed between the client and the designer.

4.15.3 Design methods

The design methods are necessarily conservative in the great majority of cases and result in structures with a great reserve of strength that can, in reality, resist higher pressure impulses than those nominally designed for, albeit with a greater degree of deformation. It should be noted that deviation from the methods outlined, for example by more sophisticated computer analysis, to take even fuller advantage of the true dynamic response of the structure, could use up this reserve, and might invalidate the basis on which the method may have proved acceptable to the approving authorities.

Care should be taken in the use of sophisticated computer aided analyses since there are a number of parameters which are extremely difficult to define and quantify accurately, e.g. physical properties such as the ultimate compressive, tensile or shear strength and stiffness can all vary to a large extent (hundreds of per cent), depending on the rate at which the material is strained.

Where, in addition to blast resistance, the FRP composite must also have an external fire rating, supplementary means of augmenting the blast resistance properties or the fire-retardant properties will need to be provided, as the mechanical properties of fire-retardant gel coats and lay-up resins are lower than those of normal resins because of the inert content required to provide the fire-retardant qualities.

4.16 FIRE DESIGN

The information in this section is taken from references 4.58 - 4.79, 4.81 and 4.83 to 4.89.

4.16.1 Fundamental requirements

Fire design has two main objectives:

- to ensure the safety of the occupants of a building, of adjacent structures and of the rescue services and to provide adequate means of escape
- to minimise damage to property by detection, control and containment of the fire and the protection of key elements.

These objectives form the basis of most national statutory regulations and building codes and the requirements of other approving authorities, including insurance companies concerned with fire safety, which often state or give guidance on how the regulations may be complied with.

However, the behaviour of structural material under actual fire conditions is difficult to predict and depends on:

- the variation of mechanical properties with temperature
- the structural form, including fixings, means of attachment to other parts of the structure, perforations, cavities and venting provisions
- the amount of fire protection and shielding
- the spatial relationship of the structural components being considered and the source of the fire
- the development of the actual fire conditions themselves.

Because of this, in most countries the fire regulations specify periods of fire resistance, as measured in a standard fire-test for load-bearing elements, depending upon the size and function of the building. They tend to concentrate on "passive" rather than active methods of protection, and the required periods of fire resistance are frequently listed in a tabular format relating periods of fire resistance to maximum compartment dimensions and building height for various types of occupancy.

In the case of FRP composite materials which are inflammable and can ignite at temperatures appreciably lower than commonly experienced in building fires, further requirements may be specified relating not only to their combustibility, e.g. heat, fire penetration or spread of flame, but to the products of combustions, e.g. smoke, and toxic and noxious fumes.

When considering repairability, with most materials the main emphasis is on estimating the residual load-carrying capacity and, hence, assessing the remedial measures, if any, that are needed to restore the structure and structural and non-structural elements to their original design for fire performance and other requirements. There are major differences in comparing FRP composites with timber, say.

FRP sections tend to be thinner than the timber sections normally used for structural purposes; in addition, with profiled sections, the specific area tends to be higher. Consequently, while charring during a real fire or under test may occur at comparable rates in the two materials and the material beneath the charred layer may not lose significant strength because of the high thermal gradient, as a result of the low thermal conductivity in each case, the residual section of uncharred FRP composite that will be left after a given period of exposure to the fire conditions would be appreciably less than the residual section of the timber. Repairs, however, to FRP sections may be more straightforward than to timber, provided that additional material may be applied to the non-moulded surface, it being difficult to achieve a high quality finish on a moulded surface which has been damaged.

4.16.2 Performance criteria

The assessment of the behaviour of FRP composites in a fire depends on a number of empirical factors that relate to the form of construction, rather than to just the materials themselves.

The factors that need to be considered in fire design include:

- fire resistance
- ease of ignition
- surface-spread of flame
- fire propagation
- fire penetration.

Other factors, related to safety though not directly to structural performance, include:

- emission of smoke
- emission of toxic and noxious fumes.

Performance criteria in each of these categories are based on the results of standard tests.

Almost every country has its own particular methods, often requiring large specimens and special equipment. The following are brief summaries of some of the tests used. The appropriate specification should be consulted for full details.

(a) Fire resistance of elements of building construction

ISO 834; BS 476 Part 20; ASTM E119

This fire test is applicable to building elements, as follows:

- flexural members (beams); stability (strength and deflection)
- compression members (columns); stability

- tension members (ties); stability.

Specimens of the form, construction, including shielding, and fire protection, as are to be used in the structure of which the element is to form part are to be loaded in the appropriate mode and are subjected to radiant heat from electric elements for the period of the test. The radiant heat is controlled so that the temperature measured by thermocouples adjacent to the test specimen follows a specified time-temperature curve.

The fire resistance of the construction is then defined as the time for which the test specimen is able:

- to sustain the applied load
- in the case of flexural members, to resist deflection during the fire test to a specified value.

(b) Fire penetration and spread of flame tests for roofing materials

BS 476: Part 3

This test also requires special equipment and generally consists of three parts: a preliminary ignition test, a fire penetration test and a spread-of-flame test. The specimen is subjected to radiant heat and a vacuum is applied to one side to simulate service conditions. A specified flame is applied to the test piece for various durations and the time for the flame to penetrate, as well as the maximum distance of flame spread, are noted. Glowing, flaming or dripping on the underside of the specimen are also taken into consideration. Results are classified as follows:

Penetration time		Spread of flame	
A.	more than 1 hour	A.	None
B.	more than ½ hour	B.	less than 21 in. (534 mm)
C.	less than ½ hour	C.	more than 21 in. (534 mm)
D.	fails preliminary	D.	more than 15 in. (381 mm) in preliminary test

The classification is prefixed by Ext. F or Ext. S according to whether the specimen was tested flat or at an inclined plane. The prefix is followed by two letters, the first showing the result of the penetration test, the second that of the spread of flame test.

If the specimen drips on the underside during the test, the letter 'X' is added to the two letter code. Thus, the best possible classification for FRP roof sheeting is Ext. S.AA.

(c) Fire propagation test for materials

BS 476: Part 6 (NFP 92501, DIN 4102, ASTM E84)

In this test, which measures ease of ignition and the rate at which materials evolve heat on combustion, specimens 225 mm x 225 mm are exposed to small gas jets about 37 mm high and to radiant heat from electric elements. The hot gases evolved from the specimen are channelled through a chimney and their temperature measured. Comparison is made with a standard non-combustible material (asbestos) and thus temperature differences at specified intervals are converted into rates of temperature rise and integrated to provide an index of performance I. Certain values for I and the sub-index are specified in the UK Building Regulations.

(d) Surface spread of flame test for materials

BS476: Part 7 (NFP 92501, DIN 4102, ASTM E84)

As with Part 3, this test requires special equipment and can be carried out only by a limited number of organisations.

The test specimen 90 cm x 22.5 cm is cut from the laminate and fixed at right angles to a radiant panel and heated to a prescribed temperature gradient. It is then ignited at the hotter end for 1 minute. The spread of flame along the specimen is recorded until extinction or for 10 minutes. Results are classified as:

Class 1	not more than 165 mm
Class 2	not more than 215 mm during first 1½ minutes and not more than 455 mm after 10 minutes or at extinction
Class 3	not more than 265 mm during first 1½ minutes and not more than 710 mm after 10 minutes
Class 4	exceeding Class 3 limits

(e) Burning rate and smoke generation

ASTM E84-76 Tunnel Test

This measures the behaviour of laminates which form the roof of a tunnel 7.62 m long and 0.51 m wide. The results on flame spread are compared with asbestos cement board (taken as 0) and red oak flooring (taken as 100). Fuel contribution and smoke can also be measured.

It must be remembered that fire tests on plastics, like fire tests generally, can be highly specific and the results are specific to the tests. In particular, the results of one type of test may not correlate directly with another. Some tests are intended mainly for screening purposes during research and development or for comparison purposes; others, such as the large scale tests, more nearly approximate actual fire conditions. Certain terms such as "self-extinguishing" and "flame spread" have meanings particular to the test in which they are used.

The principles of good design for fire safety are as applicable to composites as to other construction materials. The specific design requirements must be carefully considered, the properties of the materials taken into account and engineering judgement applied.

4.16.3 Design methods

The resins of which FRP composites are usually made are thermosetting materials. Once catalysed, an increase in temperature increases the degree of cross-linking which leads to an improvement in mechanical properties. This is why post-curing is of such benefit. However, at higher temperatures, additional chemical reactions may take place which lead to a breakdown in the polymer structure and cause degradation in these mechanical properties. The temperature at which polyester resins show signs of softening is dependent on the resin type and varies from about 55 to 150°C. The heat-distortion temperature (temperature of deflection under load), defined in BS 2782: Part 1: 102G, provides a measure of the softening temperature. As may be expected and as shown in Table 4.40, the general purpose (orthophthalic) resins perform least well at elevated temperatures, typically retaining only between 10 and 22% of their 20°C flexural strength modulus at 160°C; the heat-resistant and flame-resistant (isophthalic) resins have a 50% greater retention, and the heat and flame chemical-resistant (bisphenol) resins have about 100% better retention.

It follows that other mechanical properties of FRP laminates will degrade at these higher temperatures. However, the retention of the mechanical properties is substantially increased as the glass content is increased.

An example of the use of a fire retardant additive is antimony trioxide. A paste of this and certain chlorinated organic compounds can be made compatible with all normal resins and be easily dispersed by simple mechanical mixing. When 20% of the basic resin is replaced by such a paste, laminates with low surface spread-of-flame can be produced. Smaller percentages will give less fire resistance. Another paste dispersion, of antimony trioxide, may be used in a resin based on HET acid but can only be mixed with certain ordinary resins. Other fire retardant additives include aluminium trihydrate (ATH).

In general, though, it is not possible to achieve good fire performance and good weathering properties with the same matrix. Gel coat resins, of reasonable fire performance and with good weathering properties that can be used externally, are available but are only suitable for general industrial work, building applications, vehicle bodies and similar translucent and opaque mouldings (where a Class-2 rating, when tested to BS 476: Part 7: 1971, is acceptable). High fire-performance gel coats (which give only Class-2 rating to BS 476: Part 6L: 1968) are only suitable for use internally in building applications.

Table 4.40 Influence of temperature on short term flexural strength retention of FRP laminates, fabricated from several types of polyester resin (typical).

Polyester resin type	*% of retention of 20°C flexural strength at*		
	75°C	*130°C*	*160°C*
General purpose: high reactivity (orthophthalic)	73	19	12
General purpose (isophthalic)	-	32	-
Flame resistant (isophthalic)	-	32	-
Heat resistant (isophthalic)	85	45	18
Heat and flame resistant (isophthalic)	-	44	18
Heat, flame and chemical resistant (bisphenol)	95	57	33

It is difficult to achieve a fire resistance rating with just composite protective systems. However, fire resistance and reduced smoke and toxicity can be achieved by the use of fire barriers which include ceramic fabric, ceramic coatings, intumescent coatings, or other high-temperature foam insulative barriers. These can, in certain circumstances, achieve up to ½-hour fire resistance without resorting to non-composite, non-integral protective systems. Examples are:

- *ceramic coatings* Currently, plastics usage is limited in certain hostile environments where heat will degrade them. In many instances, these problems can be overcome by the application of a ceramic coating to improve their thermal durability. Ceramic coatings cannot, though, be sprayed directly on to engineering plastics without substrate degradation, such as warping, charring, melting or peeling of the coating due to high temperatures and impact velocity. This can be overcome by the application of a bond coat followed by a selected functional ceramic coating. The ceramic coating inhibits heat transfer to the substrate. The bond coat itself conducts heat away from "hot spots" to minimise localised degradation of the plastic substrate
- *intumescent surface coatings* These are water-based or solvent-based coatings which have been specifically designed as a protective coating against direct flaming. After the addition of a catalyst they

can be applied by brush as a flow coating. A thickness of 0.4–0.5 mm or 450–600 g/m^2 is recommended and will give a dust-free surface in one hour at 18°C or longer at lower temperatures. Applied to the reverse side of panels, it is possible to produce FRP structures which possess good weathering properties on the exterior surface and fire retardant properties on the interior surface

When a flame is applied to a cured surface coating a carbonaceous foam is formed and the inert gases produced insulate the main structure of the laminate against the flame. The foam also significantly reduces the area affected by the flame and therefore reduces the spread

When a laminate is made with a low fire hazard resin and coated with such an intumescent coating, a Class 1 rating can readily be achieved

- *other surface coatings* These comprise clear, aliphatic (non-yellowing) or pigmented two-part polyurethane or other finishing coats with good weathering qualities applied over a gel coat and/or lay-up resin formulation with good fire-retardant but poor weathering properties. Under normal circumstances, the polyurethane provides an ultraviolet resistant and weatherproof shield to the fire-retardant formulation beneath. Under fire-test, or in the event of a real fire, this finishing coat flashes off without flaming or generating enough heat to compromise the fire-test or the fire safety of the structure, thereby exposing the fire-retardant formulation.

Aside from the fire protection aspects, burning resins and insulating foams give off fumes that can be toxic, noxious or both. Resins or systems containing chloride, nitrogen or phosphorous should be avoided in places of assembly and in areas of high fire risk, as fire conditions may generate hydrogen chloride, hydrogen cyanide, phosgene and other gases. Even ordinary hydrocarbon resins may generate large quantities of carbon monoxide when burning takes place in areas of deficient oxygen.

The use of polyurethane coatings and foams used for insulation or cores can raise problems in application with toxicity of unvented isocyanate components and also of toxic fumes in the event of a fire. In areas of high risk, therefore, it may be necessary to use a mineral wool insulation.

Providing fire-retardant properties for fire-resistant structures can have an adverse effect on many other properties of FRP composites:

- *mechanical properties* Fire-retardant gel coats and lay-up matrices are inherently weaker than normal standard matrices because of either high porosity or the inert filler content required to provide the fire-retardant qualities
- *weather resistance* Such composites have poorer resistance to outdoor exposure than normal laminates. This is an important consideration because for many applications good weathering properties can be more important than low fire hazard properties

- *colour fastness* The high chlorine content of some of the fire-retardant fillers may cause pigmented matrices to fade, thereby necessitating surface paint finishes.

At high enough temperatures, such as occur in a building fire, all organic resins will burn. Nevertheless, using suitable matrices or by the use of additives, or by a combination of the two, it is possible to modify their burning behaviour so that laminates made from such matrices present a lower hazard under fire conditions. In the case of fire resistance tests, the performance requirements are likewise well defined but would be almost impossible to achieve economically in excess of, say, half-an-hour, with an unprotected composite. The measures required, though, may so alter the mechanical or weathering properties, or both, that there is little advantage to be gained. In such cases, it may be necessary to restrict the use of FRP composites to situations where fire damage is unlikely to occur, or where, if it does, it would not be a problem; otherwise extensive additional fire-protection methods may be required.

4.17 CHEMICAL ATTACK

General considerations

No additional information.

4.17.1 Acids

No additional information.

4.17.2 Alkalis

No additional information.

4.17.3 Organic solvents

No additional information.

4.18 DESIGN CHECK-LIST

No additional information.

REFERENCES

4.1 *Design Manual, Fibreglass Structural Shapes.* MMFG, Bristol, Virginia, USA.

4.2 Properties of a GRP pultruded section. Sims, G.D., Johnson, A.F., and Hill, R.D. *Composite Structures* 8 (3).

4.3 *Structural Analysis of Laminated Anisotropic Plates.* Technomic Publishing Company Inc., Lancaster, Pennsylvania 17604, USA, ISBN 87762-518-2.

4.4 Turvey, G.J. and Marshall, I.H. (eds) *Buckling and Postbuckling of Composite Plates*, Chapman Hall, London, 1994.

4.5 *Formulas for Stress and Strain,* 5th edn, McGraw-Hill Int. Book Company, New York, ISBN 0-07-085983-3.

4.6 Timoshenko, S., *Theory of Plates and Shells,* McGraw-Hill Int. Book Company.

4.7 Fibrous composite failure criteria - facts and fantasy, McDonnell Douglas Aerospace, Transport Aircraft, paper no. MDC 93K0047, presented at the Seventh Int. Conf. on Composite Structures, Paisley, Scotland, July 5-7, 1993.

4.8 *Composites Design,* 3rd edn, Think Composites, Dayton, Ohio, ISBN 0-96180900-0-0.

4.9 *Composites Design Handbook for Space Structure Applications*, Vol. 1, ESA PSS-03-1101 Issue 1/Revision 2, European Space Research and Technology Centre, Noordwijk, The Netherlands, 1986/1992.

4.10 Griffin Jr, O.H., *in Composite Materials Design and Analysis* (eds. W.P. de Wilde and W.R. Blain), p. 171, Computational Mechanics Publications, Southampton, 1990.

4.11 *Engineered Materials Handbook,* Vol. 1, *Composites,* ASM International, Metals Park, Ohio, USA, 1987.

4.12 Jones, R.M., *Mechanics of Composite Materials*, Scripta Book Company, Washington, D.C., 1975.

4.13 Agarwal, B.D. and Broutman, L.J., *Analysis and Performance of Fibre Composites*, 2nd edn, Wiley, New York, 1990.

4.14 Vinson, J.R. and Sierakowski, R.L., *The Behavior of Structures Composed of Composite Materials*, Kluwer, Dordrecht, 1990.

4.15 Saarela, O., Computer programs for mechanical analysis and design of polymer matrix composites, *Prog. Polym. Sci.*, 171-201, 1994.

4.16 CoALA College of Aeronautics Laminate Analysis, Cranfield Institute of Technology (commercial information).

4.17 LAP Laminate Analysis Program, Center for Composite Materials, Imperial College, London (commercial information).

4.18 *Composite Calculations COMPCAL, User's Manual*, Technomic Publishing Co. Lancaster, Pennsylvenia, USA, 1987.

4.19 Gillespie Jr, J.W., Shuda, L.J., Waibel, B., Garrett J.J. and Snowden J., CMAP - Composite materials analysis of plates, Report CCM 87-45, University of Delaware, Center for Composite Materials, Newark, USA, 1987.

4.20 Gillespie Jr, J.W., McCullough, R.L., Munson-McGee, S.H., Garrett, J. and Waibel, B., SMC Micromechanics model for composite materials: thermoelastic properties user's guide, Report CCM 87-29, University of Delaware, Center for Composite Materials, Newark, USA, 1987.

4.21 Ghazal, A., Aivazzadeh, S., Verchery, G. and Chu, D., *in Composite Materials Design and Analysis* (eds W.P. de Wilde, W.R. Blain), p. 201, Computational Mechanics Publications, Southampton, 1990.

4.22 Zidani, F., Aivazzadeh, S. and Verchery, G., *in Composite Materials Design and Analysis* (eds. W.P. de Wilde, W.R. Blain), p. 227, Computational Mechanics Publications, Southampton, 1990.

4.23 *Cylinder Analysis CYLAN, User's Manual*, Technomic Publishing Co., Lancaster, Pennsylvenia, USA, 1987.

4.24 Engineering Sciences Data Unit (ESDU), Estimation of the stiffnesses and properties of laminated flat plates, Item No. 83035, V1, 1989.

4.25 Engineering Sciences Data Unit (ESDU), Failure analysis of fibre reinforced composite laminates, Item No. 84018, S15, 1989.

4.26 Engineering Sciences Data Unit (ESDU), Natural frequencies of rectangular specially orthotropic laminated plates, Item No. 83036, V4, 1989.

4.27 Engineering Sciences Data Unit (ESDU), Estimation of r.m.s. strain in laminated skin panels subjected to random acoustic loading, Item No. 84008, V3, 1988.

4.28 Engineering Sciences Data Unit (ESDU), Elastic stress and strain distributions around circular holes in infinite plates of orthotropic material (applicable to fibre reinforced composites), Item No. 85001, S14, 1989.

4.29 Engineering Sciences Data Unit (ESDU), Estimation of damping in laminated and fibre-reinforced plates, Item No. 85012, V2, 1988.

4.30 Engineering Sciences Data Unit (ESDU), Natural frequencies of simply-supported sandwich panels with laminated face plates, Item No. 85037, V6, 1988.

4.31 Engineering Sciences Data Unit (ESDU), Elastic buckling of unbalanced laminated fibre-reinforced composite plates (rectangular plates of AsBtDs type, all edges simply-supported under biaxial loading), Item No. 86020, S17, 1987.

4.32 Engineering Sciences Data Unit (ESDU), Estimation of rms strain in laminated face plates of simply-supported sandwich panels subjected to random acoustic loading. Including a simplified natural frequency prediction method, Item No. 86024, V3, 1990.

4.33 Engineering Sciences Data Unit (ESDU), Elastic wrinkling of sandwich columns and beams with unbalanced laminated fibre reinforced plates, Item No. 87013, S17, 1988.

4.34 Engineering Sciences Data Unit (ESDU), Elastic buckling of cylindrically curved laminated fibre reinforced composite panels with all edges simply-supported under biaxial loading (computer program), Item No. 87025, S13, 1988.

4.35 Engineering Sciences Data Unit (ESDU), Elastic wrinkling of sandwich panels with laminated fibre reinforced face plates, (face plates of AsBoDs, AsBtDs and AsBsDs types) (computer program), Item No. 88015, S17, 1988.

4.36 Engineering Sciences Data Unit (ESDU), Natural frequencies of singly-curved laminated plates with simply-supported edges, Item No. 89011, V4, 1989.

4.37 Engineering Sciences Data Unit (ESDU), Transverse (through-the-thickness) shear stiffnesses of fibre reinforced composite laminated plates (computer program), Item No. 89013, S14, 1989.

4.38 Engineering Sciences Data Unit (ESDU), Natural frequencies of isotropic and orthotropic rectangular plates under static in-plane loading (including shear loading), Item No. 90016, V4, 1990.

4.39 Engineering Sciences Data Unit (ESDU), Delamination and free edge stresses in laminated plates subjected to constant uniaxial strain (computer program), Item No. 90021, S15, 1991.

4.40 Engineering Sciences Data Unit (ESDU), Delamination of tapered composites, Item No. 91003, S15, 1991.

4.41 Engineering Sciences Data Unit (ESDU), Thickness selection for laminated plates subjected to in-plane loads and bending moments, Item No. 92033, S14, 1992.

4.42 *Composite Experimental Mechanics Calculations CEMCAL, User's Manual,* Technomic Publishing Co., Lancaster, Pennsylvenia, USA, 1987.

4.43 Murthy, P.L.N. and Chamis, C.C., *J. Comp. Techn. Res.* **8**, 8-17, 1986.

4.44 *ASCA The Automated System for Composite Analysis, Software and User's Manual*, Technomic Publishing Co., Lancaster, Pennsylvenia, USA, 1992.

4.45 Saarela, O., LAMDA - Microcomputer software for dimensioning composite laminate and sandwich structures, Report 91-B27, Helsinki University of Technology, Laboratory of Light Structures, Otaniemi, 1991.

4.46 Menges, G. and Effing, M., SPI 43rd Annual Conference, Session 20-D/1, Anaheim, Califiornia, USA, June 1988.

4.47 Analysis/design of composite material systems (ESACOMP), ESTEC/Contract No. 9843/92/NL/PP, Nordwijk, The Netherlands, 1992.

4.48 Saarela, O., Haberle, J. and Klein, M., Composite Analysis and Design System ESAComp, CADCOMP 94, Southampton, 1994.

4.49 DAC Designing advanced composites commercial brochure, University of Zaragoza, Dept. of Mechanical Engineering, Spain, 1993.

4.50 Herrmann, A.S., Hanselka, H. and Haben, W., Faserverbundwerkstoffe am Rechner Komponieren, Kunstoffe. **6** 1992.

4.51 Hancox, N.L. and Mayer R.M., *Design Data for Reinforced Plastics: A Guide for Engineers and Designers.* Chapman & Hall, 1994.

4.52 Hogg, P.J. and Hull, D., Chapter in *Developments in GRP Technology, Vol. 1, Corrosion and Environmental Deterioration of GRP*, Ed. B. Harris, 1983.

4.53 Hollaway, L. (ed.) *Polymer and Polymer Composites*, Thomas Telford, London, 1990.

4.54 White, R.J. and Phillips, M.G., Environmental stress rupture mechanisms in glass fibre polyester laminates. *Proc. Int. Conf on Composite Materials*, 1089-99, 1985.

4.55 Pritchard, G. and Speake, S,D., Effects of temperature on stress rupture times in glass polyester composites. *Composites,* **19**, 29-35, 1988.

4.56 Yeung, Y.C. and Parker, B.E., Composite tension members for structural applications. *Composite Structures,* **4**, 1309-19.

4.57 Bulder, B.H. and Bach, P.W., Literature survey on the effects of moisture on the mechanical properties of glass and carbon plastic laminates. ECN-C-91-033, ECN, Petten, the Netherlands, 1991.

4.58 Johnson, A.F., Engineering Design Properties of GRP. British Plastics Federation Publication 215/1, 1978.

4.59 Martine, E.A., Long term tensile creep and stress rupture evaluation of uni-directional fibre glass reinforced composites. Paper 94, *SPI 48th Annual Conference,* Cincinnati, USA, 1993.

4.60 Aveston, J. and Silwood, J.M., Time dependent strength of uni-directional glass reinforced plastics in air, water and dilute sulphuric acid. Proc. *3rd Riso Int. Symp. on Metallurgy & Materials Science,* 1982.

4.61 Roberts, R.C., *Reinforced Plastics Congress Publication,* British Plastics Federation p. 145, 1978.

4.62 Steel, D.J., Technical Note 4/68, M.E.X.E., Christchurch, 1968.

4.63 ENV 1993-1-1 EUROCODE EC3: Structural steelwork, European Committee for Standardisation, 1993.

4.64 *Structural Plastic Design Manual*, Manual No. 3 ASCE, 1984.

4.65 BS 3518: Methods of fatigue testing. British Standards Institution, London.

4.66 Howe, R.J. and Owen, M.J., Cumulative damage in CSM/polyester resin laminates, *BPF 8th International Reinforced Plastics Conference,* Brighton, 1972.

4.67 Smith, T.R. and Owen, M.J., The progressive nature of fatigue damage in glass reinforced plastics, *BPF 6th International Plastics Conference,* Brighton, 1968.

4.68 Boller, K,H,, Fatigue properties of fibrous glass reinforced laminates subjected to various conditions, *Modern Plastics,* 1957.

4.69 Adam, S, and Butler, T., The static and fatigue strength of GRP compressive structure, Bath University Report 355, 1975.

4.70 Owen, M.J. and Found, M.S., The fatigue behaviour of glass fabric reinforced polyester resin under off axis loading, *Journal of Physics,* 8, 1975.

4.71 Boller, K.H., Fatigue characteristics of RP laminates subjected to axial loading, *Modern Plastics,* L1, 1964.

4.72 NFT 57-900 Réservoirs et appareíls en matíéres plastiques reinforceés: AFNOR, December 1987.

4.73 ENV 1992-1-1 Eurocode No. 2: Design of concrete structures, European Committee for Standardisation, 1992.

4.74 BS 6206: Impact performance requirements for flat safety glass and safety plastics for use in buildings, British Standards Institution, London.

4.75 BS 4618: Presentation of plastics design data, 1.2: impact behaviour, British Standards Institution, London.

4.76 BS 2782: Methods of testing plastics, British Standards Institution, London.
Method 350: Izod impact strength
Method 352D: Falling weight impact resistance
Method 353A: Multi-axial impact behaviour
Method 359: Charpy impact strength.

4.77 Thorogood, R.P., Assessment of hard body impact resistance of external walls - BRE IP 19/8, 1981.

4.78 Design data: Fibreglass Composites - Fibreglass Ltd.

4.79 *Polyester Handbook* - Scott Bader: 1980.

4.80 *Structural Analysis and Design of Nuclear Plant Facilities*, Chapter 6, Manual No. 58 ASCE.

4.81 *Structural Plastics Design Manual*, Manual No. 63, ASCE, 1984.

4.82 Process plant hazard and control building design, Chemical Industries Association.

4.83 The Building Regulations.

4.84 BS 5586: Fire precautions in the design, construction and use of buildings, British Standards Institution, London.

4.85 BS 476: Fire tests on building materials and structures, British Standards Institution, London.

4.86 BS 2782: Methods of testing plastics, British Standards Institution, London.

4.87 BS 5268: The structural use of timber, British Standards Institution, London.

4.88 Improved fire safety of composites for naval application: Sorathra, Rollhanger, Allen Hughes 1992.

CONTENTS

5 CONNECTION DESIGN

5 CONNECTION DESIGN

5.1 GENERAL

5.1.1 Definitions

The definitions given in the EUROCOMP Design Code relating to connection design are based on or taken from References 5.1, 5.2, 5.3, 5.4 and 5.5. Some of the definitions have been developed specifically for the EUROCOMP Design Code.

The hierarchical categorisation of different connections illustrated in Figure 5.1 of the EUROCOMP Design Code forms the basis of how the connection design part has been structured. It also defines the basic terminology used on each hierarchical level.

The systematic approach and the associated terminology have been introduced to avoid confusion, which easily arises from the fact that connection technology is divided into numerous groups and sub-groups with no established and consistent terminology. A commonly used term "joint type" is a good example of the ambiguous wording that is used to address anything from joint geometry to load transfer mechanism.

The primary categorisation is based on the load transfer mechanism, i.e. whether the loads are directed through an adhesive material, mechanical fasteners or both. The most descriptive title for this top level of categorisation would be "joint mechanism". As this implies an immediate reference to mechanical joints, this term is excluded from the EUROCOMP Design Code. Therefore the three main groups listed above are entitled "joint categories".

The second level, entitled "joining technique", groups the joints within each category according to how the joint is actually made, e.g. by adhesive bonding or laminating. On the final level the "joint configuration" broadly states the joint lay-out. In the design section these may be categorised even further, as in the case of lap joints into single-lap, double-lap and step-lap joints.

5.1.2 Scope

No additional information.

5.1.3 Joint classification

No additional information.

5.1.4 Joint categories and joining techniques

No additional information.

5.1.5 Joint configurations

No additional information.

5.1.6 Applied forces and moments

Further information may be obtained from Eurocode 3: Design of steel structures - Part 1.1: General rules and rules for buildings (reference 5.5).

5.1.7 Resistance of connections

Further information may be found in Eurocode 3: Design of steel structures - Part 1.1: General rules and rules for buildings (reference 5.5).

5.1.8 Design approach

No additional information.

5.1.9 Design requirements

Further information may be found in Eurocode 3: Design of steel structures - Part 1.1: General rules and rules for buildings (reference 5.5).

5.1.10 Partial safety factors

The partial safety factors given in the EUROCOMP Design Code are based on the earlier experience and general knowledge on connection design of the EUROCOMP member participants. They are not based on any specific tests or analysis.

5.1.11 Selection of joint category

Connections should be considered in the early stage of the structural design in order to take into account the connection requirements for structures and components. The most basic consideration to be made at that stage - at least provisionally - is the selection of the joint category, as this selection typically has a strong impact on the design of the individual components. If the basic considerations have not been given before the component design, the later connection design may require extensive redesign of the component and even then the final design may be far from optimum.

Several factors affecting the selection are listed in the EUROCOMP Design Code. This list should be regarded as a check list. The selection should always be based on the requirements of each individual design. Typically, only a few of the listed factors are relevant simultaneously and often one or two are more important than the others.

Tables like 5.1 and 5.2 of the EUROCOMP Design Code may assist the designer in the selection. It should be noted that these tables can only provide information on typical properties or typical behaviour of different joints. As the information has to be simplified using expressions like "yes" and "no", one has to realise the informative nature of these tables and not to consider them as definitive statements.

5.2 MECHANICAL JOINTS

5.2.1 Scope

No additional information.

5.2.2 Bolted and riveted joints in shear

5.2.2.0 Notation

A_i	Cross-sectional area of the fasteners
C	Fastener flexibility
E_1	Elastic modulus in the loading direction for the upper part of a single-lap joint
E_2	Elastic modulus in the loading direction for the lower part of a single-lap joint
E_f	Elastic modulus of the fastener
E_p	Elastic modulus in the loading direction for the centre plate of a double-lap joint
E_s	Elastic modulus in the loading direction for the side plates of a double-lap joint
F	Applied load on eccentrically loaded joint
F^c	Resolved concentric load
F_x^c	Resolved concentric load in x-direction
F_y^c	Resolved concentric load in y-direction
F_i^e	Force transferred by a fastener due to applied eccentric force
F_{ix}	Transferred fastener load in x-direction in a concentrically loaded joint
F_{ix}^c	Transferred fastener load due to concentric load in x-direction
F_{iy}	Transferred fastener load in y-direction in a concentrically loaded joint
F_{iy}^c	Transferred fastener load due to concentric load in y-direction
$F_{i\phi}^e$	Transferred fastener load in ϕ-direction due to torsional moment
F_{ix}^e	Transferred fastener load due to eccentric load in x-direction
F_{iy}^e	Transferred fastener load due to eccentric load in y-direction
l	Length of joint overlap
K_{exp}	Fastener stiffness (inverse of fastener flexibility); slope of an experimentally determined curve of R as a function of δ_{exp}
K_{ffe}	Stiffness constant of the spring element
K_{mesh}	Slope of a numerically determined curve of R as a function of δ_{mesh} using rigid springs

M Torsional moment (M=Fe)
R Load transferred by a fastener
$S_{b,crit}$ Design bearing strength
X_1 Distance from c.g. to fastener i in x-direction
Y_1 Distance from c.g. to fastener i in y-direction
a,b Empirical constants in expressions for fastener flexibility
d_{bi} Shank diameter of fastener i in a fastener group
$d_{i\phi}$ Rotation of fastener i due to torsional moment
d_k Characteristic distance from the fastener hole for predicting net-section failure
$d_{c,k}$ Characteristic distance from the fastener hole for predicting net-section (compression) failure
$d_{t,k}$ Characteristic distance from the fastener hole for predicting net-section (tension) failure
e Eccentricity of P with respect to centre of gravity of fastener group
k Stiffness constant of the fastener installation
k_r^i Normalised radial stress k_r for basic load case i
k_t^i Normalised tangential stress k_t for basic load case i
k_s^i Normalised shear stress k_s for basic load case i
r_i Radial distance from c.g. to fastener i
r_k Characteristic distance from the fastener hole for predicting bearing failure
$r - \phi$ Polar coordinate system placed at centre of gravity of fastener group
s_k Characteristic distance from the fastener hole for predicting shear failure
t_1 Thickness of the upper part in a single-lap joint
t_2 Thickness of the lower part in a single-lap joint
t_p Thickness of the centre plate in a double-lap joint
t_s Thickness of the side plate in a double-lap joint
x_i Distance from c.g. to fastener i in x-direction
y_i Distance from c.g. to fastener i in y-direction
β_i Angle between F_i^e and the x-axis
Δ Relative displacement between the joined members
Δ_{ix} Relative displacement in x-direction between joined parts at fastener i
Δ_{ix}^c Fastener deflection due to concentric load in the x-direction
Δ_{iy}^c Fastener deflection due to concentric load in the y-direction
$\Delta_{i\phi}$ Displacement in each fastener in the ϕ-direction if above
$\delta_{b,cnt}$ Design bearing strength
δ_{exp} Experimentally determined relative displacement between joined members locally at the fastener hole
δ_{fem} Predicted relative displacement between joined parts by FEM (locally at the fastener hole)
δ_{ffe} Deformation of finite spring element
δ_{mesh} Deformation of finite element mesh, locally at the fastener hole
$\varepsilon_{by\text{-}pass}$ By-passing strain at a considerable distance from the fastener hole
ε_{gross} Strain in a section a considerable distance away from the fastener hole
$\sigma_{t,\phi,k}$ Characteristic tangential strength in tension

$\sigma_{c,\phi,k}$ Characteristic tangential strength in compression
τ_{xy} Shear stress distribution in the x - y plane
$\tau_{r\phi}$ Shear stress distribution along the hole edge
τ_{xt} Shear stress distribution through the thickness of the laminate

5.2.2.1 Definitions

The fastener and the laminate are two bodies in elastic contact. As the fastener is loaded, it contacts a portion of the hole. The extent of contact varies with the load. Owing to friction within the contact zone, there are regions of slip and no slip.

Figures 5.19(b) and 5.19(a) of the EUROCOMP Design Code show the radial compressive stress distribution σ_r and the tangential stress distribution σ_ϕ, respectively, along the net-section plane at $\phi = 90^\circ$.

Figure 5.19(c) of the EUROCOMP Design Code shows the shear stress distribution, $-\tau_{sn}$, along the so-called shear-out plane. The actual stress field depends on a number of parameters such as elastic properties of the laminate and the fastener, clearance, friction and load magnitude, and loading direction (tensile or compressive).

5.2.2.2 Performance requirements

No additional information

5.2.2.3 Design requirements

Parameters affecting strength

Consider the concentrically loaded double-lap joint shown in Figure 5.1. A central composite plate member (length l, width w, thickness t, edge distance e, side distance s, spacing in the x-direction w_x and spacing in the y-direction w_y) containing four holes of equal diameter d_h is placed between two side plates (composite or metal) and fastened with four fasteners of diameter d_b. In Figure 5.2, the composite plate has been divided into four basic regions each containing a single fastener.

The strength characterisation of the central composite plate in Figure 5.1 is complicated due to the many parameters involved. The strength is affected by the geometry of the plate, material and laminate properties, fastening and loading conditions, among other factors. The parameters affecting strength may be divided into three groups (reference 5.7).

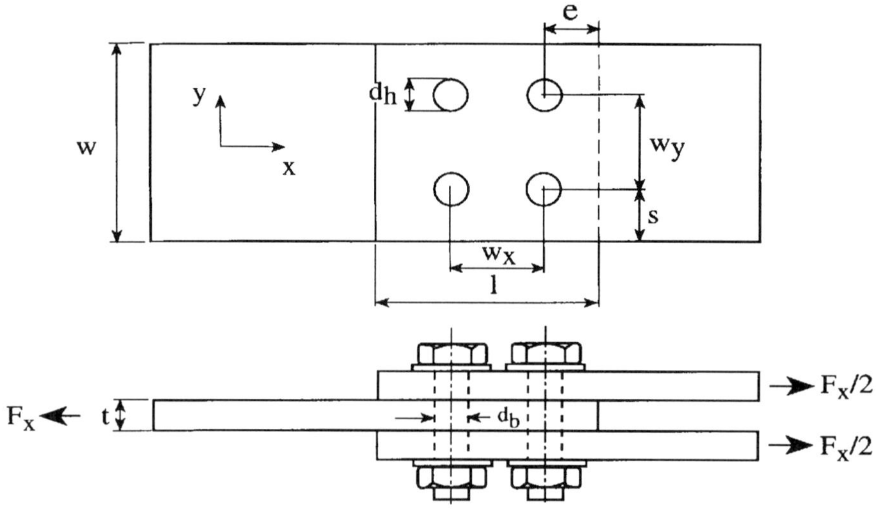

Figure 5.1 Concentrically loaded double lap joint.

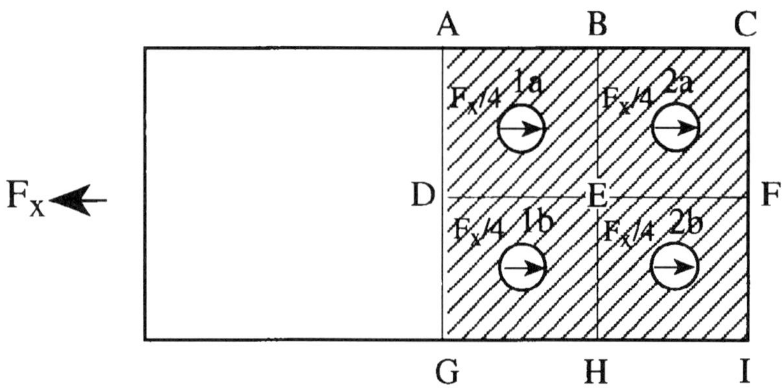

Figure 5.2 Basic regions of composite plate containing a fastener hole.

Design parameters

- geometry (width, spacings, edge distance, side distance, hole pattern, etc.)
- joint type (single lap, double lap, etc.)
- plate thickness
- loading condition (tensile, compressive, shear, etc.)

Material parameters

- fibre type and form (uni-directional, woven, fabric, etc.)
- resin type
- fibre orientation
- stacking sequence

- fibre volume fraction
- fibre surface treatment.

Fastening parameters

- fastener type (screw, fastener, rivet, etc.)
- clamping force
- washers
- fastener/hole tolerance
- fastener daimeter.

Geometry

Width, pitch and side distance: Consider the isolated region 2a (Figure 5.3) of the central plate in Figure 5.2.

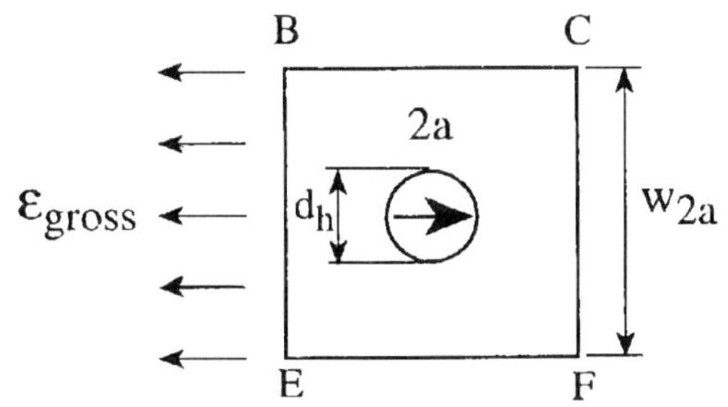

Figure 5.3 Isolated region 2a of the composite plate.

The gross-section strain in section BE is denoted by ε_{gross}. The width of the region is denoted by w_{2a}. The material is first modelled as a linear elastic material. The characteristic tensile strength and design tensile strain of the material are denoted by $\sigma_{t,\phi,k}$ and $\varepsilon_{t,crit}$ respectively. The Young's modulus of the plate in the x-direction is denoted by E_x. In the case of **bearing failure** ε_{gross} is determined by:

$$\varepsilon_{gross} = \frac{S_{b,crit}}{E_x} \frac{d_h}{W_{2a}} \tag{5.1}$$

where $S_{b,crit}$ is the ultimate bearing strength. That is, ε_{gross} is a linear function of d_h/w_{2a} for bearing failure. The slope of the curve is determined by the ratio of the bearing strength to the Young's modulus.

As the ratio of d_h/w_{2a} increases, the mode of failure changes from **bearing** to **net-section**. For large values of d_h/w_{2a} ductile material behaviour of the composite can be assumed since the size of the damage zone is a large proportion of the width. For net-section failure, ε_{gross} is determined by:

$$\varepsilon_{gross} = -\frac{\sigma_{t,\phi,k}}{E_x}\frac{d_h}{w_{2a}} + \frac{\sigma_{t,\phi,k}}{E_x} \tag{5.2}$$

Equations (5.1) and (5.2) are plotted together in Figure 5.4.

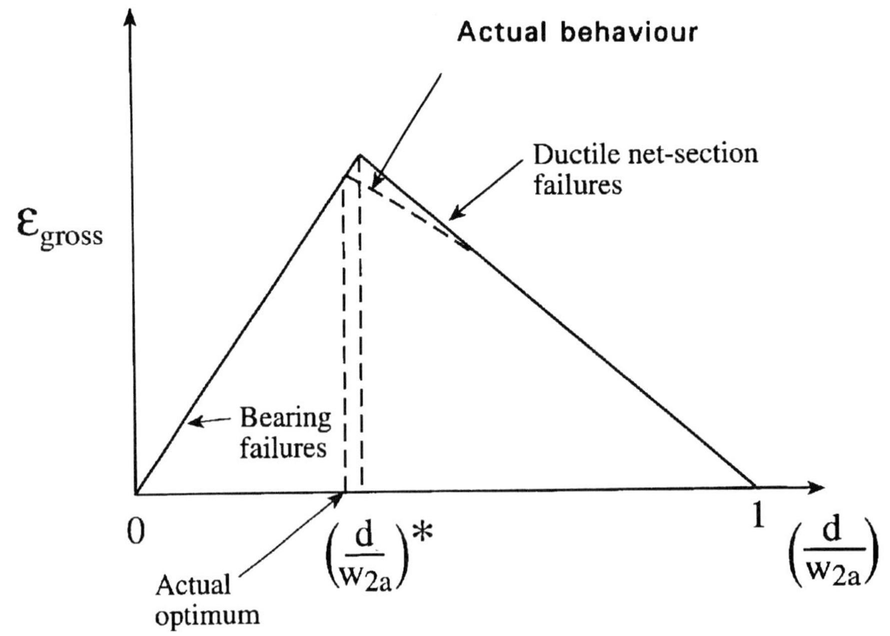

Figure 5.4 Gross-section strain as a function of diameter-to-width ratio in the bearing and net-section regions.

The gross-section strain at the point of intersection between the two straight lines is the maximum strain obtainable under the assumption of ductile material behaviour. The ratio of d_h/w_{2a} at this point is :

$$\left(\frac{d}{w_{2a}}\right)^* = \frac{\sigma_{t,\phi,k}}{S_{b,crit} + \sigma_{t,\phi,k}} \tag{5.3}$$

Using Hooke's law, Equation 5.3 is rewritten:

$$\sigma_{t,\phi,k} = E_x\ \varepsilon_{t,crit} \tag{5.4}$$

$$\left(\frac{d}{w}\right)^* = \frac{E_x\ \varepsilon_{t,crit}}{S_{b,crit} + E_x\ \varepsilon_{t,crit}} \tag{5.5}$$

As the ratio of d_h/w_{2a} is reduced, the assumption regarding ductile material behaviour becomes less accurate since the extent of the damage zone is no longer a large proportion of the width. The optimum d_h/w_{2a}-ratio for an actual glass FRP material is less than that for a ductile material provided the two materials have equal bearing strength (Figure 5.4).

Consider next the isolated region 1a (Figure 5.5) of the central plate in Figure 5.2. This region is subjected to a combination of bearing and by-pass loads. The magnitude of the transferred load expressed by ε_{gross} is affected by both the d_h/w_{1a} ratio and the ratio of bearing stress to bearing strength.

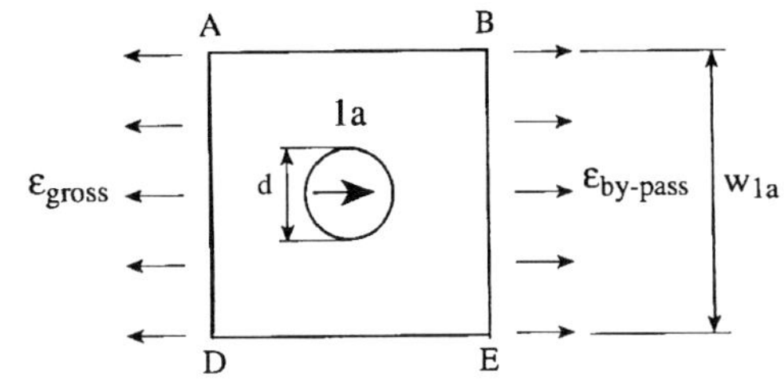

Figure 5.5 Isolated region 1a of the composite plate.

Design charts such as Figure 5.6, introduced by Hart-Smith (reference 5.8), can be used to design the structurally most efficient joint. The limiting outer envelope refers to a situation with no bearing load and only by-pass loads. The lower limit refers to a situation with only bearing loads and no by-pass loads previously discussed. This limit is defined by the straight-line (Equation (5.1)) for small d_h/w_{2a} ratios and bearing failures and by the curved line for larger d_h/w_{2a} ratios and net-section failures as the fasteners are installed closer together. For large values of d_h/w_{2a}, this curve approaches the straight line defined by Equation (5.2). The other curved characteristics in the upper left corner refer to net-section failures in multi-row joints for regions subjected to a combination of bearing and by-pass loads at progressively lower strengths as the bearing stress is increased. The roughly triangular shape adjacent to the lower part of the vertical axis defines the area in which multi-row joints fail in bearing at all fasteners.

Figure 5.6 gives information on how the strength can be maximised. This can be achieved for a single-row joint simply by using the ratio of d_h/w_{1a} which gives the highest strength. For multi-row joints, the strength can be maximised by reducing the bearing strength in the most critical region of the joint. The most critical region with respect to net-section failure is the region which is subjected to the highest by-pass loads. The strength of multi-row joints is also affected by the d_h/w_{2a} ratio.

Edge distance: A minimum value of the e/w ratio is required to develop full net-section strength of the joint. Hart-Smith (reference 5.9) presents results showing that the edge distance must be greater than or equal to the side distance for full development of the net-section strength.

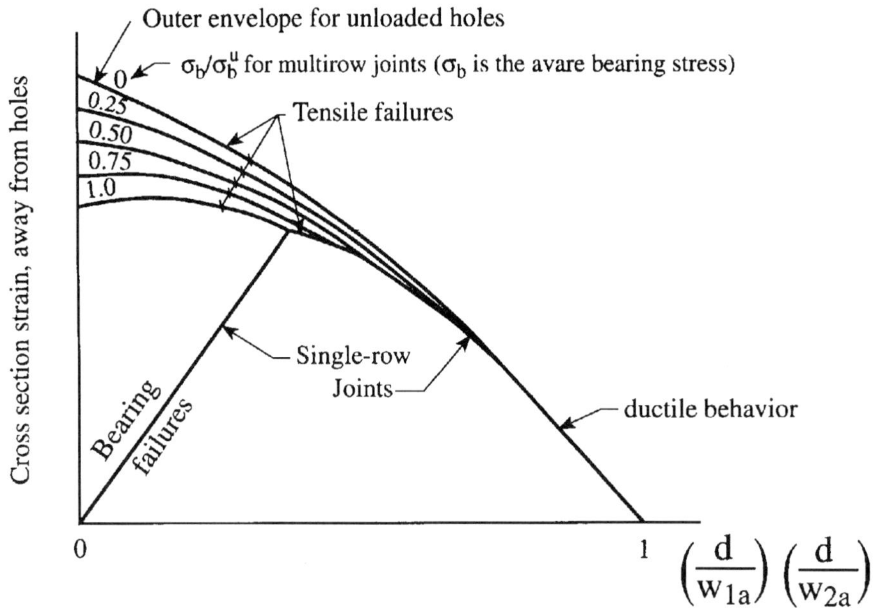

Figure 5.6 Cross-section strain as a function of diameter-to-width ratio for single and multiple-row joints.

To avoid a premature shear-out failure of the joint, a minimum value of the e/d-ratio is required. It should be emphasised that the minimum value of e/d depends on fibre orientation and the material system used. For near quasi-isotropic fibre patterns an e/d-ratio of 3 is sufficient (reference 5.8). For highly orthotropic laminates, such as pultruded laminates, shear-out failure may occur independently of the edge distance.

Member thickness: Glass fibre composites show a significant reduction in bearing strength for d_h/t-ratios greater than 1.0 (reference 5.7). A too small diameter fastener may undergo severe bending, which will reduce bearing strength, owing to loss of lateral constraint and non-uniform bearing stress through the thickness of the laminate (reference 5.8).

Eccentricity: The eccentricity in the load path in single-lap joints causes secondary bending in the laminates. This decreases the bearing strength compared with double shear configurations owing to loss of lateral constraint and non-uniform bearing stress through the thickness of the laminate.

Fibre orientation: The strength of bolted joints is significantly affected by fibre orientation. To avoid matrix-controlled failure modes, such as shear-out, there should, wherever possible, be three basic fibre orientations of 0°, 90° and +45° in a bolted laminate (reference 5.8) (the 0° direction being parallel to the load).

Stacking sequence: If the plies of a laminate are homogeneously mixed, i.e. the fibre direction changes from layer to layer, the stacking sequence has little effect on the strength of bolted joints. Bolted joints are, however, weakened if plies of the same orientation are grouped together (a blocked laminate) (reference 5.8).

Fastening parameters

Fastener type: Mechanical fasteners available are pins, screws, rivets and bolts. Pins are used for linking lug joints. Screws are not recommended for load carrying joints because they can easily be pulled out of the laminate. Care must be taken when using rivets in composite laminates. The installation process, which involves using a closing force that may not be well controlled, can result in wide variation of joint strength. The riveting operation may cause excessive damage to the laminate and have an adverse effect on its strength. Generally, solid rivets produce stronger joints than hollow rivets. With bolts, the lateral constraint is better controlled than with rivets, resulting in higher strength and less scatter of test results. Sufficient clamping pressure is obtained by tightening the bolt. Care must be taken to avoid over-tightening the bolt, as this may cause damage to the laminate. On external surfaces, countersunk fasteners and rivets may be required. The countersunk angle should be as large as possible (reference 5.7).

Clamping force: For unconstrained specimens, failure due to bearing has been shown to occur through buckling or brush-like failure (reference 5.10). Examination of failed regions of constrained specimens has indicated that failure has occurred by initiation of cracks at the hole edge and subsequent crack propagation to the edge of the constrained region, where the mode has tended to revert to one of local instability and delamination. Hart-Smith reported (reference 5.8) that there is often no visible damage at all in the high bearing stress area under the washer, the damage having been initiated by delamination at the edge of the washer.

In a pin-loaded specimen (no constraint), failure occurs immediately adjacent to the contact points of the bolt and laminate. In a laterally constrained specimen, on the other hand, the failure cannot occur in the constrained region and instead occurs at the washer edge. The stress level at the washer edge is lower than the stress level at the hole edge. Consequently, a laterally constrained specimen can be subjected to a higher load than a pin-loaded specimen before failure occurs.

In a laterally constrained specimen, some load is transferred by frictional forces at the interface between the washer and the laminate. Thus, the load is effectively distributed over a larger area than in a pin-loaded specimen, resulting in a higher bearing strength. In finger-tightened specimens, the applied load transferred by frictional forces is relatively small. The higher strength of a finger-tight specimen, compared to that of a pin-loaded specimen, is thus due to the dominance of the first mechanism. In torqued specimens, relatively higher loads are transferred by frictional forces between the washer and the laminate. Therefore, the higher strength of a clamped specimen, compared to that of a finger-tight specimen, is to a large extent due to the effective distribution of the load over a larger area (the second mechanism). Furthermore, the higher the clamping torque, the more the out-of-plane mechanisms are restrained from occurring, resulting in an additional strength increase.

Washers: The bearing and pull-out strength of composite laminates containing fastener holes can be improved by clamping the laminate well between large, stiff and tight-fitting washers using protruding heads wherever possible (reference 5.8).

5.2.2.4 Design methods

(a) *Fastener load distribution - simple method*

Determination of the load distribution in the joint is a statically indeterminate problem which depends on a number of factors:

- elastic properties of members being joined
- elastic properties of fasteners
- geometry of joined members
- joint type (single lap, double lap, etc.)
- action of applied load
- clamping force
- friction between joined members
- clearance between bolts and connected parts.

The influence of member stiffness (elasticity and geometry) on the bolt load distribution is illustrated by a simple example. Figures 5.7 and 5.8 show schematically a single-lap joint with uniform lap thickness loaded in tension. The upper and lower parts represent the joined members and the layer between represents a row of fasteners. When the load is applied, the members deform concentrically and the fasteners in shear. If the members were rigid, the fasteners would transfer equal amounts of the load, and the shear deformation would be equal in all fasteners (Figure 5.7).

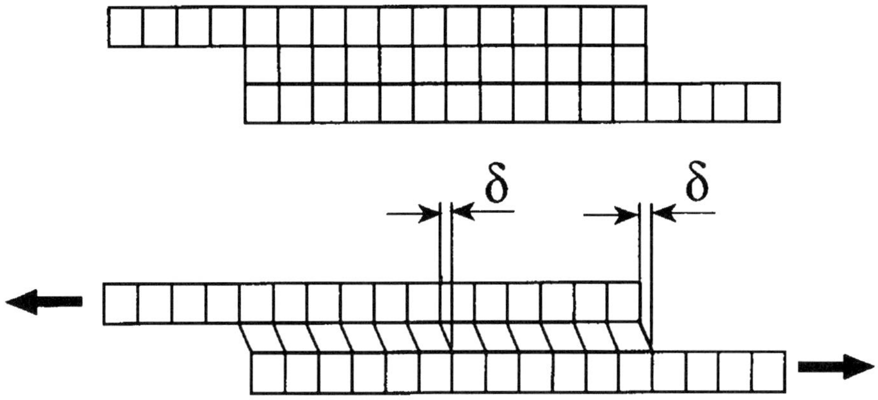

Figure 5.7 Deformation of a single-lap joint with rigid members.

In reality, however, the members are elastic (Figure 5.8). The upper and lower members will deform continuously along their lengths. The relative displacement between the members is maximum at the ends of the overlap (Figure 5.8). This means that fasteners which are located at the ends of the overlap transfer a greater amount of the load than fasteners located at the center of the overlap.

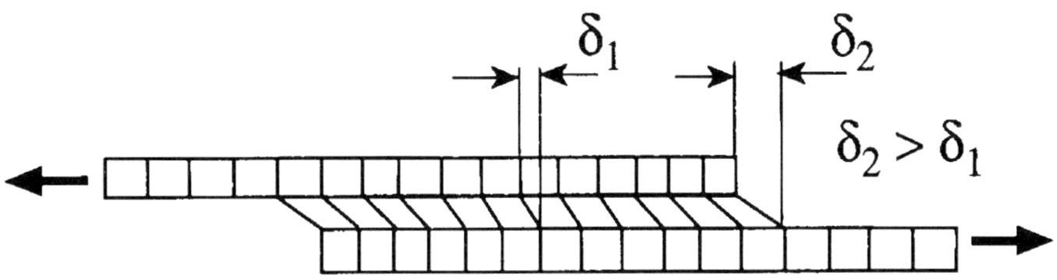

Figure 5.8 Deformation of a single-lap joint with elastic members.

To determine the load distribution in a bolted connection accurately, taking into account the elastic properties of the joined members and the fasteners explicitly, it would be necessary to use a numerical method such as the finite element method (FEM). To greatly simplify the calculation of fastener load distribution it is assumed that the members are macroscopically rigid and that the elasticity is limited to local areas in the vicinity of the fasteners. It may furthermore be assumed that the load versus deformation response of an individual fastener is linear. Thus, the effect of member stiffness on fastener load distribution is taken into account by means of special correction factors.

(i) *Concentrically loaded lap joints*

Consider two plates with uniform thickness (Figure 5.9) which are joined together by a certain number (n) of fasteners and subjected to a uniaxial tensile or compressive load F_x. The cross-sectional area of the fasteners is denoted by A_i. As the load is applied, the plates and fasteners deform and the load is distributed among the fasteners.

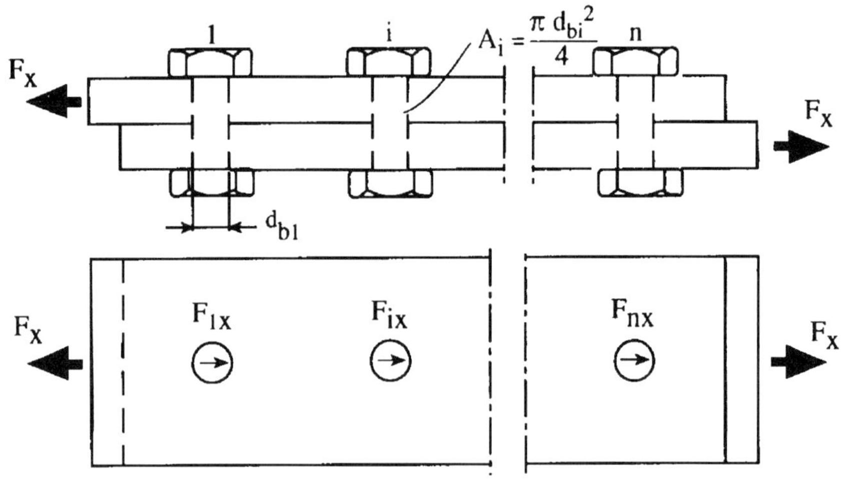

Figure 5.9 Concentrically loaded single-lap joint.

The load F_x yields a relative displacement between the joined members. The relative displacement between the connected parts in the vicinity of the fastener holes is denoted by Δ_{ix}. Since the connected parts are assumed to be also macroscopically rigid, all Δ_{ix} are identical. It is also assumed that the force transmitted by a fastener is proportional to the cross-sectional area of the fasteners A_i and the relative displacement Δ_{ix}:

$$F_{ix} = kA_i \, \Delta_{ix} \tag{5.6}$$

where k is a constant depending on the stiffness properties of the joined members and bolt material, among other factors. The constant k is assumed to be identical for all fasteners.

Equilibrium gives:

$$\sum_{i=1}^{n} F_{ix} = F_x \tag{5.7}$$

Equations (5.6) and (5.7) give the load transmitted by each fastener:

$$F_{ix} = \frac{F_x}{\frac{1}{A_i}\sum_{i=1}^{n} A_i} \tag{5.8}$$

(ii) *Eccentrically loaded lap joints*

Figure 5.10 shows two plates which are connected by a group of fasteners. The connection is subjected to an applied load F with a line of action passing outside the centre of gravity (c.g.) of the fastener group. The origin of both a Cartesian coordinate system (x - y) and a polar coordinate system (r - ϕ) is placed at c.g. The cross-sectional area of the fasteners is denoted by A_i. The eccentricity of F with respect to c.g. is denoted by e. As the load is applied, the plates and the fasteners deform and the load is distributed among the fasteners. The load transferred by each fastener is denoted by F_i^e. The angle between F_i^e and the x-axis is denoted by β_i. The problem is to predict the magnitude and direction of the load transferred by each fastener.

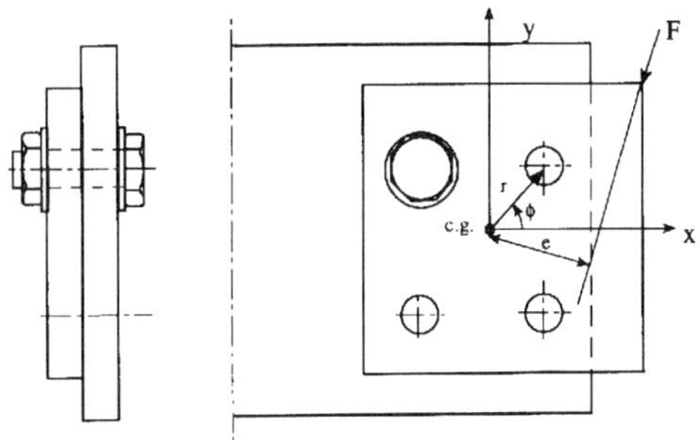

Figure 5.10 An eccentrically loaded connection.

It is assumed that the rotation of the connection takes place about c.g. The eccentric load is resolved into a concentric load, F^c, acting through c.g. (see Figure 5.11) and a torsional moment, M=Fe (Figure 5.12).

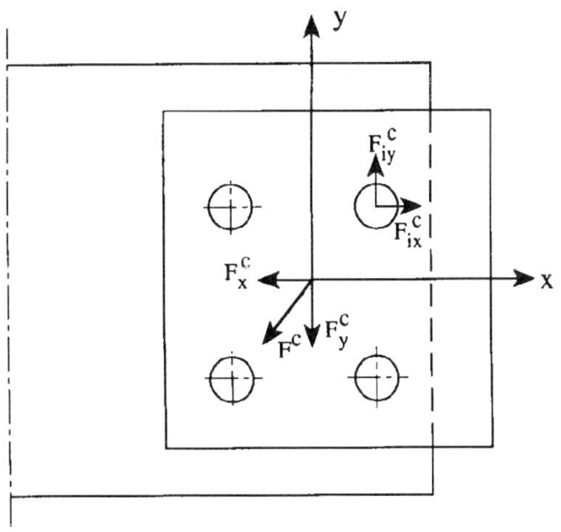

Figure 5.11 Fastener load distribution due to concentric load.

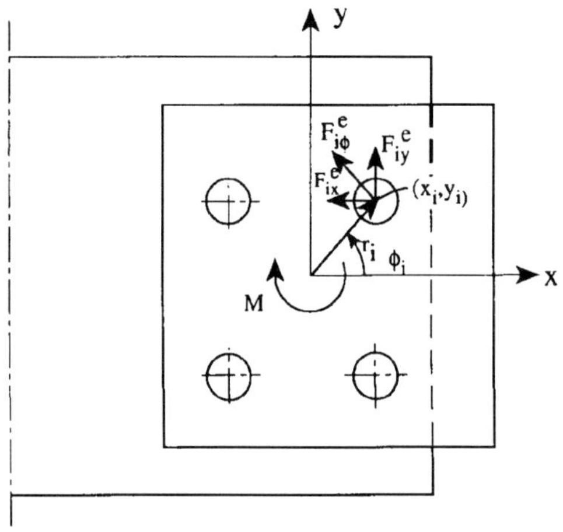

Figure 5.12 Fastener load distribution due to torsional moment.

The moment is assumed to cause loads in the fasteners that vary linearly with the distance from the centre of gravity. The load in any fastener is determined by taking the vector sum of the load components (the concentric load F^c and the moment M).

The load F^c yields fastener deflections $\Delta_{ix}{}^c$ and $\Delta_{iy}{}^c$ in the x- and y-directions, respectively. The deflections in all fasteners are assumed to be identical. The force transmitted by a fastener i is calculated from:

$$F^c_{ix} = kA_i \Delta^c_{ix} \tag{5.9}$$

$$F^c_{iy} = kA_i \Delta^c_{iy} \tag{5.10}$$

where k is a constant depending on the stiffness properties of the joined members, among other factors. The constant k is assumed to be identical for all fasteners.

Equilibrium gives:

$$\sum_{i=1}^{n} F^c_{ix} = F^c_x \tag{5.11}$$

$$\sum_{i=1}^{n} F^c_{iy} = F^c_y \tag{5.12}$$

Equations (5.9) to (5.11) give:

$$F_{ix}^{c} = \frac{F_x^{c}}{\frac{1}{A_i}\sum_{i=1}^{n} A_i} \tag{5.13}$$

$$F_{iy}^{c} = \frac{F_y^{c}}{\frac{1}{A_i}\sum_{i=1}^{n} A_i} \tag{5.14}$$

The moment is assumed to rotate the connection so that the load $F_{i\varphi}^{e}$ in each fastener is acting perpendicularly to the radius r_i of rotation. The displacement in each fastener is:

$$\Delta_{i\phi} = r_i d_{\phi i} \tag{5.15}$$

where $d_{\phi i}$ is the rotation of the plate. The force in each fastener is expressed as:

$$F_{i\phi}^{e} = kA_i r_i \Delta_{i\phi} \tag{5.16}$$

The fastener loads in the x- and y-directions are:

$$F_{ix}^{e} = F_{i\phi}^{e} \sin\phi_i \tag{5.17}$$

$$F_{iy}^{e} = F_{i\phi}^{e} \cos\phi_i \tag{5.18}$$

The moment equilibrium gives:

$$\sum_{I=1}^{n} F_{i\phi} r_i = M \tag{5.19}$$

Equations (5.15) to (5.19) give:

$$F_{ix}^{e} = \frac{-MA_i y_i}{\sum (x_i^2 + y_i^2) A_i} \tag{5.20}$$

$$F_{iy}^{e} = \frac{MA_i x_i}{\sum (x_i^2 + y_i^2) A_i} \tag{5.21}$$

The resulting force in each fastener, due to the concentric load and the torsional moment, is:

$$F_i^e = \sqrt{(F_{ix}^c + F_{ix}^e)^2 + (F_{iy}^c + F_{iy}^e)} \qquad (5.22)$$

As previously discussed, the applied load is unevenly distributed between the fasteners, depending on the stiffness properties (elasticity and geometry) of the members being joined. This is not taken into account by the presented analytical approach. The analytical results must therefore be corrected.

The special purpose finite element program BOLTIC (reference 5.11) was used to determine the fastener load distribution for a number of realistic joint configurations. Figure 5.13 shows the computed results. In BOLTIC, the fasteners are represented by spring elements which are given a representative stiffness. In this context, an empirical relation for fastener flexibility according to Huth (reference 5.12) was used. Constants in the Huth expression valid for carbon FRP joints were used, since no data are available for glass FRP joints. The elastic constants of the analysed laminates are given in 5.2.2.5 *Experimental data.*

Figure 5.13 (a) shows the bolt load distribution for two equally thick (3.9 mm) glass-fibre laminates joined together by 2, 3 and 4 tandem rows of fasteners. In the two-row joint (w_x=40 mm and 120 mm, respectively) the load is evenly distributed among the two rows. In the three-row joint (w_x=40 mm and 60 mm, respectively) the fasteners at the end of the overlap transfer about 7% higher bolt load than that of an even distribution, whereas the bolt in the middle transfers a load 14% less than that of an even distribution. In the four-row configuration (w_x=40 mm) the bolts at the end of the overlap transfer about 20% higher load than that of an even distribution. Hence, the two bolts in the middle transfer about 20% less load than that of an even distribution.

Figure 5.13 (b) shows the bolt load distribution for two glass-fibre laminates of equal thickness (7.8 mm) joined by 2, 3 and 4 tandem fastener rows. In the two-row joint (w_x=40 mm) the fastener rows transfer an equal amount of the load. In the three-row joint (w_x=40 mm) the end bolts transfer about 4% higher load than an even distribution. Consequently, the middle bolts transfer about 8% less load than that of an even distribution. Finally, the end and middle bolts of the four-row configuration (w_x=40 mm) transfer about 10% higher and lower load, respectively, than that of an even distribution.

Figure 5.13 (c) shows the bolt load distribution for a 3.9 mm thick glass-fibre laminate joined to an equally thick steel plate by 2, 3, and 4 tandem fastener rows. For all configurations the bolts at the end of the overlap transfer substantially higher load than the bolts in the beginning of the overlap.

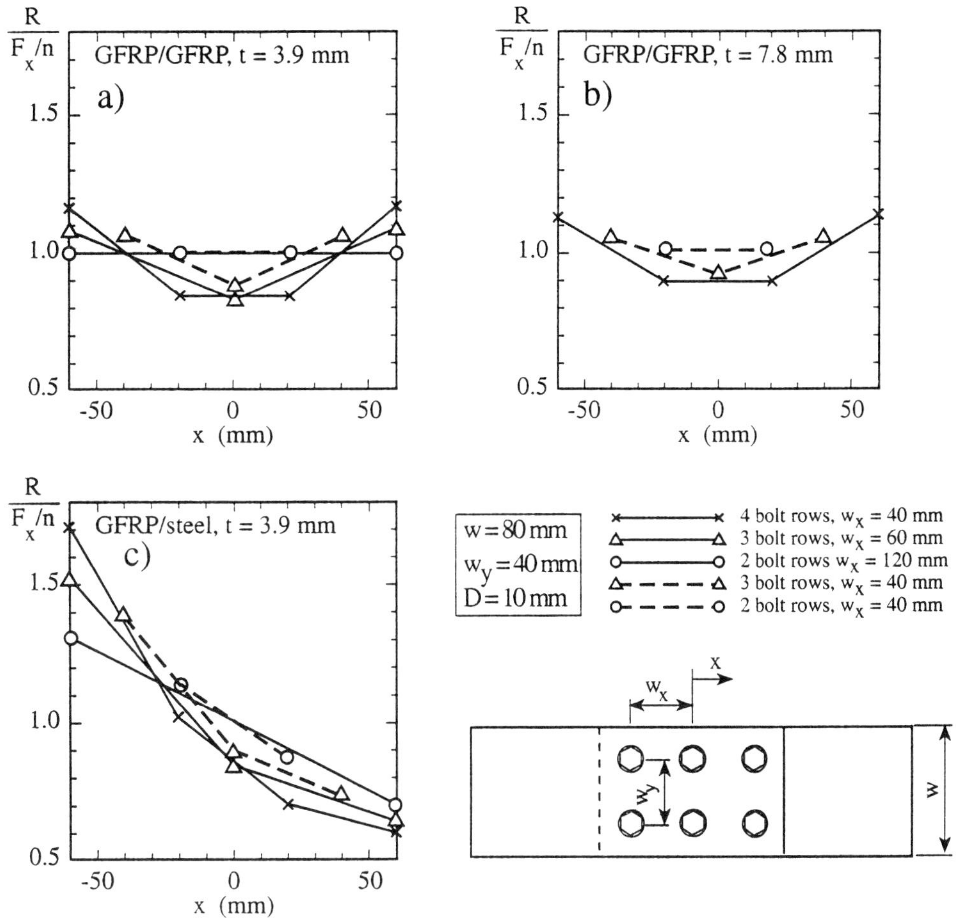

Figure 5.13 Influence of member stiffness and bolt spacing on load distribution in multiple-row single-lap joints: (a) two 3.9 mm glass FRP members, (b) two 7.8 mm glass FRP members, (c) one glass FRP member and one steel member.

(b) *Advanced method*

A major limitation of the simple method of determining the fastener load distribution is that the stiffness (elasticity and geometry) of the joined members is not explicitly taken into account. Another limitation of the simple method is that the specific elasticity properties of the analysed member is not considered when determining the fastener hole stress distribution. These limitations can effectively be overcome by adopting a numerical method such as the finite element method (FEM) for evaluation of the load distribution.

The advanced method contains the following steps:

Step 1 (source analysis): determination of load distribution (fastener load distribution and far-field load distribution)

Step 2 (target analysis): determination of fastener hole stress distributions

Step 3 (failure analysis).

The advanced method can be executed using any general purpose FE-code with the capability of modelling anisotropic structures and contact problems. However, the disadvantage of using a general purpose FE-code is that the modelling work may be very time consuming and costly. To dramatically reduce modelling time and make the advanced method a practical and cost-effective alternative to the simple method, a special purpose FE-code BOLTIC has been developed by Eriksson, Bäcklund and Möller (reference 5.11).

(i) *Load distribution analysis*

Consider the web-flange splice shown in Figure 5.14. The fasteners are subjected to shear. Assume that the flange splice transfers the bending moment through tensile and compressive forces and that the web splice carries the transverse shear. The composite beams are made from fibre reinforced composite material. As the joint is loaded, the flanges and web move and the bolts contact portions of the holes. As the load is increased, failure of the joint may occur in different modes, i.e. net-section, bearing, and shear out.

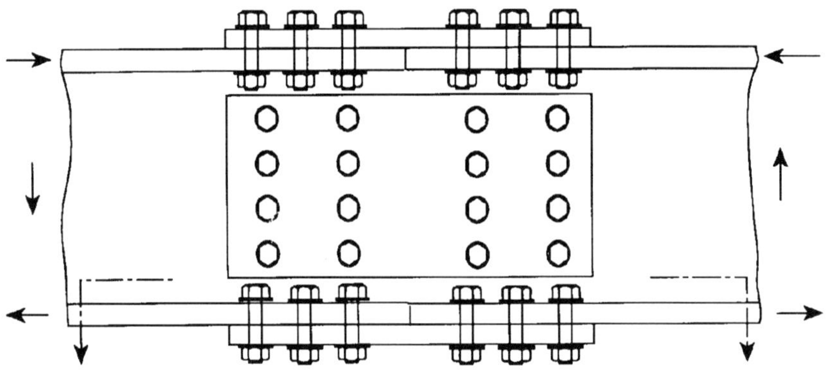

Figure 5.14 Connection of two I-beams.

Figure 5.15 shows the lower flange splice and Figure 5.16 shows an isolated region of the lower flange splice containing a single fastener. Assume that the flange splice is stiff with respect to bending, so the bending deformations are small in comparison to the in-plane deformations. The load transferred by the fastener is denoted by R. The relative displacement between the members at some distance away from the hole consists of local and global parts. The local part δ_{exp} is made up of bending and shearing deformations of the bolt and, in addition, by deformation of the hole edge. The global part is related to the elastic deformations of the members away from the fastener hole. Tate and

Rosenfeld (reference 5.13) proposed a linear approximation for δ_{exp}:

$$\delta_{exp} = CR \tag{5.23}$$

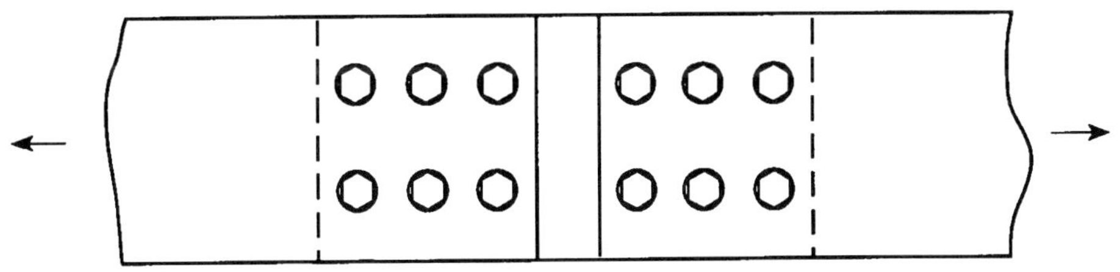

Figure 5.15 Lower flange splice.

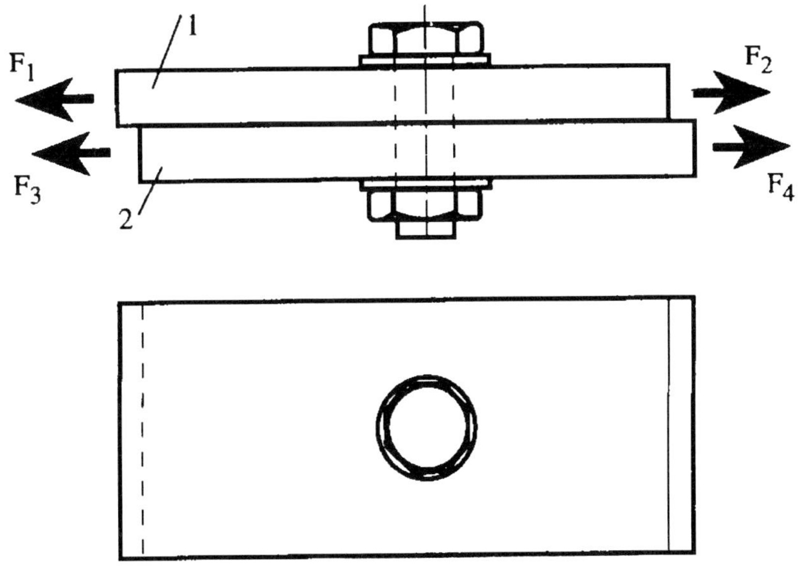

Figure 5.16 Isolated region of lower flange splice.

Jarfall (reference 5.14) suggested the constant C be called the fastener flexibility. The inverse of the fastener flexibility is denoted as the fastener stiffness K_{exp}:

$$K_{exp} = \frac{1}{C} = \frac{R}{\delta_{exp}} \tag{5.24}$$

Figure 5.17 shows a finite element model of the isolated region. The members are modelled by plane stress elements. The fasteners and holes are not modelled explicitly but are represented by spring elements which are connected to node points on the members. As the load is applied, the members and springs deform. The relative displacement between the members, measured with respect to the connected points, is denoted by δ_{fem} and is made up of the deformation of the spring element δ_{ffe} and the local deformation of the mesh, δ_{mesh} .

$$\delta_{fem} = \delta_{ffe} + \delta_{mesh} \tag{5.25}$$

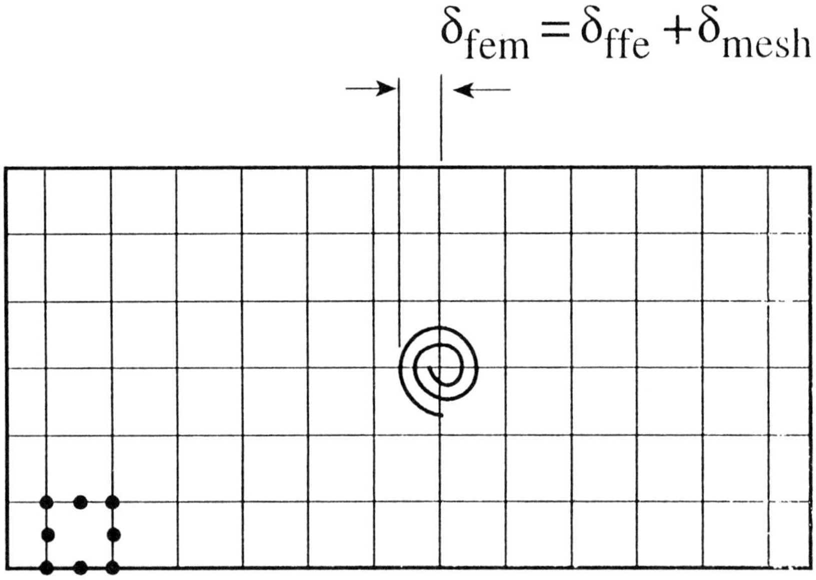

Figure 5.17 Finite element model of an isolated region of lower flange splice.

In turn, δ_{mesh} depends on the element type, mesh density and mesh configuration. To predict accurately the joint displacements and load distribution, Jarfall [5.14] pointed out that the stiffness constants of the spring elements shall be determined so that:

$$\delta_{exp} = \delta_{ffe} + \delta_{mesh} \tag{5.26}$$

Equation (5.26) may be rewritten as:

$$\delta_{ffe} = \delta_{exp} - \delta_{mesh} \tag{5.27}$$

Division by R gives:

$$\frac{\delta_{ffe}}{R} = \frac{\delta_{exp} - \delta_{mesh}}{R} \tag{5.28}$$

Assuming that the load versus displacement response of the joint is approximately linear, the transferred loads are linear functions of each of the displacements δ_{ffe} , δ_{mesh} , and δ_{exp} :

$$R = K_{ffe}\,\delta_{ffe} \tag{5.29}$$
$$R = K_{mesh}\,\delta_{mesh} \tag{5.30}$$
$$R = K_{exp}\,\delta_{exp} \tag{5.31}$$

where:

K_{ffe} is the stiffness constant of the spring element
K_{mesh} is the slope of a numerically determined curve of R as a function of δ_{mesh} using a rigid spring
K_{exp} is the slope of an experimentally determined curve of R as a function of δ_{exp} .

Equations (5.28) to (5.31) give:

$$K_{ffe} = \frac{K_{exp}K_{mesh}}{K_{mesh} - K_{exp}} \tag{5.32}$$

To determine K_{ffe} from Equation (5.32), K_{mesh} and K_{exp} must be known. K_{mesh} is determined from a finite element analysis of the source model using rigid spring elements. K_{exp} is usually unknown and must be estimated from analytical or semi-empirical models. There exist several semi-empirical expressions for the fastener flexibility (the inverse of the fastener stiffness K_{exp}) which are used in the industry. Those presented by Huth (reference 5.12) are valid for both metallic and composite (carbon FRP) members. The fastener flexibility for a single-lap joint from (reference 5.12) is:

$$C = \left(\frac{t_1 + t_2}{2d_b}\right)^a b\left(\frac{1}{t_1E_1} + \frac{1}{t_2E_2} + \frac{1}{t_1E_f} + \frac{1}{2t_2E_f}\right) \tag{5.33}$$

and for a double lap joint:

$$C = \left(\frac{t_p + t_s}{2d_b}\right)^a \frac{b}{2}\left(\frac{1}{t_pE_p} + \frac{1}{t_sE_s} + \frac{1}{2t_pE_f} + \frac{1}{4t_sE_f}\right) \tag{5.34}$$

In Equations (5.33) and (5.34) the indices 1 and 2 refer to the upper and lower members, respectively, of a single-lap joint; and s and p refer to the side and centre plates, respectively, of a double-lap joint. The lap thickness is denoted by t, elastic modulus of members and fasteners by E and E_f, respectively, and fastener diameter by d_b. The empirical constants a and b depend on the material of the members and the type of fastener. For bolted metallic members (a=2/3, b=3.0), for riveted metallic members (a=2/5, b=2.2), and for bolted carbon FRP members (a=2/3, b=4.2). It is recommended that the values valid for carbon FRP members be used for glass FRP joint composites until specific data are available.

(ii) *Detailed stress analysis*

Stress distributions in the vicinity of the bolt holes, which are required for failure analysis, are determined by a series of detailed FE-analyses of regions containing single bolt holes.

For single-lap joints, all stresses as calculated above should be increased by 100% (C_m = 2.0). For double-lap joints and similar joints which may reasonably be assumed to be symmetrically loaded, only σ_r, as calculated above, should be increased by 20% (C_m = 1.2).

These increases are to allow for non-uniform stress distributions through the laminate due to non-symmetry, bolt-bending and loss of lateral restraint.

(iii) *Failure analysis*

During increasing loading of a composite plate with a hole, a damage zone forms and grows in the most stress-intense region at the edge of the hole. "Damage" includes the collective effects on the material of micromechanical failure mechanisms, such as fibre breaking, fibre microbuckling, fibre and matrix shear failure, and delamination. Since fibre reinforced composites have some capability of redistributing stresses, failure criteria based on the stress concentration at the hole edge are generally conservative for composites. One possible way to improve the accuracy is to evaluate stresses at a characteristic distance ahead of the fastener hole. Whitney and Nuismer (reference 5.15) first adopted this principle and presented the well-known point stress criterion (PSC) and the average stress criterion (ASC). The models of Nuismer et al. are simple to apply and are therefore attractive to designers. A specific failure mode is associated with a specific stress component and a characteristic distance. It is assumed that failure has occurred at some point around the hole circumference when a specific stress component at some characteristic distance from the hole edge reaches a specific unnotched strength value (associated with the specific failure mode) of the laminate.

The failure criterion parameter unnotched strength in tension, compression, or shear at any location and direction around the hole circumference is predicted on the basis of the laminate (extensional) stiffness equation in conjunction with a critical strain value (design value) in tension, compression or shear, respectively. Failure of the unnotched laminate is assumed to have occurred when the strain exceeds the critical value (design value).

Net-section and bearing failure can be determined from the stress values either at the hole edge (conservative design) or at a characteristic distance ahead of the hole, if such a distance has been determined experimentally. Shear-out failure can be determined from the maximum shear stress along the shear-out plane or from the shear stress at a characteristic distance from the hole edge, if such a distance has been determined experimentally.

Net-section failure is evaluated around the hole circumference at points located along radial lines through hole boundary node points (Figure 5.18).The most critical point defines the location of failure. Failure at any of the points is assumed to occur when the tangential stress σ_ϕ at either the hole edge (conservative design) or a characteristic distance d_k from the hole edge reaches the unnotched strength $\sigma_{\phi,k}$ of the laminate in the direction considered. Both tensile and compressive net-section failures are considered. Hence, unnotched strengths in tension ($\sigma_{t,\phi,k}$) and compression ($\sigma_{c,\phi,k}$) in conjunction with characteristic dimensions in tension ($d_{t,k}$) and compression ($d_{c,k}$) are required to evaluate net-section failure. The unnotched tensile and compressive strengths are predicted from:

$$\sigma_{t,\phi,k} = E_\phi \epsilon_{t,crit} \tag{5.35}$$

$$\sigma_{c,\phi,k} = E_\phi \epsilon_{c,crit} \tag{5.36}$$

where $\varepsilon_{t,crit}$ and $\varepsilon_{c,crit}$ are the critical strains (design values) in tension and compression, respectively, and E_ϕ is the elastic modulus in the tangential direction of the hole (ϕ-direction) from (reference 5.16):

$$\frac{1}{E_\phi} = \frac{1}{E_x}\sin^4\phi + \left(\frac{1}{G_{xy}} - 2\frac{\nu_{xy}}{E_x}\right)\sin^2\phi\cos^2\phi + \frac{1}{E_y}\cos^4\phi \tag{5.37}$$

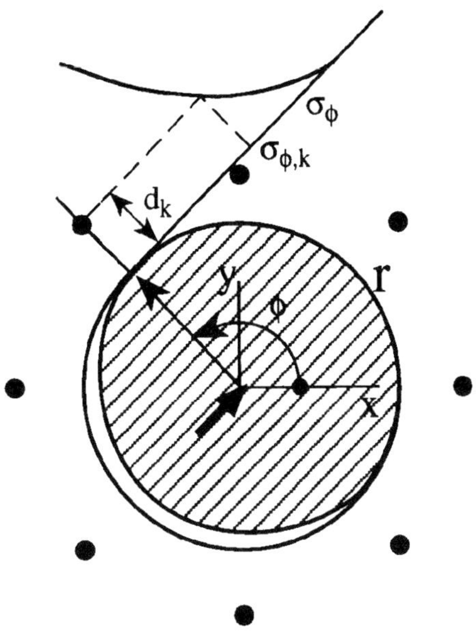

Figure 5.18 Evaluation of net-section failure around the hole circumference.

Bearing failure is evaluated around the loaded part of the hole circumference at points located along radial lines through hole boundary node points (Figure 5.19). Bearing failure occurs at any of the selected points when the radial compressive stress σ_r reaches the unnotched compressive strength $\sigma_{r,k}$ of the laminate at either the hole edge (conservative design) or a characteristic distance r_k from the hole edge. The unnotched radial compressive strength is predicted from:

$$\sigma_{r,k} = E_r \, \varepsilon_{c,crit} \tag{5.38}$$

where E_r is the radial elastic modulus in the direction considered (see reference 5.16).

$$\frac{1}{E_r} = \frac{1}{E_x}\cos^4\phi + \left(\frac{1}{G_{xy}} - 2\frac{\nu_{xy}}{E_x}\right)\sin_2\phi\cos^2\phi + \frac{1}{E_y}\sin^4\phi \tag{5.39}$$

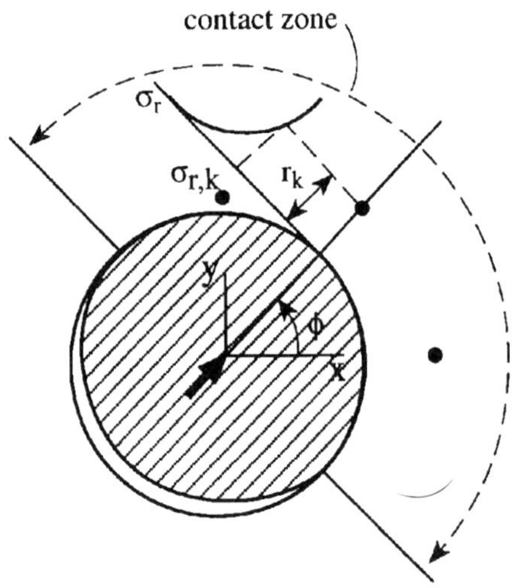

Figure 5.19 Evaluation of bearing failure around the loaded half of the hole circumference.

Shear-out failure is evaluated along shear-out planes (s - n planes), which originate from opposite points on the hole boundary where the contact between bolt and laminate ends. The shear-out planes are parallel to the direction of the principal bolt load (Figure 5.20). It is assumed that failure occurs when the shear stress $\tau_{sn,s}$ reaches the unnotched shear strength $\tau_{sn,k}$ of the laminate, at some characteristic distance s_k away from the hole edge. The unnotched shear strength is predicted from:

$$\tau_{sn,k} = G_{sn}\varepsilon_{s,crit} \tag{5.40}$$

where $\varepsilon_{s,crit}$ is the critical (design) shear strain and G_{sn} is the shear modulus in the s - n plane see (reference 5.16):

$$\frac{1}{G_{sn}} = \frac{1}{G_{xy}} + \left(\frac{1+\nu_{xy}}{E_x} + \frac{1+\nu_{yx}}{E_y} - \frac{1}{G_{xy}}\right)\sin^2 2\phi \qquad (5.41)$$

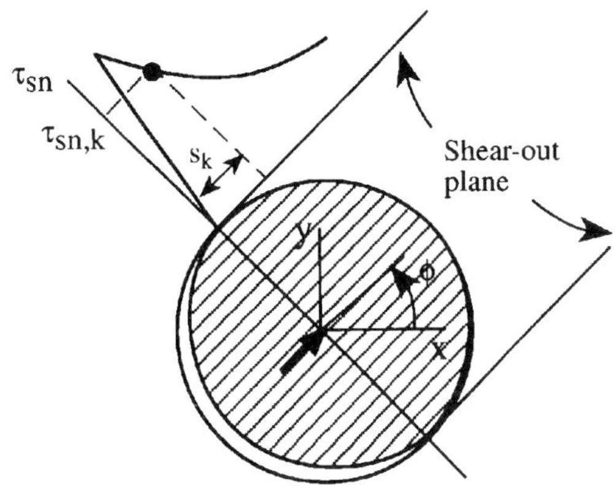

Figure 5.20 Evaluation of shear-out failure along shear-out planes (s - n planes).

It should be emphasised that the characteristic distances ($d_{t,k}$ and $d_{c,k}$ for net-section tensile and compressive failure, respectively; r_k for bearing failure; and s_k for shear-out failure) vary depending on the ply-orientation. Thus, the characteristic distance varies around the hole boundary since the ply orientation varies around the hole. Hence, a great amount of testing would be necessary to determine experimentally these distances for all directions considered. To reduce the amount of testing, a limited number of data points (at least three) could be used to establish an empirical relation between the characteristic distance and the associated unnotched strength. The characteristic distance of a laminate of the same material system but with a different ply orientation and, therefore, a different value of the unnotched strength could be found from this empirically determined relation. The characteristic distance r_k depends on the degree of lateral constraint (clamping force), since bearing strength is strongly affected by the degree of lateral constraint. Hence, depending on the degree of lateral constraint (pin-loaded, finger-tight, or clamped), different values of r_k are required.

(c) *Illustrative worked example*

(This analysis is carried out using ultimate strength data and without including partial safety factors.)

The plate was made by hand lay-up. Details are given in 5.2.2.5 *Experimental data.*

Consider the concentrically loaded double-lap connection shown in Figure 5.21. A composite plate (over-lap length l, width w, thickness t, edge distance e, side distance s, spacing in the x-direction w_x and spacing in the y-direction w_y) containing four bolt holes of equal diameter d_h is placed between two steel plates and fastened with four finger-tightened steel bolts of diameter d_b. The geometry of the joint is presented in Table 5.1.

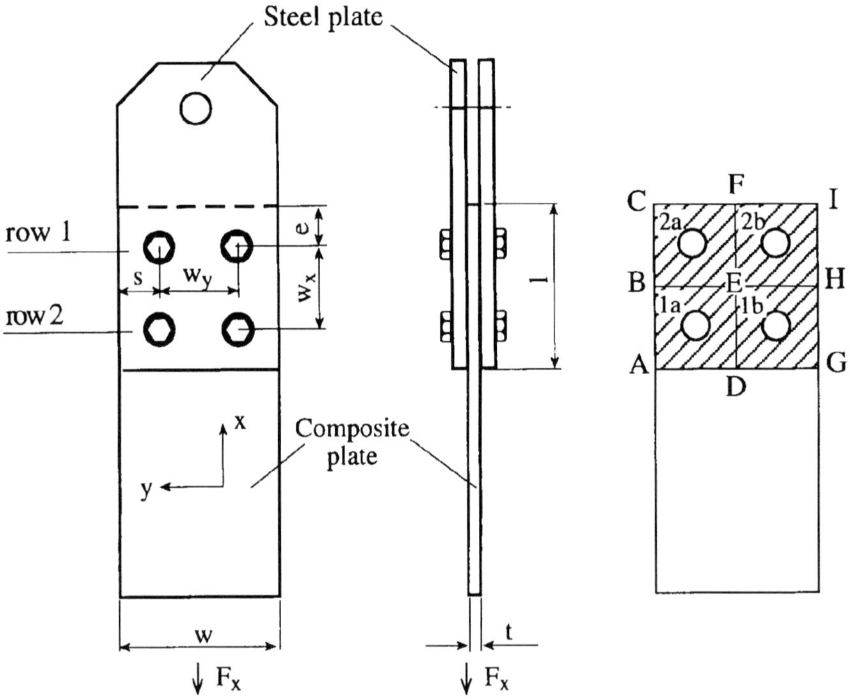

Figure 5.21 Concentrically loaded double-lap joint.

Table 5.1 Geometry of joint (dimensions in mm).

l	w	t	e	s	w_x	w_y	d_h	d_b
320	140	3.9	40	40	60	60	10.0	9.9

The plate is subjected to a uniaxial tensile load F_x. During loading, the composite plate deforms until failure occurs. The load displacement curve of the specimen is shown in Figure 5.22. According to the performance requirements (see 5.2.2.2 in the EUROCOMP Design Code), on-set of non-linear load deflection behaviour is a serviceability limit-state which in this particular case occurs at 43 kN (see Figure 5.22) The analysis is carried out according to the simple method presented in the EUROCOMP Design Code as well as the advanced method presented here in this section.

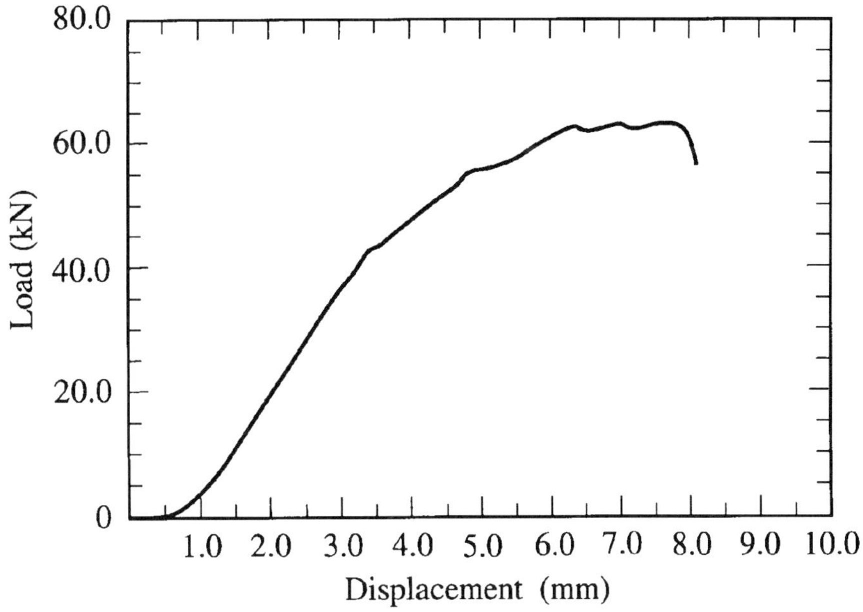

Figure 5.22 Load displacement curve of the concentrically loaded double-lap joint shown in Figure 5.21.

(i) *Simple method*

(a) *Determination of fastener shear load distribution*

The connection is a concentrically loaded double lap joint containing two rows of fasteners of equal diameter. The composite plate is fastened to two steel plates. According to section 5.2.2.4 in the EUROCOMP Design Code, the first row of fasteners transfers a greater amount of load than the second row. The first and second fastener rows transfer $0.57F_x$ and $0.43F_x$, respectively.

(b) *Determination of by-pass load distribution*

According to 5.2.2.4 of the EUROCOMP Design Code, the by-pass load distribution is determined from the conditions of equilibrium. A free body diagram of parts ABGH and BCHI (Figures 5.23 (a) and (b)) gives that the load acting on section BH is $0.57F_x$. Consider next a free body diagram of part 1a (Figure 5.2). Owing to symmetry, it is assumed that $0.5F_x$ is acting on sections AD and DG. The load acting on section BE and EH is $0.285F_x$. The load acting on section BE of part 2a and section EH of part 2b is $0.285F_x$.

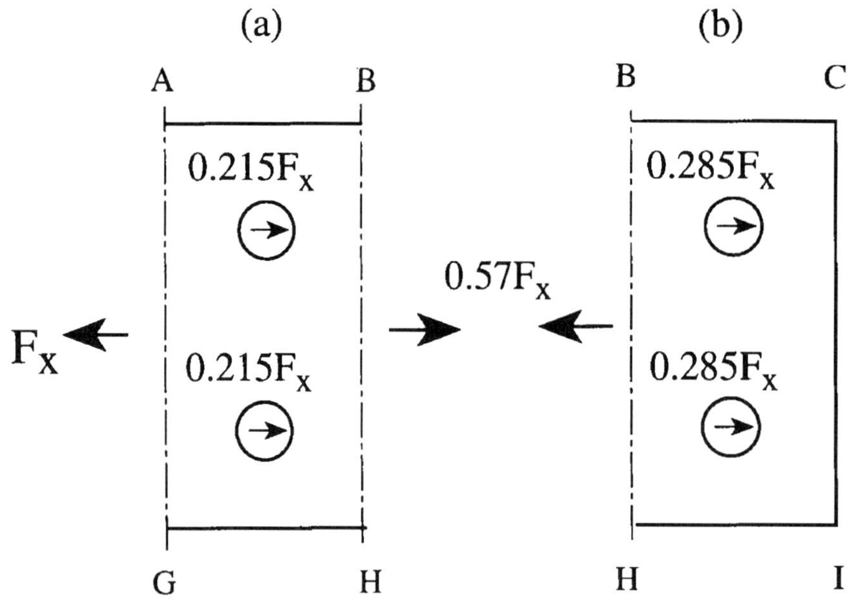

Figure 5.23 Free body diagram of parts ABGH (a) and BCHI (b).

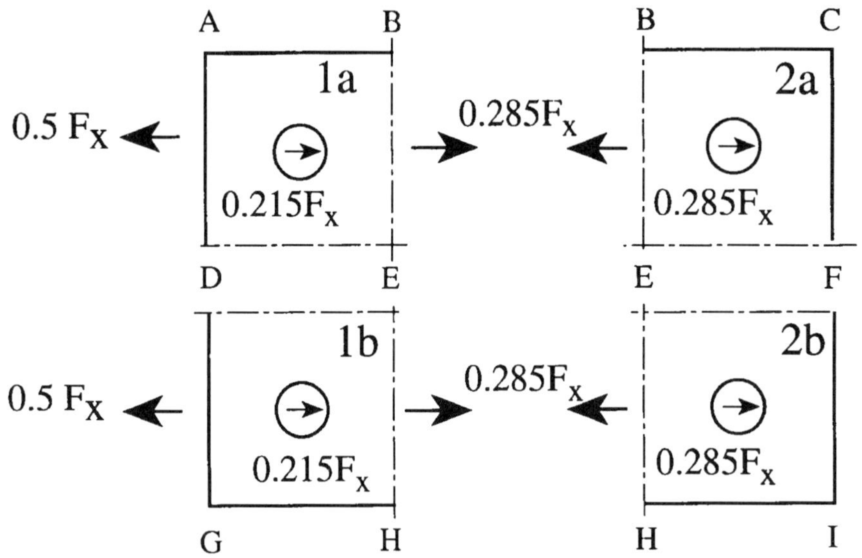

Figure 5.24 Free body diagram of parts 1a-2b.

Owing to symmetry, parts 1a and 1b are identical and so are parts 2a and 2b. Parts 1a and 1b are made up of the basic load cases 1 and 3 (Figure 5.25). Parts 2a and 2b are made up of basic load case 1 (Figure 5.26).

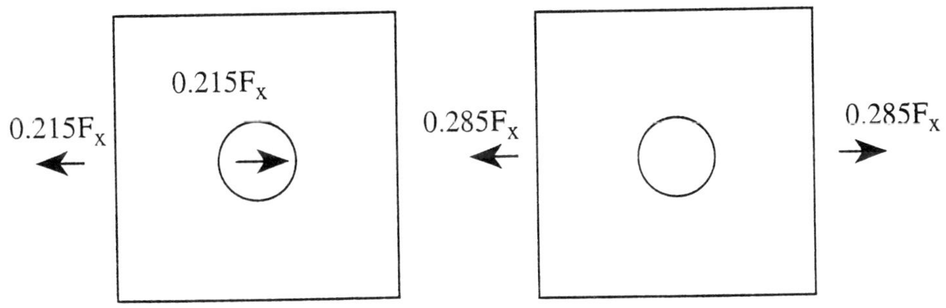

Figure 5.25 Basic load cases 1 and 3 making up the load case for parts 1a and 1b.

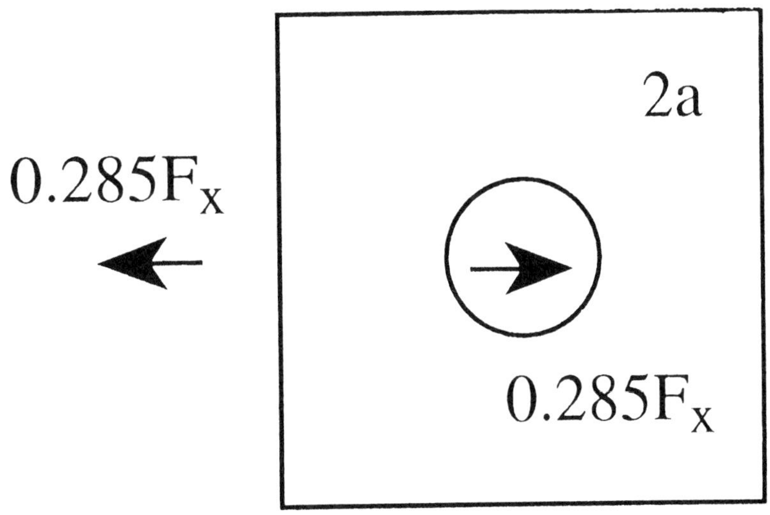

Figure 5.26 Basic load case 1, making up the load case for parts 2a and 2b.

(c) *Determination of normalised stress distribution around the fastener hole*

The stress distributions are determined from section 5.2.2.4 in the EUROCOMP Design Code. To secure a conservative design, the upper boundary stress level is taken as the actual stress level in the laminate studied. The stress charts generated for the 35% glass content laminate is used since the glass content of the studied laminate (42%) is close to 35%. The normalised stress values (k_r^i, k_t^i, and k_s^i) for basic load case 1 and 3 are listed in Tables 5.2 to 5.5.

Table 5.2 Normalised radial stress k_r^1 = minus $-\sigma_\phi/S_b$ and tangential stress $k_t^1 = \sigma_\phi/S_b$ along the hole boundary at 0°, 45° and 90° for basic load case 1.

Location	k_r^1	k_t^1
0°	1.35	1.60
45°	1.10	0.65
90°	0.00	2.80

Table 5.3 Normalised shear stress $k_s^1 = -\tau_{xy}/S_b$ along the shear-out plane at different locations x/d from the hole boundary for basic load case 1.

x/d	k_s^1
0.2	0.59
0.3	0.62
0.4	0.60

Table 5.4 Normalised tangential stress $k_t^3 = \sigma_\phi/\sigma_N$ along the hole boundary at 0°, + 45° and 90° for basic load case 3.

Location	k_t^3
0°	−1.90
45°	1.50
90°	5.10

Table 5.5 Normalised shear stress $k_s^3 = -\tau_{xy}/\sigma_N$ along the shear-out plane at different distances x/d from the hole boundary for basic load case 3.

x/d	k_s^3
0.2	0.80
0.3	0.60
0.4	0.40

(d) *Compliance check*

The failure analysis is carried out according to section 5.2.2.4 of the EUROCOMP Design Code. The net-section, bearing and shear-out failure modes were evaluated. For this reason, the ultimate strength values $\sigma_{t,\phi,k}$ $\sigma_{c,\phi,k}$ and $\tau_{sn,k}$ were calculated.

The ultimate strain in tension was determined experimentally. The results were 2.0% in the x-direction (tangential direction at 90°) and 1.4% in the y-direction (tangential direction at 0°). The ultimate strain at 45° was not determined experimentally. The lowest of the strain values measured in the x- and y-directions (1.4%) is taken for the ultimate tangential strain at 45°. The calculated strength values are presented in Table 5.6.

The ultimate strain in compression was determined experimentally. The results were 1.6% in the x-direction (radial direction at 0°) and 1.82% in the y-direction (radial direction at 90°). The ultimate compressive strain at 45° was not determined experimentally. The lowest of the strain values measured in the x- and y-direction (1.6%) is taken for the critical radial strain at +45°. The calculated ultimate compressive strength values are presented in Table 5.6.

The ultimate shear strain was determined experimentally. The ultimate shear strain in the x - y plane was 3.3%. The unnotched shear strength in the x-y plane is presented in Table 5.6.

Table 5.6 Elastic moduli E_ϕ, E_r and G_{sn}, and ultimate strength $\sigma_{t,\phi,k}$, $\sigma_{r,k}$ and $\tau_{sn,k}$ at 0°, 45° and 90°.

Location	E_ϕ (N/mm^2)	E_r (N/mm^2)	G_{sn} (N/mm^2)	$\sigma_{t,\phi,k}$ (N/mm^2)	$\sigma_{r,k}$ (N/mm^2)	$\tau_{sn,k}$ (N/mm^2)
0°	7353	19360	vacant	103	310	vacant
+45°	11946	11946	vacant	167	191	vacant
90°	19366	7353	5029	387	132	166

Net-section failure analysis of parts 2a and 2b

The tangential stress distribution around the bolt holes of parts 2a and 2b is obtained from:

$$\sigma_{\phi,s} = k_t^1 S_b \tag{5.42}$$

The value of k_t^1 at 0°, +45° and 90° is presented in Table 5.2. S_b is the average bearing stress calculated from:

$$S_b = \frac{0.285F_x}{d_h t} \tag{5.43}$$

Equations (5.42) and (5.43) give:

$$\sigma_{\phi,s} = k_t^1 \frac{0.285F_x}{d_h t} \tag{5.44}$$

Net-section failure occurs when the tangential stress ($\sigma_{\sigma,s}{}^1$) is equal to the ultimate strength ($\sigma_{t,\phi,k}$) in the direction considered. The net-section failure load is predicted from:

$$P_x = \frac{\sigma_{t,\phi,k} d_h t}{0.285 k_t^1} \tag{5.45}$$

Bearing failure analysis of parts 2a and 2b

The radial stress distribution around the bolt holes of parts 2a and 2b:

$$\sigma_{r,s} = -k_r^1 S_b \tag{5.46}$$

The values of k_r^1 at 0°, +45°, and 90° are presented in Table 5.3. Equations (5.43) and (5.46) give:

$$\sigma_{r,s} = \frac{k_r^1 0.285F_x}{d_h t} \tag{5.47}$$

Bearing failure occurs when the radial stress ($\sigma_{r,s}{}^1$) is equal to the ultimate compressive strength ($\sigma_{r,k}$) in the direction considered. The bearing failure load is thus predicted from Equation (5.48).

$$F_x = \frac{\sigma_{r,k} d_h t}{0.285 k_r^1} \tag{5.48}$$

Shear-out failure analysis of parts 2a and 2b

The shear stress distribution ($\tau_{sn,s}$) along the shear-out planes (x - y planes) of parts 2a and 2b is obtained from Equation (5.49).

$$\tau_{sn,s} = -k_s^1 S_b \tag{5.49}$$

The values of k_s^1 are listed in Table 5.3. Equation (5.43) and (5.49) give:

$$\tau_{sn,s} = -k_s^1 \frac{0.285F_x}{d_h t} \tag{5.50}$$

Shear-out failure occurs when the shear stress ($\tau_{sn,s}^1$) is equal to the ultimate shear strength ($\tau_{sn,k}$). The shear-out failure is thus predicted from Equation (5.51):

$$F_x = \frac{\tau_{sn,k} d_h t}{0.285 k_s^1} \tag{5.51}$$

The results predicted from Equations (5.45), (5.48) and (5.51) in the directions 0°, +45° and 90° are summarised in Tables 5.7 and 5.8.

Table 5.7 Predicted net-section and bearing failure for parts 2a and 2b.

Location	*Net-section* F_x *(kN)*	*Bearing* F_x *(kN)*
0°	8.8	31.4
+45°	27.0	23.8
90°	18.9	∞

Table 5.8 Predicted shear-out failure for parts 2a and 2b along shear-out plane at ϕ = 90°.

Location (mm)	*Shear-out* F_x *(kN)*
0.2	38.5
0.3	36.6
0.4	37.8

Net-section failure analysis of parts 1a and 1b

Parts 1a and 1b are made up of basic load cases 1 and 3. The tangential stress distribution around the bolt holes of basic load case 1 is given by Equation (5.44). The tangential stress distribution around the bolt hole of basic load case 3 is:

$$\sigma^{3}_{\phi,s} = k_t^{3}\sigma_N \qquad (5.52)$$

where

$$\sigma_N = \frac{0.215F_x}{wt\ /2} \qquad (5.53)$$

Equations (5.52) and (5.53) give:

$$\sigma^{3}_{\phi,s} = k_t^{3}\frac{0.215F_x}{wt\ /2} \qquad (5.54)$$

Equations (5.44) and (5.54) give:

$$\sigma^{1}_{\phi,s} + \sigma^{3}_{\phi,s} = F_x\left(\frac{k_t^{1}0.285}{d_h t} + \frac{k_t^{3}0.43}{wt}\right) \qquad (5.55)$$

Net-section failure occurs when the tangential stress ($\sigma_{\varphi,s}{}^{1} + \sigma_{\phi,s}{}^{3}$) is equal to the ultimate tensile strength ($\sigma_{t,\phi,k}$) in the direction considered. The net-section failure load is predicted from Equation (5.56):

$$F_x = \frac{\sigma_{t,\phi,k}}{\dfrac{k_t^{1}0.285}{d_h t} + \dfrac{k_t^{3}0.43}{wt}} \qquad (5.56)$$

Bearing failure analysis of parts 1a and 1b

The contribution to the radial stress distribution from basic load case 1 is given by Equation (5.46). Any contribution to the radial stress distribution from basic load case 3 is neglected.

Bearing failure occurs when the radial stress ($\sigma_{r,s}{}^{1}$) is equal to the ultimate compressive strength ($\sigma_{r,k}$) in the direction considered. Thus bearing failure load is predicted from Equation (5.57):

$$F_x = \frac{\sigma_{r,k}d_h t}{0.215K_r^{1}} \qquad (5.57)$$

Shear-out failure analysis of parts 1a and 1b

The shear stress distribution from basic load cases 1 and 3 is given by Equations (5.49) and (5.58), respectively:

$$\tau^3_{sn,s} = k^3_s \sigma_N \tag{5.58}$$

Equations (5.53) and (5.58) give:

$$\tau_{sn} = k^3_s \frac{0.285F_x}{wt/2} \tag{5.59}$$

Equations (5.50) and (5.59) give:

$$\tau^1_{sn,s} + \tau^3_{sn,s} = F_x\left(-\frac{0.215k^1_s}{d_h t} + \frac{0.57k^3_s}{wt}\right) \tag{5.60}$$

Shear-out failure occurs when the shear stress ($\tau_{sn,s}{}^1 + \tau_{sn,s}{}^3$) is equal to the ultimate shear strength $\tau_{sn,k}$. The shear-out failure load is predicted from Equation (5.61):

$$F_x = \frac{\tau_{sn,k}}{-\frac{0.215k^1_s}{d_h t} + \frac{0.57k^3_s}{wt}} \tag{5.61}$$

The results predicted by Equations (5.55), (5.57) and (5.60) in the directions 0°, +45° and 90° are summarised in Tables 5.9 and 5.10.

Table 5.9 Predicted net-section and bearing failure for parts 1a and 1b.

Location	*Net-section* F_x *(kN)*	*Bearing* F_x *(kN)*
0°	10.1	41.6
+45°	28.1	31.5
90°	15.8	∞

Table 5.10 Predicted shear-out failure for parts 1a and 1b along shear-out plane at ϕ = 90°.

Location (mm)	*Shear-out* F_x *(kN)*
0.2	69.2
0.3	59.5
0.4	57.4

(ii) *Advanced method*

In this section the advanced design method is used to analyse the joint. A special purpose finite element program (BOLTIC), which can be used to analyse a joint in accordance with the procedure set out in the EUROCOMP Handbook has been developed reference (5.11). The stress analysis is carried out in two steps. In the first step (source analysis) the far field stress distribution in the composite is determined. In the second step (target analysis) a series of detailed stress analyses of regions containing a single bolt hole is performed to determine the stress distribution around the bolt holes. Far field stress distributions, obtained in the source analysis, are automatically transferred to the boundaries of the target models. The failure analysis includes evaluation of the failure modes net-section, bearing and shear-out according to simple point stress criteria.

Step 1: Source analysis

Figure 5.27 shows a finite element model of the joint. The members are modelled by 8-node Serendipity elements under the assumption of plane stress conditions. The two steel plates of thickness h are modelled by one steel plate of thickness 2h for an accurate overall stiffness representation. The fasteners and holes are represented by finite spring elements which are connected to node points on the members. The stiffness constants of the spring elements (K_{ffe}) are determined according to the procedure described in section 5.2.2.4 of the EUROCOMP Handbook.

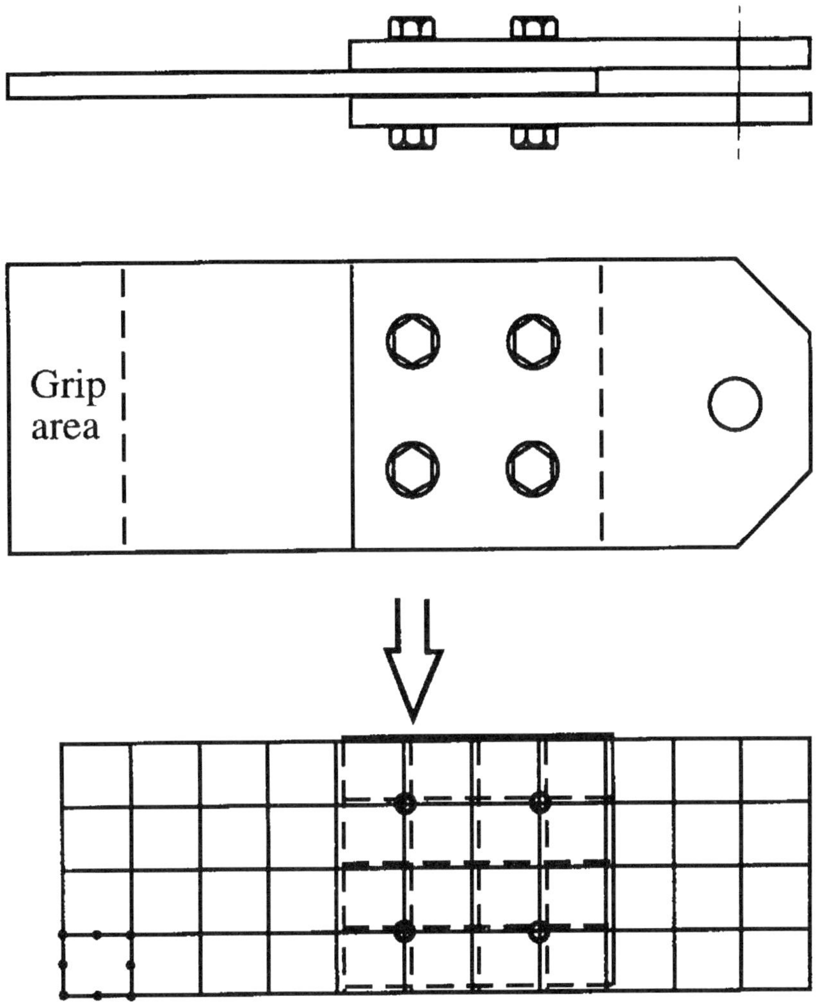

Figure 5.27 Finite element model of concentrically loaded double-lap joint (source model).

Step 2: Target analysis

The composite member plate is divided into rectangular regions (1a-b and 2a-b) (target regions) containing a single fastener hole (Figure 5.28). Owing to symmetry, regions 1a and 1b are identical, as are regions 2a and 2b. The target regions are modelled by linear triangular plane stress elements. Such elements are used by BOLTIC to facilitate automatic mesh generation and mesh refinement. Loads obtained in the source analysis are applied to the edges of the target models. The frictionless contact between the bolt and laminate is taken into account.

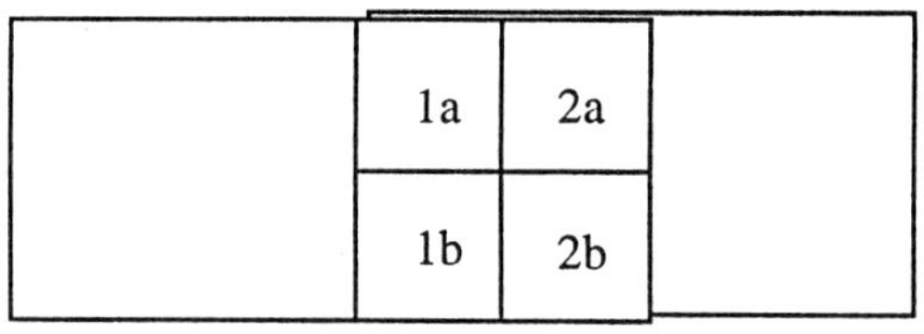

Figure 5.28 Four target regions containing single fastener holes.

Step 3: Failure analysis

The basic failure modes net-section, bearing, and shear-out were evaluated according to section 5.2.2.4(b)(iii) of the EUROCOMP Handbook. Parameters required for the failure analysis were determined experimentally and are listed in 5.2.2.5 *Experimental data.* The predicted results for target regions 1a and 1b were 91 kN for net-section failure and 40 kN for bearing failure. For target region 2 the net-section failure load was 200 kN and the bearing failure load was 52 kN.

5.2.2.5 *Experimental data*

Details of the hand lay-up composite plate tested are given in Table 5.11. The elastic properties of the plate were determined experimentally and are also presented in Table 5.14.

Two laminates were made by resin transfer moulding (RTM). Details are given in Tables 5.12 and 5.13. The elastic properties were again determined experimentally and are given in Tables 5.15 and 5.16.

Table 5.11 Properties of hand lay-up laminate.

Reinforcement	5 layers Devold DBT 800 glass fibre fabric
Matrix	NORPOL 20M-80 polyester
Laminate configuration	5 layers Devold fabric with 50% of the fibres oriented in 0° direction and 50% in ± 45° direction. The 0° direction is parallel to the x-direction
Fibre volume content	42% (estimated from laminate weight)
Fabrication method	Hand lay-up, room temperature curing
Producer	Karlskronavarvet AB, Karlskrona, Sweden

Table 5.12 Properties of RTM laminate 1.

Reinforcement	Ahlstrom multiaxial (0/90 and ±45) "non-crimp", Ahlstrom M113 chopped strand mat (CSM); Vetrotex U-812 continuous filament mat
Matrix	Neste polyester G300-2120
Hardener	Akzo Butanox M-50 (1.75% by weight)
Laminate configuration	[CSM, 0/90, CFM, 0/90, CFM, 0/90, CFM, 0/90, CSM.] The 0° direction is parallel to the x-direction
Fibre volume content	Approximately 50% by weight
Producer	Neste Composite Technology, Helsinki, Finland

Table 5.13 Properties of RTM laminate 2.

Reinforcement	Ahlstrom multiaxial (0/90 and ±45) "non-crimp", Ahlstrom M113 chopped strand mat (CSM); Vetrotex U-812 continuous filament mat
Matrix	Neste polyester G300-2120
Hardener	Akzo Butanox M-50 (1.75% by weight)
Laminate configuration	[CSM, 0/90, CFM, ±45, CFM, ±45, CFM, 0/90, CSM.] The 0° direction is parallel to the x-direction
Fibre volume content	Approximately 50% by weight
Producer	Neste Composite Technology, Helsinki, Finland

5.2.3 Bolted and riveted joints in tension

5.2.3.0 Notation

No additional information.

5.2.3.1 Definitions

No additional information.

5.2.3.2 Performance requirements

No additional information.

5.2.3.3 Design requirements

No additional information.

Table 5.14 Elastic properties, strength, and failure criterion parameters for the glass fibre polyester system Devoid DBT 800/Norpol 20 M-80.

Definition	Symbol	Value
Elastic properties		
Initial tangent modulus (x-direction) (N/mm^2)	E_x	19 366
Initial tangent modulus (y-direction) (N/mm^2)	E_y	7353
In-plane shear modulus (chord modulus) (N/mm^2)	G_{xy}	5029
Poisson's ratio	ν_{xy}	0.5
Strength data		
Ultimate tensile strain[1] (x-direction) (%)	$\varepsilon_{xt,ult}$	2.0
Ultimate tensile strain[1] (y-direction) (%)	$\varepsilon_{yt,ult}$	1.4
Ultimate compressive strain[1] (x-direction) (%)	$\varepsilon_{xc,ult}$	1.6
Ultimate compressive strain[1] (y-direction) (%)	$\varepsilon_{yc,ult}$	1.8
Ultimate shear strain[1] (x - y plane) (%)	$\varepsilon_{xy,ult}$	3.3
Characteristic dimensions (net-section failure)		
Tension failure in x-direction (mm)	$d_{t,k,x}$	2.8
Tension failure in 45° direction (mm)	$d_{t,k,45}$	not available
Tension failure in y-direction (mm)	$d_{t,k,y}$	2.7
Compression failure in x-direction (mm)	$d_{c,k,x}$	2.4
Compression failure in 45° direction (mm)	$d_{c,k,45}$	not available
Compression failure in y-direction (mm)	$d_{c,k,y}$	6.2
Characteristic dimensions (bearing failure)		
x-direction	$r_{k,x}$	1.3
45° direction	$r_{k,45}$	not available
y-direction	$r_{k,y}$	not available
Characteristic dimensions (shear-out failure)		
x - y plane	$s_{k,xy}$	not available

1 Calculated from ultimate strength and initial tangent modulus (conservative estimation)

Table 5.15 Elastic properties, strength and failure criterion parameters for RTM laminate 1.

Definition	Symbol	Value
Elastic properties		
Initial tangent modulus (x-direction) (N/mm^2)	E_x	15 077
Initial tangent modulus (y-direction) (N/mm^2)	E_y	14 390
In-plane shear modulus (chord modulus) (N/mm^2)	G_{xy}	2290
Poisson's ratio	ν_{xy}	0.23
Strength data		
Ultimate tensile strain[1] (x-direction) (%)	$\varepsilon_{xt,ult}$	1.8
Ultimate tensile strain[1] (y-direction) (%)	$\varepsilon_{yt,ult}$	1.6
Ultimate compressive strain[1] (x-direction) (%)	$\varepsilon_{xc,ult}$	1.6
Ultimate compressive strain[1] (y-direction) (%)	$\varepsilon_{yc,ult}$	1.6
Ultimate shear strain[1] (x - y plane) (%)	$\varepsilon_{xy,ult}$	4.0
Characteristic dimensions (net-section failure)		
Tension failure in x-direction (mm)	$d_{t,k,x}$	3.7
Tension failure in 45° direction (mm)	$d_{t,k,45}$	2.0
Tension failure in y-direction (mm)	$d_{t,k,y}$	2.8
Compression failure in x-direction (mm)	$d_{c,k,x}$	not available
Compression failure in 45° direction (mm)	$d_{c,k,45}$	not available
Compression failure in y-direction (mm)	$d_{c,k.y}$	not available
Characteristic dimensions (bearing failure)		
x-direction	$r_{k,x}$	1.8
45° direction	$r_{k,45}$	not available
y-direction	$r_{k,y}$	not available
Characteristic dimensions (shear-out failure)		
x - y plane	$s_{k,xy}$	4.7

1 Calculated from ultimate strength and initial tangent modulus (conservative estimation)

Table 5.16 Elastic properties, strength and failure criterion parameters for RTM laminate 2.

Definition	*Symbol*	*Value*
Elastic properties		
Initial tangent modulus (x-direction) (N/mm^2)	E_x	13 340
Initial tangent modulus (y-direction) (N/mm^2)	E_y	12 170
In-plane shear modulus (chord modulus) (N/mm^2)	G_{xy}	3230
Poisson's ratio	ν_{xy}	0.30
Strength data		
Ultimate tensile strain[1] (x-direction) (%)	$\varepsilon_{xt,ult}$	1.6
Ultimate tensile strain[1] (y-direction) (%)	$\varepsilon_{yt,ult}$	1.5
Ultimate compressive strain[1] (x-direction) (%)	$\varepsilon_{xc,ult}$	1.4
Ultimate compressive strain[1] (y-direction) (%)	$\varepsilon_{yc,ult}$	1.4
Ultimate shear strain[1] (x - y plane) (%)	$\varepsilon_{xy,ult}$	4.2
Characteristic dimensions (net-section failure)		
Tension failure in x-direction (mm)	$d_{t,k,x}$	2.0
Tension failure in 45° direction (mm)	$d_{t,k,45}$	1.1
Tension failure in y-direction (mm)	$d_{t,k,y}$	1.9
Compression failure in x-direction (mm)	$d_{c,k,x}$	not available
Compression failure in 45° direction (mm)	$d_{c,k,45}$	not available
Compression failure in y-direction (mm)	$d_{c,k,y}$	not available
Characteristic dimensions (bearing failure)		
x-direction	$r_{k,x}$	3.7
45° direction	$r_{k,45}$	not available
y-direction	$r_{k,y}$	not available
Characteristic dimensions (shear-out failure)		
x - y plane	$s_{k,xy}$	4.7

1 Calculated from ultimate strength and initial tangent modulus (conservative estimation)

5.3 BONDED JOINTS

5.3.1 General

In these EUROCOMP documents the bonded joints are assumed to be bonded by using either a proper adhesive or a laminating resin. At least one of the adherends to be joined is expected to be of fibre reinforced plastic (FRP). As the existing analytical models to describe the adhesive stresses and strains are developed for plate-to-plate connections, only plate-to-plate connections may be analysed using the analytical models. These models include several simplifying assumptions; thus the models can only give approximations of the actual stress or strain state in the adhesive. The accuracy of the models has however, been found, to be adequate for most cases. The applicability of the existing models is often limited. Joints falling outside the applicability ranges of the models may only be estimated using these models. As a result, design by testing is often the only acceptable method for final dimensioning of bonded joints.

The design conditions given in the EUROCOMP Design Code for bonded joints are:

- allowable shear stress of the adhesive is not exceeded
- allowable tensile (peel) stress of the adhesive is not exceeded
- allowable through-thickness tensile stress of the adherend is not exceeded.

The first two conditions are relatively straight forward. They are applied only to the isotropic adhesive material and are to ensure that neither of these two loading modes exceed the load-bearing capacity of the adhesive. In the case of isotropic adherends, typically only these two loading conditions need to be considered.

However, the layered structure of most FRP adherends has a major further impact on the conditions mentioned above. First, layered FRP materials are relatively weak in the through-thickness direction, which makes them vulnerable to through-thickness tensile loads. The third condition is to ensure that this failure mode does not become critical.

Secondly, the layered structure causes interlaminar shear stresses at each layer interface together with other in-plane shear stresses. Therefore, it should also be verified that the allowable in-plane shear stress of the adherend is not exceeded. This is not included, however, in the conditions above, as the interlaminar shear failure is typically preceded by a through-thickness tensile failure. The required value of the interlaminar shear strength is also seldom available, and there are no standardised test procedures to determine such a value. As a result, bonded joint induced interlaminar shear stresses are not calculated in the design procedures presented.

It should be noted that the material strength values needed for designing against the three failure modes listed above often are difficult to obtain. Often the material manufacturers are unable to provide such values and the mechanical test methods required for the determination are difficult to perform. It may be envisaged, however, that in the future the situation will be improved as the need for providing the designer with such data is more widely recognised and the whole area of designing and analysing adhesively bonded FRP joints is better established.

5.3.1.1 Stresses and strains

The primary loading modes of bonded joints shown in Figure 5.25 of the EUROCOMP Design Code based on (reference 5.21) are:

- peel
- tensile
- shear
- cleavage.

These loadings are generally used to describe the loading of the adhesive material. Similar descriptions for adhesive loading modes can also be found in references 5.22 and 5.23.

Shear is the preferred loading mode giving the highest joint strength in real applications. In principle, tensile loaded joints are also very effective, but joint constructions that give a pure tensile loading to the adhesive are impractical to construct in real structures.

Peel and cleavage loadings should always be avoided. When their presence can not be avoided, their effects should be minimised. Peel loads are mainly produced by out-of-plane loads acting on a thin adherend or by the eccentricity of the in-plane loads. When the adherend is thick and stiff, the out-of-plane loads typically produce cleavage loads. The critical areas in the joint with respect to peel and cleavage loadings are the ends of the overlap.

Bonded joints are generally loaded simultaneously by more than one of the above mentioned loadings. In order to achieve high joint efficiencies shear loading should be favoured and other loading modes should be minimised.

5.3.1.2 Failure modes

Possible locations of failure initiation and critical strengths for the three primary failure modes given in the EUROCOMP Design Code are indicated in Figure 5.26 of the EUROCOMP Design Code based on reference 5.22.

The most widely used joint configurations are single- and double-lap and single- and double-strap joints (Figure 5.33, EUROCOMP Design Code). These joints have stress concentrations at the ends of the overlap when loaded with in-plane tensile, compressive or shear loads. The adhesive shear and peel stresses have their maximum values at the end of the overlap. Typical shear stress and shear strain distributions of a double-strap joint are shown in Figure 5.27 of the EUROCOMP Design Code which is based on reference 5.24. A typical peel stress distribution of a single-lap joint is shown in Figure 5.28 of the EUROCOMP Design Code which is based on reference 5.25.

The predominant failure mode depends primarily upon the following parameters, using the notation shown in Figure 5.29:

- lap length L
- eccentricity in lap and strap joints
- adhesive thickness t_a
- fibre orientations in the layer adjacent to the adhesive
- adhesive mechanical properties
- adherend mechanical properties.

Although stress concentrations are a problem mainly with lap and strap joints, it should be noted that stress concentrations are present also in scarf and step-lap joints. However, in scarf joints the stress concentrations are irrelevant when the scarf angle is low, typically less than 20°. In step-lap joints the stress concentrations generally have to be taken into account only at the ends of the outermost steps.

5.3.1.3 Effect of joint geometry on joint strength

The joint design procedure consists of two primary tasks:

- determination of the joint lay-out
- analysis.

The joint lay-out is defined to cover the following subjects:

(a) joining technique
(b) joint configuration
(c) adhesive selection

(d) joint geometry, including:
- overlap length
- strap lay-up and thickness
- adhesive layer thickness.

(e) methods for reducing peak stresses.

The analysis includes the calculation of the stresses induced by the applied loading. In addition to these two primary tasks, manufacturing aspects, such as surface treatments and the bonding process, have also to be considered.

The potential maximum joint strength, i.e. the maximum load-bearing capacity of the joint, depends significantly on the joint configuration. An indication of this can be seen in Figure 5.34 of the EUROCOMP Design Code, which is based on reference 5.24. Highest joint efficiencies can only be reached using scarf or step-lap configurations. However, in most cases lap and strap configurations can provide a sufficient strength, and as these configurations are simpler to manufacture than scarf and step-lap joints they are primarily used.

The main problem with lap and strap joints is the eccentricity of the load paths. This eccentricity generates stress concentrations which reduce the overall load-bearing capacity of the joint. Peel loadings producing through-thickness stresses in the adherends are particularly harmful. Methods to minimise the peel stresses are given in the EUROCOMP Design Code clauses and Figure 5.29 taken from reference 5.17.

Although peak peel stresses can be reduced using long overlaps, the maximum load-bearing capacity can not be increased significantly by increasing the lap length once a certain minimum lap length has been exceeded. This is because the maximum shear stress (shear stress peaks at the ends of the overlap) remains constant after a certain lap length has been exceeded and this shear stress becomes the limiting factor of the joint strength.

5.3.1.4 Adherends

Typically the structure and the dimensions of the members to be joined are fixed before the joint design procedure is commenced, i.e. the joint has to be designed using adherends which have a structure determined by requirements other than those arising from the joint design. However, some important factors concerning adherends should always be considered when designing bonded joints between composite adherends.

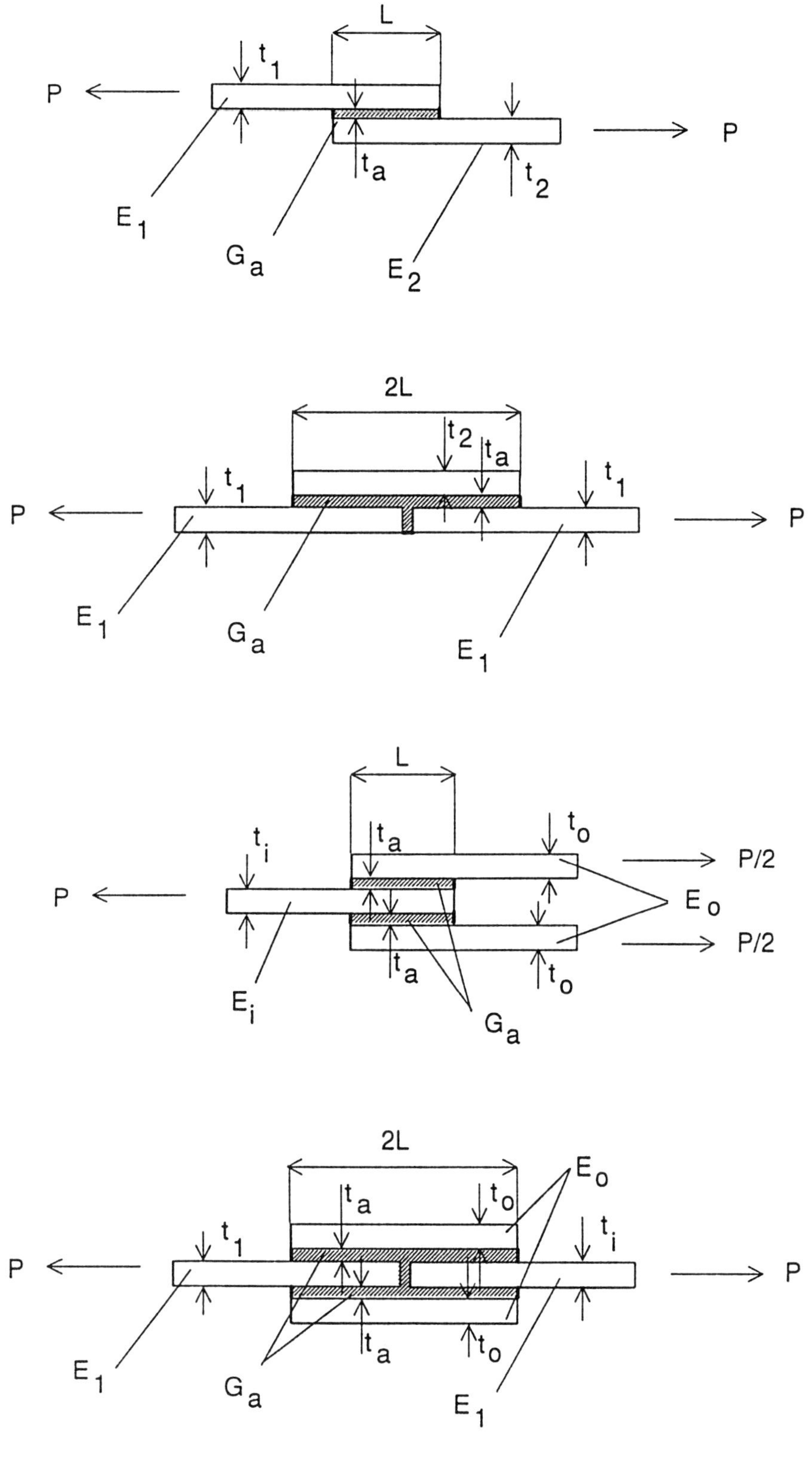

Figure 5.29 Notation of lap and strap joints.

The relative weakness of FRP materials in the through-thickness direction may often become an unacceptable restriction to an adequate joint strength. In such a case it may be practical to tailor the laminate structure locally in order to increase the through-thickness strength of the laminate. The through-thickness strength may be increased for example by stitching the reinforcement layers together, by using three dimensional reinforcements or, as a minimum requirement, by paying special attention to the quality of the layer interfaces during the manufacture of the adherends.

The fibre orientation on the bond surface should be parallel to the primary loading direction. Otherwise it is likely that the joint will fail due to adherend failure at a relatively low loading level. Owing to a great variety of possible laminate structures and applications, it is unrealistic to require that the fibre orientation on the bond surface and the primary loading direction should be parallel in all connections. This requirement is essential for primary structural connections (see 5.1.1 for definition) and for other connections it is a recommendation. The requirement allows the use of fabrics, woven rovings and uni-directional reinforcements on bond surfaces when at least half of the fibres are parallel to the loading direction. In primary structural connections mats are not allowed to be used on bond surfaces.

Adherends of dissimilar materials

The primary problem in joining dissimilar materials is the difference in thermal expansion. When the joint is used at any temperature other than the cure temperature of the adhesive, the adhesive will be stressed owing to the relative movements of the adherends. These stresses may become high enough to cause the joint to fail without external loads. It is possible to calculate these stresses analytically together with the stresses from the mechanical loading. Procedures for single-lap and double-lap joints are presented in 5.3.5. When the members to be joined have dissimilar thermal expansion behaviour, ductile adhesives are preferred.

5.3.1.5 Surface treatments

A proper surface treatment is probably the most important single factor in the process of ensuring the reliability and durability of a bond. Therefore, the importance of the surface treatment prior to bonding cannot be over emphasised. One of the basic design assumptions is that no adhesive failure is encountered. To guarantee that this assumption will be valid on all occasions, all bond surfaces shall be treated according to the EUROCOMP Design Code.

The required methods are, in fact, relatively simple to perform. For many cases roughening the surface by using an abrasive paper is sufficient. For higher quality and consistency this may be replaced by grit blasting, the grit material typically being fine sand or alumina. The grit size and type, and the blast pressure should be selected so as not to damage the outer fibres. Where relatively large amounts of surface material are to be removed, including non-structural surface mats, a rotary abrasive disc may be used. However, great care has to be taken not to damage the primary load-bearing fibre reinforcements. Solvent degreasing has to be performed before and after the abrasion.

The use of a peel ply on the bond surface during the component manufacture is recommended to prevent surface contamination. Typically it also provides a good surface, like abrasion, that needs no further treatment before bonding, except for primary structural connections. This requires, however, that the peel ply has truly shielded the surface from any contamination and is removed only just before the bonding is commenced. The peel ply should not leave any threads of fabric on the bond surface either. All these points need to be checked before accepting the surface for bonding without further treatment.

The use of a peel ply is also beneficial because it provides a smooth surface as a base for further treatments. The coarse fibre pattern of certain mats and woven rovings can significantly lower the bondline quality. Removing the fibre pattern by machining or abrading is not recommended as it may seriously damage the fibres. If the use of peel ply is impractical, the reinforcement layers on the bond surfaces should be selected so that they provide a reasonably smooth surface for bonding.

Surfaces of metallic materials have to be prepared specially for bonding. Owing to the wide variety of metallic materials no single general surface treatment method can be specified. Suitable methods may be found from the instructions of the adhesive or component manufacturer or from several text books.

The bonding process should be performed as soon as possible after the members have been treated. This is to avoid contamination of the surfaces, including effects of moisture. If the bonding is not performed soon after the pretreatment, the surfaces shall be protected against contamination and possibly dried prior to bonding.

5.3.2 Design principles

5.3.2.1 General principles

There are certain fundamental rules in the bonded joint design that have to be followed throughout the design. These rules, or principles, also form the basis to how the different joining techniques are dealt with in the EUROCOMP Design Code and, more generally, how the bonded joint sections of the Code are structured. The main joint design principles are listed below, followed by further explanations:

- joint design covers only the design of the actual joint; in addition the adherend through-thickness strength is checked against the joint-induced internal loadings
- detailed analysis of other than plate-to-plate joints requires decomposing the joints into series of plate-to-plate joints
- three alternative design approaches exist:
  1. analytical
  2. numerical
  3. design by testing
- perfect adhesion is always assumed
- the joint finally manufactured is identical to the joint originally designed.

It is important to realise that the bonded joint design may only ensure that the joint is capable of withstanding the external loads assumed in the design. Further on, in the case of certain configurations with highly eccentric load paths, such as lap and strap joints, it is also checked that the internal loadings induced by the joint do not exceed the through-thickness strength of the adherend. The joint design procedures do not prevent the designer from exceeding the load-bearing capacity of any of the components to be joined. The most clear example of this is the simple formula given for the scarf joint resistance (see 5.3.5.6), where the joint resistance increases to infinity with decreasing scarf angles. This formula should be taken merely as an indication of when the joint resistance reaches that of the basic material, thus making the design component critical. It is obvious that in a constant thickness joint under uniform tensile loading any resistance beyond the component resistance is physically impossible. Ensuring sufficient component strength under external loads is part of the component design process. In this particular case (scarf joint) the EUROCOMP Design Code actually warns the designer from falling into this 'trap'. However, the same applies to all the cases where analytical formulae are given, although not always separately stated. The only joint design approach that also gives a good indication of the component strength is design by testing, provided that the load arrangement is truly representative of that experienced in the final overall structure.

All the existing formulae for bonded joint design are given for relatively simple plate-to-plate configurations. As can be seen, especially in the case of lap and strap joints, even this leads to rather complicated equations. Providing any formulae for the joints comprising different sections (or like) would be impractical as the number of different combinations would be infinite and the formulae would be extremely complicated. However, the plate-to-plate design procedures may be employed in all design cases by breaking down the joint into a series of individual plate-to-plate joints. Such decomposing is a common engineering practice and is based on the fundamental rules of structural mechanics, although sometimes with certain simplifications. For example, an intersecting joint of two I-beams should be designed to have the flanges connected by single or double straps to transfer the flange in-plane loadings (resulting from axial and bending loads), whilst the webs should be connected as a tee joint using bonding angles for the shear load transmission.

The different design approaches listed above are effectively the same as given in 5.1.8 of the EUROCOMP Design Code. The analytical methods may be classified as being either rigorous or simple or, in some cases, as mere design guidelines. The rigorous procedures, when given, are intended to be fully comprehensive closed-form methods enabling a detailed analysis of the joint. The simple methods are either semi-empirical alternatives to the rather laborious rigorous methods or simple one-equation checks against the most critical condition. The design guidelines given for some joining techniques and joint configurations are mainly targeted towards initial estimations of cases requiring experimental verification due to the lack of appropriate design procedures. The numerical methods may be applied to practically any design case, including bonded joints between FRP members. However, the detailed use of these methods is an extremely wide subject, thus being beyond the scope of the EUROCOMP Design Code. The basic principles to be followed when modelling bonded joints for numerical analysis are discussed in 5.3.2.2, but are useful only to those already familiar with such techniques. All joints may be designed by testing sufficiently representative units to verify the load-bearing capacity under the loads used in testing. For complex joints this is the most reliable way of the design, provided that the tests are carefully performed, and for many joints it is the only alternative.

Another important point which the designer has to realise, is that the level of adhesion cannot be verified with calculational methods. Therefore, perfect adhesion is always assumed in the analysis, i.e. it is assumed that the failure occurs within the adhesive or in one of the adherends, not at the interface. Making this assumption requires proper surface treatment in order to ensure that a good level of adhesion is achieved in the final bond. This is a reasonable requirement, as with thermoset plastics and most metals this may be reached with relatively simple methods, when carefully applied. It is worth noting that preventing the adhesion from failing is not only to validate the design formulae. Good adhesion also increases the load-bearing capacity of a bonded joint, and improves the consistency of the overall joint quality and the predictability of the joint behaviour.

A behaviour of a bonded joint may be highly sensitive to changes in certain geometrical factors (dimensions, shapes, etc.). This fact is often utilized in the design, for example to maximize the load-bearing capacity or to bring added conservatism to the design. Achieving the improved performance requires, however, that the actual joint manufactured is equal to the joint originally designed. This applies to both standard and non-standard designs. Methods to ensure this include specifying appropriate installation/manufacturing tolerances for bonded joints and issuing clear working instructions when other than standard practice is called for. The most critical parameters typically are the bondline thickness and lap length. For example, if a certain bondline thickness has been specified, a deviation to any direction may reduce both the strength and durability of the joint. Other features requiring special attention include, for example, the various stress relief methods discussed in 5.3.1.3 (tapers, fillets, etc.).

5.3.2.2 Finite element analyses of bonded joints

In complex structures or loading conditions a finite element (FE) analysis of a bonded joint may be a useful tool in the assessment of the overall behaviour of the joint area. This global analysis, even when a very simple three dimensional model is used, may help the designer to understand the loadings acting in the joint and to identify the crucial elements in the joint. Typical factors that can be evaluated by using the global analysis have been listed in the EUROCOMP Design Code. In this context the word 'global' refers to a model or analysis of a structure or component where the joint is taken into account but is represented by a very simplified model, for example by a continuous structure with double plate thickness at the joint, the adhesive being neglected.

It should be noted that the global model should be employed only for the purpose discussed above. It should not be used to study any of the design details or the load-bearing capacity of the joint, although it is often believed that by increasing the precision of the global model it may also be used to analyse details. However, even at its best, the results given by the global analysis on details may be highly erroneous or misleading, as the precision would still be totally insufficient when compared with what is required for a reliable detailed analysis. It is good practice to pay no attention to the absolute values of the joint stresses given by the global analysis, for example at the stress concentrations, but to concentrate on matters of a more general nature, such as the location of maximum stresses and their probable causes, large deformations, etc.

It is essential that the flexibility of the connections is taken into account in the design. If it is not incorporated into the global FE analysis, it should be at least considered separately so that the designer can be sure of understanding it correctly. The main effect in terms of flexibility comes from the adhesive or resin. It is obvious that the behaviour of a joint with a thick layer of ductile adhesive differs from that of a laminated joint with a thin layer of matrix resin between the joint members. Other factors to be considered are, for example, the flexibility of the members to be joined (including possible stiffeners in the joint area).

Based on the global analysis, a more accurate model may be employed on details that require further investigation. For this purpose, a two dimensional or axisymmetric model of the joint area is often an optimum choice. Typically, the following requirements should be set for a reliable model of a connection:

- the model has to allow non-linear behaviour

- the model must have increased mesh density at the points of stress concentration (joint ends); for example, for bondline thicknesses of between 0.1 and 0.5mm, from four to six second order elements are required in the through-thickness direction
- the modelling of the adherends (shapes, etc.) should be realistic
- sharp corners should be avoided in the model, the bondline has to have a minimum fillet with dimensions equal to the bondline thickness
- rectangular elements should be preferred
- the correct numerical functioning of the high aspect ratio elements should be ensured, as these cannot be avoided, owing to the highly varying mesh densities.

When increasing the mesh density in the model, particularly in the bondline and its immediate vicinity, one has to maintain a good sense of balance between the modelling of the bondline and adherends. Usually the adherend is modelled even in a high density mesh as a homogeneous material which is not an exact representation of a composite. Very high accuracy somewhere in the model, in the bondline for example, would require the FRP adherends be modelled to the level of individual fibres, which would be impractical.

The detailed analyses should concentrate on enhancing the understanding of the joint behaviour or the effects of joint details in relative terms and not used to determine the level of the load-bearing capacity. Aspects such as the adherend or fillet geometry may have a significant effect on the overall joint performance, and the detailed analysis can be a powerful tool in studying these effects. However, when making detailed optimization of the joint geometry the practical and achievable production tolerances need to be considered.

The practical problem the designer is often faced with is the lack of material data available for the analysis, especially on the adhesives. The usefulness of carrying out an FE analysis has always to be judged against the accuracy of the input data. Also the real joint/adherend geometry should be accurately represented by the model. A detailed FE analysis based on insufficient or unreliable data may easily create an unjustifiable trust in the results and is often a waste of time. Furthermore, the loadings have to be fed into the analysis with the same level of precision as the analysis in general is carried out. This is where the global model may prove particularly useful.

5.3.3 Selection of joining technique

No additional information.

5.3.4 Adhesives

5.3.4.1 Classification and characterisation of adhesives

As a wide range of adhesives exists, a clear classification is essential in selecting the adhesive most suitable for the application. The adhesives are typically classified according to the following factors:

- type of adhesive
- curing process activation
- requirements for curing process parameters
- form of adhesive.

Most of the adhesive families have either a thermoset or thermoplastic base. This is also the primary and the most traditional way of categorising adhesives, although within some adhesive families, such as polyurethanes, both thermoset and thermoplastic adhesives may be found. Thermoset adhesives form bonds that are essentially infusible and insoluble after curing and they typically have a much higher load-bearing capability than thermoplastic adhesives. Thermoplastic adhesives are fusible, soluble, soften when heated and their creep resistance is lower than that of the thermoset adhesives. The most common thermoset adhesives are epoxies, phenolics and polyurethanes, while the most widely used thermoplastic adhesives include acrylics (including anaerobics, hot melts and cyanoacrylates) and thermoplastic polyurethanes. A brief description of these adhesives (both thermoset and thermoplastic) is given below from reference 5.20 and 5.28.

Categorisations based on curing activation, process requirements and adhesive form are important as these factors determine among other things the required tooling for application and cure, which further on determine matters like the appropriate site for the bonding process. These requirements may, for example, exclude the possibility of preparing joints on-site. All these categorisations are also significant cost factors, both directly and indirectly.

Adhesive descriptions

Epoxy adhesives are thermosetting resins which cure by polymerisation. They are available either as two-part mixtures (resin plus hardener) or as a one-part resin. Two-part systems cure when mixed (sometimes accelerated by heat), while one-part materials require heat to initiate the curing reaction. The properties of epoxies vary with the type of curing agent and resin used. Epoxies generally have high cohesive strength, are resistant to oils and solvents, and exhibit little shrinkage during the cure.

Epoxies provide strong joints and their excellent creep properties make them particularly suitable for structural applications, but the unmodified epoxies have only moderate peel and low impact strength. These properties can be improved by modifying the resin, to produce more flexible materials which have an improved resistance to brittle fracture. These adhesives include combinations such as epoxy-nylon, epoxy-polyamide, epoxy-polysulphide, epoxy-polyurethane and epoxy-phenolic. However, real toughness is only obtained in the so-called toughened adhesives, in which resin and rubber are combined to form a finely dispersed two-phase solid during curing.

Epoxies are used for structural bonding of metals, glass, ceramics, wood, concrete and plastics, including thermosetting reinforced plastics. Typical service temperatures for epoxies range from –55° to +120°C and for epoxy novolacs up to +250°C.

Phenolics Phenol and resorcinol formaldehyde adhesives cure by condensation polymerisation with the elimination of water, and therefore require high curing pressures. They are normally available as two-component systems consisting of a paste resin and a liquid hardener. Traditional uses include wood bonding and plywood fabrication, but nowadays phenolics, especially modified phenolics, are also used in structural bonding of metals and plastics.

Modified phenolics are materials whose resistance to brittle failure has been enhanced by the addition of a more ductile component without fundamentally reducing the good environmental characteristics of the basic phenolic resin. Typical adhesives are based on nitrile phenolics, neoprene phenolics, epoxy phenolics, and vinyl phenolics. Typical service temperatures range from –60 to +90°C and for phenolic epoxies up to +250°C.

Polyurethanes Thermosetting polyurethane adhesives can be used for structural applications. They are normally two-component systems based on an isocyanate resin, the second component being one of a number of hardeners. Some thermoplastic solvent-based polyurethanes are available.

The structural polyurethanes can be regarded as durable, load-bearing adhesives, with adequate water resistance and high tolerance to oil and fluids. Applications include bonding of metals, rubbers, plastics and fabrics. Typical service temperatures range from –75 to +80°C.

Acrylics are thermoplastic resins based on acrylates or derivatives such as amides and esters. Acrylics are available as emulsions, solvent solutions, and monomer - polymer mixtures (one or two components) with a catalyst. Emulsion - solvent types set by evaporation and absorption of a solvent. Polymer mixtures set through polymerisation by heat, ultra-violet radiation and/or the action of a chemical catalyst.

Mainly toughened acrylics are used in structural applications. Toughened acrylics comprise three main classes:

1. Two-part adhesives in which the polymerisation catalyst is applied separately to the substrate prior to joint assembly. Anaerobic

peroxides may be incorporated to speed curing. Amines or nitrogen-free primers prepare the substrate and initiate the curing process.

2. Two-part adhesives which cure by incorporating a catalyst just before use to effect radical polymerisation.

3. One-part anaerobic systems, including ultra-violet activating compositions.

Acrylic adhesives cure by addition polymerisation reactions, which provide a rapid setting of the adhesive once the cure has initiated.

Some typical features of the above-mentioned families of adhesives are listed in Tables 5.17 and 5.18 taken from reference 5.20. Because each family includes several different adhesives it is not possible to give exact property descriptions. The descriptions given above and in Table 5.17 should be taken as typical and relative properties enabling initial comparisons between the main adhesive families to be made.

Table 5.17 Advantages and limitations of some widely used chemically reactive structural adhesives (reference 5.20).

Epoxy	*Modified phenolic*	*Polyurethane*	*Modified acrylic*
Advantages			
High strength Good solvent resistance Good gap-filling capabilities Good elevated temperature resistance Wide range of formulations Relatively low cost	High strength Good hot water and weathering resistance Good elevated temperature resistance Wide range of formulations Relatively low cost	Varying cure times Tough Excellent flexibility even at low temperatures One or two component Cure at RT or at elevated temperature Moderate cost	Good flexibility Good peel and shear strengths No mixing required Bonding of dirty (oily) surfaces possible Room temperature cure Moderate cost
Limitations			
Exothermic reaction Exact proportions needed for optimum properties Two-component formulations require exact measuring One-component formulations often require refrigerated storage and elevated-temperature cure Short pot life	High-pressure high-temperature cure Two-component formulations Limited shelf life	Both uncured and cured are moisture sensitive Poor elevated-temperature resistance May revert with heat and moisture Short pot life Special mixing and dispensing equipment required	Low hot-temperature strength Toxic Flammable Odour Limited open time Dispensing equipment required

5.3.4.2 *Adhesive mechanical properties*

The most important adhesive mechanical properties are the shear and tensile characteristics. The adhesive shear strength is one of the factors that determines the ultimate strength of a well designed bonded joint while the shear modulus determines the stiffness of the joint. In addition, good tensile strength is needed to resist peel stresses. However, even today most of the adhesive manufacturers are unable to provide the information on the adhesive mechanical properties needed in the analytical models. Some data on mechanical properties are available in research reports published in journals and conference proceedings, but usually the data need to be obtained by testing. The testing is often complicated to perform and special fixtures and transducers are needed. In the future, designers should encourage their adhesive suppliers to provide the most essential design data automatically.

Adhesives have either ductile or brittle shear behaviour, as shown in Figure 5.32 of the EUROCOMP Design Code. Ductile adhesives are preferred in structural joints because of their better resistance to peel and impact loads. However, brittle adhesives typically have better creep and environmental resistance than ductile adhesives and are therefore used in applications where these factors are critical.

Adhesives, as all plastics, are viscoelastic materials combining characteristics of both solid materials like metals and viscose substrates like liquids. Typically, the adhesive shear stress vs. shear strain curve is non-linear. This behaviour is characteristic especially for thermoplastic adhesives and modified thermosetting adhesives. Thermosetting adhesives are, by their basic nature, more brittle than thermoplastic adhesives but, as discussed earlier, are often modified for more ductile material behaviour.

Adhesive mechanical properties are affected by temperature, moisture and chemicals. Increased temperature and moisture content plasticise the adhesive, reducing the stiffness, i.e. shear and tensile modulus, and typically reduce the corresponding strengths as well (see Figure 5.30). Low temperatures make adhesives stiffer and more brittle, i.e. increase the moduli and decrease the failure strength and strain. Typical, indicative values of some of the properties are given in Table 5.18 (based on reference 5.20).

Table 5.18 Typical properties of some chemically reactive structural adhesives (from reference 5.20).

Property	*Epoxy*	*Modified phenolic*	*Polyurethane*	*Modified acrylic*
Service temperature range, °C	–55 to 120	–55 to 90	–155 to 80	–70 to 120
Impact resistance	Poor	Fair	Excellent	Good
Tensile shear strength, N/mm²	15	15	15	26
T-peel strength, N/m	< 525	2500	14,000	5250
Creep resistance	Excellent	Excellent	Poor	Poor
Heat cure or mixing required	Yes	Yes	Yes	No
Solvent resistance	Excellent	Excellent	Good	Good
Moisture resistance	Excellent	Excellent	Fair	Good
Gap limitation, mm	None	None	None	0.75
Odour	Mild	Mild	Mild	Strong
Toxicity	Moderate	Moderate	Moderate	Moderate
Flammability	Low	High(liquid)/Low(film)	Low	High

5.3.4.3 Adhesive selection

Factors to be considered in adhesive selection are listed in the EUROCOMP Design Code. Further information on the influence of these factors is given below in the relevant order.

Some adherend - adhesive material combinations may have adhesion problems due to chemical incompatibility. The adhesive manufacturer can provide information on which adhesives may be used with each matrix material.

A typical environmental factor which effectively limits the use of possible adhesives for a certain application is the service temperature. In outdoor applications the existence of high humidity may also restrict the number of possible adhesives. Environmental factors such as temperature, humidity and chemicals may have a severe degrading effect on the adhesive properties and, in particular, the creep properties.

The applied loading and the joint geometry influence the loading of the adhesive. Some adhesives are more durable under certain loading modes than others.

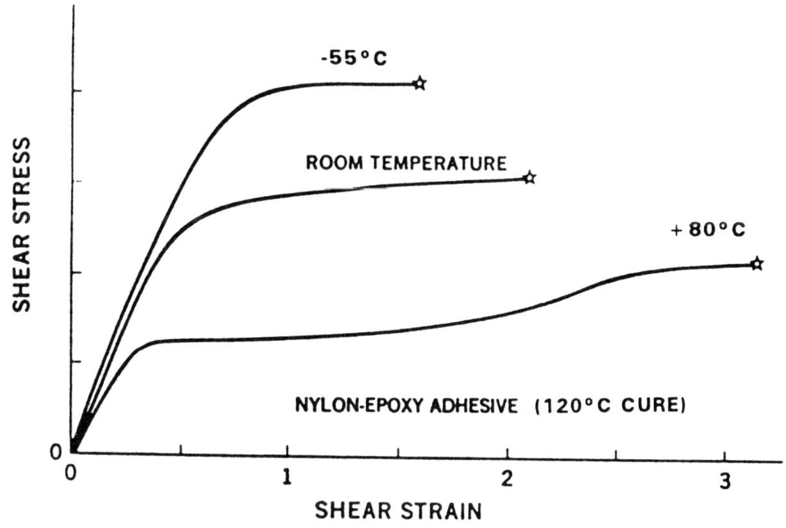

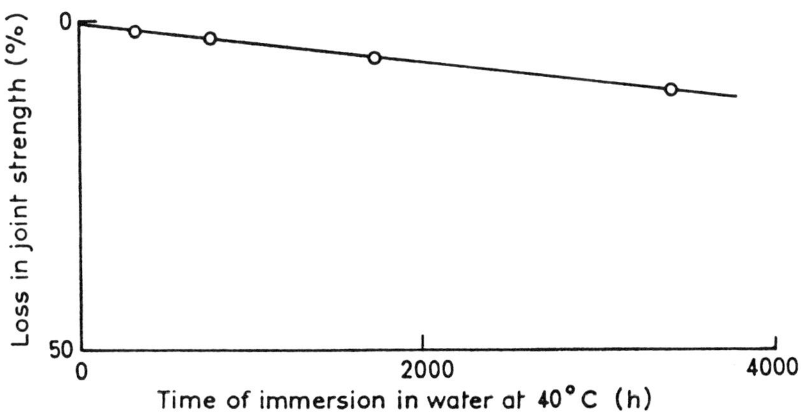

Figure 5.30 (upper) Effect of temperature on adhesive shear behaviour (from reference 5.18). (lower) Effect of immersion in water of FRP/two-part acrylic adhesive/FRP joints on the tensile lap-shear strength (from reference 5.19).

When the bonding is performed in a factory, the environmental parameters, such as temperature and humidity, may be controlled. If the bonding is performed on the construction site, environmental conditions can typically only be monitored, not controlled. Two-component adhesive systems should be used in on-site bonding as, typically, they do not require elevated temperature cure. On-site bonding becomes impractical if the areas to be bonded are large. Elevated temperature cure requires an external heat source. In a factory this is typically an oven and on-site a radiator, a hot air blower or a heating blanket. The last three may also be used in a factory environment but maintaining the same level of temperature control as in an oven may not be possible (see also 6.2.6).

Cost factors should be based on the total costs of the whole bonding process, including materials, preparation, application, curing, service maintenance and repair. In this comparison the adhesive material cost may become less important or even insignificant. Some adhesives require elevated temperature cure with simultaneous application of pressure. The application of such adhesives is often more expensive than the use of room temperature adhesives. In addition to curing considerations, the required surface treatments and the application of adhesive should be included in the cost estimates.

Special requirements on the adhesives may include factors such as impact strength, radiation tolerance and recyclability.

In general, ductile adhesives should be used in structural bonding in preference to brittle adhesives. The former are better able to withstand tensile, peel, and cleavage stresses and have a better impact resistance. The main disadvantage of ductile adhesives is their tendency to creep.

The adhesive selection process, like any material selection process, is complicated to construct in such a way that (1) all essential factors are included and (2) it gives the best candidates for further evaluation. Realising this, Lees (reference 5.27) has stated that currently any selection process can do no more than suggest one or more generic types of adhesives that are worthy of more detailed examination.

The starting point for any successful adhesive selection procedure is a comprehensive and clear specification. Bearing that in mind, one possible selection procedure is given in the following. In this procedure, the different requirements set for the adhesive are categorised according to their nature. For simplicity, it is suggested the requirements be divided into two different groups.

The first group can be described as "absolute requirements", i.e. the requirements that have to be fulfilled without any compromises. Typical examples of these are temperature resistance or resistance to chemical or other types of environmental attack. Also certain mechanical properties may become absolute requirements if, for example, a minimum value is set for the property in the specification. Any adhesive failing to meet any of the absolute requirements leads to that particular adhesive being dropped from of the selection process. If none of the adhesives satisfies all the absolute requirements, it is obvious that either the requirements that have been set are unrealistic and have to be reconsidered or adhesive bonding is not the correct method for the case studied.

Typical absolute requirements include the following:

- material compatibility
- operating temperature
- exposure to water or chemicals
- type of loading (fatigue, long-term static, impact)
- curing temperature.

The second group could be called "relative requirements" as these are the requirements where each property has to be compared with and balanced against other properties, making compromises where needed. Examples of requirements falling into this category are listed below. This phase of the selection process is sometimes called "target evaluation". The grouping of the requirements needs to be carefully considered. No relative requirement should be given the status of an absolute requirement, as this may easily lead to the early exclusion of a potential adhesive.

Target evaluation can be assisted and made more efficient by introducing tools for the selection process that lead to a systematic and neutral approach. A typical method would be one where each requirement is evaluated with care against the properties of each adhesive, giving numerical values (within a specified range) to the adhesives according to their performance. Some requirements can be emphasised by using weight factors. After the evaluation the values are summed and the adhesive or the adhesives getting the highest ranks are further evaluated. In the target evaluation phase the adhesives to be considered have to be specified to the trade name level. The evaluation should not be based on typical properties of the adhesive family.

Typical target requirements are

- material compatibility
- creep/peel/fatigue/impact resistance
- environmental resistance
- gap-filling properties
- working time
- adhesive/preparation/application/curing costs
- manufacturing aspects.

In reference 5.28 an alternative adhesive selection procedure is presented. The method is based on the elimination of adhesives that do not fulfil the requirements.

5.3.5 Adhesively bonded joints

5.3.5.1 General

No additional information.

5.3.5.2 Selection of joint configuration

The selection of the joint configuration is affected by several factors, such as:

- loadings
- required load-bearing capacity
- manufacturing costs

- aesthetics.

Various joint configurations give different potential load-bearing capacities, as shown in Figure 5.34 of the EUROCOMP Design Code based on reference 5.24. However, the most efficient joints are typically expensive to manufacture and the bonding process requires greater attention than the use of simpler configurations like lap or strap joints. Typically, a well designed double-lap or -strap joint gives adequate strength.

5.3.5.3 Design of lap and strap joints

The design procedure for lap joints is divided into two procedures according to their accuracy and sophistication:

1. simplified procedure
2. rigorous procedure.

The use of these methods depends on the status of the design, i.e. preliminary or final design, and the classification of the connection (see 5.1.3, EUROCOMP Design Code). The rigorous procedure should be used for all primary structural connections. The simplified procedure may be used for the preliminary design of primary structural connections or for the final design of secondary and non-structural connections. The simplified procedure can be employed with only the minimum amount of adhesive data, i.e. the lap shear strength. Whenever the maximum joint efficiency is required the rigorous procedure should be used, as it is calibrated to give higher allowable joint loads than the simplified procedure.

The simplified design procedure is based on test results obtained by standardised lap shear tests. These results are generally available for all structural adhesives. The lap shear strengths are transformed to joint resistance by using reduction factors. Owing to the inherent inaccuracy of the method, a separate coefficient C_m (coefficient of design method inaccuracy) is introduced in order to ensure that the allowable stresses given by the simplified method never exceed those given by the rigorous procedure.

The rigorous design method is based on generally accepted closed-form models. The adhesive behaviour in the models is assumed to be linearly elastic. Only the formulae used in the calculation of the temporary maximum joint resistance require the complete shear stress - shear strain curve or the elastic - plastic model of the adhesive to be known. As adhesives typically have a non-linear shear behaviour, using only the linear part of the stress - strain curve brings added conservatism to the models with respect to the actual joint resistance.

It should to be noted that lap and strap joints are treated identically. As the numerical methods are based on lap joint analysis, applying these same methods gives equal or conservative results for strap joints, because strap joints are generally stronger than lap joints.

5.3.5.4 Simplified design procedure for lap and strap joints

With the simplified design procedure, the joint resistance can be calculated based on the adhesive behaviour in lap shear specimens. The shear stress - strain behaviour of the adhesive need not be known. The lap shear data can be based on tests performed according to ISO 4587-1979, ASTM D 3163 or corresponding test arrangements. The lap shear data are, with many adhesives, the only relevant strength data that the adhesive manufacturers are able to provide. The simplified design procedure has been developed specifically for the EUROCOMP Design Code.

In this procedure, the joint failure load per unit width is determined from the tests results. The lap length in the joint to be designed is taken as twice the lap length in the test specimen, but not less than 50 mm. This is to give additional safety to the design because of the uncertainties in this design procedure. Typically, the increase in lap length does not affect the magnitude of the peak shear stress but reduces the peak peel stress, and also improves the damage tolerance of the joint. The simplified design procedure may only be used if the essential joint parameters of a lap shear specimen and the actual joint are identical within reasonable accuracy.

5.3.5.5 Rigorous design procedure for lap and strap joints

The rigorous procedure utilises analytical models to determine the stress state in the adhesive. In the immediate vicinity of the through-thickness stress peak in the adhesive, the adherends are expected to experience the same stress. In the analysis the joint design is satisfied when the following conditions are met:

- $$\tau_{0\,max} \leq \tau_{0\,allowable}$$
- $$\sigma_{0\,max} \leq \sigma_{0\,allowable}$$
- $$\sigma_{0\,max} \leq \sigma_{z\,allowable}$$

where the subscript 0 refers to the adhesive and subscript z refers to the through-thickness direction in the adherend.

The design approach chosen for the rigorous procedure limits the allowable adhesive shear stress to the elastic zone only. It has been considered to be a good design practice not to induce stresses in the adhesive which exceed this limit. However, this limitation leads, especially in the case of ductile adhesives, to high margins of safety in respect to the joint failure load. This is because the shear strength of adhesively bonded joints depends on the strain energy to failure per unit bond area of the adhesive layer, rather than on any of the individual properties such as peak shear stress.

In addition, the design approach determines the material parameter, adhesive shear stress at upper limit of elastic behaviour τ_e, to be divided by the partial material safety factor γ_m, which further reduces the allowable joint load. This is clarified in Figure 5.31, where the total area $A_{failure}$ under the adhesive shear stress - shear strain curve represents the work to failure and the area under the design shear stress A_{design} represents the allowable work as determined by the design procedure. The effect of this is that at least in the case of a ductile adhesive the design joint load will be very small compared with the joint failure load, typically only a few per cent. For joints with brittle adhesive this ratio is considerably greater, the load bearing capacity thus being better utilised.

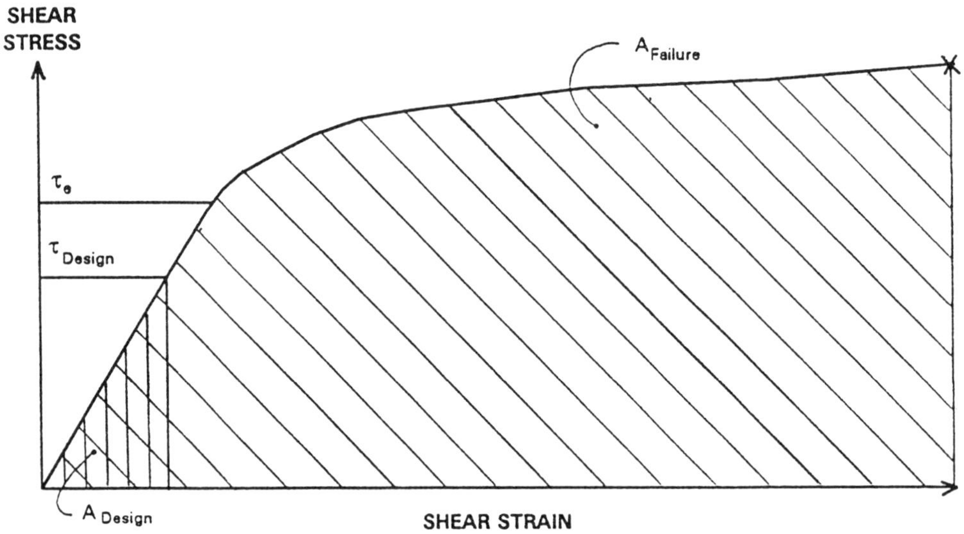

Figure 5.31 Ultimate load-bearing capacity of a bonded joint ($A_{Failure}$) and the load-bearing capacity utilised according to the design procedure (A_{Design}).

The design is based on the maximum shear stress which occurs at the end of the joint. However, the shear stress decreases rapidly within a short distance from the joint end. It should also be noted that this peak stress occurs immediately after a load is applied, but owing to the viscoelastic nature of polymer adhesives, this peak flattens in the course of time. This behaviour is shown in Figure 5.32, based on the work of Groth (reference 5.36). This viscous behaviour further increases the margin of safety in the design approach.

Despite the great conservatism, the design method applied is considered to be the best available for the use in construction applications. Another method which, in addition, takes into account the plastic part of the adhesive stress - strain curve has been presented by Hart-Smith. His design approach is described in Figure 5.33, where the lap length is determined so that the plastic zones are dimensioned to carry the ultimate adherend load and the elastic zone in the middle stays as a reserve. An elastic - plastic model (see Figure 5.34) is used for the adhesive characterisation. The length of the elastic zone is determined so as to keep the minimum shear stress below a specified level (10% of τ_p). This is to provide resistance against fatigue and creep and also to increase the damage tolerance of the joint. This method gives a higher load-bearing capacity of the joint than the design method primarily used in the EUROCOMP Design Code, but as it takes into account the plastic behaviour of the adhesive, it cannot be used as a general design method for all bonded joints.

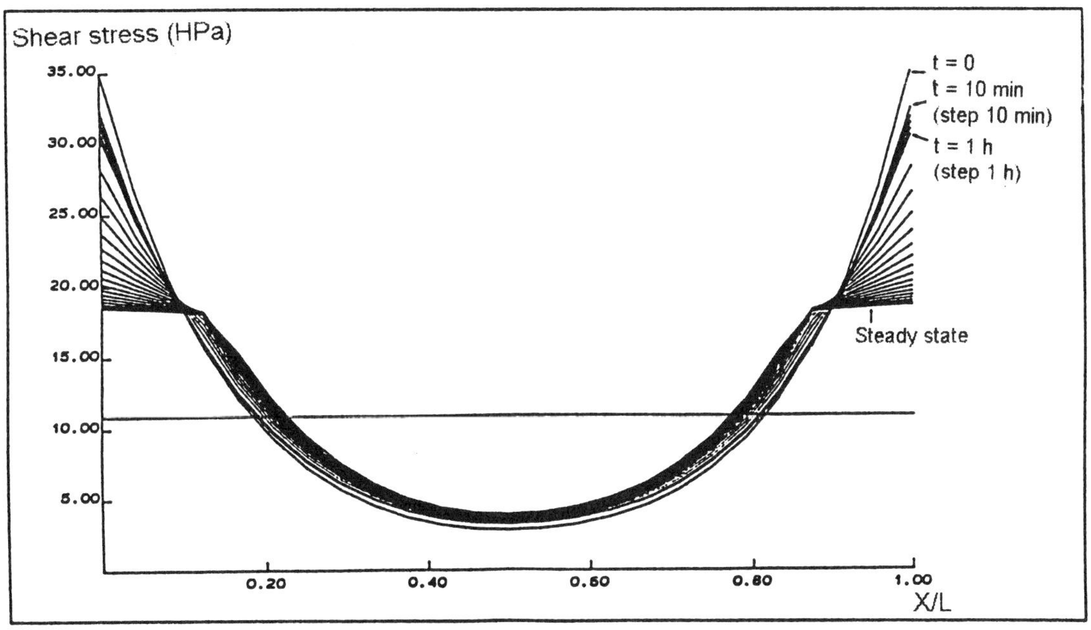

Figure 5.32 Adhesive shear stress distribution as a function of time in a single-lap joint using a viscoelastic-viscoplastic adhesive model, based on Groth (reference 5.36).

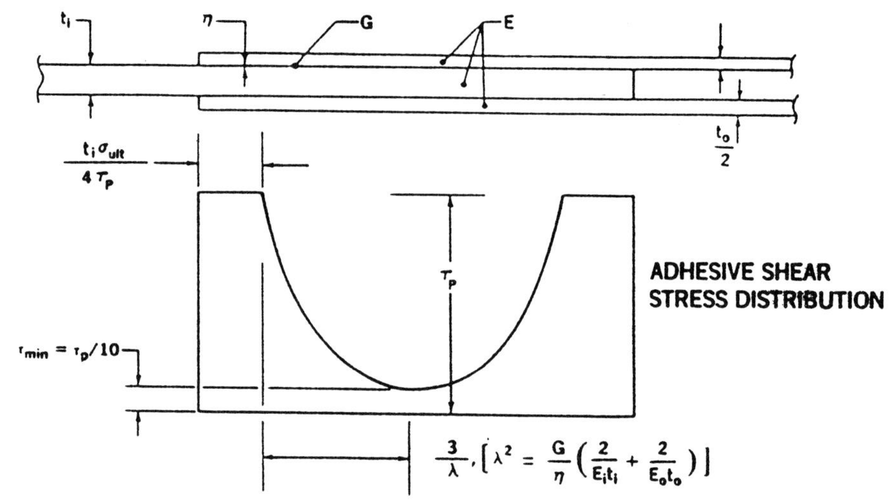

Figure 5.33 Design of a double-lap joint according to Hart-Smith (reference 5.26).

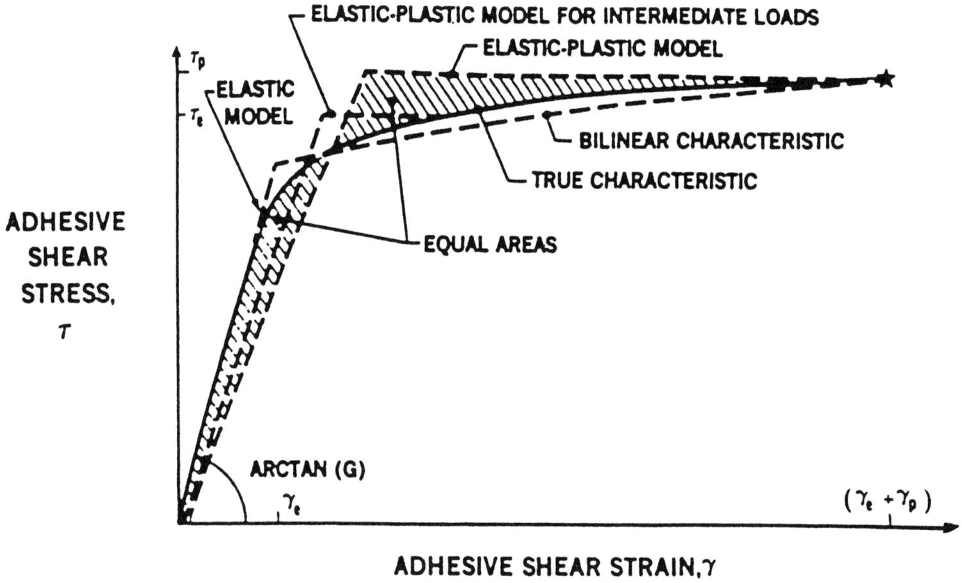

Figure 5.34 Models of adhesive behaviour from (reference 5.24).

(a) *Tensile shear loading*

The rigorous procedure given in this chapter is based on the analysis by Goland and Reissner (reference 5.29). Other closed-form analytical models have been published by several writers. Many of these models are extensions of the Goland and Reissner analysis. Details of some of these models is given below.

- Adams and Wake (reference 5.17) considered the adhesive stresses and adherend direct stresses acting across the width of an adhesive joint. They included the adherend shear stresses but neglected the effects of bending, thus excluding peel stresses. They used a linear elastic analysis.

- Hart-Smith (references 5.25, 5.26, 5.30 and 5.31) has conducted extensive studies of bonded joints using the elastic - plastic model for the adhesive. He has covered the analysis of lap, strap, scarf and step-lap joints. He has modified the load eccentricity induced peel stress approach by using a modified bending stiffness. He has studied the effects of non-uniform adhesive thickness, adhesive non-uniform moisture absorbtion and defects in the bondline. He has also included thermal stresses in his models.
- The ESDU/Grant method (references 5.32 and 5.33) uses a non-linear model for the adhesive shear behaviour in double-lap joints. The analysis is available in the form of a computer program and on ESDU (Engineering Science Data Unit) data sheets.

The following approach, in which adhesive maximum shear stress and peel stress are calculated, is based on (references 5.17, 5.18, 5.29 and 5.34).

Volkersen presented his "shear lag" analysis in 1938. It is based on the following assumptions:

- adhesive deforms only in shear
- adherend deforms only in tension
- shear stresses are constant across the adhesive thickness
- materials are linearly elastic.

The joint configuration is a single-lap joint with adherends of equal thickness t. Note that the bending moment due to the load path eccentricity is not taken into account. Joint rotation due to bending of the adherends is also neglected.

Goland and Reissner took the bending behaviour into account by using a bending moment factor, k, which relates the bending moment on the adherend at the end of the overlap, M_0, to the in-plane loading by the relationship:

$$(M_0 = kP\frac{t}{2}) \qquad (5.62)$$

where P is the applied load and t is the load eccentricity (i.e. half the sum of the adherend thicknesses). If the load on the joint is very small, there is no rotation of the overlap. Then the line of action of the load is as shown in Figure 5.35(a), passing close to the edge of the adherends at the ends of the overlap. In this case, therefore, $M_0 \approx Pt/2$ and $k \approx 1.0$. As the load increases, the overlap rotates, bringing the line of action of the load closer to the centre-line of the adherends (i.e. the load eccentricity $< t$), as shown in Figure 5.35(b), and thus reducing the value of the bending moment factor.

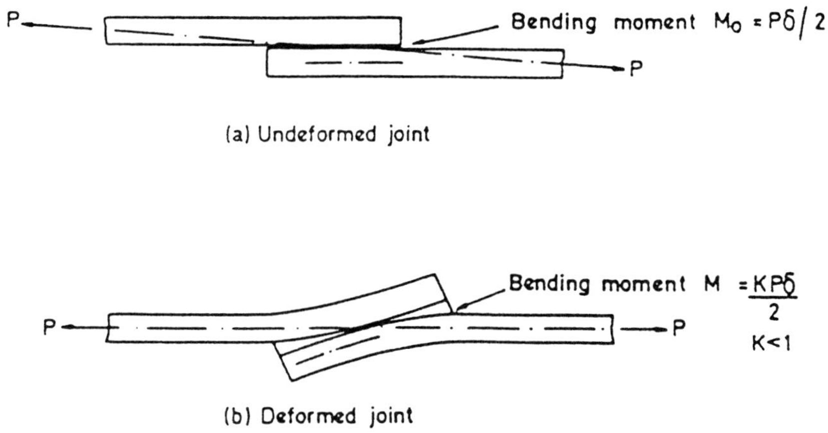

Figure 5.35 Goland and Reissner bending moment factor (reference 5.29).

Limitations (reference 5.27) The range of validity is given by Goland and Reissner (reference 5.29) as:

$$\frac{t_1 G_a}{t_a G_1} < 0.1, \quad \text{and} \quad \frac{t_1 E_a}{t_a E_1} < 0.1$$

where E, G and t are Young's modulus, the shear modulus and the thickness and the subscripts 1 and a refer to the adherends and adhesive respectively.

The above mentioned validity requirements are exceeded with the materials typically used. However, Lubkin and Reissner (1956) have shown that the bounds of the theory are conservative but gave no indication of the true bounds.

The solutions for the adhesive shear stress, τ_x, and the normal stress,σ_0, are:

$$\tau_x = \frac{\sigma t}{8c}\left[\frac{\beta c}{t}(1 + 3k)\frac{\cosh\frac{\beta x}{t}}{\sinh\frac{\beta c}{t}} + 3(1 - k)\right] \qquad (5.63)$$

and

$$\sigma_0 = \frac{\sigma t^2}{c^2 R_3}\left[\left(R_2\lambda^2\frac{k}{2} - \lambda k' \cosh\lambda \cos\lambda\right)\cosh\frac{\lambda x}{c}\cos\frac{\lambda x}{c} + \left(R_1\lambda^2\frac{k}{2} - \lambda k' \sinh\lambda \sin\lambda\right)\sinh\frac{\lambda x}{c}\sin\frac{\lambda x}{c}\right] \tag{5.64}$$

where

$$c = \frac{L}{2}$$

$$\sigma = \frac{P}{t}$$

$$\beta = \sqrt{8\frac{G_a}{E}\frac{t}{t_a}}$$

$$\lambda = \frac{c}{t}\left(\frac{6E_a t}{E t_a}\right)^{\frac{1}{4}}$$

$$R_1 = \cosh\lambda \sin\lambda + \sinh\lambda \cos\lambda$$

$$R_2 = \sinh\lambda \cos\lambda - \cosh\lambda \sin\lambda$$

$$k = \frac{\cosh(u_2 c) \sinh(u_1 L)}{\sinh(u_1 L) \cosh(u_2 c) + 2\sqrt{2}\cosh(u_1 L) \sinh(u_2 c)}$$

$$k' = k\frac{c}{t}\left(3(1-\nu^2)\frac{\sigma}{E}\right)^{\frac{1}{2}}$$

$$u_1 = 2\sqrt{2}\,u_2$$

$$u_2 = \frac{1}{\sqrt{2}\,t}\sqrt{3(1-\nu^2)\frac{\sigma}{E}}$$

and

P	is load per unit width
L	is length of the overlap
t	is adherend thickness = $t_1 = t_2$
E	is adherend tensile modulus = $E_1 = E_2$
G_a	is adhesive shear modulus
t_a	is adhesive layer thickness
E_a	is adhesive tensile modulus
ν	is adherend Poisson's ratio.

The maximum value of τ_x is found at the end of the joint:

$$\tau_{0\,max} = \frac{\sigma t}{8c}\left[\frac{\beta c}{t}(1 + 3k)\coth\frac{\beta c}{t} + 3(1 - k)\right] \tag{5.65}$$

Increasing the lap length beyond the limit $\beta c/t \approx 25$ has no effect on the magnitude of the maximum shear stress, which remains as:

$$\tau_{0\,max} = \frac{\sigma}{8}(1 + 3k)\sqrt{8\frac{G_a t}{E t_a}} \tag{5.66}$$

The maximum value of the peel stress acts also at the joint end and is:

$$\frac{\sigma_{0\,max}}{\sigma}\left(\frac{c}{t}\right)^2 = \lambda^2\frac{k}{2}\,\frac{\sinh(2\lambda) - \sin(2\lambda)}{\sinh(2\lambda) + \sin(2\lambda)} - \lambda k'\,\frac{\cosh(2\lambda) + \cos(2\lambda)}{\sinh(2\lambda) + \sin(2\lambda)} \tag{5.67}$$

For long overlaps ($\lambda > 2.5$) this becomes:

$$\frac{\sigma_{0\,max}}{\sigma} = \frac{k}{2}\sqrt{6\frac{E_a}{E}\frac{t}{t_a}} + k'\frac{t}{c}\sqrt{6\frac{E_a}{E}\frac{t}{t_a}} \tag{5.68}$$

In the design procedure the lap length is determined according to the assumption that the lap length has no effect on the magnitude of the maximum shear stress, when $(\beta c)/t \geq 25$. Figure 5.37 of the EUROCOMP Design Code has been constructed in accordance with this limit value.

The procedure presented is valid for single-lap joints and it is assumed to be valid, with reasonable accuracy, for single-strap joints.

The design procedure for double-lap and double-strap joints is based on the approach by Hart-Smith (reference (5.26). This is based on the Goland and Reissner theory which has been modified to include adhesive plastic behaviour. Adhesive shear stress - strain behaviour is modelled using an elastic - plastic model, where the actual adhesive behaviour is described using a simplified but quite accurate model, (see Figure 5.34). Hart-Smith's analysis, in which the adhesive behaves elastically through the whole joint (fully-elastic analysis), is valid for double-lap joints, but the same analysis is also assumed to be valid with a reasonable accuracy for double-strap joints. The notation for double-lap joints is given in the EUROCOMP Design Code (Figure 5.38).

The Hart-Smith approach for the adhesive shear stress at the end of overlap can be expressed as follows:

$$\tau_{0\,max} = \frac{\lambda P}{4}\left[\frac{\cosh(\lambda c)}{\sinh(\lambda c)} + \Omega\frac{\sinh(\lambda c)}{\cosh(\lambda c)}\right] \tag{5.69}$$

where

$$\lambda^2 = \frac{G_a}{t_a}\left(\frac{1}{E_o t_o} + \frac{2}{E_i t_i}\right)$$

$$\Psi = \frac{E_i t_i}{2 E_o t_o}$$

and

Ω = the greater of $(1 - \Psi)/(1 + \Psi)$ or $(\Psi - 1)/(1 + \Psi)$
P = load per unit width

The maximum peel stress according to Hart-Smith is:

$$\sigma_{0\,max} = \tau_{0\,max}\left(\frac{3 E_a'(1 - \nu^2)\, t_o}{E_o t_a}\right)^{\frac{1}{4}} \tag{5.70}$$

where
E_a' = effective adhesive transverse modulus
ν = outer adherend Poisson's ratio
t_a = adhesive layer thickness.

The effective adhesive transverse modulus E_a' takes into account the fact that adhesive properties are different in bulk and film forms. In a bonded joint the adhesive layer is a film. When loaded using through-thickness tensile or compressive loads the adherends restrict the deformations of the adhesive, unlike when the bulk form of the adhesive is being similarly loaded. These restrictions in deformations tend to increase the adhesive modulus measured between the relatively stiff adherends. This effective modulus is difficult to establish theoretically.

An analytical relationship for adhesive effective modulus and bulk modulus can be given for the case where tubular bars are bonded together and loaded in their axial direction. (see references 5.17 and 5.19). It is assumed that the adherends are infinitely rigid i.e. the radial and hoop strains in the adhesive and adherends are zero. Then the effective tensile modulus is:

$$E_a' = \left[\frac{(1 - \nu_a)}{(1 + \nu_a)(1 - 2\nu_a)}\right] E_a \tag{5.71}$$

where ν_a is the adhesive Poisson's ratio and E_a is the adhesive tensile modulus determined from the bulk material.

The temporary maximum joint resistance (bond shear strength) can be calculated by assuming that the adhesive failure shear strain is encountered at the end of the overlap. For a double-lap configuration the maximum load is specified by the lesser of the values given by the following pair of equations (reference 5.30):

$$P = \sqrt{2\,\tau_p\,t_a\left(\frac{\gamma_e}{2} + \gamma_p\right)2\,E_i\,t_i\left(1 + \frac{E_i\,t_i}{E_0\,t_0}\right)} \qquad (5.72)$$

$$P = \sqrt{2\,\tau_p\,t_a\left(\frac{\gamma_e}{2} + \gamma_p\right)2\,E_0\,t_0\left(1 + \frac{E_0\,t_0}{E_i\,t_i}\right)} \qquad (5.73)$$

where τ_p is the plastic adhesive shear stress, γ_e is the elastic adhesive shear strain, γ_p is the plastic adhesive shear strain. Other symbols are defined in the EUROCOMP Design Code (Figure 5.39).

(b) *Compressive loading*

The design procedure is based on the Hart-Smith approach (reference 5.26). He has stated that the design of a joint in compressive shear loading is identical to that in tensile shear loading. Therefore the equations given for a tensile shear loaded joint are also valid for the analysis of a compressive shear loaded joint. He has also stated that peel stresses are always harmful at the end of the joint for tensile shear loading but are typically not of concern for compressive shear loading. It is again assumed that the method described for double-lap joints can be applied also to double-strap joints.

(c) *In-plane shear loading*

The shear loads acting in the plane of a plate type adherend produce similar adhesive loadings to tensile shear loading (see Figure 5.39, EUROCOMP Design Code). The design procedure for in-plane shear loading is based on the Hart-Smith approach (reference 5.26). The analysis of an in-plane shear loading on a double-lap joint or a single-lap joint restrained against a transverse deflection proves to be governed by differential equations of basically the same form as in the case of tensile shear loading. Figure 5.39 (EUROCOMP Design Code) gives the geometry and notation used in the analysis of a double-lap joint. As the same differential equations govern the behaviour of both tensile shear and in-plane shear loaded double-lap joints, Equations (5.69) and (5.70) can be used to calculate the maximum adhesive stresses. However, the parameters for tensile loading have to be replaced with the parameters for in-plane shear loading, as indicated in Table 5.6 of the EUROCOMP Design Code.

It is assumed that the method described for double-lap joints can be applied also to double-strap joints.

(d) *Combined loading*

Hart-Smith (reference 5.26) has presented the situation where the joint is loaded by tensile or compressive and in-plane shear loadings simultaneously. Joint failure will occur when the bondline displacement resultant caused by these actions exceeds the capacity of the adhesive. If the tensile (compressive) loading develops a maximum bondline displacement of $t_a(\gamma_{t,c})_{max}$ and the shear loading induces an orthogonal displacement of $t_a(\gamma_s)_{max}$ at the same end of the joint, the displacement at failure is:

$$t_a(\gamma_e + \gamma_p) = \sqrt{[t_a(\gamma_{t,c})_{max}]^2 + [t_a(\gamma_s)_{max}]^2} \qquad (5.74)$$

Generally, as the adhesive seems to be mainly a strain limited material the best failure criterion for a combined loading would be some strain based failure criterion such as the maximum strain criterion.

Russell (reference 5.35) has presented a method to analyse adhesively bonded joints under generalised in-plane loading. His approach is based on the Hart-Smith method using a non-linear adhesive model.

5.3.5.6 Design of scarf joints

The design procedure for scarf joints is based on the presentation in (reference 5.21. It gives the following procedure for a single-taper scarf joint. The geometry and notation used in the analysis is given in the EUROCOMP Design Code (Figure 5.40).

The tensile load P produces an average shear stress τ_m in the adhesive layer:

$$\tau_m = \frac{P\cos\theta}{L} = \frac{N\cos\theta}{t}\sin\theta \qquad (5.75)$$

and the maximum shear stress is

$$\tau_{0\,max} = K\,\tau_m \qquad (5.76)$$

where

$$K = \sqrt{\frac{1}{4}\left[\tan\theta\,(1-\psi) - \frac{E_a}{E}\cot\theta\right]^2 + 1}$$

and

$$\psi = \nu_a - \nu_{xz}\frac{E_a}{E} \quad .$$

In general, $\theta < 20°$ and

$$\frac{E_a}{E} \ll 1$$

then, $K \approx 1$ and $\tau_{o\,max} \approx \tau_m$.

The procedure for the symmetrical double-taper scarf joint is identical to that for the single-taper scarf joint case except that the required bond length is half the bond length of the single-taper joint.

5.3.5.7 Design of butt joints

Butt joints are capable of transferring compressive and in-plane shear loads and also, to a reasonable extent, tensile loads. Butt joints are, however, vulnerable to the cleavage loads induced by bending. As in actual applications the avoidance of bending loads is difficult or impossible, the use of butt joints in load-bearing connections is not recommended. Instead a strap, scarf or lap joint configuration should be used, as shown in Figure 5.41 of the EUROCOMP Design Code. However, when it is supported against bending deformations, the butt joint configuration can be used at least in non-critical connections.

5.3.5.8 Design of step-lap joints

Step-lap joints share features in common with both double-lap joints and scarf joints. The scarf joint represents the mathematical limiting case of a step-lap joint with an infinite number of steps. (see reference 5.30).

Reference 5.21 gives the following procedures for calculating the joint strength of step-lap joints:

- for three or fewer steps the joint can be analysed as a single-lap joint, taking each lap as a single-lap of thickness t_i, (see Figure 5.42 of the EUROCOMP Design Code)
- for four or more steps the analysis of a double-lap joint can be carried out by taking the step thickness so that the outer adherend thickness t_o of the double-lap joint equals the step thickness t_i and the inner

laminate thickness t_i equals two step thicknesses, i.e. $2t_i$ (Figure 5.43, EUROCOMP Design Code).

It is important to note that in the case of non-homogeneous layered materials like FRP, the tensile stress distribution within the adherends may be highly uneven, thus causing the step loads to be different also. This is particularly evident in a 0°/90° laminate, where some steps coincide with 0° plies and some others with 90° plies. When determining the step loads the Designer should always check the fibre orientations within each step and at each step shear surface in order to avoid underestimating any step loads by unjustifiable averaging. Once the proportional loads carried by each step have been determined, it is obvious that they should be given such values that the sum of the step loads equals or equates with the applied external load.

In addition to the above mentioned, Hart-Smith (reference 5.31) has given three practical limitations for the optimisation of bonded step-lap joints:

1. The outer steps of any adherend should not have an excessive L/t ratio to ensure that it does not become overloaded by the higher-than-average load transfer near the ends of the overlap.

2. The outermost steps should not be too thick in order to prevent premature failure by joint induced through-thickness tensile stresses.

3. Increasing the lengths of any or all of the steps does not increase the joint strength indefinitely. This phenomenon is identical to that of increasing the lap length in a double-lap joint.

According to Hart-Smith the adherend stiffnesses should be balanced as much as the relative strengths of the adherends permit (i.e. product Et should be kept constant). This needs to be considered only when two dissimilar materials are joined together.

If the joint strength has to be increased, it is preferable to increase the number of steps and to decrease the incremental step thickness to improve the shear strength. An increase in step lengths alone is not usually sufficient.

Adams (reference 5.17) has analysed step-lap configurations with carbon FRP adherends. He has investigated the loads transferred by butt and shear surfaces. Figure 5.36 shows the adhesive shear stress distributions in 3-step joints between un-idirectional carbon FRP and aluminium adherends. For a small butt spacing d, the maximum shear stress at the ends of the steps is less than the average applied shear stress. This is because at least half of the applied load is transferred by the adhesive between the butt faces, giving stress concentrations of the order of 10 in these regions. The proportion of the load transferred by shear, rather than by the butt faces, is increased considerably by increasing the thickness of the bondline between the butt faces.

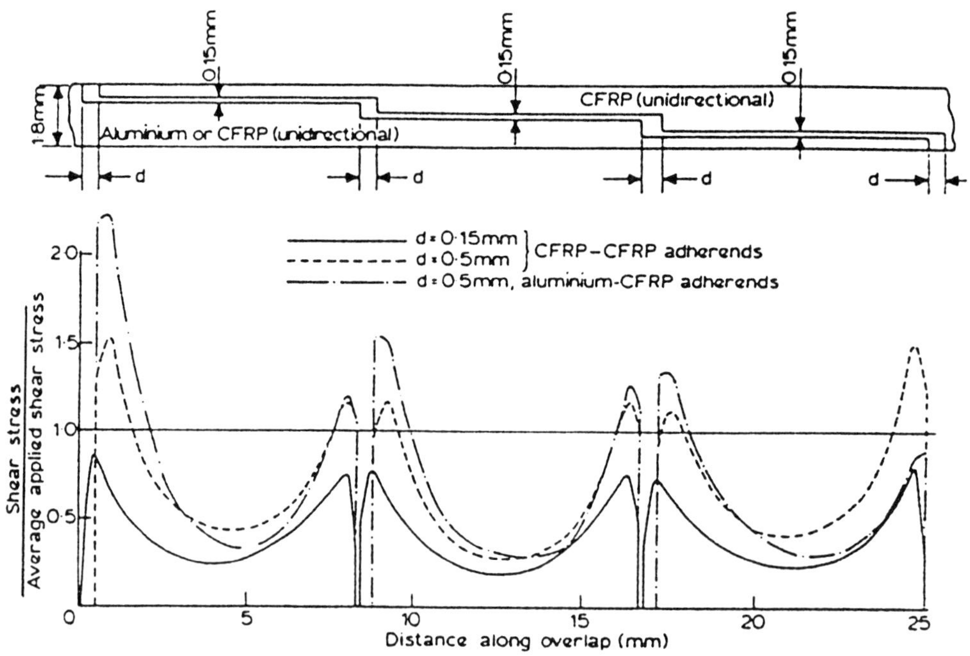

Figure 5.36 Adhesive shear stress distribution in step joints based on (reference 5.17).

When elastic-plastic adhesive behaviour is assumed, the analysis predicts that the adhesive will become plastic first between the butt faces at each end of the step-lap joint. At higher loads, the proportion of the load transferred by the butt faces is reduced as the joint efficiency is increased.

5.3.5.9 Design of angle joints

The main object of the joint is to enable smooth transfer of loads between the two panels meeting at the joint. The joint is formed by bonding two pre-formed angles on the inside and outside of the two panels to form the overall angle connection. The load transmission is done entirely through the adhesive layer and the bonding angles. The potential failure modes are dominated by the adhesive performance in tensile, tensile shear and peel loadings.

This particular joint configuration does not lend itself to simple, closed-form analytical modelling. Simple guidelines may be provided for an initial estimate, but the final design has to be based on testing and/or numerical analysis.

Published literature in this area is very limited. One study (reference 5.37) examined structural testing of different types of bonded connections (for detailed discussion see 5.3.5.10). One of the main conclusions to emerge from this study was that weakest link in FRP structures is at the secondary bonding connections. Because of this, (pure) bonded angle joints should be used only after careful verification. The preferable method of joining should involve bonding and bolting, for instance.

5.3.5.10 Design of tee joints

In terms of failure modes, design and analysis this joint is practically identical to the angle joints discussed above. The only difference is that the joint is formed by bonding two pre-formed angles attaching either side of the "web" piece, i.e. the leg of the tee, to the "flange" piece (the base of the tee) to form the overall tee connection.

Reference 5.37 examined structural testing of different types of bonded connections. The study was concerned with the bonding of different types of stiffener/plate connections. One type of connection dealt with the use of a bonding angle to connect a stiffener to plating. The strengths of such connections were evaluated in four-point bending tests on the overall specimen. It was reported that most failures occurred at the bonding surface between the stiffener and the plating. Further, it was stated that bondline thickness, the quality of the bond and other fabrication imperfections were difficult to measure; because of this, residual stresses in the bonded structure could not be predicted. Also in this case one of the main conclusions was that the weakest link in FRP structures is at the secondary bonding connections.

Another recent work (reference 5.38) makes a passing reference to connections between top-hat sections and plating, which enable transmission of loads in two orthogonal directions similar to tee connections. The example cited concerns a design which utilises a shaped (or cross-cut) aluminium extrusion or composite pultrusion to distribute the load over a much greater area of plating than would otherwise have been possible. It has been reported that when a toughened adhesive is used as a laminating resin in the critical areas, there is a substantial decrease in the dimensional accuracy required. Ductility of the adhesive helps to minimise stress levels in the region of the joint. The capacity of the toughened adhesive appears to prevent premature catastrophic failure.

Reported evidence of the actual use of such tee joints in "real" structures is limited. Because of this, (pure) bonded tee joints should be used only after careful verification. The preferable method of joining should involve bonding and bolting, for instance.

5.3.5.11 Manufacturing aspects

General

It is important to realise that structural bonding is critical and demanding work, comparing well with welding as a task requiring special skills and knowledge. Because of this, only properly trained personnel should be allowed to perform adhesive bonding.

Polymer composite materials absorb water from humid air. This water may degrade the bond, especially when elevated temperature cure is required. Water or humidity on the adherend bond surface may seriously reduce the adhesion compared to that with a dry surface condition. Therefore, the dryness of the laminates that are to be bonded has to be ensured. However, if room temperature cure is used and drying of laminates is considered impractical, an adhesive appropriate to this condition has to be selected. Particular care should be excercised when performing surface treatments under on-site conditions.

Surface treatment is one of the most important factors affecting the joint strength of an adhesively bonded joint. An incorrect or improper surface treatment typically leads to a premature adhesive failure (i.e. adhesion fails) and the joint strength remains modest. In the case of FRP adherends the failure of the adhesion can be avoided by employing proper surface treatments. For FRP adherends a degrease - abrasion -degrease surface treatment has been found to be sufficient. For metallic adherends some chemical surface treatment is typically required before bonding. These treatments depend on the metallic material used and are typically given by adhesive manufacturers or may be found in representative text books. Once treated, the surface has to be protected from further contamination.

Special attention has to be paid also to the environmental conditions during bonding, particularly when on-site. In addition to the adhesion problems associated with the absorbed moisture in the adherends, high relative humidity may have an unfavourable effect on the adhesion properties of the adhesive. Temperature effects on the viscosity, working time and cure of an adhesive are similar to those experienced with lamination resins (see Chapter 6) and need to be considered separately. To avoid problems, the environmental conditions should be those specified by the adhesive manufacturer or the component manufacturer. During the bonding process air escape routes have to be provided for the air entrapped in the bondline, as it may significantly reduce the strength and durability of the joint.

Adhesives do not have any load-carrying capability prior to cure. Therefore, components to be bonded have to be supported during curing by jigs or even by bolts, if appropriate. The curing temperature has to be within the limits given by the adhesive manufacturer or otherwise specified. When the adhesive is cured at the prevailing environmental temperature, it is sufficient to monitor only the temperature of the surrounding air. When additional heat is applied during the cure, both the adhesive and component surface temperatures have to be monitored.

Some further manufacturing aspects specific to certain joint configurations are discussed below.

Scarf joints

A scarf joint reaches its maximum strength when the scarf tip is sharp. However, in actual applications this kind of a sharp tip is difficult to manufacture. Also the bondline thickness becomes extremely difficult to control close to the tip. Therefore, it is required that the tip should have a butt end with a finite thickness of at least 1.0 mm.

Step-lap joints

Step-lap joints are more complicated and more expensive to manufacture than lap joints. Requirements for the geometrical tolerances of the joint are also much more demanding.

In the EUROCOMP Design Code, two possible methods are given for the manufacture of the steps of a step-lap joint. The steps may be produced either by machining the cured laminate or by producing the steps during the laminate manufacture. Of these two methods, machining gives geometrically more precise steps than lamination. It is, however, extremely difficult to machine a step on to the surface of a specific layer. Especially in thick laminates with low fibre volume content, the absolute distance of a specific layer from the laminate surface may vary significantly. Therefore, the recommendation that the fibre orientation on the bond surface is to coincide with the principal loading direction is sometimes hard to fulfil. Furthermore, it is to be noted that machining may damage the fibres on the step shear surface.

Producing the steps in the course of the lay-up process provides control on the fibre orientations on the bond surfaces, but the dimensional accuracy of the steps is not as good as with machined steps, unless the steps are laminated against a specially manufactured mould surface. If this is not the case, light machining of the laminated steps often gives the most favourable result.

5.3.6 Laminated joints

5.3.6.1 General

Laminated joints are formed by laminating fibre layers on cured composite parts, often by hand lay-up. The resin used is typically identical to that used in the components to be joined. However, different resins may also be used for fabricating laminated joints, for example epoxy on polyester parts.

The mechanical behaviour of laminated joints can be characterized by using two different assumptions:

- laminated joints are adhesively bonded joints where the adhesive is replaced by a matrix resin

- laminated joints form a solid structure combining the joint and the adherends.

The first case includes the assumption that there is a distinguishable layer between the adherend and the on-laminated joint. However, there is no specific layer that can be said to be the adhesive layer; thus the thickness of the adhesive layer is not accurately definable. This leads to the second assumption, that the joint strength is dependent on the internal strength of the laminate. In reality, the performance of laminated joints indicates that the first assumption is closer to the truth.

A typical laminated joint is a butt joint on to which doublers are laminated, the result being a double-strap joint. The following features distinguish this laminated joint from the corresponding adhesively bonded double-strap joint:

- no specific adhesive layer
- shape of the doubler (no sharp corners)
- cure shrinkage of the doubler
- doubler manufacturing process (typically hand lay-up for laminated joints).

Also, polymer resins used as matrices in FRP materials are more brittle than typical adhesives. Generally, matrix resins have no plastic behaviour and thus local peak loads cannot be redistributed by plasticity. However, when doublers are laminated, it is relatively easy to taper their ends.

Various matrix resins can have compatibility problems, as the adhesion between dissimilar resins may be poor. It has been found that sometimes even the use of polyester resin in a laminated joint of cured polyester does not give satisfactory adhesion. Therefore, the compatibility of matrix resins should always be examined or verified by testing.

A laminated joint is typically a weak link in the structure for the following reasons:

- there is a discontinuity in the mechanical properties in the joint
- the adherend and the doubler have different thermo-mechanical properties
- typically a relatively thick layer of pure polymer is formed on a bond surface.

The discontinuity in the mechanical properties of the adherend and the laminated doubler can be due to:

- resins
- fibre orientation
- fibre volume fraction
- manufacturing process.

Also the cure shrinkage of the doubler can produce internal stresses that decrease the joint strength.

The design procedure presented in the EUROCOMP Design Code comprises two phases:

1. Preliminary design, which will be performed according the corresponding procedures for adhesively bonded joints. Matrix resin properties are used instead of adhesive properties. The adhesive layer thickness is evaluated by assuming a particular value.

2. The preliminary design, which is verified using component tests.

It is practically impossible to define the actual bondline thickness. Therefore one has to make an estimate of the thickness. In most cases it is beneficial for the joint strength to have a thin bond layer. However, an unrealistically low estimate of the bond layer thickness should not be made. A realistic minimum value is 0.1 mm.

5.3.6.2 Joint configurations and applied loads

No additional information.

5.3.6.3 Design of lap and strap joints

No additional information.

5.3.6.4 Design of scarf joints

No additional information.

5.3.6.5 Design of step-lap joints

No additional information.

5.3.6.6 Design of angle joints

The main purpose of the angle joint is to facilitate a smooth transfer of loads between the two panels meeting at the joint. The joint is formed by laminating strips of reinforcing cloth (or overlaminates) on the inside and the outside of the two panels to form a double boundary angle connection. The load transmission is done entirely through the reinforcing plies and the filleting resin in the corner of the connection. The make-up of the boundary angles needs close attention. Two criteria governing the overall performance are the stiffness and strength of the joint. Traditional approaches in design have tended to equate the gross properties of the boundary angles to those of the base laminates, i.e. the in-plane properties of the two laminates is equivalent to that of the weaker of the two members being joined. Because, in most cases, the material used on the overlaminates is the same as that in the base laminates, the thickness of the overlaminate is specified as a function of the thinnest member being joined.

The two elements carrying the load are the overlaminate and the filler fillet. In order to achieve maximum efficiency in the joint, the overlaminate and fillet should be designed such that they reach their individual failure limits at the same value of external load. If the overlaminate is relatively thick, then it is stiff as well. If, for instance, a flexural load is applied to the joint, high through-thickness stresses develop in the radiused corner of the joint and delamination develops in the plies. This generally occurs when the boundary angle is in tension. As the load is increased, further layers delaminate until the overlaminate disbonds from the filler fillet or from one of the base laminates. This has the potential for causing the unsupported filler fillet to develop a crack and thus fail. This mode of failure is similar to that in tee joints.

It is important to point out here that such angle joints have not yet seen widespread use in other applications. The knowledge base, i.e. papers and articles in the open literature, from which design guidance can be obtained is limited. Consequently, it is especially important here that explicit modelling is done to fully understand all potential failure mechanisms.

The influence of the design variables defining the joint and their efficiency in enabling smooth load transfer between the two base members must be carefully accounted for through specimen testing. Such testing should preferably be supplemented by detailed finite element analysis to understand the internal response of the material constituents and the associated margins to failure in each case. The approach to be adopted here can be similar to that for tee joints, as outlined in 5.3.6.7. See, for example, reference 5.39 for further details of test procedures and numerical modelling schemes.

5.3.6.7 Design of tee joints

The primary function of the tee joint is to transmit flexural, tensile and shear loads between two sets of panels meeting at the joint. The joint itself is formed by laminating strips of reinforcing cloth (or overlaminates) either side of the joint to form a double (or boundary) angle connection. The load transmission is done entirely through the reinforcements in the plies and any filleting resin within the boundary angle. The details of the reinforcements and resin therefore need close attention. Traditionally, this has been the only criterion considered - that is, the sum of the in-plane properties of the two overlaminates is equivalent to the weakest member being joined. Because, in most applications, the material used in the overlaminate is the same as that used in the structure, the thickness of the overlaminate is specified as a function of the thickness of the base structural members (i.e. the leg and top of the tee) intersecting at the joint.

However, it is also recognised that the overall flexibility of the joint is as important as its strength. This is valid if the stress raising effect of the "hard point", created by the presence of the leg, is to be avoided. Thus the designer is faced with apparently contradictory requirements: that of strength (requiring a thick overlaminate) and that of flexibility (requiring a thin overlaminate).

A recent study (reference 5.39) has made a comparison between the following two cases:

1. a "traditional" joint with thick overlaminate of small radius
2. a "new" joint with a flexible fillet resin (such as urethane acrylate with a high strain to failure value) coupled with a thin overlaminate of large radius.

It was shown that the joint in case 1 was stronger than the one in case 2 by a margin of 50%, though the deflection was much larger.

The apparent contradiction mentioned above is illustrated by considering the failure modes associated with such joints. The two elements carrying the load are the overlaminate and the filler fillet. For both the elements to be exploited efficiently, they should reach their failure stresses (both in-plane and through-thickness in the case of the overlaminate) at the same load. If the overlaminate is relatively thick, then it is stiff as well. As the load is applied to the joint, high through-thickness stresses develop in the radiused corner of the joint and delamination develops in the plies. This generally occurs in the boundary angle which is in tension. As the load is increased, further layers delaminate until the overlaminate disbonds from the filler fillet. When this occurs the unsupported fillet develops a crack originating at the location of the disbond between it and the overlaminate. During this process, the in-plane stress in the overlaminate is generally well within its failure value. Similarly, immediately prior to the disbonding of the overlaminate from the fillet, the stress in the fillet is also lower than the failure value of the material. Figure 5.37 illustrates the failure modes associated with joints having thick boundary angles, thin boundary angles and no boundary angles (i.e. only filleting resin). This last mentioned, i.e. a joint held together with filleting resin only, is not normally allowed for structural use; it is shown here for illustrative purposes only.

Clearly, to exploit the materials used fully, it is necessary to reduce the through-thickness stress, whilst allowing the in-plane and fillet stresses to increase. Any design method must be able to quantify these stresses and to determine the effect of design variables on them in an explicit manner. One approach is to conduct a systematic, parametric study to check the influence of varying different design parameters outlined below (see for example reference 5.40).

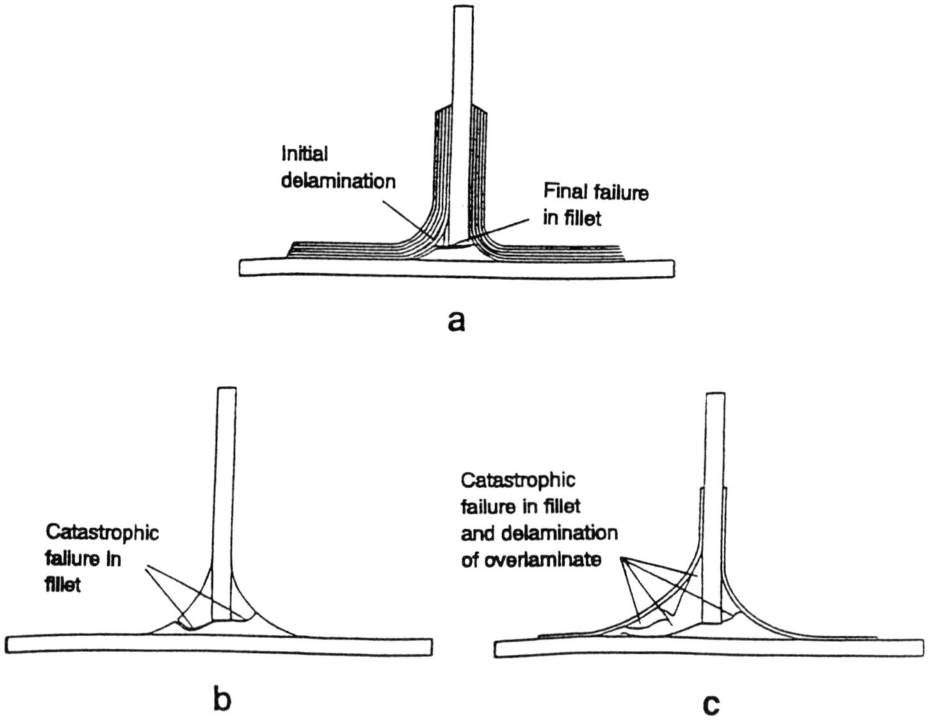

Figure 5.37 Typical failure modes for different tee joint configurations: (a) thick boundary angle; (b) no boundary angle; (c) thin boundary angle.

Table 5.19 Influence of material and geometry variations on the behaviour of tee joints.

	Feature	*Response*		
		In-plane stress in laminate	Through-thickness stress in laminate	Principal stress in fillet
Geometry	Increasing thickness of overlaminate	Decreases	Increases	Decreases
	Increasing radius: overlaminate	Decreases	Decreases	Decreases
	Increasing radius: pure fillet	-	-	Approx. same
Material	Impact of overlaminating on a fillet	-	-	Decreases
	Impact of using flexible fillet resin	Decreases	Decreases	Decreases

The influence of design variables defining the joint needs, as far as possible, to be accounted for in both physical specimen testing and finite element analyses. Recent studies (references 5.39, 5.40) have identified the most dominant design variables affecting joint performance to be the thickness of the overlaminate, radius of the fillet and the failure strain of the filleting resin. The effects of these variables are summarised in Table 5.19. Generally, choice of design variables should be such as to make the joint as flexible as possible while keeping the stresses low. As an illustration, a joint should use filleting resin with a high strain to failure value and the boundary angles should be as thin as possible with a large radius.

5.3.6.8 Manufacturing aspects

Basically similar manufacturing principles may be applied to laminated joints as to adhesively bonded joints. However, some additional points should be considered, while some others become irrelevant. These additional points include the following:

- The same manufacturing requirements should be applied to laminated joints as to fabricating laminated structures in general.
- In certain cases no surface treatment prior to joining is required. This is the case when the parent laminate is in the state where the formation of chemical bonds across the bondline is still possible. This requires a low degree of cure in the laminate and the time limit for this has to be defined separately for each resin and application.
- The joint should have a temperature resistance similar to that of the parent laminate or what is required from consideration of the whole structure. This often means that either the whole structure or the joint has to be post-cured at elevated temperature. When no elevated temperature performance or special environmental resistance is required, post-cure may not be needed.

5.3.7 Moulded joints

5.3.7.1 General

The three moulding processes listed in the EUROCOMP Design Code are only typical examples of possible moulding processes. However, these methods cover the field reasonably well. Of the methods listed, RTM members have the highest strength potential, as they may consist at least partially of continuous fibres and have a relatively high fibre volume content. SMC members have only chopped fibres (typical length 25 mm) and typically have a low fibre volume content. Both the RTM and SMC processes use thermoset plastics. Injection moulded reinforced thermoplastics members typically have the lowest strength of these three types of joint members. This is because of the relatively low strength of the thermoplastic material and the low fibre length (about 5-10 mm at maximum). They also possess the inherent problem of poor adhesion of the thermoplastics when traditional surface treatments are applied.

5.3.7.2 Joint configurations and applied loads

Moulded joints are typically used in frames or in truss or lattice type structures, where they may be subjected for example to shear load at the interface due to axial tensile loads in one of the members to be joined, i.e. axial tensile forces pulling one of the members out of the moulded joint. As an example, Table 5.20 gives tested axial pull-out loads in a truss structure made of pultruded tubular sections (glass fibre/polyester) with RTM and SMC mouldings in the leg/diagonal joints. The values presented should only be considered as indicative for moulded joints made using different materials and manufacturing methods. Any generalization of the data is not allowed.

Table 5.20 Axial pull-out load of a tubular FRP section in the case of a moulded joint of a truss structure.

Tube diameter (mm) outer/inner		*Pull-out load at failure (kN)*	
		RTM member	*SMC member*
35/32	leg	16.7	15.6
20/17	diagonal	12.2	11.9
50/46	leg		
32/28	diagonal	13.6	
65/61	leg		
32/28	diagonal	16.0	17.7

5.3.7.3 Design

In principle, thermosetting moulded joints (RTM, SMC) are related to laminated double-lap joints, enabling similar types of design procedures to be used for preliminary dimensioning. However, these procedures should not be used for the final design of load-carrying structures, owing to certain basic differences between these joining techniques.

When practical, the methods described in 5.3.1.3 to reduce joint peak stresses should be applied to moulded joint design.

Injection moulded reinforced thermoplastic joints should only be used in lightly loaded non-structural connections because they typically give a weak adhesion to almost all materials. Therefore, the load transfer from the joint component to the injection moulded reinforced thermoplastic joint member should be based on mechanical interlocking and not on the adhesion.

The design aspects considered above can be applied to the preliminary design of a moulded joint, but the final design should always be based on testing.

5.3.7.4 Manufacturing aspects

Similar principles can be applied to the manufacture of moulded joints as to adhesively bonded and laminated joints, but owing to the special manufacturing techniques involved some additional points should be considered. Three major aspects are discussed below.

Each moulding process requires the use of correct process parameters. The most important ones are injection/moulding pressure, temperature and time. These mutually dependant parameters have to be defined experimentally for each application, as they have a significant effect on the mechanical properties of the joint.

Material escape during the process leads to the formation of voids and other defects and results in inadequate filling of the mould. Both the outer edges of the mould and the ends of the members to be joined need to be sealed as they typically are pultruded sections with open ends. This can be, for example, by using plugs or by pre-assembling the joint using a non-structural sealant type adhesive. The sealing required also depends on the process parameters, especially on the temperature in relation to the rate of applied pressure.

The moulding methods discussed typically require the use of internal release agents. These agents weaken the adhesion and therefore lower the load-bearing capacity of the joint. The use of these agents should be minimised, but can seldom be entirely avoided.

5.3.8 Bonded insert joints

5.3.8.1 General

In the EUROCOMP Design Code bonded inserts are defined as components adhesively bonded to a composite member that enable later mechanical attachment of this member to another member through a bonded insert (typically metallic) (see Figure 5.31 of the EUROCOMP Design Code). It is important to understand the basic difference between the bonded inserts, which rely solely on the adhesion for their attachment to the parent component, and embedded inserts, where the attachment to the composite member is based on mechanical interlocking or entrapment. The inserts of the latter type are often positioned and temporarily bonded during installation. This is mainly for convenience, for example to hold the inserts while the structure is being assembled. The transfer of the service loads in the completed structure is based on mechanical interlocking (insert flanges behind the plate or between the plies, etc.). For this type of insert joint, the design procedures for mechanically fastened joints should be used.

5.3.8.2 Joint configurations and applied loads

Bonded inserts are typically used with pultruded sections or with similar types of components having closed cross-sections. The inserts may be bonded either on the outer or inner surface of the section. The sections are typically loaded in axial tension or compression or in bending, resulting in shear loadings (primarily) in the bonded joint.

Well designed bonded insert joints can transfer high loads. The maximum possible axial load transmitted by a bonded insert joint is of the same order of magnitude as the ultimate strength of the composite member/section, with failure primarily taking place in the section material. Achieving load-bearing levels similar to that of the section however, requires, careful design and manufacture of the joint.

Bonded inserts are also commonly used in sandwich structures, most typically in panels. The insert is bonded into the core and to one or both skins of the sandwich. Loads are then directed to the joint through the mechanical fasteners (i.e. the insert) in their axial direction, causing out-of-plane loading of the sandwich component. Sandwich structures typically have thin skins and therefore the loads from the insert are mainly transferred to the core. If the insert is loaded in the in-plane direction of the sandwich member, the connection is categorised as an embedded insert and it should be designed as a mechanical connection.

5.3.8.3 Design

No additional information.

5.3.8.4 Manufacturing aspects

In addition to the general principles which apply to bonded joints, there are certain manufacturing aspects specific to bonded insert joints. The insert is usually of metallic material and therefore it often requires a chemical surface treatment prior to the bonding. Once the insert is ready to be bonded, it has to be verified that the whole bond area is covered by the adhesive after the insert has been installed. This problem needs to be solved by developing production methods that guarantee the quality and consistency of the bonding.

5.3.9 Cast-in joints

5.3.9.1 General

No additional information.

5.3.9.2 Joint configurations and applied loads

No additional information.

5.3.9.3 Design

The strength of a cast-in joint has to be determined only when adhesive/resin failure leads to a total failure of the structure or when the serviceability limits are exceeded due to adhesive/resin failure. This is because in many applications the adhesive/resin is primarily used for purposes other than load transfer and in these cases the bond strength is not a design consideration.

The preliminary design of a cast-in joint may be performed using the design procedures for adhesively bonded double-lap joints. Typically a cast-in joint design becomes a component design problem instead of a joint design problem. The final design of cast-in joints always has to be based on testing equivalent structures.

Typical applications of cast in-joints may result in relatively thick bond layers. In the design process, account should be taken of the fact that the adhesive/resin layer thickness may not exceed the maximum value given by the adhesive/resin manufacturer.

5.3.9.4 Manufacturing aspects

In addition to those given in the context of adhesively bonded joints, the following aspects, arising from the very nature of thermosetting polymers, should be recognised, especially when the adhesive/resin layer is thick:

- the joint and the components may be damaged if the temperatures induced by the exothermic cure reaction are too high
- the joint quality may be affected by the significant cure shrinkage.

The effect of the detrimental exothermic reaction is assisted by the low thermal conductivity of the adhesive/resin itself and the FRP adherends. If the dimensions cannot be changed, the heat build-up should be limited, for example, by adding fillers into the adhesive/resin and thus reducing the mass of thermoset polymer in the adhesive/resin. Depending on the type of filler, the thermal conductivity may also be improved. In any case, the adverse impact of the above points should be reduced by the design or other appropriate precautions.

5.3.10 Defects and quality control

Defects

The most common defects encountered in bonded joints are inadequate adhesion or disbonds and flaws and/or porosity. Examples of other possible defects are variations in the adhesive thickness and undercuring.

The importance of the defects depends on factors such as:

- the number and size of the defects
- the possible consequences of the defects
- whether the defects are evenly spread over the bonded area or concentrated at critical areas
- whether or not the defects are an indication of a degradation process
- whether or not the defects are indicative of material deficiencies.

Any bonding defect leads to a redistribution of loads through the adhesive layer. This would suggest an increase in peak stresses at the points of discontinuity in the adhesive layer. However, the actual effect is smaller than expected and, in most cases, only an imperceptible increase in stress occurs. It should be noted that this statement is valid only when the flaw size is proportionally small compared to the joint size.

Adhesion defects are either local or extend over the whole bond area. They are generally due to poor surface treatment, lack of preconditioning (such as drying) or material incompatibility. In the case of defective adhesion over the whole bond area, a sudden adhesive rupture can occur even at relatively low load levels. When the adhesion defect is local, a redistribution of loads occurs as with flaws. This load redistribution is shown in Figures 5.38 to 5.42 for tensile loaded double-lap joints. These stress distribution calculations can be performed by using, for example, the computer code A4EI (reference 5.42).

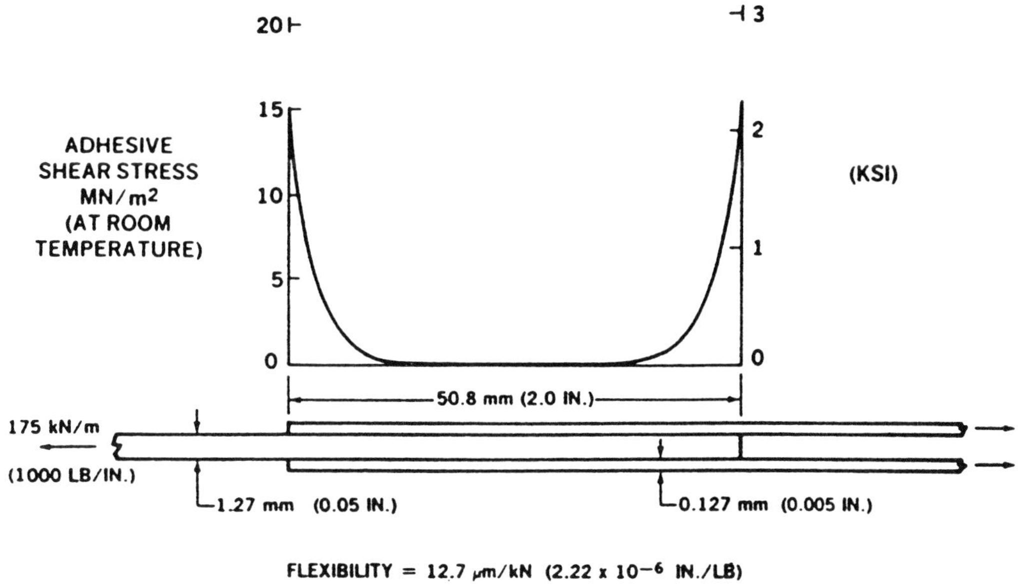

Figure 5.38 Adhesive shear stress distribution in intact bonded joints (based on reference 5.31). (Copyright ASTM: reprinted with permission).

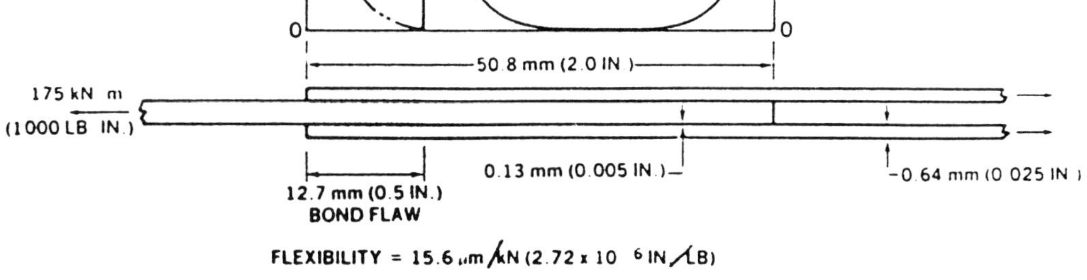

Figure 5.39 Adhesive shear stress distribution in flawed bonded joints (based on reference 5.31). (Copyright ASTM: reprinted with permission).

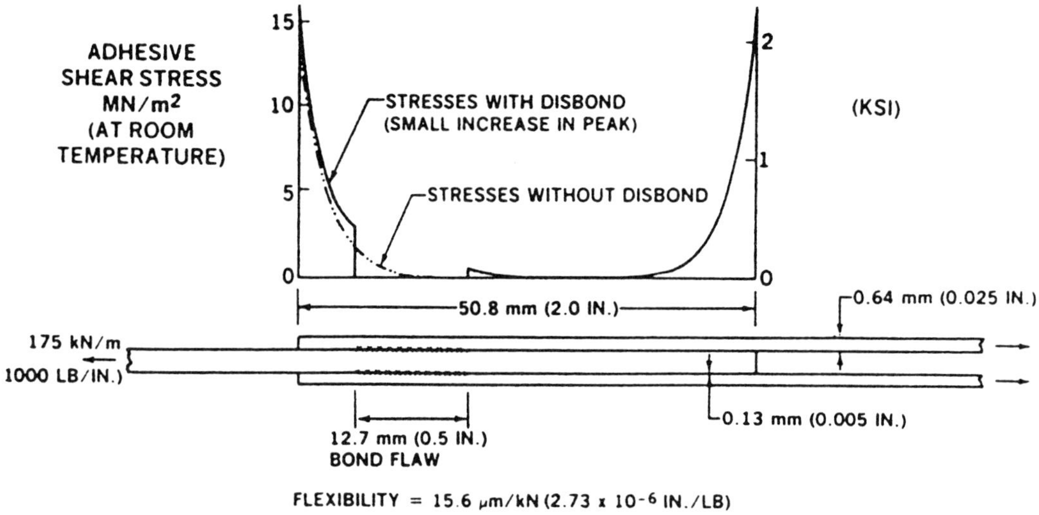

Figure 5.40 Adhesive shear stress distribution in flawed bonded joints (based on reference 5.31). (Copyright ASTM: reprinted with permission).

Experience has shown that the best results in the bonding of composites using ductile epoxy adhesives are obtained when the adhesive layer has a thickness ranging from 0.1 to 0.25 mm. Often, this optimum thickness cannot be achieved in practice.

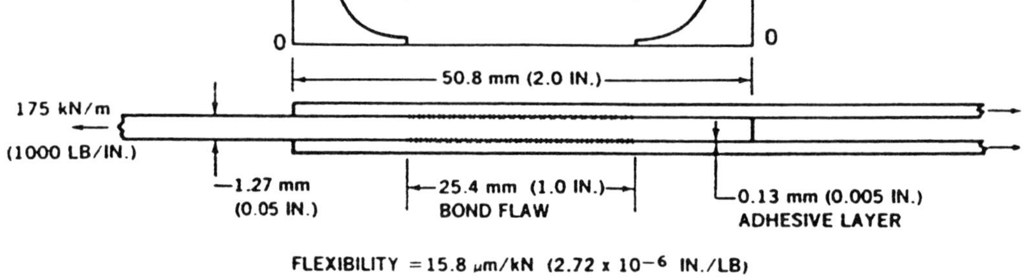

Figure 5.41 Adhesive shear stress distribution in flawed bonded joints (based on reference 5.31). (Copyright ASTM: reprinted with permission).

The variation in adhesive layer thickness is most critical at the ends of the overlap. When pressure is used during the bonding process, the adhesive tends to flow out from the joint ends owing to adherend bending. The effect of adhesive layer thickness on the adhesive shear stress at the ends of the overlap is shown in Figure 5.42.

Undercuring of the adhesive is due to insufficient time and/or temperature for curing. Other possible reasons for undercuring are a wrong adhesive component ratio or inadequate mixing, or alternatively a presence of water or chemicals during bonding.

Quality control

Quality control of the bonding should concentrate on the following points:

- preparation of the adherends for bonding
- mixing and handling of the adhesive
- application of the adhesive
- curing.

The following factors have to be considered when adherends are prepared for bonding:

- surface contamination
- adherend surface humidity
- water content of the composite adherend
- environmental factors.

The following factors concerning the adhesive and the application of the adhesive have to be considered:

- the ratio between the adhesive/resin and the hardener (two-component adhesives)
- adequate mixing (two-component adhesives)
- pot life
- ageing of the adhesive
- storage conditions
- environmental conditions during adhesive application.

During the curing process the following factors have to be considered:

- curing temperature and time
- applied pressure
- environmental conditions.

The minimum requirement for controlling the bonding process is to make an adhesive casting from every batch. These castings should be cured identically to the corresponding adhesively bonded joint. The adequacy of the curing state may then be verified from the castings. This verification can be performed using thermal, thermo-mechanical or mechanical inspection. Typical methods, in corresponding order, are differential

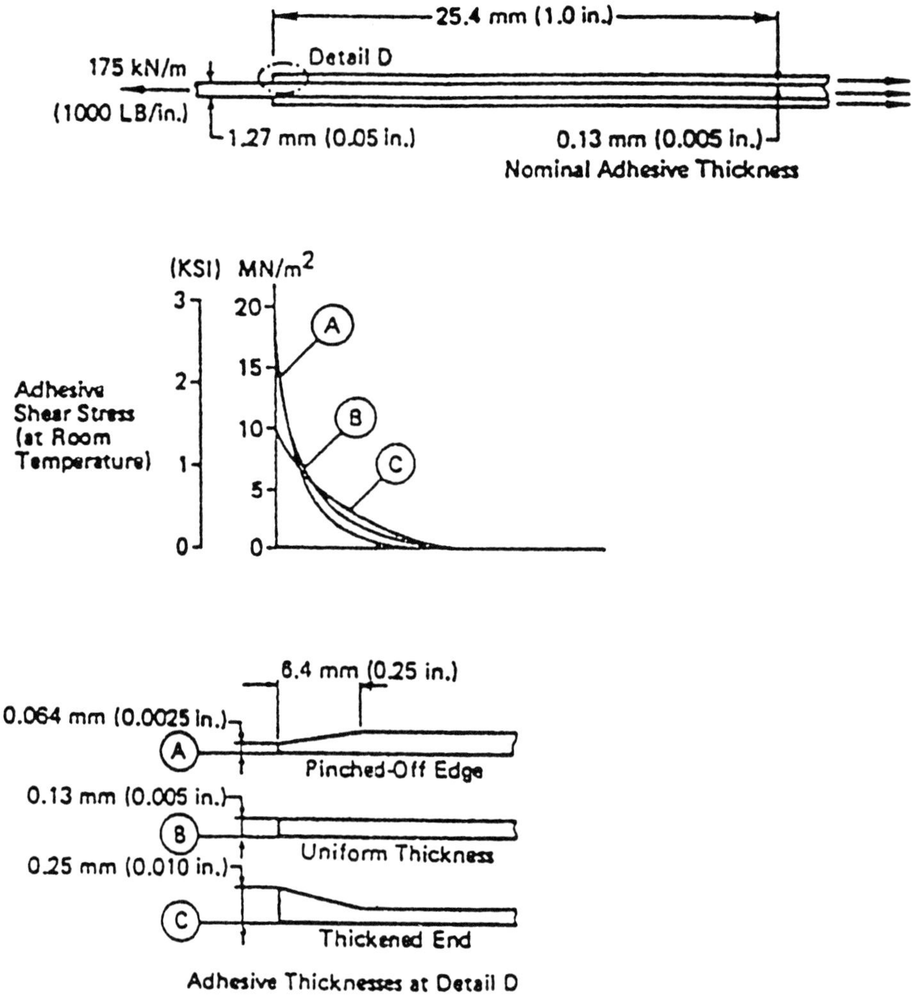

Figure 5.42 Variation of peak adhesive shear stress with the thickness of adhesive at the ends of the overlap (based on reference 5.31. (Copyright ASTM: reprinted with permission).

scanning calorimetry (DSC), differential mechanical thermal analysis (DMTA) and a hardness test, such as Barcol hardness (see 7.2.1). Whenever possible, it is recommended that a control sample be bonded for mechanical testing using surface treatments and adhesives identical to the actual joint. This control sample could be two plates (200 mm x 200 mm) bonded together using a 12.5 mm overlap and having a single-lap configuration. This sample may then be cut into 25.0 mm wide test specimens and loaded in tension according to ISO 4587: 1979 or a corresponding national standard. The results may be used as quality control data.

Non-destructive inspection methods

The non-destructive inspection methods appropriate for detecting defects in bonded joints are briefly summarized below from (references 5.21 and 5.44).

Ultrasonic: In this method, sound waves ranging in frequency from 1 to 10 MHz are used to find changes in thickness or detect porosity and delamination or unbonded areas. There are three ultrasonic systems:

- **pulse-echo reflector:** To detect flaws and/or delaminations
- **transmission:** To detect flaws
- **resonant frequency:** To detect unbonded areas.

Sonic emission: This method applies sonic vibration to find flaws or microflaws, porosity, undercured or/and unbonded areas and variation in density.

Infrared: This method is used to find delaminations, unbonded areas and porosity.

Penetrant liquid: This method is used to detect flaws, porosity and delaminations.

Tapping: This method is used to detect relatively large unbonded areas.

5.3.11 Repairability

In the case of bonded joints it is particularly important that the reason for the failure is investigated as stated in the EUROCOMP Design Code. It is reasonably common that the primary reason for the failure is not any overloading, although it may easily be the first interpretation, as the real cause of failure may be built-in to the design. As discussed in the context of the individual design procedures, bonded joints are often designed with insufficient or inaccurate data, especially on internal loadings and material properties. Also several assumptions are typically made in the course of the design which can not be verified. As an addition to these, reliable data on the loads to which the whole structure or component is subjected may not be available (environmental loads in particular). The result of all this may be that, although the design has been carried out following the procedures correctly, it is inherently erroneous. This problem is, of course, of more general nature but needs to be specifically addressed here as all the inaccuracies are often accumulated in the bonded joint design. Therefore, it is important that the reason for the failure is carefully considered before any actions are taken, as the repair can hardly be successful if the joint remains based on erroneous design. If the reason is found to be any of those listed in the EUROCOMP Design Code, redesign is necessary.

In addition to the design problems discussed above, bonded joints are also particularly vulnerable to poor workmanship, most typically insufficient surface treatment or unsuitable environmental conditions during bonding. Further potential sources for defects are given in the manufacturing sections of each joining technique. None of these should become a problem if a proper and detailed manufacturing specification is made and is followed throughout the bonding process. Experience has shown that often this is not the case. Therefore, if the design of a failed connection is found to be satisfactory, the bonding instructions should be reviewed in order to check that they are correct, sufficient and practical and that they have been followed in practice. It is obvious that unless the necessary modifications are made to the instructions for the repair or the manufacture of the replacement, no improvement in the performance can be guaranteed.

Additional repair instructions for some specific joining techniques are given in the following:

Bonded insert joints

The following repair instructions may be given for two typical cases:

- When a bonded insert has separated from a composite section the damaged part of the structure shall be replaced by an intact structure. This may result in discarding the whole damaged member. It may also be feasible to cut off the damaged section and to bond the insert into the intact part of the section
- When a bonded insert has separated from a sandwich structure the damaged part of the structure shall be rebuilt. Usually this means cutting off the damaged part of the core material and removing an adequate amount of the skin material. A new core block has to be bonded into the existing core and the skin has to be repaired by laminating or bonding a skin patch.

Cast-in joints

If the instructions given for the repair of the other bonded joints were to be followed, cast-in joints could only be repaired by disassembling and rebonding or by using externally bonded doublers. It may be assumed, however, that often cast-in type of joints do not rely to the adhesive forces to the same extent as the other bonded configurations. The primary function of the cast adhesive may be to provide mechanical locking for the intersecting member with zero tolerances. In such a case simpler repair procedures may be acceptable. The most obvious one would be to grind off as much of the cast adhesive as possible and re-cast the joint. The best possible surface preparation should be carried out, although it is clear that the original level of adhesion cannot be maintained. Therefore, it is necessary to verify that the level of adhesion to be achieved corresponds with the load-bearing requirements. If the adhesive joint needs to be fully loaded, then simpler repair methods cannot be accepted.

5.3.12 Example: An analysis of a plate-to-plate bonded joint using the simplified procedure

Description of the joint

The single-lap joint is loaded by an in-plane (characteristic) tensile load per unit width of 6 kN/m.

Material descriptions

Adherends RTM manufactured glass-reinforced polyester laminates having the following mechanical properties:

$E = E_1 = E_2 = 15$ kN/mm^2
$\sigma_{z,k} = 20$ N/mm^2
$\nu = 0.3$

Adhesive A ductile epoxy adhesive having the following mechanical properties:

$G_a = 371$ N/mm^2
$E_a = 950$ N/mm^2
$\tau_{0,k} = 10$ N/mm^2
$\sigma_{0,k} = 30$ N/mm^2

The design loading is a short-term permanent loading only. The joint will be located at an indoor environment at room temperature. The adhesive is applied manually using a spatula and the adhesive layer thickness is controlled using copper wires. The adhesive and adherend mechanical values have been determined by testing.

The following values for partial safety factors for materials may be obtained from 2.3.3.2 and 5.1.10 in the EUROCOMP Design Code for the adhesively bonded connection designed by testing:

$\gamma_{m,1} = 1.25$
$\gamma_{m,2} = 1.25$
$\gamma_{m,3} = 1.0$
$\gamma_{m,4} = 1.0$
$\gamma_m = 1.56$

The following value for the partial safety factor for the action may be obtained from 2.3.3.1, Table 2.2 of the EUROCOMP Design Code (Permanent actions only, unfavourable effect):

$$\gamma_f = 1.35$$

Step 1 A series of lap shear tests have been performed under laboratory conditions using the following geometric parameters:

t = 4.0 mm
L = 12.5 mm
W = 25 mm (width of the joint)

The average lap shear strength was determined to be 11 N/mm^2.

Calculate the ultimate joint load per unit width

$$N_u = 11\ \text{N/mm}^2 \times 12.5\ \text{mm} = 137.5\ \text{N/mm} = 137.5\ \text{kN/m}$$

Step 2 The lap length of the actual joint shall be taken as

L is the greater of 2 x 12.5 mm or 50 mm

$\Rightarrow$ L = 50 mm

Step 3 Calculate the maximum joint resistance, Equation (5.8) in the EUROCOMP Design Code

$$N_{t,\,Rd} = \frac{N_u}{y_m} = \frac{137.5\ \text{kN/m}}{1.56} = 88.1\ \text{kN/m}$$

Step 4 Applied load versus joint resistance, Equation (5.9) in the EUROCOMP Design Code

C_m = 10 (see 5.3.5.4 of the EUROCOMP Design Code, Step 4)

$$N_{t,Sd} = F_k\ \gamma_f\ C_m = 6 \times 1.35 \times 10 = 81.0\ \text{kN/m} < 88.1\ \text{kN/m} = N_{t,Rd}$$

The design condition is satisfied with respect to the joint load.

5.3.13 Example: An analysis of a plate-to-plate bonded joint using the rigorous procedure

Description of the joint

The single-strap joint shown in Figure 5.43 is loaded by an in-plane (characteristic) tensile load per unit width of P_k = 60 kN/m. The joint has the following thicknesses:

$t_1 = t_2 = 4$ mm
$t_a = 0.5$ mm

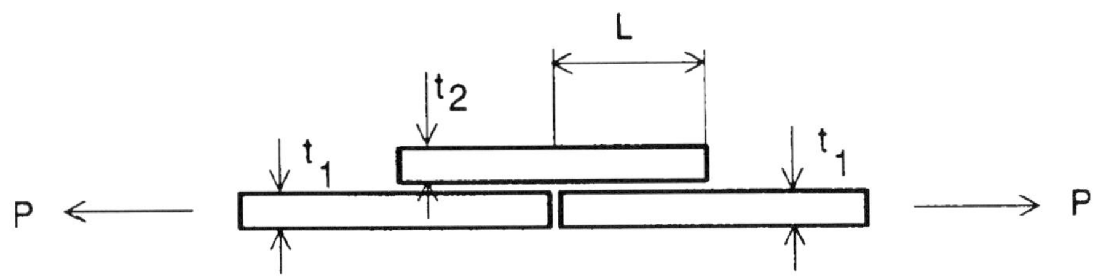

Figure 5.43 Single-strap joint in the example of rigorous procedure.

Material descriptions

Adherends RTM manufactured glass-reinforced polyester laminates having the following mechanical properties:

$E = E_1 = E_2 = 15$ kN/mm^2
$\sigma_{z,k} = 20$ N/mm^2
$\nu = 0.3$

Adhesive A brittle epoxy adhesive having the following mechanical properties:

$G_a = 371$ N/mm^2
$E_a = 950$ N/mm^2
$\tau_{0,k} = 11$ N/mm^2
$\sigma_{0,k} = 30$ N/mm^2

The type of loading will be a short-term permanent loading only. The joint will be located in an indoor environment at room temperature. The adhesive is applied manually and the adhesive layer thickness will be controlled. The adhesive and adherend mechanical values have been produced by testing.

The following values for partial safety factors for materials may be obtained from 2.3.3.2 and 5.1.10 of the EUROCOMP Design Code:

for the adherends $\gamma_{m,\ ADHEREND} = 1.65$

for the adhesive $\gamma_{m,1} = 1.25$
$\gamma_{m,2} = 1.25$
$\gamma_{m,3} = 1.0$
$\gamma_{m,4} = 1.0$

$\Rightarrow \quad \gamma_{m,\ ADHESIVE} = 1.56$

The following value for the partial safety factor for the action may be obtained from 2.3.3.1 Table 2.2 of the EUROCOMP Design Code (Permanent action only, unfavourable effect):

$$\gamma_f = 1.35$$

The rigorous procedure

Step 1 Define the lap length using Equation (5.10) and Figure 5.37 both in the EUROCOMP Design Code

$$\frac{\beta}{t} = \sqrt{\frac{8\ G_a}{E\ t\ t_a}} = \sqrt{\frac{8 \times 371}{15000 \times 0.004 \times 0.0005}} \approx 315$$

$$\Rightarrow c = 77 \text{ mm} \Rightarrow L = 2\ c = 154 \text{ mm}$$

Step 2 Calculate the maximum adhesive shear stress using Equation (5.11) of the EUROCOMP Design Code

$$\tau_0 = \frac{\sigma}{8}(1 + 3k)\sqrt{8\frac{G_a\ t}{E\ t_a}}$$

where

$$\sigma = \frac{60 \times 1.35}{0.004} \text{ kN/m}^2 = 20.3 \text{ N/mm}^2$$

$$u_2 = \frac{1}{\sqrt{2} \times 0.004}\sqrt{3(1 - 0.3^2)\frac{20.3}{15000}} \text{ 1/m} \approx 10.73 \text{ 1/m}$$

$$u_1 = 2\sqrt{2}\ u_2 \text{ 1/m} \approx 30.35 \text{ 1/m}$$

$$k = \frac{(\cosh(10.73 \; x \; 0.077) \; x \; \sinh(30.35 \; x \; 0.154)}{M + N}$$

$$where\ M = \sinh(30.35 \; x \; 0.154) \; x \; \cosh(10.73 \; x \; 0.077)$$

$$N = 2\sqrt{2} \; \cosh(30.35 \; x \; 0.154) \; x \; \sinh(10.73 \; x \; 0.077)$$

$$Hence\ k = 0.34$$

$$\Rightarrow \tau_0 = \frac{20.3}{8} (1 + 3 \text{ x } 0.34) \sqrt{8 \; \frac{371 \text{ x } 0.004}{15000 \text{ x } 0.0005}} \; \text{N/mm}^2 = 6.5 \; \text{N/mm}^2$$

Step 3 Magnitude of the shear stress

$$\tau_{o,\ ALLOWABLE} = \frac{T_{0,k}}{\gamma_{m,\ ADHESIVE}} = \frac{11 \; \text{N/mm}^2}{1.56} = 7.1 \; \text{N/mm}^2$$

$$\tau_{0,max} = 6.5 \; \text{N/mm}^2 < \tau_{0,\ ALLOWABLE} = 7.1 \; \text{N/mm}^2$$

The design condition is satisfied in respect of/or respect to the adhesive shear stresses.

Step 4 Calculate the maximum peel stress

Define λ from Equation (5.14) of the EUROCOMP Design Code

$$\lambda = \frac{c}{t} \left(\frac{6 \; E_a \; t}{E \; t_a} \right)^{\frac{1}{4}} = 14.34 > 2.5$$

Calculate the maximum peel stress from Equation (5.15) of the EUROCOMP Design Code

$$\frac{\sigma_{0,max}}{\sigma} = \frac{k}{2} \sqrt{6 \; \frac{E_a}{E} \; \frac{t}{t_a}} + k' \; \frac{t}{c} \sqrt{6 \; \frac{E_a}{E} \; \frac{t}{t_a}}$$

where

$$k' = k \; \frac{c}{t} \sqrt{3 \; (1 - \nu^2) \; \frac{\sigma}{E}} = 0.40$$

and then

$$\frac{\sigma_{0,max}}{\sigma} = 0.31 \quad \Rightarrow \quad \sigma_{0,max} = 6.3\ \text{N/mm}^2$$

Step 5 Magnitude of the peel stress

$$\sigma_{0,\ \text{ALLOWABLE}} = \frac{\sigma_{0,k}}{\gamma_{m,\ \text{ADHESIVE}}} = \frac{30\ \text{N/mm}^2}{1.56} = 19.2\ \text{N/mm}^2$$

$$\sigma_{z,\ \text{ALLOWABLE}} = \frac{\sigma_{z,k}}{\gamma_{m,\ \text{ADHEREND}}} = \frac{20\ \text{N/mm}^2}{1.65} = 12.2\ \text{N/mm}^2$$

and

$$\sigma_{0,max} < \sigma_{0,\ \text{ALLOWABLE}}$$

$$\sigma_{z,max} < \sigma_{z,\ \text{ALLOWABLE}}$$

The design condition is satisfied with respect to the adhesive peel and adherend out-of-plane stresses.

5.4 COMBINED JOINTS

5.4.1 Bonded-bolted joints

A bonded-bolted joint combines adhesive bonding and mechanical fastening. Functions of both elements and possible applications are discussed in this section.

Generally, combining bonding and bolting will not improve the joint strength compared to that of a well-designed undamaged bonded joint. Since a structural adhesive provides a much stiffer load path than the fasteners, the load is carried by the adhesive almost entirely and no load sharing between the adhesive and the bolts occurs. As a result, the strengths of bonded and bolted joints cannot be simply combined in order to get the bonded-bolted joint strength, and the design of bonded-bolted joints should be performed according to the procedures used in the design of adhesively bonded joints.

A typical shear stress or strain distribution in an adhesively bonded lap joint is shown in Figure 5.38. If fasteners are placed in conventional positions (i.e. metal bolted connections), they are located in the area of minor shear stresses. Therefore, when the joint is undamaged the fasteners will be only moderately loaded.

Fasteners will be significantly loaded only in a damaged joint. If, as an example, debonding takes place at the end of an overlap, as shown in Figure 5.44, the outermost fastener will be loaded. Then the edge distance e has to be long enough to prevent local shear-out failure.

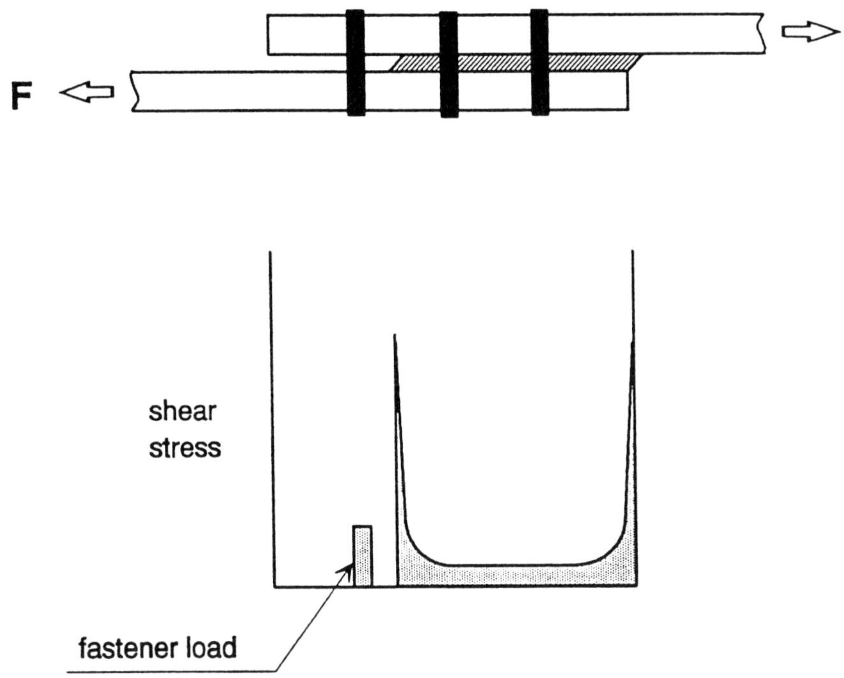

Figure 5.44 Load distribution in a bonded-bolted joint with a joint end debond.

All this would suggest that using bolts in a bonded joint is of little practical benefit. However, there are some cases when the combination of bonding and bolting can be justified.

Bolting may be advantageous in the repair of defective bonded joints and in the in-service repair of damaged structures. In both cases bolts transfer significant loads in the debonded areas and reduce peak stresses in the adhesive, since they enable the adhesive to be stressed to a higher level than before the repair. Bolts can also provide the clamping pressure required in bonding. This is often beneficial in repairs.
In lightly loaded bonded-bolted joints the bolts can provide a fail-safe load path in the event of adhesive failure. In heavily loaded structures mechanical fasteners can effectively prevent the spreading of debonding or delaminations from local damage or initial faults.

Peel stresses exist at the ends of adhesively bonded lap joints, especially in single-lap joints. To reduce these undesirable stresses the use of mechanical fasteners near the joint ends is sometimes recommended. As high peel stresses occur only in the immediate vicinity of the joint ends, the fasteners have to be located very close to the joint ends and cannot provide a fail-safe load path in the case of debonding. Therefore, the use of mechanical fasteners to reduce peel stresses is not recommended here. Instead the bonded joint should be correctly designed to reduce the peel stresses to an acceptable level.

When mechanical fasteners are used with bonding, a better damage tolerance is obtained. However, the weight of the joint structure is also increased, joining becomes more complicated and expensive, and even reductions in joint strength are possible. All these aspects have to be considered when a bonded-bolted joint is to be used instead of a bonded joint.

There are cases where the adhesive is not used primarily to transfer loads but is there to act as a sealant. These joints are typically lightly loaded. Therefore, commonly optimum lap lengths are not applied. Using bonded-bolted joints can then provide a fail-safe structure, especially in fatigue loading. Typically the failure of the joint initiates at the edge of the adhesive as a result of the combination of peel and shear stresses. A crack or cracks grow until the adhesive ceases to carry any load. At this stage the sealing effect is also lost. The bolts can take the moderate joint load and the structure will not fail totally. An increase in joint flexibility is encountered.

It may be possible to design a structure where service loads are carried by the adhesive and the ultimate loads by bolts, for example in the case of bond failure. Typically, the adhesive provides a stiffer load path with smaller deformations than the bolts. When deformations are the critical factor at the serviceability limit state, a bonded joint could be designed to fulfil these requirements. The same joint could then be designed as a bolted joint in respect to the ultimate limit state requirements. The final configuration and geometry of the joint would be determined according to the limit state design, which is more conservative with respect to both bonded and bolted joint designs.

There are no simple analytical methods for analysing bonded-bolted joints. The static strength of an undamaged bonded-bolted joint can be analysed identically as for the corresponding bonded joint. When the effect of defects needs to be evaluated, computer codes or finite element (FE) packages are required. Hart-Smith (reference 5.45) has developed a FORTRAN code called A4EK for analysing intact and flawed bonded-bolted step-lap joints with linearly elastic adherend deformations.

Finite element methods can also be used to analyse bonded-bolted joints. A high stress concentration exists in angular corners of the adherends. This phenomenon is identical to that encountered in the analysis of bonded joints. Non-linearities in the adhesive and adherends have to be considered. With an FE analysis a more detailed picture of the behaviour of a bonded-bolted joint may be reached than with closed-form analytical methods. However, perfoming FE analyses can be time consuming and their accuracy is dependent on the accuracy of the model, as discussed in 5.3.2.2.

5.4.2 Bonded-riveted joints

No additional information.

REFERENCES

5.1 *Standard Terminology of Adhesives*, ASTM D 907 - 82, Volume 15.06, 1987.

5.2 Skeist, I. (ed.) *Handbook of Adhesives*, 2nd edn, Van Nostrand Reinhold, New York, 1977.

5.3 *Engineered Materials Handbook,* Volume 1, Composites, ASM, USA, 1987.

5.4 Lubin, G., *Handbook of Composites,* Van Nostrand Reinhold Company, New York, USA, 1982.

5.5 Eurocode 3: Design of steel structures - Part 1.1: General rules and rules for buildings, Reference No. ENV 1993-1-1 : 1992 E, CEN.

5.6 *Structural Plastics Design Manual,* ASCE Manuals and Reports on Engineering Practice No. 63, American Society of Civil Engineers, 1984.

5.7 Matthews, F.L. and Godwin, E.W., A review of the strength of joints in fibre reinforced plastics, *Composites,* July 1980.

5.8 Hart-Smith, L.J., Design and analysis of bolted and riveted joints in fibrous composite structures, Douglas paper 7739, presented to International Symposium on Joining and Repair of Fibre-reinforced Plastics, Imperial College, London, 1986.

5.9 Hart-Smith, L.J., Mechanically-fastened joints for advanced composites - phenomenological considerations and simple analyses, Douglas paper 6748A, presented to Fourth Conference on Fibrous Composites, Structural Design, San Diego, California, 1978.

5.10 Collings, T.A., The strength of bolted joints in multi-directional CFRP laminates, Aeronautical Research Council, Current paper 1380, 1977.

5.11 Eriksson, I., Bäcklund, J. and Möller, P,. Design of multiple-row bolted composite joints under generalised loading, accepted for publication in *Composites Engineering Journal.*

5.12 Huth, H., Influence of the fastener flexibility on the prediction of load transfer and fatigue life of multiple-row joints, in *Fatigue in Mechanically Fastened Composite and Metallic Joints,* (ed. John Potter), ASTM STP927, American Society for Testing and Materials, Philadelphia, 1986, pp. 221-250.

5.13 Tate, M.B., and Rosenfeld, S.J., Preliminary investigation of the loads carried by individual bolts in bolted joints, NACA TN1051.

5.14 Jarfall, L., Shear loaded fastener installations. Report KHR-3360t Saab-Scania AB, Linkoping, 1986.

5.15 Whitney, J.M., and Nuismer, R.J., Stress fracture criteria for laminated composites containing stress concentrations. *Journal of Composite Materials,* **8**, 253-265, 1974.

5.16 Jones, R.M., *Mechanics of Composite Materials,* Scripta Book Company, Washington, D.C., 1975.

5.17 Adams, R.D. and Wake, W.C., *Structural Adhesive Joints in Engineering,* Elsevier Applied Science Publishers., 1984.

5.18 Matthews, F.L. (ed.), *Joining of Fibre-reinforced Plastics,* Elsevier Applied Science Publishers., Great Britain, 1987.

5.19 Kinloch, A.J., *Adhesion and Adhesives* (Science and Technology), Chapman and Hall, 1987.

5.20 *Engineered Materials Handbook,* Volume 3, Adhesives and Sealants, ASM International, Metals park, Ohio, USA, 1990.

5.21 *Composites Design Handbook for Space Structure Applications,* Volume 2, European Space Agency, ESA PSS-03-1101, Issue 1, Nordwijk, The Netherlands, June 1988.

5.22 Shields, J., *Adhesives Handbook,* 3rd edn, Butterworth, Great Britain, 1984.

5.23 *Structural Plastics Design Manual,* ASCE Manuals and Reports on Engineering Practice No. 63, American Society of Civil Engineers, 1984.

5.24 Hart-Smith, L.J., Lecture material, Three day course on joining of composite materials, 27 - 29 November 1990, The Royal Institute of Technology, Stockholm, Sweden.

5.25 Hart-Smith, L.J., Adhesive bonded single lap joints, NASA CR-112236, NASA Contractor Report, Langley, USA, 1973.

5.26 Hart-Smith, L.J., Adhesive-bonded double-lap joints, NASA CR-112235, NASA Contractor Report, Langley, USA, 1973.

5.27 Lees, W.A., Adhesive selection, Papers from a one day seminar organised jointly by the *International Journal of Adhesion and Adhesives* and RAPRA Technology Ltd., 7th July 1988, Butterworth.

5.28 Waterman, N.A., Ashby, M.F. and (ed), *Elsevier Materials Selector,* Elsevier Applied Science, London, 1991

5.29 Goland, M. and Reissner, E., The stresses in cemented joints, *Journal of Applied Mechanics,* March 1944.

5.30 Hart-Smith, L.J., Analysis and design of advanced composite bonded joints, NASA CR-2218, NASA Contractor Report, Langley, USA, 1974.

5.31 Hart-Smith, L.J., Further developments in the design and analysis of adhesively-bonded structural joints, in *Joining of Composite Materials,* (Ed. K.T. Edward), ASTM STP 749, American Society for Testing and Materials, 1981.

5.32 ESDU, Shear stresses in the adhesives in bonded joints. Single step double lap joints loaded in tension., Engineering Science Data Unit Report No. 78042, London, 1978.

5.33 ESDU, Inelastic shear stresses and strains in the adhesives bonding lap joints in tension or shear, Engineering Science Data Unit Report No. 79016, London, 1979.

5.34 Anderson, G.P., Bennett, S.J. and DeVries, K.L., *Analysis and Testing of Adhesive Bonds,* Academic Press, New York, 1977.

5.35 Russell, S.G., Nonlinear analysis of an adhesive-bonded joints under generalised in-plane loading, *AIAA Journal,* **28,** (No.12), December 1990.

5.36 Groth, H., Shear lag analysis and testing of adhesive joints with viscoelastic and viscoplastic adhesives, *Proceedings of Nordic Composites and Sandwich Meeting*, Royal Institute of Technology Stockholm, Nov. 19-20 1991, Organised by Nordic Industrial Fund and KTH.

5.37 Lu, X.S. and Jin, D., Structural design and tests of a trial GRP hull, *Marine Structures,* 3, 1990, pp 133-148.

5.38 Lees, W.A. (ed.), *Adhesives and the engineer,* Mechanical Engineering Publications., London, 1989.

5.39 Shenoi, R.A. and Hawkins, G.L., Influence of material and geometry variations on the behaviour of bonded tee connections in FRP ships, *Composites,* **23**, (No.5), September 1992, pp 335-345..

5.40 Hawkins, G.L. and Shenoi, R.A., A parametric study to determine the influence of geometric variations on the performance of a bulkhead to shell plating joint, *Proceedings of the Ninth International Conference on Composite Materials,* Madrid, 12-16 July, 1993, Vol IV, pp97-104.

5.41 Brander, T.J., Bonded joint tests for EUROCOMP, Report LLS-94-T103, Helsinki University of Technology, Laboratory of Lightweight Structures, Otaniemi, 1994.

5.42 *Adhesive Bonding Handbook for Advanced Structural Materials,* European Space Agency, ESA PSS-03-210, Issue 1, Nordwijk, The Netherlands, 1990.

5.43 ESDU, Guide to the use of Data Items in the design of bonded joints, Engineering Science Data Unit Item No. 81022, London, UK, 1981.

5.44 Cowley, P. and Adams, R.D., Defect types and non-destructive testing techniques for composites and bonded joints. *Materials Science and Technology*, **5,** 413-425. 1989.

5.45 Hart-Smith, L.J., Design methodology for bonded-bolted composite joints, USAF Contract Report AFWAL-TR-81-3154, Vol. 1, February, 1982.

CONTENTS

6 CONSTRUCTION AND WORKMANSHIP

6 CONSTRUCTION AND WORKMANSHIP

6.1 OBJECTIVES

6.1.1 Design and coordination

The purpose of these clauses is to ensure that the design of the FRP components is properly coordinated with that of the structure or building of which they are to form part and that any assumptions made in the design are carried through the manufacturing EUROCOMP Design Code, assembly and erection processes. It is not the purpose of the Code to assign specific responsibilities to individual members of the design and construction team. This should, however, be dealt with in the Specification for the project.

6.1.2 Proposed method of fabrication and erection

These clauses apply whatever the function of the FRP components, the materials used in their manufacture and the method and workmanship used in the manufacture.

Recommendations and procedures covered in the text that follows are mainly concerned with small scale production systems, either permanent or temporary, off site and on site. Whilst the principles are applicable to all work and processes involving FRP, they are not intended to be self-sufficient or unique.

6.1.3 Production of mock-ups and samples

No additional information.

6.2 MANUFACTURE AND FABRICATION

6.2.1 Quality of work

Where special properties are required, these should be obtained by selecting special resins rather than by use of additives by the fabricator. Significant exceptions are:

- mixing of pigments with resin where the quality required is not sufficient for premixing by the manufacturer
- where thixotropic properties are required for local vertical surfaces (e.g. returned edges) within a panel, when small quantities of silica flour, or similar approved material, can be added
- wax to prevent air inhibition of flow coat resins.

6.2.2 Workshop conditions

The general "housekeeping" in the workshop should be of the highest standard possible. The presence of scrap materials, empty drums, discarded resin mix containers, rags, trimming off-cuts, etc. not only contributes to poor working conditions but also increases the risk of contaminations of work in progress. In particular, contamination of the gelcoat arising from poor housekeeping can markedly influence the performance of a laminate exposed to weathering conditions.

Tables for cutting and tailoring glass fibre reinforcements should possess a clean surface and be sufficiently large to avoid the risk of the reinforcement trailing on the shop floor and becoming contaminated.

The moulds should be clean and dry and positioned in an area free from draughts, dust and dampness. As far as practical the mould should be distant from any areas where sanding or cutting is in progress, particularly in the gelcoat stage, when dust contamination can be most serious.

Fabrication workshops should be kept at near constant temperature of 21°C, to reduce and control curing and thermal movements of plastic materials during the manufacture of structural elements and units. A temperature range of ±3°C about a chosen mean internal factory temperature is suggested.

At all times the relative humidity should be such that condensation on moulds and materials is avoided. The relative humidity of an atmosphere is influenced by temperature changes, falling with increase in workshop temperature. For this reason, the materials should be allowed to reach workshop temperature after transfer from stores, before being put into use.

Exposure of glass fibre reinforcement to the workshop atmosphere before acclimatisation can result in the condensation of moisture from the atmosphere upon the fibre. Reinforcements should be retained in their plastic wrappings until required for both cutting and use.

Unless moulds are to be specially heated, the temperature should be as near as possible to that of the workshop. Accordingly, where moulds have been stored outside, sufficient time should be allowed for the mould to reach the workshop temperature. Failure to allow sufficient time to acclimatise may result in condensation on the mould and/or undercure of the gelcoat.

As a guide, a 200 litre drum of resin may require 48 hours or even longer to acclimatise to the workshop temperature when transferred from outside or from a cold store.

Adequate lighting levels to achieve manufacture and joints of fabricated elements and inspection of the finished product should be provided in workshops. The appropriate Factory Acts in force regarding lighting levels of illumination and glare must be met. A level of illumination of

500 lumens/m^2 falling on the work surface should be suitable for most operations.

Provision should be made for adequate equipment to measure accurately both the quantity of resin and the amounts of catalyst and accelerator added. Separate equipment for measuring the latter materials is essential.

Resin containing fillers, pigments or additives must be thoroughly stirred before use in case any settling has occurred during storage.

Equipment for blending resins with catalyst and accelerator should ensure that such blending is uniform. However, over vigorous blending which results in air becoming dispersed in the resin mix should be avoided. In particular very high speed propeller mixers which create a vortex and aerate the resin should not be used.

Suitable release agents should be applied to the mould. In the case of wax release agents, a heavy build up should be avoided and all wax applied should be well polished. If a polyvinyl alcohol release agent is employed, care should be taken to ensure that this is adequately dried before proceeding to the application of the gelcoat. In all cases, it is essential that the manufacturer's instructions and recommendations be followed.

All equipment should be regularly cleaned and serviced, and frequent checks made to ensure the liquid components are being mixed and applied in the correct proportions.

6.2.3 Manufacturing accuracy

Clause 6.2.3 P(1) in the Eurocomp Design Code is a general good practice requirement. However, by itself it will give only limited control over the appearance of the finished product, it being necessary to consider inaccuracies in the building structure and in erection of the FRP units.

The requirement that at least two-thirds of measurements fall within one-third of the permissible deviations is based on a statistical approach; in practice the degree of accuracy to be aimed for should be far higher than might be assumed from the figures in table 6.1 of the EUROCOMP Design Code.

Smaller tolerances on the cut lengths than those given in the table can be achieved but should be specified in advance to the fabricator.

Reinforced plastic can be cut on site using a diamond-tipped cutting disc. It is recommended that the blade's peripheral velocity is 35 to 40 m/s.

Wet and dry joints filled with adhesive resin should have a filled gap size that reflects the properties of the jointing detail and adhesive type, with regard to filling the gap and minimising subsequent deformation. Generally, smaller filled gaps between joined parts are to be preferred, though larger gaps filled with adhesive may perform satisfactorily. Any filled joint greater than 12 mm in width should be justified by tests.

Fabricated units should be kept as large as possible, provided that they can subsequently be transported and erected on site. Several smaller moulds producing units to suit the construction programme may dictate the size of the modular unit.

Consideration should be given to methods of controlling costs in the manufacturing process, e.g. to the re-use of forms, dies and other equipment so as to standardize methods of producing polymeric composites for further structures.

Where members may become distorted during manufacture, consideration should be given to the units being completely fabricated within a jig, and removed only when the adhesive has reached the stipulated strength.

For plastics to be competitive with other structural materials (i.e. timber, steel, reinforced concrete, and aluminium) mechanised production of standard simple geometric units should be used, which, when joined together, produce the designed structure.

Variations to the basic unit should be avoided where possible. Many small variations will add to the overall cost.

6.2.4 Gelcoats to external surfaces (applies also to resin flowcoats)

The gelcoat must be as free from entrapped air as possible, with the catalyst mixed in homogenously.

The gelcoat may be applied by any suitable method that will yield the specified thickness.

Gelcoat resins as supplied by the resin manufacturer should not be diluted with acetone or other inert solvents either to render them sprayable or for any other reason.

If the gelcoat is too thin it may not fully cure and the glass fibre pattern may show through from the backing laminate. If it is too thick, it may crack or craze and will be more sensitive to impact, particularly from the reverse side of the laminate. A gelcoat of uneven thickness will cure at different rates over its surface. This causes stresses to be set up in the resin which may lead to crazing and in the case of pigmented gel coats, a patchy appearance. A thin coat (or thin parts) will be very prone to colour fading.

The required thickness is usually applied in a single coat but occasionally double gel-coating may be necessary, e.g.:

- on vertical mould surfaces, to prevent "sagging";
- where the chosen colour is not very opaque. (This problem can often be overcome by using a pigmented laminating resin.)

The tests for determining wet film thickness are primarily intended for paint, but can also be used for the much thicker gel coats. A wheel gauge or a comb gauge may be used; neither will give more than an approximate measurement, hence the fairly wide range of acceptable readings.

Unlike paint films, polyester resins are thermosetting, so there is little difference between the wet and dry film thicknesses (about 5%). As a rough guide 500-650 g/m² of gel coat mixture will give the recommended thickness of 450-550 microns).

6.2.5 Laminating

A lightweight mat, not greater than 300 g/m², may be used behind the gel coat in order to minimise the risk of air entrapment at the back of the gelcoat. This applies to both hand laminating and spray depositing techniques.

Subsequent laminate plies should be maintained as free of air inclusions as possible and should provide a good bond to foregoing plies. If laminating is interrupted overnight or over the weekend, the recommendations of the resin supplier should be sought with regard to any surface preparation which may be necessary before lamination recommences.

When processes other than contact lamination are used, similar criteria will apply, i.e., elimination of air, temperature control and gelcoat bonding.

If, unavoidably, the gelcoat must be left before laminating starts, then the mould should be protected to prevent contamination of the gelcoat.

Glass fibre is laid on the gelcoat and wetted out with resin. It is important that the laminate adheres to the gelcoat and therefore the gelcoat should still be tacky when laminating commences (gelcoated moulds should not be left overnight). It is important that each layer of the wet laminate is properly consolidated to remove air by using a roller.

Further layers of mat or woven fabric are added until the face skin is complete. 450 g/m² layers of chopped strand mat are commonly used. Lighter weights may be used to laminate complex or fine details. In general any finished FRP skin should have not less than 900 g/m² of glass fibre. Each additional layer must be added whilst the preceding layer is still tacky, otherwise delamination may occur in use.

Any fixings to the rear of a single skin construction are applied and overlaminated, and followed by ribs (if any) held against the laminate by weights.

A layer of tissue glass fibre may be applied to smooth the internal hand finished surface. A flow coat of unreinforced resin is then applied to the whole of the rear of the skin as the internal equivalent of the gelcoat.

Core materials and sandwich plies must be placed on to an adequate layer of resin and then firmly and evenly weighted to ensure full contact with the outer skin until bonded. Care needs to be taken to provide escape routes for entrapped air. They may be held against the wet resin by using a vacuum or pressure bag, press or autoclave instead of weights - this is a very efficient method. In situ foaming of core material can be used instead of ready foamed slabs, especially on difficult curves or for gap filling.

Inserts such as fixings or light fittings can be located within the core if necessary. Small timber, metal or solid laminate blocks may, for example, be inserted for subsequent screw fixings.

The rear of the core material is coated with resin and further layers of glass/resin added as required, ensuring that the inner and outer skins are balanced to avoid bowing of the unit. Finally a flow coat is applied to the whole of the rear skin.

6.2.6 Curing

For optimum performance, it is important that the laminate should be properly cured in accordance with the instructions and recommendations of the resin manufacturer, as these may be different for different resin types, manufacturers and applications. The following information should therefore be used only for guidance purposes.

Curing should be carried out, generally within the temperature range of 16°C to 20°C for a period of approximately seven days before leaving the controlled environment of the workshop. The cure temperature should never fall below 10°C. When the temperature exceeds 20°C, a shorter maturing period may be employed. Consultation with the resin supplier is essential to establish the extent of reduction of the period of cure.

Should space considerations make it necessary to remove the laminate to the outside before adequate cure is achieved, then it is necessary to give consideration to post-cure at elevated temperature and to maintain proper jigging throughout the cure cycle. In general post cure at 40°C for 16 hours should prove sufficient.

The resistance of a laminate to colour change on exposure is strongly dependent on the degree of cure before leaving the workshop. Exposure of adverse weather conditions, cold or moisture, before it is thoroughly cured will have an adverse effect on both the laminate strength and its weather resistance.

Premature release from the mould before cure is well advanced may lead to the appearance of a fibre pattern in the gelcoat and residual distortion, particularly of large panels.

6.2.7 Records

No additional information.

6.3 DELIVERY AND ERECTION

6.3.1 Information to be provided

This clause is likely to be relevant on most jobs to ensure that the FRP manufacturer is closely involved in the on-site operations. Even where the manufacturer is a nominated sub-contractor, he may make informal arrangements for use of the main contractor's or other local labour.

6.3.2 Preparation of FRP units for transportation

FRP units must be adequately secured during transportation to prevent movement.

To prevent over-stressing of large units during handling and transportation temporary stiffening supporting jigs should be provided where necessary.

FRP units being generally low in weight should not present lifting problems in the factory or on site. Careful handling of FRP units at all stages of movement must be observed in the factory, during transport, and during erection on site. Large units should be housed in cradles during transporting to the site from the factory.

Placing large units in containers reduces the risk of damage during transport by rail and sea. Transporting units as open deck cargo by ship is not recommended.

Temporary supporting cradles should be strong and stiffly cross-braced to prevent distortion or impact damage during movement or lifting of a unit. A painted steelwork framework is suggested for the supporting cradle with good racking resistance and high stiffness. However, care must be taken in the design of cradles made of materials other than the fabricated plastic composite material. For example, differential temperature movement between a steel or aluminium support cradle and the FRP unit being housed within the metal framework of the cradle must be allowed for.

The design of the fabricated units should include lifting points and brackets where possible to assist with lifting. Where required, to reduce the effects of damage to plastic elements and finishes during movement, lifting procedure should be agreed with the fabricator, providing special lifting harnesses, polyester and nylon slings, and rubber protection to metal slings, etc.

6.3.3 Protection

Units should be stored on a flat level surface free from sharp protrusions, stones, bricks, etc. and adequately supported to prevent local damage. Whilst in the storage area, the units should be protected from impact damage such as by fork lift trucks or cranes. Units should be secured to prevent movement due to wind load, and temporary anchors should be provided where necessary.

Degradation of stored polymeric composites should be minimised, especially in poor environments. Protection of materials and manufactured units should be provided, with separation between surfaces to avoid direct contact or frictional damage. Wrapping, packaging and crates may be used as separators.

Where materials are not designed for external environments all stacked units should be protected from the elements by waterproof and opaque coverings. The bottom layer should be placed on packings, such as clean untreated softwood timber, in order to keep the products clean and dry and prevent scratching to finishes.

It may be necessary to limit the number of products in a stack, to avoid overstressing of products and high contact stresses causing damage.

6.3.4 Accuracy of erection

Lifting fixed points and brackets provided in the factory could be used for further lifting of units on site. These lifting points may remain in the final structure, provided that agreement is given by the designer.

Lifting fixings must not be added to the completed factory unit on site without the agreement of the designer.

A method statement should be prepared and agreed with the designer and a copy kept on site. This erection method statement should be prepared by the erection contractor and include a programme. Use of temporary braces, ties and ropes should be listed in the Method Statement. Erection drawings should be supplied to site where necessary. The sequence of the erection of the units should also be given.

Temporary roof covering to protect the structure and operatives locally from the elements (e.g. rain) should be provided while joints are being made and until the curing of the adhesive is complete.

Connected parts should be drawn carefully together (e.g. where bolted connections are specified). Gaps remaining after tightening that will affect the integrity and load-carrying capacity of a joint must be adjusted where possible or given remedial treatment to the satisfaction of the design engineer.

Gaps remaining in completed adhesive joints should not be permitted. Defective adhesive joints/members containing gaps should be replaced or given remedial treatment (see section 5.3).

A site plan showing the datum level and setting-out grid lines should be produced and attached to the Method Statement for erection.

A suitable level hardstanding area free of standing water for the erection crane should be provided for the crane supplier, to suit the crane manufacturer's recommendations. Details of underground services and overhead cables and other site obstructions should be obtained for the erection contractor before work commences. Extra care should be taken with long carbon composite members that are lifted near live exposed electrical wires, as these composite plastics, like many metals, will conduct electricity.

Requirements for any temporary works, temporary bracing and propping should be given in the Erection Method Statement (see 6.1.2). The stage when temporary bracings/supports can be removed, and the erected structure is able to accept all loadings, including wind, must be defined in the Method Statement.

All panels should be placed in position in such a way as to ensure that loads are distributed as intended in design, e.g. using prepared anchor blocks or metal brackets.

6.4 CONNECTIONS

6.4.1 Bolted connections

FRP cannot be welded or easily shaped after fabrication. Screws are not recommended if the primary role of the joint is structural, because both self-tapping screws and tapped-hole screws cause extensive damage to the laminates. If screws are to be used for structural applications, it should be possible to cast metal blocks into the composite at the time of manufacture or to mount metal blocks on the composite and then to drill and tap into them in the conventional way.

Because the process of forming the rivets involves applying a lateral pressure to the composite, there may be cases in which the riveting operation causes damage to the laminate, especially when using blind rivets. The clamping force exerted by the rivets also affects the strength of the joint. If countersunk screws are to be used to ensure a smooth external surface, they must not be allowed to cut completely through the laminate. This will generally imply a restriction on the minimum thickness of the composite. Care must be taken in tightening joints to avoid driving washers into and crushing the laminate, particularly in the case of sandwich panels. But, provided excessive over-tightening of the joint does not take place, no damage should be done to the composite during assembly.

It is important to locate accurately the relative positions of the holes in the composite materials to be joined, so that high stresses leading to premature failure during assembly are avoided, either by accurate templates, clamping the units to be bolted together and drilling them in the workshops or drilling and bolting as one operation on site.

Long term creep in the structure could reduce the initial tightening of bolted joints and this should be considered in the design.

Large washers are sometimes required from design considerations, and the designer and fabricator must provide sufficient spacings between bolts that lapping of washers is eliminated.

Bolt hole sizes should not be specified smaller than necessary and the use is of large washers recommended.

Where close tolerance bolts are specified, holes must not be enlarged beyond the specified size.

6.4.2 Adhesively bonded and laminated joints

It is desirable to have the free end of the adherend in an adhesive joint tapered to reduce the stress concentration and avoid premature failure by induced peel stresses.

Adhesive joints tend to form a fillet of adhesive spew, owing to the squeezing of the adherents during bending. The existence of a spew fillet can reduce the stresses at the free end appreciably, because of a stress relief mechanism, so unless otherwise agreed it is desirable to maintain a form of spew fillet.

The suggested procedure for bonding is as follows:

(1) remove grease and dust
(2) roughen the surface
(3) remove grease and dust
(4) spread the adhesive
(5) apply compression (this can be combined with screwed and riveted fastenings) using a compressive stress of about 0.1 N/mm^2
(6) allow adhesive to harden and cure in accordance with manufacturer's recommendations.

6.5 SEALANT JOINTS

All joints must be sealed against the ingress of moisture. Water may have an effect on the glass - resin bond, causing composites to fail well below their full strength. Chlorinated water (e.g. swimming pools) can bleach FRP pool liners, and weaken the bond of the composite. Moisture can cause deterioration of composites. Water absorption by resins may cause swelling and increases in thickness. Short term water absorption mainly affects the surfaces of FRP laminates in the case of sandwich panels and long term immersion in water leads to plasticising. Moreover, the water tends to fill any voids in the laminate and this can cause blisters at fibre - resin interfaces.

The compressive force in sealed joints is applied by means of bolts and the sealing material could be either a compressed foam or various forms of sealants and gaskets.

Compressible sealant strips can be used, provided the flanges are sufficiently rigid not to bow between bolting points.

The necessary properties of a sealant are:

- a good adhesion with the joint
- permanent elasticity
- low rate of hardening
- low rate of shrinkage
- durability.

The selection of sealants must be discussed in detail with sealant manufacturers. The final choice will, however, be a compromise, as no one product has all the above mentioned attributes.

6.6 REPAIR OF LOCAL DAMAGE

Repairs may be made by building up the section thicknesses at the defect as if making a laminated joint. Thicknesses of repair material should be at least equal to the thickness of the parent defective tube, bar, plate, etc. and in no case less than 3 mm. The procedure should be as set down in section 5.3 for bonded or laminated joints.

Bolted, screwed or riveted joints are difficult to repair and should be replaced. Consideration may be given to cutting out the defective joint and a length of the member and replacing them with a new joint and profiled length to make a splice at a place of low stress.

Minor repairs may be acceptable if the units are not located in a prominent position.

Even if repairs are made using the same materials as in the manufacture of the unit, the colour may not match because the unit will have a moulded finish, and the repair will have a coated finish. Colour/texture variations may be disguised by coating the whole unit with a two-pack polyurethane system matched to the correct colour. However, a painted finish will normally be less durable than a moulded gel coat, and periodic repainting will be necessary. Colour change through time would make colour matching increasingly difficult.

6.7 MAINTENANCE

6.7.1 Inspection

The importance of site protection to glass FRP components to avoid impact damage and damage due to rough handling cannot be overstressed (see 6.3.3). Much damage could be avoided by:

- asking the tenders to list the methods of storage and handling on site
- determining the minimum clearance requirement between the main structure and the scaffolding
- requesting that one of their operatives be present to supervise the unloading on site.

It is essential to establish a firm programme of erection (see 6.3.4). Especially important are corner units, infill sections and part sections which may have to be produced from an adaptation of existing moulds, by insertion of blocking pieces into the mould. Such mouldings should necessarily be left until last to prevent witness marks on flashings being evident on subsequent mouldings.

Although successful repairs can be made to FRP units, they will detract from the overall visual appearance and will be costly.

6.7.2 Washing

The responsibility for cleaning the FRP units after installation is often overlooked and should be clearly stated in the contract documents. A Method Statement should be included with tenders.

6.7.3 Painting

The painting of plastics is not straightforward. It can sometimes even damage the material. General guidance on choice of paint and methods of cleaning and painting is offerai but the advice of a reputable paint supplier should always be sought before commencing painting or refurbishing.

Before painting it should be ensured that the paint is compatible in order to achieve acceptable adhesion and durability. Given a suitable paint, plastics will generally provide a more stable base for painting than timber or metal.

Thermal movement of the panels should be considered when specifying paint colour. Plastic panels usually have a higher coefficient of thermal expansion than other building materials.

If existing installations in pale colours are painted in a darker shade (black in the extreme), thermal movement in sunlight will increase. This may place excessive strain on joints and fixings or cause permanent distortion or, in the case of thin laminates, matrix softening. Conversely excessive thermal movement in dark-coloured plastics will be reduced by painting them white.

It is recommended that non-yellowing (aliphatic) two-pack polyurethane paints of proven durability to weathering be used as finishing paints for FRP, to obtain maximum adhesion and life between repaints. These are available in a full range of colours and gloss levels and also in versions giving fine-textured finishes. The latter are a useful means of disguising an uneven surface produced by repair work and, unlike free-flowing types, will cover fine hairline cracks. These finishing paints normally adhere well if applied directly to FRP, but special primers (e.g. epoxide or moisture-curing polyurethane types) are recommended by some paint manufacturers. Brushable or roller-applied formulations are best for on-site use. Spray application would raise problems associated with the toxicity of the unreacted areas by wind-borne overspray. Freshly applied two-pack polyurethanes are particularly sensitive to damage by rain until drying and curing are substantially complete, when a very high degree of water resistance develops. Hence the weather must be judged carefully.

Where circumstances suggest the use of a more conventional type of finishing paint with better tolerance of site application conditions than the above, silicone alkyd enamels should be considered.

6.8 HEALTH AND SAFETY

Working with reinforced plastic profiles (i.e. FRP) can cause irritation to the eyes and skin by contact from dust and cuttings.

Personal protective equipment and clothing must be worn, including facemasks and gloves, when cutting and grinding with these materials. Additionally, when large quantities of cuttings and dust are produced mechanical extraction of particles must be provided in the factory or on site.

Air-borne contaminants must be prevented from escaping beyond the working area and affecting the general workforce or public. A high standard of personal hygiene must be practised by the workforce and adequate washing facilities must be provided.

Certain resins, seals, paints, dust and adhesives are causes of dermatitis. To avert this hazard, contact with the skin must be avoided. All loose materials must be handled carefully to avoid splashing. Suitable plastic gloves and correct barrier creams on forearms and hands should be used.

People working plastic, toxic and glass fibre materials should wash their hands before eating, smoking or using the lavatory. Gloves should be well washed before they are removed.

Individuals that become skin-allergic to the materials from repeated exposure should be given medical treatment for skin dermatitis before continuing work. All the appropriate factory and site working conditions acts in force must also be met.

Adequate ventilation must be provided at all times when using volatile materials (e.g. resins, paints and solvents) and in confined spaces mechanical extraction of fumes must be provided together with fresh air in the workplace in accordance with statutory requirements and good practice. Smoking should be prohibited in and around the workshops.

Substances that have a fibrous form present a potentially serious hazard to health if the fibres are of a certain shape (e.g. in the case of glass fibre) - see above for personal protective equipment and clothing which must be supplied to the workforce.

REFERENCES

(Note: These references have been used as the basis for the guidance in this chapter: they are not specifically cited in the text).

6.1 Glass reinforced plastics panel cladding/features (National Building Specification: H41 (1989))

6.2 Guidance notes for the construction of reinforced plastics cladding panels (British Plastics Federation: 287/1 (1981))

CONTENTS

7 TESTING

7 TESTING

7.1 GENERAL

Testing within the content of this Chapter includes all the various physical checks that may be required during production of the components and their fabrication into the finished structure, as well as on the structure itself. They may be purely visual checks, dimensional measurements, or chemical or physical tests to determine various specified properties. In general the purpose of the tests will be to check that the structure and its component parts conform with the initial specification.

While the primary role of testing is likely to be for compliance purposes, it will also be required as a part of the development of new materials, elements or structures. The very wide range of combinations of resins and fibres that can be envisaged means that the range of properties of the resulting composites will be far greater than would be available with conventional construction materials. While simple methods may be used to estimate some composites' properties, based on a knowledge of the individual fibre and resin properties, the results may be inaccurate. The more complex the laminate the less accurate the estimate. Hence there will be a need for significant testing, when new combinations of resins and fibres are being developed. There will also be considerable benefits to the designer, as Chapter 2 requires lower partial safety factors to be used when measured material properties are available.

When the properties of a particular laminate are well understood, less testing will be required when developing a new element. Load tests will be largely to confirm that the form of analysis used when considering the behaviour of the element was valid. It may also be appropriate to use load testing when the loading on the finished structure is complex or when the structural response is not fully understood. An example of the latter might be when the effect of the stiffness of the connections, and hence the degree of restraint, is unclear.

Although in the presentation of the EUROCOMP Design Code the two different reasons for carrying out testing, namely for compliance and for development testing, are treated separately, the actual tests that may be required will be the same in both cases. One significant difference may be in the number of samples tested in respect of each situation. For compliance testing, particularly in the case of the manufacture of a standard component, relatively few samples may be required at any one time, though they will be taken at regular intervals to check on the consistency of the process.

For development testing, it is likely that a greater number of samples will be tested at a single time, so that a good indication of the properties of a particular laminate, component or structure may be determined.

Figure 7.1 in the EUROCOMP Design Code illustrates the design and construction process and indicates the areas in which testing may be required, either for compliance purposes or as part of the design or development process. The point of entry into this figure will depend on the nature of the project and on the amount of data available on the material or components being used.

The amount of testing that is to be carried out for compliance purposes will generally be specified in the EUROCOMP Design Code or in agreed Standards to which it refers. However, it is essential that all parties are fully aware what testing is required at each stage of the construction process and that this should be agreed. Sufficient tests will be required to give all parties confidence in the applicability of the results to the completed structure. Again the number and type of tests must be agreed by all parties.

Testing, as said earlier, is to check that the structure conforms with the specification. Thus the quality of the testing must be such that the results can be accepted with confidence. This means that tests must be carried out in accordance with agreed Standards by suitably trained staff. This can be achieved by using ISO (International Standards Organisation) Standards and approved laboratories.

Published ISO Standards cover most of the tests that may be required both on the basic materials, glass, resins, etc., and on the finished product. In the following sections the ISO Standards applicable to the various stages of the production process are listed along with other European National Standards where they are exactly equivalent to the ISO ones. It should be noted that the list does not include all the standards that may be required in a particular situation. Many are subject to revision and will eventually become Euronorms.

Many ASTM (American Society for Testing and Materials) Standards are used for testing. However, it has been shown that the values they give for certain properties are different from those given by ISO tests. Hence the ASTM test methods should only be used where no equivalent ISO tests have been published. However, for convenience, a list of some of the relevant ASTM tests is given at the end of this chapter because they have been so widely used in the past.

The testing methods given in ISO and International Standards generally use relatively small and thin specimens. While these may be representative of the material used in some sectors of the aerospace industry, for which many of the tests were developed, they may not be strictly applicable to the thicker units that will be required for structures designed in accordance with the EUROCOMP Design Code. Thus while they may be appropriate for compliance testing, where the main purpose of the test is to ensure consistency, the results may not be strictly applicable when used as part of development testing. With complex laminates it may be preferable to use techniques based on the ISO ones but using larger test specimens, more representative of the real structure. To minimise differences, the larger test specimens should be of the same shape and have the same aspect ratios (for example ratio of length to thickness) as the standard specimens.

All testing work, along with the whole production process, should be carried out by suitably qualified staff under an approved quality scheme. The relevant ISO Standards are:

ISO 9000:1987	Quality Management and Quality Assurance Standards - Guidelines for selection and use
ISO 9001:1987	Quality systems - Model for quality assurance in design/development, production, installation and servicing
ISO 9002:1987	Quality systems - Model for quality assurance in production and installation
ISO 9003:1987	Quality systems - Model for quality assurance in final inspection and test
ISO 9004:1987	Quality management and quality system elements - Guidelines

[These are equivalent to EN29001/3.]

Testing should, where possible, be carried out by suitable laboratories or test houses having an accreditation under the ISO Standards listed above or National schemes such as BS 5750 or NAMAS in the UK. In all cases the test methods to be used should be agreed in advance by all concerned, along with the number of specimens to be tested, the test conditions, etc.

Guidance on the preparation and conditioning of test specimens is given in the following ISO Standards:

ISO 291:1977	Plastics - Standard atmospheres for conditioning and testing

[This approximates to BS 2782: Part 10: Method 1004 and to EN62.]

ISO 1268:1974	Plastics - Preparation of glass fibre reinforced, resin bonded, low-pressure laminated plates or panels for test purposes [see also BS 2782: Part 9: Methods 920A to 920C: 1977]
ISO 3167:1983	Plastics - Preparation and use of multipurpose test specimens [see also BS 2782: Part 9: Method 931A: 1988]

7.2 COMPLIANCE TESTING

7.2.1 During production

As stated earlier, the purpose of compliance testing is to ensure that the final product conforms to the specification in every aspect. The first step in this process is to check the constituent materials, i.e. the resins, fibres, fillers, etc., to ensure that they have the correct properties, appropriate to the production process that is to be used.

Similarly when the fibres are formed into woven mats, multi-axial fabrics and the like, compliance tests will have to be carried out to ensure that they are in accordance with the specification. These will be mainly visual and geometric checks as the checks on the properties of the glass fibres themselves will have been carried out already.

Where a woven or multi-axial fabric is produced with a specific application in mind, its location in the finished laminate may be known in advance. Where appropriate the manufacturer of the fabric should positively identify the location, for example by the inclusion of coloured threads.

Compliance tests on the laminates are chiefly intended as a check on the manufacturing process, as the checks on the constituent materials will have been carried out already.

When the manufacturer makes a standard range of products, he will normally supply values for the properties commonly required in the design process. Regular determination of these properties will check all aspects of the production process as well as, indirectly, the properties of the materials used.

When a strength for a standard element is specified by the manufacturer, tests will be required to demonstrate that the units conform. Generally, sufficient tests will be required so that a characteristic strength, i.e. one below which not more than one in twenty results fall, may be determined.

Visual checks are an important part of the process of testing for compliance. This applies both to the manufacturing processes and to the finished product. As detailed later, regular checks, carried out at an agreed rate and in accordance with an agreed plan, will ensure that a consistent standard is maintained.

The type of testing that is required will depend upon the nature of the structure of which the product will form a part and the intended use. For example, ultimate load tests would be inappropriate for an element that will only be lightly loaded. Conversely deflections may be critical and hence the stiffness must be known accurately.

The number of stages of the manufacturing process, from raw materials to finished product, will influence the amount and type of tests that may have to be carried out.

The following ISO Standards cover the properties of the glass fibre and of the rovings. The supplier of the glass should carry out the necessary tests and should supply the material with a certificate of conformity. In general the composites manufacturer will not be required to carry out any additional testing unless the particular specification to which the work is being carried out is more restrictive than that used by the supplier. Even in this case it may be preferable for the supplier to carry out the necessary tests and then certify that the product conforms to the manufacturer's specification.

ISO 1889:1987	Textile glass - Continuous filament yarns, staple fibre yarns, textured yarns and rovings (packages) - Determination of linear density
ISO 2078:1985	Textile glass - Yarns - Designation
ISO 2797:1986	Textile glass - Rovings - Basis for a specification
ISO 3341:1984	Textile glass - Yarns - Determination of breaking force and breaking elongation
ISO 3375:1975	Textile glass - Determination of stiffness of rovings
ISO 3598:1986	Textile glass - Yarns - Basis for a specification

The following ISO Standards cover the properties of mats and woven or multi-axial fabrics:

ISO 2113:1981	Textile glass - Woven fibres - Basis for specification
ISO 2559:1980	Textile glass - Mats (made from chopped or continuous strands) - Basis for a specification
ISO 3342:1987	Textile glass - Mats - Determination of tensile breaking force
ISO 3374:1990	Textile glass - Mats - Determination of mass per unit area
ISO 3616:1977	Textile glass - Mats - Determination of average thickness, thickness under load and recovery after compression
ISO 4602:1978	Textile glass - Woven fabrics - Determination of number of yarns per unit length of warp and weft
ISO 4603:1978	Textile glass - Woven fabrics - Determination of thickness

ISO 4605:1978	Textile glass - Woven fabrics - Determination of mass per unit area
ISO 4606:1979	Textile glass - Woven fabric - Determination of tensile breaking force and breaking elongation by the strip method

Tests dealing with the mechanical properties of the mat or fabric may be required by the manufacturer of the FRP product. However, they will generally be more appropriate as part of the testing associated with the development of a new laminate or a new manufacturing process (see 7.3).

The following ISO Standards cover the properties of the fibre-based materials used in the manufacturing process. All materials should be supplied with a certificate of conformity and hence, in general, no further testing will be required for compliance purposes, unless the specification specifically calls for it.

ISO 1886:1990	Reinforcement fibres - Sampling plans applicable to received batches
ISO 8604:1988	Plastics - Prepregs - Definitions of terms and symbols for designations
ISO 8605:1989	Textile glass reinforced plastics - Sheet moulding compound (SMC) - Basis for a specification
ISO 8606:1990	Plastics - Prepregs - Bulk moulding compound (BMC) and dough moulding compound (DMC) - Basis for a specification

Similarly, the resins, catalysts, etc., should be supplied with a certificate stating that they conform to the relevant ISO Standards, of which those relevant to the manufacture of the laminate are listed below. With a certificate of conformity, testing by the manufacturer of the laminate will generally be unnecessary. However, some testing may be required when developing a new process, or using new materials (see 7.3).

ISO 584:1982	Plastics - Unsaturated polyester resins - Determination of reactivity at 80°C (conventional method)
ISO 1628:1984	Determination of viscosity number [see also BS 2782: Part 7: Method 732A to F: 1991]
ISO 1675:1985	Plastics - Liquid resins - Determination of density by the pyknometer method [see also BS 2782: Part 6: Method 620E: 1980]
ISO 2555:1989	Plastics - Resins in the liquid state or as emulsions or dispersions - Determination of apparent viscosity by the Brookfield Test method

ISO 3219:1977	Plastics - Polymers in the liquid emulsified or dispersed state - Determination of viscosity with a rotational viscometer working at defined shear rate [see also BS 2782: Part 7: Method 730B: 1978 (1991)]
ISO 6186:1980	Plastics - Determination of pourability

The amount, location and orientation of the fibre in the finished product, and its location, will have a significant effect upon the final properties. For continuous processes, a continuous supply of reinforcement material should be ensured and monitored on a regular basis. For batch and single processes, a suitable system should be operated to ensure that the correct amount of reinforcement is used. One approach is to divide the material into suitable packages, with one package being used for each finished unit, rather than using a continuous supply and relying on the operator to determine the required quantity.

In all processes, care should be taken to ensure that any significant faults in the reinforcement material, which should have been identified by the supplier, are avoided or suitably compensated for.

Where a reinforcement fabric includes markings to identify its location in the finished product, for example by the use of coloured tape, such markings should ideally be visible at the end of the production process.

As well as measuring the dimensions of the finished product, determination of the density of a suitable sample will provide a check on the overall production process. This is covered by:

ISO 1183:1987	Plastics - Methods for determining the density and relative density of non-cellular plastics [see also BS 2782: Part 6: Methods: 620A to D:1991]

Visual inspection of the finished product will give a good indication of the quality. Defects will generally be due to some error in the production process or problems with the constituent materials. In continuous processes such as pultrusion minor adjustments can be made to rectify the defects subsequently.

The assessment of defects is highly subjective. The specification should state clearly what defects are permitted and at what frequency. The acceptable level will depend upon the use to which the product will be put. Some examples of defects are given in:

ASTM 2563:	Classifying visual defects in glass-reinforced plastic laminate parts

This includes photographs showing typical surface defects and descriptions of the causes of the defects. The document also defines four levels of visual acceptance. For the first three, the maximum size or number of defects is defined, while for level IV the permissible defect must be fully defined in the product drawing or in the contract. It may be that different levels of acceptance will apply to different parts of an element or structure, a more severe check being required for critical areas. The document also covers briefly the repair of defects, stating that acceptable methods must be agreed and specified. It should be stressed that the document deals only with inspection without the aid of magnification. There would appear to be no suitable ISO standard at present.

Dimensional checks should be carried out using standard measuring equipment, that has been calibrated under an approved scheme. For standard products, simple go/no-go templates may be used to check that the units are within the required tolerances. The allowable tolerances, given in Table 6.1 of the EUROCOMP Design Code, are based on current practice and not on International or National Standards.

Local tests on surface hardness are a good check on the overall quality of the production process. The following ISO tests give suitable methods:

ISO 868:1985 Plastics and ebonite - Determination of indentation hardness by means of a durometer (Shore hardness)
[see also BS 2782: Part 3: Method 365B: 1992]

ISO 2039-1:1987 Plastics - Determination of hardness - Part 1: Ball indentation method
[see also BS 2782: Part 3: Method 365D: 1991]

ISO 2039-2:1987 Plastics - Determination of hardness - Part 2: Rockwell hardness
[see also BS 2782: Part 3: Method 365E: 1992]

ISO 6603-1:1985 Plastic - Determination of multi-axial impact behaviour of rigid plastics - Part 1: Falling dart method
[see also BS 2782: Part 3: Method 359A: 1991]

ISO 6603-2:1989 Plastic - Determination of multi-axial impact behaviour of rigid plastics - Part 2: Instrumented puncture test

In addition there is the following British Standard test method which has no ISO equivalent:

BS 2782: Part 10: Method 1001:1977, Measurement of hardness by means of a Barcol impressor
[equivalent to EN 59]

When the product is a standard item, for which the manufacturer provides basic information for use in design, tests should be carried out to determine the properties, as discussed in 7.3.2. The data so obtained will not only give the designer properties based on a statistical analysis of the results but also act as part of the quality control process.

Where the product may be considered as a structure in its own right, the loadings that will be applied to it in service may be fully defined. In this case compliance testing will require the testing of representative samples to the specified loadings. An example would be FRP pipes, for which the ISO Standard defines the method for checking resistance.

ISO 1167:1973	Plastic pipes for the transport of fluids - Determination of the resistance to internal pressure

Other specifications include testing requirements for bending and point loads, which again should be checked on representative samples.

With filament winding of a large structure it may be appropriate to use the technique to form small, trial specimens. These can be tested either as a control measure for the process or as a test for the complete structure (see 7.3.5).

ASTM 2585:	Preparation and testing of filament wound pressure vessels

describes such a technique.

7.2.2 On site

The amount of testing that has to be carried out on site as part of the constructor's overall quality control process will depend on the nature of the structure and the construction process. If the units are delivered to site with a certificate confirming that their dimensions are within the specified tolerances, only critical dimensions need to be checked on site. If the units are subsequently joined to form an assembly, prior to being installed in the final structure, additional overall dimensional checks will be required.

Testing for the dimensional accuracy of components on site may be more difficult to carry out than testing at the place of manufacture, because of adverse weather conditions for example. In this case it may be more appropriate for the checks to be carried out by, or on behalf of, the end-user at the place of manufacture. Failing that, the components should be delivered with a certificate of conformity stating that the dimensions are within the specified tolerances.

The allowable tolerance will depend on the nature of the finished structure or on the importance of the particular component. Similarly the accuracy required for any holes and fixings will depend on the importance of the connection. Adjustments on site, for example elongating holes, may lead to a significant reduction in the capacity of connections between components.

The components should have been checked for defects by the manufacturer, who should issue a certificate stating that they conform with the agreed specification. However, they should be checked again on delivery to site, and before connection or installation in the finished structure, to ensure that no further damage has occurred during transport or storage.

Where possible, connections should be made under factory conditions. Where it is necessary to make connections on site, it may be necessary to test some, in accordance with 7.3, to check that they comply with the specified behaviour. This will be particularly important when the efficiency of the joint depends to a large extent on workmanship or where the ambient conditions are particularly unfavourable.

Ideally, the tests should be carried out on actual connections in the finished structure, but this may not be practicable. As an alternative, special connections suitable for testing may be made up, using the same techniques and under realistic conditions. Such connections can be tested to their ultimate capacity, while only factored service loadings should be applied to the connections in the complete structure.

7.3 TESTING FOR DESIGN AND VERIFICATION

7.3.1 Laminates

It is unlikely that the situation will arise in which a new resin and a new fibre are being used in a new manufacturing process or technique. Thus generally there will be prior knowledge of some aspects of the laminate, and this will aid in developing the test programme. This will also aid in the interpretation of the results, which should thus be applicable to a wider range of parameters than those actually considered in the tests.

Testing may be carried out either on samples cut from a larger element or on specially prepared samples, made by the same process and under the same conditions. In the latter case it is necessary to ensure that the sample is, as far as possible, representative of the material in the larger unit. This is particularly important where the data is required for the design of connections.

The nature of the mechanical loading, either static, dynamic or long term, will depend on the purpose of the testing and on the existing knowledge of the laminate properties.

Similarly, when the laminate will be subjected to significant environmental loadings, testing may be required to determine its long term response.

As with all testing it is essential to plan the work and to agree the plan with all concerned. The nature of the specimens tested will depend upon the particular circumstances, but the plan should have clear objectives and also state clearly the limitations of the testing.

The number of specimens to be tested will generally be stated in the Standard being used (see below). However, Mottram (reference 7.1) has suggested that the specified numbers are generally inadequate and that as many as 20 specimens may be required to determine a particular property. Then the characteristic strength can be taken as the mean value less 1.64 standard deviations.

Testing should be in accordance with ISO Standards as listed below, where possible. However as discussed in 7.1, the specimens may not be suitable for testing laminates, consisting of multiple layers of reinforcement, which are considerably thicker than the standard specimens. Similarly problems can arise when the fibres are not uniformly arranged in the cross-section, for example where individual rovings are used rather than continuous mats. Here the width of the standard specimen may be less than the spacing between the rovings, leading to the possibility of highly unrepresentative results. In this case it may be more appropriate, with the agreement of all parties, to use specimens of the same geometry but with larger, more representative dimensions. Ideally some attempt should be made to correlate the results with those from the standard specimens, although this may be difficult when through-thickness effects are significant.

ISO Standards covering the determination of the basic properties of the FRP laminate are listed below (other tests are listed in Appendix 7.1). These cover the elastic, short term properties and may either be used for the testing of a new product or as part of the compliance testing process described in 7.2.

ISO 178:1975 Plastics - Determination of flexural properties of rigid plastics
[equivalent to EN 63]
[see also BS 2782: Part 3: Method 335A: 1978]

ISO 3268:1978 Plastics - Glass reinforced materials - Determination of tensile properties
[equivalent to EN 61]

ISO 3597:1977 Textile glass reinforced plastics - Composites in the form of rods made from textile glass rovings - Determination of flexural (cross-breaking) strength

ISO 4585:1989 Textile glass reinforced plastics - Determination of apparent interlaminar shear properties by short-beam test

ISO 4899:1982 Textile glass reinforced thermosetting plastics - Properties and test methods

For laminates that will be subjected to high levels of stress in service, standard tests should be carried out to determine the creep properties, as follows:

ISO 899:1981 Plastics - Determination of tensile creep

ISO 6602:1985 Plastics - Determination of flexure creep by three point loading

Where the laminate will also be subjected to an unusual environment, for example high ambient temperature or aggressive atmosphere, the creep tests should be carried out under appropriate conditions.

Where the laminate may be subjected to impact, the following tests will be applicable for determining the local response of the material. However, the overall behaviour of the finished structure, including the response of connections, may also have to be considered, as outlined in 7.3.3 and 7.3.5.

ISO 179:1982 Plastics - Determination of Charpy impact strength of rigid materials
[see also BS 2782: Part 3: Method 359: 1984]

ISO 180:1982 Plastics - Determination of Izod impact strength of rigid materials
[see also BS 2782: Part 3: Method 350: 1984]

It has been shown that different test methods can give different values for a given property. This may be due to size effects, end effects or other causes. However, provided a particular test is used throughout a testing programme, these differences should not cause concern. In general, the properties of the laminate will be required so that the properties of the complete component can be assessed. Measuring the properties of the laminate will be one stage in the development of the component or the complete structure. Ideally tests should also be carried out on the component, as described in 7.3.2. Then the analytical model, which predicts the behaviour of the component on the basis of the material properties of the laminate, can be refined. However, the methods used for determining the laminate properties must be clearly stated.

Standard test methods will define the environment in which the specimens are to be tested. When the conditions in the final structure are unusual, for example a combination of a high stress in an aggressive environment, specimens should be tested under appropriate conditions. Alternatively it may be appropriate in this example to test in the aggressive environment, but at a range of stresses, in order to determine limiting conditions for use in service.

Testing will generally be restricted by time and cost to a limited number of parameters. It may be possible to use analytical techniques, based on previous experience of similar materials or manufacturing techniques, to widen the range of parameters. Here a knowledge of the properties of the fibre and resin may be required, for example, so that other materials can be considered. But as with testing the laminate, consistent test methods must be used throughout.

Again because of the restricted time available for testing, accelerated testing will be required to determine long term properties. However, the results should be treated with caution unless they are consistent and have a predicable trend, say are linear against the logarithm of time. In addition some correlation must be made with real-time tests to validate the accelerated testing.

7.3.2 Components

Where the component is the finished item itself, and will not be joined to other components so as to form a more complex structure, testing of the component will indicate its behaviour in service. An example might be a simple plank which will be used in a simply supported condition. However, if the ends are built in, providing a moment connection, tests on the component will not fully represent its final condition and hence 7.3.5 will apply.

Testing of components will be required because it will not, in general, be possible to predict accurately the overall behaviour simply on the basis of knowledge of the behaviour of the individual laminates, owing to out-of-plane and other effects.

Testing of components will generally be one stage in an overall programme, either developing a new product or part of the testing of a complete structure, or as described in 7.3.5. Thus there will be some prior knowledge, either of similar materials or of similar products, which can be used to extend the information gained from the tests.

By checking the structural behaviour of the complete component, one is able to check the combined behaviour of the individual laminates, when their properties are known. Alternatively, it gives a direct measure of the response when the properties of the individual laminates are not known or where their interaction is not understood.

The analysis of a complex cross-section may be difficult and is likely to be inaccurate when the material properties are not uniform throughout. In this case, testing of the component will give a direct measure of the structural behaviour, as well as being used to check the method of analysis.

In general, testing of components will concentrate on the structural behaviour, considering the response in bending, tension and so on. Tests for resistance to corrosive attack and other environmental loadings will probably be best carried out on suitable laminate specimens. Exceptions will be where the structural response and the environment interact, such as corrosive attack on a stressed component. In addition, the response may be influenced by the geometry of the member, or its size, such as the response to fire.

As with all testing, a clear test plan must be drawn up and agreed by all parties at the start of the programme. The size of the programme will depend on the nature of the testing. If it is part of the development of a standard component, a large number of tests will be required which cover a number of different aspects of the structural response. On the other hand, if it is part of testing a complete structure, it may only be necessary to carry out a few tests under the appropriate loadings.

Relevant ISO Standards covering compliance testing are listed in 7.2 of this EUROCOMP Handbook. In general, they cover the behaviour of test samples, either specially made or cut from the component, and not the behaviour of the component itself. (Some component tests are included in ASTM methods, see appendix to this chapter.) Thus it will be necessary to agree appropriate tests to be carried out, bearing in mind the size of the component and its intended use. As discussed before, all testing should be carried out under an approved quality scheme.

Design processes will generally consider the effect of individual loadings and then superimpose them to determine the total response. Thus it is appropriate to test components under simple loading, so that the response can be readily understood. However, care should be taken to ensure that the loadings are in fact those assumed. For example, when testing a component in bending, the end supports must be free to move, to avoid applying a longitudinal force.

An appropriate loading pattern may be defined when the end use for the component is known. For example, the loading may lead to a defined relationship between bending and shear. The true boundary conditions may be more difficult to simulate. If, for example, the end fixity is uncertain, it may be more appropriate to test some specimens with free ends and some with fixed, to identify the two extremes of the response.

Standard tests are carried out under standard conditions of temperature and environment. These will be appropriate when the testing is part of the development of a component for general use. However, if the end-use of the component is known, appropriate conditions should be used during the test. This is particularly important when considering long term effects, such as creep.

The behaviour of a component will depend not only on the materials from which it is made, but also on the variability of the manufacturing process. Thus it is not possible to define exactly how many test results are required in all situations. Three tests under a given loading should be taken as the minimum. The more variable the results obtained the more tests will be required. Statistical methods may then be used to determine the required property of the component to the required level of accuracy. The minimum number of tests to be carried out should be agreed in the test programme.

As discussed earlier, one purpose of component testing will be to determine the overall response and to compare it with the predicted response based on a knowledge of the behaviour of the individual laminates. Thus, having tested the component, it is necessary to check that the material of which it is made is in fact the same as had been assumed. This is best determined by testing samples cut from the component after test, avoiding those parts that have been significantly stressed. If for some reason this is not possible, because the method of testing has stressed all the component, then the test samples might be taken from other similar components. An example might be when using pultruded sections. It may safely be assumed that units from the same production run as the component actually tested have the same properties and hence they may be used as samples for materials testing.

If additional units are not available, it may be possible to obtain data from the manufacturer, provided that he has sufficient experience of the materials and the manufacturing technique used. While the data so obtained may not be directly applicable to the components actually tested, it should aid in the wider interpretation of the results, giving an indication of the variability that might be expected if similar components were tested.

Accelerated testing will generally be the only method for obtaining long-term results within a reasonable timescale. However, as discussed in 7.3.1, some correlation with tests under the true conditions is essential.

Standard tests will specify the load to which the component is to be subjected during environmental testing. This may be appropriate for compliance testing but is not appropriate when the testing is part of the development for a standard component. Where the loading in the final structure can be completely defined, the correct level of stress should be applied. Otherwise tests should be carried out at a range of stresses, covering those likely to be encountered in service.

Depending on the nature of the component, it may be appropriate to load it to the service load for which it is designed, for example to determine its stiffness. Alternatively the component may be loaded to failure and the resulting collapse load compared with that assumed in design, to determine the actual factor of safety achieved. Loading the component only to its design ultimate load will give more limited information; it will indicate whether the component is capable of carrying the load but will not give a direct measure of the margin of safety associated with collapse.

7.3.3 Connections

The EUROCOMP Design Code covers a range of possible methods of forming a connection between two members. By implication, testing will only be required when the connection plays a significant structural role. It is unlikely that testing will be required for connections that carry only a nominal load.

Testing of connections will be required when a new technique is being developed. This may be required when new materials, or new structural forms, are to be connected. Alternatively, an existing design of connection may be adopted to take account of a new adhesive or a new fabrication technique. As indicated in 7.3.1, testing of the laminates may be required to develop data for use in the design process. In addition, testing will be required when no suitable analytical method exists. Here the recommended method of design is by testing.

Connections are designed primarily to carry certain types of loads and may be significantly less efficient under other types of loading. For example, a connection designed mainly for shear may have a low resistance to loads in other directions. Testing will therefore generally consist of loading the connection, either statically or under repeated loading, in a particular direction to represent its in-service conditions. Where it is likely that the effects of environmental loads will have a significant effect on the behaviour of the connection, testing should be carried out under suitable conditions.

As with all the testing discussed in this chapter, a clear test plan must be drawn up and agreed by all concerned at the start of the programme. It is particularly important that the specification for making the connections should be agreed, including all stages of the preparation of the components, curing of any resins and so on. The specification should be the same as is recommended for general use.

The range of tests and the number of samples of each type to be tested will depend on prior knowledge of the likely behaviour of the connection. Appendix 7.2 lists appropriate tests for the adhesive and the connection. Testing of the complete connection will not be covered by ISO or other National Standards. Hence the test method must be chosen carefully, so that it adequately represents the loadings that will be applied to the connection in service.

Because the efficiency of the connection will generally depend to a large extent on the workmanship, care must be taken that the process is representative of that which will be found in practice. For example, drilled holes should be within specified tolerances, but not necessarily to tighter tolerances, while surface preparation, adhesive application and curing should be carried out under realistic conditions. Ideally the skill of the personnel making the connection should be similar to that encountered in practice.

If the testing is part of a general development programme, consideration should be given to testing a range of skill levels and a range of variations in making the connection, such as different ambient conditions.

When connections are tested as part of the testing of an assembly or a complete structure (see 7.3.4 and 7.3.5) the pattern of loading will be known. The test arrangement should accurately represent these loadings and the environmental conditions should be similar. It will be necessary to judge which arrangement of test loads and which size of specimen is the most economic solution. For example, a long length of component on either side of the connection will help to reduce end effects, but will require a large test facility. On the other hand, a shorter length of component, with a more complex loading system will be easier to handle and lead to less material being used.

When the testing is part of a development programme, the intention will be to obtain data on the behaviour under clearly defined loading. This may be, for example, direct tension or pure shear. The testing should ensure that the pattern of loads applied to the connection represents these loadings as accurately as possible. Additional testing will be required to study the effects of combinations of load on the capacity of the connection.

Also the environmental conditions used during the testing of the connection should be representative of those in the actual structure, if known. Alternatively, if it is thought that the environment could have a significant effect on the behaviour of the connection, it may be appropriate to test nominally identical connections under a range of likely conditions.

Where possible the connection should be subjected to mechanical loads, as well as the environmental loads, to simulate the true in-service conditions. As before, it may be appropriate to use a range of loads to cover the range of likely in-service conditions if they are not well understood.

The number of connections to be tested under any particular loading condition will depend on the scatter of results obtained; sufficient tests will be required to enable statistical methods to be used to set a level of confidence to the results.

The loadings to be carried by the connection should correspond to those in the final structure, if these are known. In this case the same factors as used when testing the complete structure (see 7.3.5) are appropriate. In general a factored service load should be used.

Loading to the design ultimate, or a multiple thereof, will give confidence that the connections in the complete structure are capable of performing satisfactorily. Unless a linear behaviour up to collapse of the connection can be assumed, such tests give little indication of the ultimate margin of safety. However, they confirm that premature adhesive failure will not occur.

If the testing is part of a development programme, the connections may be loaded to failure: this may be defined as a maximum load capacity or may depend on limiting deflections or rotations. Once a failure load has been determined, a safe service load for the connection may be determined by applying a suitable safety factor.

When the behaviour of a particular connection is critical to the behaviour of the structure, for example when failure would lead to a disproportionate collapse, higher load factors should be used when testing as part of the assessment of a complete structure.

Information is required on the properties of the components so that the results from the connection tests can be evaluated. The same approach should be adopted as discussed in 7.3.2, with test samples being taken from lightly stressed areas away from the connection. The main purpose should be to check that the material in the region of the connection is representative of that elsewhere or in other similar components. With standard components, historical data from the manufacturer giving the range of values achieved in practice will aid interpretation and allow a degree of confidence to be placed on the results.

Similarly, when the joint is bonded, the adhesive should be tested to check its strength. Where possible this should again be compared with historical data.

7.3.4 Assemblies

It may be necessary, or convenient, to connect together a number of components at one location to form a unit and then to move the unit to its final location. This may be at the construction site, for example when units are assembled at ground level and then lifted into position, or may involve transport over a longer distance, for example from the factory to the site. In all cases, steps should be taken to minimise unnecessary additional stresses, due to lifting and transport, by the use of temporary supports and braces, as detailed in Chapter 6. Where this is not possible, some additional testing on the assembly may be required.

Testing of an assembly may be treated as a limited case of the testing of a structure, and hence 7.3.5 applies. However the loadings will be far more limited, being only those appropriate to transport and lifting. Particular attention should be paid to situations in which the loadings on the assembly are in a direction different from those in the final structure. For example lifting a beam at mid-span will introduce hogging bending throughout, while in the final structure, under a uniform loading, it would be in sagging throughout. More severe stresses may also occur when an assembly is transported, for reason of size, on its side rather than vertically.

It is important to remember that permanent deformations of the assembly may lead to additional loads on the completed structure, for example by introducing additional moments by displacing a column from its assumed line of action. Where appropriate, the assembly should be tested with a view to checking critical dimensions.

7.3.5 Structures

7.3.5.1 Purpose of testing

There are a number of reasons why it may be necessary to test a complete structure or part of a structure. These include situations in which the design assumptions and the analysis methods used do not accurately represent the complete structure. In addition tests may be required when there are doubts about the materials or the workmanship.

It may also be necessary to test the complete structure when it has been repaired following damage, for example due to overload or fire.

Finally, testing may be carried out when it is considered that the rules for calculation in the EUROCOMP Design Code would lead to uneconomic results. One example might be when the effects of multi-axial loading lead to permissible stresses that are higher than the simple uni-axial ones generally considered.

However, it should be remembered that load testing, particularly of a complete structure, is complex and expensive. For all but the simplest structures it should only be undertaken as a last resort, when the results of tests on specimens or components removed from the structure have proved inconclusive.

7.3.5.2 Load tests to substantiate design

The objectives of this type of testing are to define the load-carrying capacity and the deformations of the whole or a part of a structure and hence to demonstrate that it satisfies the limit state requirements. The test specimen used should be truly representative of the finished structure, or a part thereof. The size of specimen that can be tested will depend on the available facilities, the magnitude of the loads and reactions, and how accurately they can reflect the situation in the completed structure.

The designer should specify the loading conditions, the programme of tests and their objectives. In particular the loading stages and their duration should be specified, and whether or not various test specimens are to be subjected to different load combinations, to ascertain the most critical conditions.

There are a number of approaches to planning a test programme. The major choice, which will be governed by costs on many cases, is between:

(a) a large number of simple tests on components or structures, subjected to a wide range of loadings to cover the entire range of structural behaviour

(b) a limited number of more complex tests with a limited range of parameters, the test results being then used to predict the behaviour under other loading conditions

(c) tests on scale models of complex structures, when experience has shown that the effects of scale are small or can be allowed for in the interpretation of the results.

In all cases the magnitudes and combinations of the various loads and reactions should correspond with those used in the design process. However, judgement may be necessary to select the most relevant values and combinations to ensure that the scale and cost of the testing is not excessive in relation to the size of the project.

To obtain as much information as possible from the test, component parts should be tested whenever possible. By applying simple loading, for example pure bending, the complex loading patterns that occur in the real structure can be avoided. In addition, the uncertainties induced by connection behaviour are eliminated.

While the results of tests on component parts will obviously be directly applicable to the particular structure being tested, they may not be truly representative of practice. Data should be obtained from the manufacturer, to give an indication of the variability in the values of a given property. Hence the designer may assign a level of confidence to his test results and make them more generally applicable.

The degree to which it is appropriate to take samples from the structure and the value of the material properties so obtained will depend on the nature of the particular programme of work. For a simple structure, such as a frame made up of standard pultruded sections, the basic material properties will be of little benefit apart from providing a check that the sections are within specification. (However, even this would probably be better done by testing full components.) For a more complex structure, say a shell of variable thickness, a knowledge of the localised properties could be essential.

As with the testing of component parts, data from the manufacturer, giving the range of values likely, will aid the interpretation of the results and allow the designer to assign a level of confidence.

Where the behaviour of the connection can have a significant influence on the behaviour of the structure, for example owing to slip or rotation, it may be appropriate to test individual connections. It is not likely that they can be removed from the structure after test, unless there are typical connections that have been only lightly stressed. It may be more appropriate to make special test specimens, under the same conditions as those in the complete structure.

7.3.5.3 Load tests of completed structures or parts of structures

As indicated earlier, load testing of the complete structure should generally be considered as the last resort when other methods of assessment have either failed or given inconclusive results. General guidance on the appraisal of structures (built with traditional materials) is given in reference 7.2, which also includes guidelines for testing. With a small, simple structure in which the loads and reactions are understood clearly, it may be practical to load the complete structure. In general, however, it will only be possible to load a representative part of the structure and use the data so obtained to predict the behaviour of the whole.

In general, loading will be to the service load, multiplied by a small factor. Care should be taken to ensure that all dead loads are adequately taken into account, including finishes, partitions and so on.

When loading part of a structure, it is important to consider the effects of load sharing. Ideally loads should be applied to a representative portion of the structure, which should be isolated from the remainder. Generally this will not be possible without removing connecting elements, which could affect the subsequent behaviour of the whole structure. However, it may be possible to estimate the proportion of the applied load that is transferred to other parts of the structure, for example by applying different patterns of load to the portion under test.

All load testing can be potentially dangerous, particularly for existing structures whose behaviour may not be clearly understood beforehand. An experienced engineer should be in charge of the entire preparation and execution of the test and, in particular, to be responsible for safety. The engineer should be aware of the signs of impending collapse and should ensure that adequate provisions are made for supporting the structure in the event of failure. The possibility of damage to adjacent parts of the structure, owing to collapse of the portion under test, should be taken into account.

A test plan must be drawn up and agreed by all parties before work begins. Of particular importance is the instrumentation that should be used. There are various methods for measuring the surface strains on the components and the deformations. The choice will depend on the type of structure, access to critical locations during the test, the provision of a fixed frame of reference and, above all, on the purpose of the test. The major limitations are likely to be time and cost. It may be appropriate to concentrate on the overall behaviour of the structure, for example by mainly measuring deflections under load, and to determine the local behaviour of components or connections as a separate exercise, using the methods in the previous sections.

The sequence of loading will again depend on the nature of the structure. However, the EUROCOMP Design Code recommends incremental loading, to the factored service loads, followed by unloading and subsequent reloading. The number of increments should be chosen such that the overall length of the test is not so long that creep unduly influences the results. At each increment the structure should be allowed to settle for 5 minutes, or until deflections are sensibly constant, before readings are taken. The purpose of the first loading is chiefly to allow the measuring equipment to bed down. The deflections, and the strains if appropriate, measured during the second loading should be compared with those measured during the first loading, to give an indication of the amount of settling down. Provided on unloading for the second time the deflections and stresses return to the values they had at the start of second loading, to within 5% of the range recorded, the values under the second loading should be taken as the response of the structure.

If the additional residual deflections after the second loading are significant, i.e. more than 5% of the range, the structure should be reloaded until repeatable values are obtained. Should it be impossible to achieve repeatable values, then the structure is being overloaded, leading to damage in some components or in some connections leading to slip.

As with all the testing described in this chapter, it is important to gather as much information as possible on the materials and the component parts of the structure. Where possible, samples should be removed from the structure and tested in accordance with 7.3.1. Similarly it may be possible to remove components or complete connections, in which case they could be tested in accordance with 7.3.2 and 7.3.3 respectively. However, this will only be possible when bolted connections without bonding are used, so that the units can be reinstated once they have been tested.

Alternatively, if the structure is formed of standard components, such as pultruded sections, tests may be carried out on similar components or data obtained from standard information sheets supplied by the manufacturer, provided that it is agreed that they are representative of those in the complete structure.

7.4 ADDITIONAL TESTS FOR SPECIAL PURPOSES

ISO Standards include tests which address the durability of the FRP product, under either the action of the environment or biological action, or due to abrasion. They are listed below:

ISO 846:1978 — Plastics - Determination of behaviour under the action of fungi and bacteria - Evaluation by visual examination or measurement of change in mass or physical properties

ISO 877:1976 — Plastics - Determination of resistance to change upon exposure under glass to daylight
[see also BS 2782: Part 5: Method 560A: 1977]

ISO 4582:1980 — Plastics - Determination of changes in colour and variation in properties after exposure to daylight under glass, natural weathering or artificial light
[see also BS 2782: Part 5: Method 552A: 1981]

ISO 4607:1978 — Plastics - Methods of exposure to natural weathering
[see also BS 2782: Part 5: Method 550A: 1981]

ISO 4611:1987 — Plastics - Determination of the effects of exposure to damp heat, water spray and salt mist
[see also BS 2782: Part 5: Method 551A: 1988]

ISO 4892:1981 — Plastics - Methods of exposure to laboratory light sources
[see also BS 2782: Part 5: Method 540B: 1982]

ISO 6601:1987 — Plastics - Friction and wear by sliding - Identification of test parameters

ISO 9352:1989 Plastics - Determination of resistance to wear by abrasive wheels

Where fire is a major design consideration, testing may be required. The following ISO Standards consider the behaviour of the material itself but do not address the structural consequences:

ISO 181:1981 Plastics - Determination of flammability characteristics of rigid plastics in the form of small specimens in contact with an incandescent rod

ISO 871:1980 Plastics - Determination of temperature of evolution of flammable gases (decomposition temperature) from a small sample of pulverized material

ISO 1210:1982 Plastics - Determination of flammability characteristics of plastics in the form of small specimens in contact with a small flame

ISO 4589:1984 Plastics - Determination of flammability by oxygen index

To determine the structural response, full scale fire tests would be required, either on individual components or on parts of the complete structure.

REFERENCES

7.1 Mottram J.T., Compression strength of pultruded flat sheet material, *Journal of Materials in Civil Engineering*, 6 (2), 185-200, 1994.

7.2 *Appraisal of Existing Structures*, Institution of Structural Engineers, July 1980.

APPENDIX 7.1 ASTM STANDARDS

These standards are widely used for testing. However, they should only be used within the context of the EUROCOMP Design Code when no suitable ISO Standard exists. Some of the relevant test methods are listed here, chiefly for reference purposes.

(a) *Tests on fibres and high modulus fibres*

D2343-67 Tensile properties of glass fibre strands, yarns and rovings used in reinforced plastics

D3800-79 Density of high modulus fibres

D4018-81 Tensile properties of continuous filament carbon and graphite yarns, strands, rovings and tows

(b) *Tests during construction*

D1505-85 Density of plastics by the density gradient technique

D2563-70 Classifying visual defects in glass-reinforced plastic laminate parts

(c) *Tests on laminates*

D256-90B Impact resistance of plastic and electrical insulating materials

D638-90 Tensile properties of plastics

D695-90 Compressive properties of rigid plastics

D790-90 Flexural properties of unreinforced and reinforced plastics and electrical insulating materials

D2344-84 Apparent interlaminar shear strength of parallel fibre composites by short beam method

D3039-76 Tensile properties of fibre - resin composites

D3410-87 Compressive properties of unidirectional or crossply fibre - resin composites

D3518-91 In-plane shear stress-strain response of unidirectional reinforced plastics

D3846-79 In-plane shear strength of reinforced plastics

(d) *Tests on components*

D2290-87 Apparent tensile strength of ring or tubular plastics and reinforced plastics by split disk method

D2585-68 Preparation and tension testing of filament wound pressure vessels

D4475-85 Apparent horizontal shear strength of pultruded reinforced plastic rods by the short beam method

D4476-85 Flexural properties of fibre reinforced pultruded rods

(e) *Long term tests*

D3479-76 Tension-tension fatigue of oriented fibre - resin matrix composites

APPENDIX 7.2 TEST METHODS FOR ADHESIVELY BONDED JOINTS

Mechanical strength tests relating to the adhesive characterisation and strength of bonded joints are listed below:

Adhesive characterisation

ASTM D 3983-81 - Measuring strength and shear modulus of nonrigid adhesives by the thick adherend tensile lap specimen [thick adherend test]

Joint strength

Tensile shear loading

ISO 4587: 1979 - Adhesives - Determination of tensile lap-strength of high-strength adhesive bonds [single lap]

ASTM D 1002-72 - Strength properties of adhesives in shear by tension loading (metal-to-metal) [single lap]

ASTM D 3163-73 - Determining the strength of adhesively bonded rigid plastic lap-shear joints in shear by tension loading [single lap]

ASTM D 3165-73 - Strength properties of adhesives in shear by tension loading of laminated assemblies [single lap]

Tensile loading [butt joint]

ISO 6922: 1987 - Adhesives - Determination of tensile strength of butt joints

ASTM D 897-76 - Tensile properties of adhesive bonds [metal-to-metal]

ASTM D 2095-72 - Tensile strength of adhesives by means of bar and rod specimens

Peel loading

ASTM D 1781-76 - Climbing drum peel test for adhesives

ASTM D 1876-72 - Peel resistance of adhesives (T-peel test)

ASTM D 3167-76 - Floating roller peel resistance of adhesives

Cleavage loading

ASTM D 950-82 - Impact strength of adhesive bonds

ASTM D 1062-78 - Cleavage strength of metal-to-metal adhesive bonds

Durability

ASTM D 1151-84 - Effect of moisture and temperature on adhesive bonds

ASTM D 1828-70 - Atmospheric exposure of adhesive-bonded joints and structures

ASTM D 2918-71 - Determining durability of adhesive joints stressed in peel

ASTM D 2919-84 - Determining durability of adhesive joints stressed in shear by tension loading

ASTM D 3672-79 - Adhesive-bonded surface durability of aluminium (wedge test)

Creep

ASTM D 1780-72 - Conducting creep tests of metal-to-metal adhesives (tensile shear loading)

ASTM D 2293-69 - Creep properties of adhesives in shear by compression loading (metal-to-metal)

ASTM D 2294-69 - Creep properties of adhesives in shear by tension loading (metal-to-metal)

Fatigue

ASTM D 3166-73 - Fatigue properties of adhesives in shear by tension loading (metal-to-metal)

CONTENTS

8 QUALITY CONTROL

8 QUALITY CONTROL

8.1 SCOPE AND OBJECTIVES

Quality control is essential at all stages of the design, fabrication and erection process to ensure that the finished structure is in accordance with the specification and should thus be able to fulfil the role for which it was designed. This chapter aims to outline the various control measures that will or may be required: the actual choice of the level and degree of control will depend on the nature of the structure and its intended purpose.

8.2 CLASSIFICATION OF THE CONTROL MEASURES

For the purpose of this chapter, three different types of control are identified, namely internal, external and conformity. The first two refer to control carried out by the organisation performing the particular task or control by an independent authority. The third is that control which is required to ensure that the product, in all aspects, conforms with the specification. As such it will generally be carried out directly by the client or by an independent authority on the client's behalf.

8.2.1 Internal control

It is essential that the designer, fabricator, supplier and contractor all exercise their own internal quality control, which will cover the particular aspect of the process with which they are concerned. In general this should be in accordance with ISO 9000 to 9003, as appropriate, as detailed in Chapter 7 of this EUROCOMP Handbook and in Table 8.1 of the EUROCOMP Design Code. The standards require agreed procedures for carrying out any particular process, which are audited on a regular basis both internally by the organisation itself and externally by the certifying authority.

8.2.2 External control

The client may require additional control of all, or certain parts, of the manufacturing and construction process. This will be particularly likely when the structure may be considered to be outside the general scope of the EUROCOMP Design Code, because of the importance of structure, the environment or the loadings. The control may consist of an independent audit of the existing quality control system, with particular emphasis on those parts that are of particular relevance to the project. In addition, the independent authority may require further quality control checks. These will be over and above the existing internal control measures and should be treated separately from them.

8.2.3 Compliance control

It is important that all stages in the production process should be carried out in accordance with the agreed specifications. This will generally consist of testing at various stages, as detailed in 7.2, and the maintenance of accurate records. The amount of testing that is required at any stage will be as defined in the EUROCOMP Design Code or as required by the client or the independent authority.

8.3 VERIFICATION SYSTEMS

The amount of control that may be required will depend on the nature of the structure being built, or even on the behaviour of particular elements or connections. The control required for the design, manufacture and assembly of a component that is critical to the overall behaviour of the structure, for example a highly stressed member controlling the stability, will be more severe than for a member of minor importance.

8.4 CONTROL OF DIFFERENT STAGES

No additional information.

8.5 CONTROL OF DESIGN

The EUROCOMP Design Code requires that the design should comply with the appropriate National administrative procedures. In the United Kingdom, these will include the Building Regulations.

8.6 CONTROL OF COMPONENT MANUFACTURING

8.6.1 Objectives

The aim of control of the manufacturing process is to ensure that the finished product conforms with the original specification. The intention is to ensure a uniform quality of the product, covering both the materials and the workmanship. Testing, as detailed in Chapter 7 of the EUROCOMP Design Code, will be required along with an overall quality plan.

8.6.2 Elements of component manufacture

As outlined in Chapter 7 of the EUROCOMP Design Code, checks are required on the basic materials, such as the glass and resin, during the course of manufacture and on the finished product. To avoid repetition, it will generally be sufficient for materials to pass from one stage to the next, e.g. glass rovings to woven mat to composite manufacture, with a certificate of conformity, which will state that the product conforms to the relevant specification.

Different verification systems are appropriate for different manufacturing processes. For a continuous process, the main checks are on the supply of materials and the operation of the machinery such that the product is of a uniform quality. There is only a limited workmanship aspect once the process has been set up and is running, Conversely for single products there is likely to be a large element of workmanship and hence control will be concentrated in this area.

8.6.3 Initial tests

For all products it is essential to carry out preliminary tests to demonstrate that the fibres, resins, etc., are suitable for the particular equipment or manufacturing process being used. For standard products or processes this may be on the basis of past experience.

At all times, only materials that have been shown to be suitable for the particular end-use of the product should be used.

8.6.4 Checks during manufacture

The manufacture of products should be under an approved quality plan, in accordance with the relevant ISO Standards or Euronorms. Checks should be in accordance with 7.2.1.

Part of any quality control system is the keeping of adequate records, which should be available to all parties. These are particularly valuable when the manufacturer makes standard products, as they record the variability of the properties over a period of time. They also give confidence when a standard product is to be adapted for a particular application, by changing the shape, the materials or otherwise, that the properties will be as predicted.

8.7 CONTROL OF COMPONENT DELIVERY

As part of the quality control process, all components delivered to site must be accurately identified. The delivery ticket should contain as much detail as possible of the manufacture of the unit and its subsequent curing and storage. Some checks will be required on site, as detailed in 7.2.2, particularly in the case of damage during transit, but in general the components should be delivered with certificates confirming that they conform to the specification.

All delivery notes should be stored, along with all other relevant information on the components.

8.8 CONTROL OF ASSEMBLY

The assembly must be in accordance with the specification and the detailed drawings. Any significant variations must be carefully recorded and the data made available to all, including the future owners of the structure. This is particularly important when modifications are carried out on site, for example elongation of fixing holes, that could effect the long term strength or behaviour of the structure.

Care should be taken to ensure that any connections are made in accordance with the specification. Again any significant variations from the specification should be recorded.

A record must be kept of the results of any tests carried out on components, connections, assemblies or parts of the finished structure. These should be available to all parties.

8.9 CONTROL AND MAINTENANCE OF COMPLETED STRUCTURE

The owner of the structure should be provided with a detailed programme which will cover any checks and maintenance that may be required. It should detail suitable repairs that may be used in the event of damage as well as routine work such as the replacement of surface coatings. In addition it should state clearly the use for which the structure is intended, the loadings, environment, etc., so that the necessary corrective or remedial action can be taken in the case of misuse.

Part 3 Test Reports

CONTENTS: TEST REPORTS

TESTING OF GLASS REINFORCED PLASTIC STRUCTURAL PANELS

BONDED JOINT TESTS

TESTS ON TUBULAR MEMBERS

TESTS ON NOMINALLY PINNED CONNECTIONS FOR PULTRUDED FRAMES

TESTING OF A PULTRUDED GRP PORTAL FRAME

THE TESTING OF GLASS REINFORCED PLASTIC (GLASS FRP) STRUCTURAL PANELS TO BE USED ON THE WINTERBROOK BRIDGE ENCLOSURE FOR THE EUROCOMP PROJECT

by

W.M. Banks

Department of Mechanical Engineering
University of Strathclyde
75 Montrose Street
Glasgow G1 1XJ
Telephone: 0141-552 4400 Ext. 2321
Fax: 0141-552 5105

CONTENTS

TESTING OF GLASS FRP STRUCTURAL PANELS

1 INTRODUCTION

Five glass fibre reinforced plastic (glass FRP) panels were supplied by ANMAC of Nottingham to the University of Strathclyde to be tested. The panels are to be used in the Winterbrook Bridge enclosures. They are to be hung below the bridge and used as walkways for inspection purposes. The main advantage of using composite materials such as glass FRP over traditional materials for this kind of application is that they display not only an attractive combination of mechanical properties and lightness, but also environmental resistance, and consequently reduced through-life costs, which is a principal consideration.

The panels, of overall dimensions 1.85 m x 1 m, are reinforced by closely spaced longitudinal glass FRP stiffeners, together with widely spaced transverse small glass FRP frames. Tests were carried out at two levels:

- on a local scale, where the elastic and ultimate properties of the different structural parts of the panel were determined and compared with theoretical models using micromechanics and macromechanics - as detailed in the EUROCOMP Design Code
- on a full scale, the panels in bending using actual support conditions. Pull-out tests on the support hangers as well as compression tests on the central longitudinal stiffeners were completed.

A simple theoretical model based on composite beam theory associated with the traditional strength of the materials was developed to describe the behaviour of the panels in the elastic range. A three-dimensional finite elements analysis on the whole panel was also performed to simulate its flexural behaviour.

2 STRUCTURAL CHARACTERISTICS

The panels are reinforced by closely spaced longitudinal stiffeners, which provide strength and stability under conditions of longitudinal bending, together with widely spaced transverse small frames, which give intermediate support to the longitudinals and provide rigidity and strength under transverse loads (see Figure 1(a) and (b)).

In terms of structural components, the panel (made by hand lay-up) is made basically of two distinct entities for which the nature and the arrangement of the layers are specific. Details are given in Figures 2 and 3.

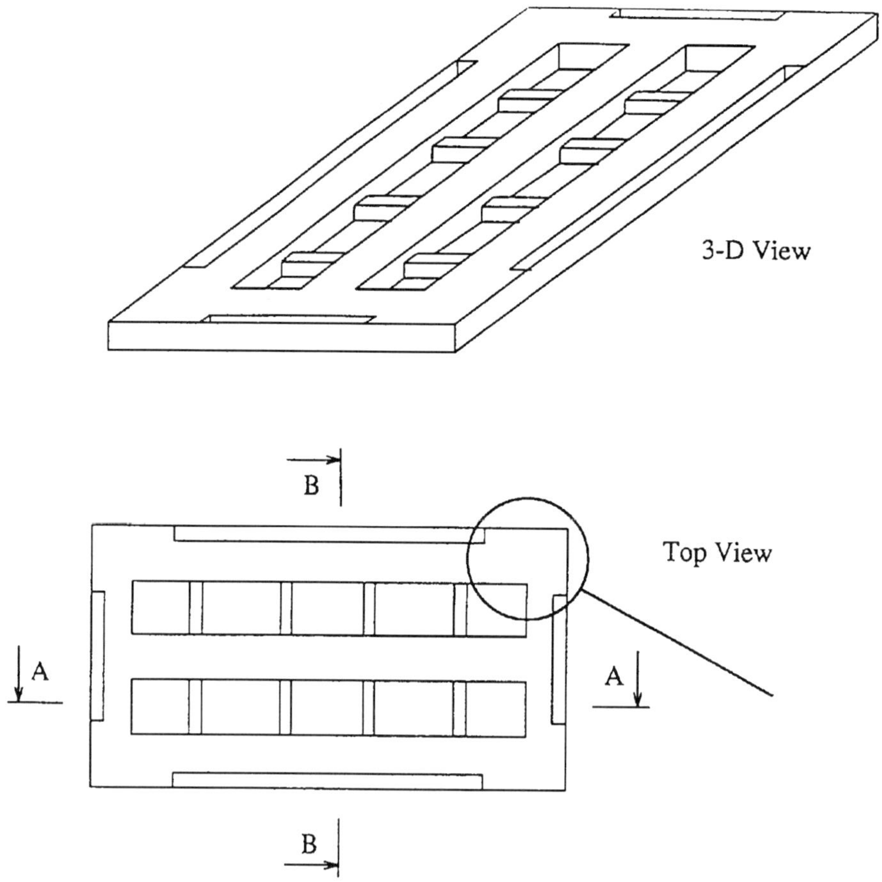

Figure 1a Overview of the panel.

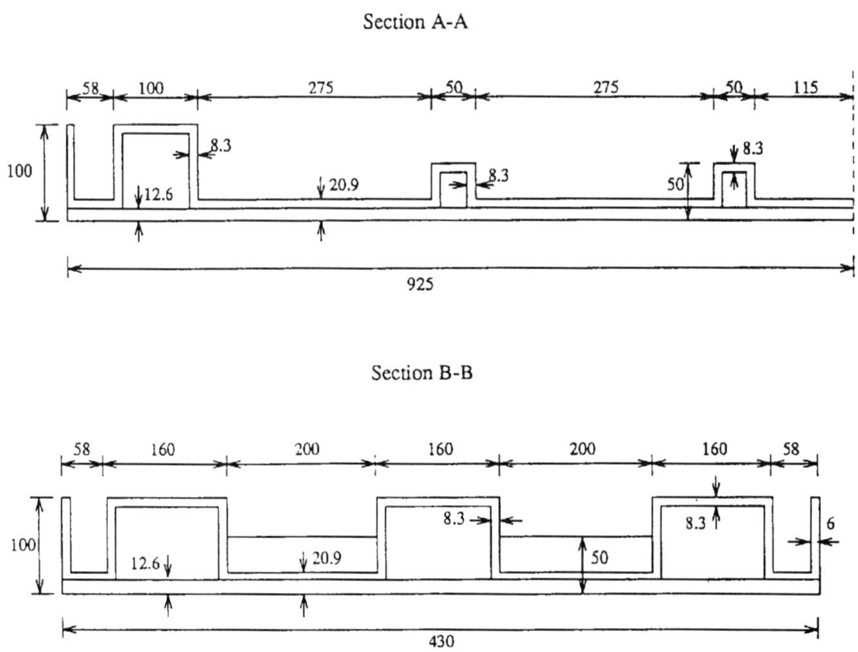

Figure 1b Longitudinal and transverse sections.

(a) Laminate stiffeners are reinforced by a combination of 800 g/m^2 glass woven rovings with a fibre weight fraction (W_f) of 0.55 and 450 g/m^2 chopped strand mat with a fibre weight fraction of 0.33.

- transverse and longitudinal plain channel stiffeners: [CSM_5/WR/CSM_5], t = 8.3 mm (designated L1)
- edges: [CSM_4/WR/CSM_4], t = 6 mm (designated L2)

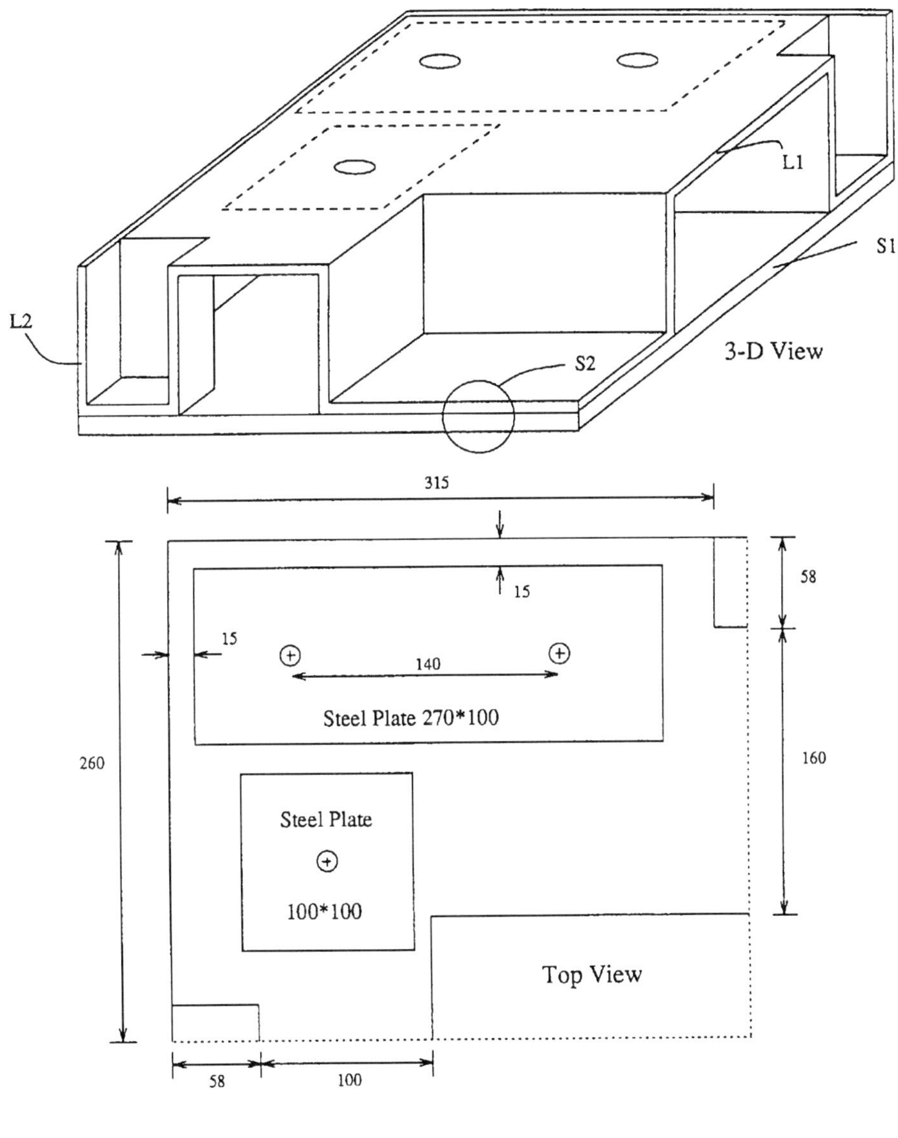

Figure 2 Corner detail.

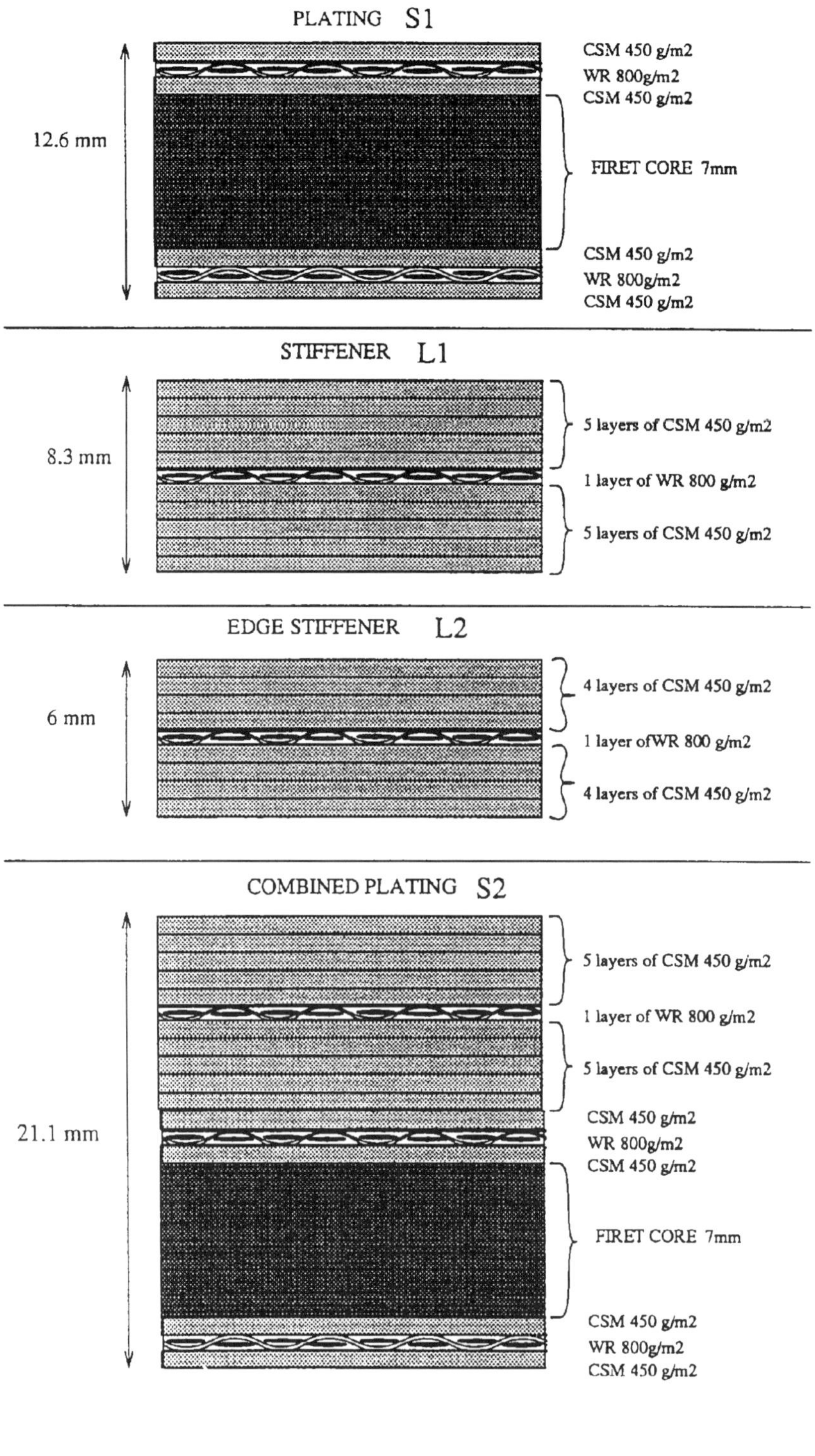

Figure 3 Detail of the layers.

(b) sandwich construction

- the twin-skinned plating consisting of a glass/polyester skin plus a core material: [CSM/WR/CSM/COREMAT/CSM/WR/CSM], t = 12.6 mm (designated S1)
- a mixed part where the laminate reference S1 and L1 are 'stacked' one on to the other: [CSM_5/WR/CSM_6/WR/CSM/COREMAT/CSM/WR/CSM], t = 21.1 mm (designated S2).

3 MATERIAL CHARACTERISTICS

3.1 Raw materials

The resin is an isophthalic polyester resin. Its principal advantage, besides low cost, lies in its rapid and controllable cure chemistry and therefore in its ease of use within hand lay-up fabrication processes. There are several types of polyester but isophthalic resins offer the most attractive combination of mechanical properties and environmental resistance (i.e. reduced corrosion and maintenance costs, and good resistance to attack by chemicals). Moreover, previous studies on effects of immersion and weathering (exposure to wind, rain, sun) show that the performance of glass FRP laminates are generally good, especially with isophthalic polyester resin.

In the manufacturing of the panel, sandwich construction was used, combining glass FRP faces with a light core material. It offers an alternative to stiffened single-skin construction for the plating of the panel, thus increasing its thickness and its flexural properties.

3.2 Monolayer and laminate properties

The properties for the basic layers of CSM and WR constituting the different parts of the panel were computed using micromechanics as detailed in the EUROCOMP Design Code. More specifically, elastic constants of a monolayer of CSM were derived using the theory of composites with randomly orientated continuous fibres, and those of WR were computed using the concept of "ply efficiency" equal to 0.5 for a bi-directional balanced cloth ply. Table 1 shows the respective proportions and properties of the glass fibres and resin used in the fabrication of the panels and the subsequent elastic constants for a monolayer of CSM and WR.

Table 1 Mechanical properties and thickness of CSM and balanced WR.

	Resin	*Fibres*	*CSM 450 33%*	*WR 800 55%*
Young's modulus (N/mm^2)	3,500	72,000	9,991	21,193
Shear modulus (N/mm^2)	1,287	30,000	3,735	2,920
Poisson's ratio	0.36	0.2	0.3374	0.1332
Density (g/m^3)	1.53	2.55		
Fibre weight fraction (%)			33	55
Fibre volume fraction (%)			22.8	42.3
Weight of reinforcing fabric (kg/m^2)			0.45	0.8
Thickness (mm)			0.7736	0.7415

The properties of the multilayer laminates referenced L1, L2, S1, S2 are derived from macromechanics. The direct and shear elastic constants are calculated from the direct and shear rigidities A_{ij} and the flexural elastic constants are determined from the modified flexural rigidities D_{ij}. The theoretical average elastic constants of the different parts of the panel are summarised in Table 2.

Table 2 Theoretical values of elastic properties of laminates.

	Thickness (mm)	E_x (N/mm^2)	E_y (N/mm^2)	G_{iy} (N/mm^2)	υ_i	W_{mtot} (%)	E_f (N/mm^2)
L1	8.48	11,037	11,037	3,664	0.3058	64.9	10,795
L2	6.93	11,268	11,268	3,648	0.2993		
S1	11.58	5,805	5,657	1,414	0.2439	59.3	11.205
S2	20.06	8,025	7,941	2,365	0.2809		9.251

4 TESTS FOR EVALUATION OF LAMINATE PROPERTIES

In order to confirm theoretical estimates of stiffness necessary for a finite element analysis, experimental determination of laminate properties are required. Measurement of laminate thickness and fibre and resin content as well as mechanical properties are also required.

4.1 Fibre and resin content

Fibre and matrix volume and weight fractions are normally determined by destructive tests on small representative coupons of laminates. In the case of glass FRP, a burn-off test is likely to be the most effective method.

A small coupon, whose weight and volume have been measured accurately, is heated (to 625°C) under controlled conditions until the polymer matrix has vaporised. The weight of residual glass reinforcement is then subtracted from the original coupon weight to obtain the matrix weight. The burn-off test was carried out following the British Standard BS 2790 : Part 10 Method 1002.

The resin contents, by weight, were as follows:

L1 Theoretical 64.9% Experimental 55.4%
S1 Theoretical 59.3% Experimental 57.8%

4.2 Tensile tests (based on BS 2782 : Method 320f : 1976)

Measurements of Young's modulus and tensile strength were carried out by using long specimens as detailed in the British Standard. Glass FRP tabs were bonded to the ends of test specimens in order to avoid premature failure at the test machine grips.

The results are summarised in Table 3.

Table 3 Comparison of the experimental and theoretical values of tensile properties.

Property	Unit	*Theory* (N/mm^2)	*Experiment* (N/mm^2)	*Difference* (%)
Tensile	L1	11,037	10,953	0.8
modulus	S1	5,805	6,589	11.9
	S2	8,025	7,465	7
Tensile	L1	113	104	7
strength	S1	60	76.6	21
	S2	55	66.8	18
Flexural	L1	10,795	10,271	4.8
modulus	S1	11,205	11,187	0.8
	S2	9,251	6,530	30
Flexural	L1	187	148	21
strength	S1	89	73.4	18
	S2	98	69.2	29
Compressive	L1	143	117	18
strength	S1	77	78.9	3
	S2	82	100	18

- *Tensile failure mode*
  The tensile strength of cross-ply or fabric-reinforced laminates is intermediate between the longitudinal and transverse strengths of a uni-directional composite having the same volume fraction. Failure involves a combination of the processes observed for uni-directional laminates. In the case of L1, S1 and S2 initial damage, having the form of resin cracking and fibre debonding, is found to occur at tensile strains of between 20 and 50% of the ultimate value and is associated with a reduction of up to 40% in the Young's modulus, which is typical of marine-type glass FRP structures. For this reason tensile strains under normal design loads are usually confined to between 20 and 30% of ultimate values.

- *Comparison of theoretical and experimental values*
  The tensile test gives mean values for the tensile moduli which are broadly in agreement with values predicted by theory. The values match very well especially for the laminate L1. For the sandwich construction S1 and S2 a slight difference appears, perhaps because of their intrinsic structure. Theory therefore provides good estimates for tensile moduli.

  In terms of strength, the values obtained are in the range of the generally accepted values for this kind of material. Usually the tensile strength of a CSM (33%) laminate is around 100 N/mm^2, and of a WR (55%) laminate around 250 N/mm^2. The tensile strength of the core material is around 3 N/mm^2. By a simple rule of proportionality, the strength of the laminate can be very roughly calculated.

- *Comparison of strength and moduli between L1, S1 and S2*
  The elastic moduli in a composite is directly related to the amount of reinforcement.

  As expected, the tensile modulus of L1 is greater than the tensile modulus of S1 and of S2 because of the very low modulus in tension characterising the core material. The intermediate value for S2 between S1 and L1 is certainly due to the fact that compared with S1, S2 has one more layer of WR, which has a greater tensile modulus than CSM. Also the total weight fraction of fibres for S2 is greater than for S1.

  As regards strength, S1 and S2 have the lowest values, because of the very low strength of the core material, which tends to weaken the structure. Moreover, the strength of S2 is half that of S1. In this case the lay-up is not symmetrical, thus inducing complex behaviour.

4.3 Flexural test (based on BS 2782 : Method 335a : 1993)

The flexural rigidity and strength of a laminate are evaluated by a three-point bending test on parallel strips of material. Specimen dimensions are important to avoid significant shear deformation and interlaminar shear stress effects. It should be borne in mind that the flexural rigidity and strength of non-homogeneous laminates will depend strongly on the location of individual reinforcing plies relative to the neutral axis. This is less significant in laminates formed entirely of CSM or WR reinforcement of the same material, but apparent flexural properties can still be influenced by surface anomalies, e.g. a thick gel coat. The test specimen supported as a beam is deflected at a constant rate at the mid-span until the specimen fractures.

The results are summarised in Table 3.

- *Flexural failure mode*

  Flexural failure involves a combination of the processes occurring in tensile and compressive failures; it starts usually on the outer plies in tension of the laminate and spreads to the underlying plies. Its progressive nature leads to strengths which are slightly higher than the tensile or compressive strength.

  For a sandwich construction the failure starts within the core material, and the crack which appears grows with increase of load, till it reaches the end of the specimen. Then the load is carried by the skin of the specimen. If the skin is very thin compared with the thickness of the core, final failure occurs a few seconds after, as is the case for S1. If, on the other hand, the skin is very thick, as for S2, the curve obtained for the flexural stress is in two parts; one corresponding to the behaviour of the sandwich construction before the failure of the core material, and a second part corresponding to the behaviour of the skin only.

- *Comparison of theoretical and experimental values*

  As for the tensile test, the flexural test results are in general agreement with the values predicted theoretically, except for S2; but the large difference here can be attributed essentially to its particular failure mode, which is induced by its geometry (i.e. thick upper skin and non-symmetric sandwich construction).

  In terms of strength, the values obtained are in the range of the generally accepted values for this kind of material. Usually the flexural strength of a CSM (33%) laminate is around 77 N/mm^2, and of a WR (55%) laminate around 295 N/mm^2. The flexural strength of the core material is around 6 N/mm^2. By a simple rule of proportionality, the strength of the laminate can be very roughly calculated. This yields the same kind of values as those obtained experimentally, except for the sandwich construction S2.

- *Comparison of flexural strength and moduli between L1, S1 and S2*

  As expected S1 has a greater flexural stiffness than L1 and S2, and S2 has a slightly better flexural stiffness than L1. This is due to the presence of the core material, which increases the thickness and consequently the flexural rigidity.

In terms of strength, L1 still has the best value, whereas the values for the sandwich constructions are lower, because of the very low strength of the core material.

4.4 Compression tests (based on BS 2782 : Method 345a : 1993)

Compressive strength is much more complicated to evaluate, the main problem being avoidance of buckling effects. Tests can be carried out on short specimens, or preferably on long specimens having a short, central gauge length and extended glass FRP or metal tabs. For this latter type, compressive load is applied through slip conical or wedge grips. The test specimen is compressed along its major axis at constant speed until the specimen fractures. The load sustained by the specimen is measured during this procedure.

Typical results are summarised in Table 3.

- *Compression failure mode*
  Failure of a laminate under compressive load acting in the fibre direction is likely to involve microscopic buckling of individual fibres, which act as cylindrical beam-columns on an elastic foundation provided by the surrounding matrix. Compressive failure of glass FRP laminates is strongly influenced by imperfections, including:

  - imperfect fibre alignment
  - fibre non-contiguity
  - deficient fibre-matrix adhesion
  - voids, particularly where these occur at the fibre-matrix interface.

- *Comparison of compressive strength and moduli between L1, S1 and S2*
  As expected L1 has the highest compressive strength, followed by S2 and S1. However, if the test is to be used as an effective means of comparison of the compressive strength between the different parts of the panel, the values must be taken with care because of the possible buckling behaviour (load eccentricity due to the particular geometry of the specimen) which could occur before fracture.

  The values are in the anticipated range when taking into account the compressive strength of CSM laminates (137 N/mm^2) WR laminates (213 N/mm^2) and the core materials (22 N/mm^2).

4.5 Validity of theoretical methods

Theoretical methods of evaluating moduli for glass FRP laminates are difficult to verify experimentally, partly because accurate estimates of moduli for fibre and resin constituents are seldom available and partly because of the scatter of experimental data, attributable in part to inherent material variability but possibly also to inadequate testing techniques. While no final conclusions can be drawn regarding the accuracy of the theoretical methods used, it is evident that the computed results are broadly in agreement with measured data.

5 STRUCTURAL PERFORMANCE: OVERALL TESTS ON THE GLASS FRP PANELS

Three different tests were carried out on the panels for this programme:

- overall bending tests under actual conditions of support (two panels were tested)
- local tests on the corners (pull-out tests on one panel)
- compression test on the central longitudinal stiffener of the panel (two panels were tested).

5.1 Bending tests

An overall view of the bending test is given in Figure 4. The panels were tested under actual conditions of support. This means that the specimen was hung at the four corners below a static frame with brackets and rods exactly as happens in practise. In fact two panels were tested for two different types of loading.

- *Three-point loading test*
  The panel was tested under three equal static loads applied at equal distances along the length of the specimen, as shown in Figure 5. Hence, at the centre of the panel a load P was applied to the structure as well as two loads of the same magnitude 325 mm away from the centre of the panel. The loads were applied at the required points with small jacks (manifold system: three single acting cylinders). Each load was a static point load, and in order to spread the load across the breadth of the panel, load spreaders (square hollow tubular sections 80 mm x 80 mm x 5 mm) were fitted on to the panel.

- *One point loading test*
  Only one static load was applied at the centre of the panel and distributed across the breadth of the panel by the spreader.

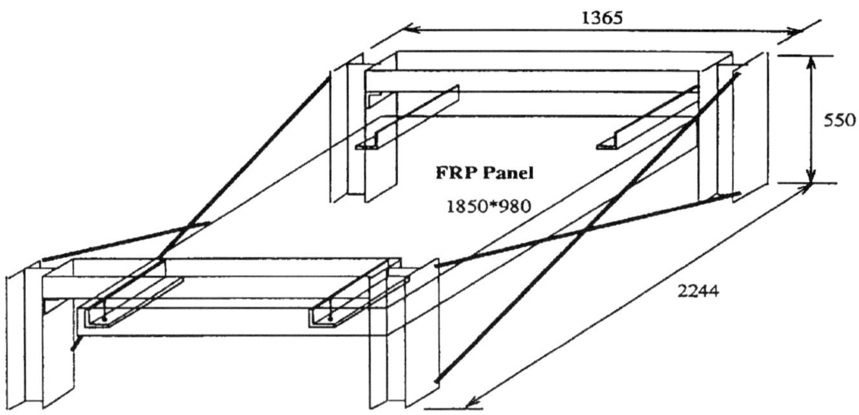

Figure 4 Overall bending test: frame supporting the panel.

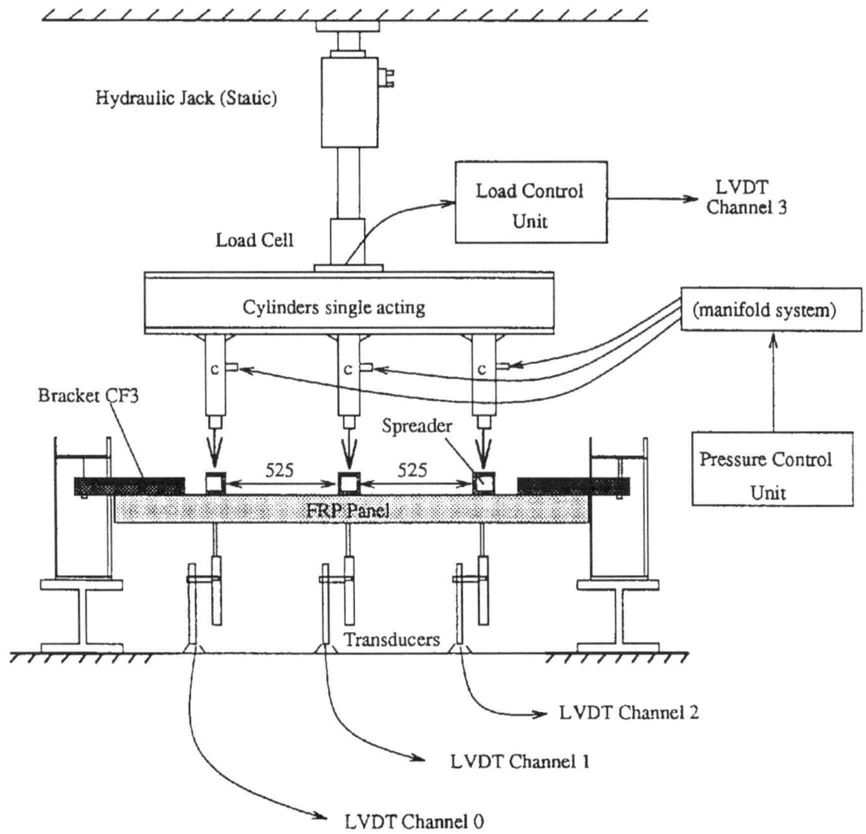

Figure 5 Three point load test set-up.

The single acting cylinders pushed against the panel, and the overall pressure was transmitted to the load cell of the hydraulic jack. The load was increased by increments until the panel reached its ultimate load. The deflection was measured from the bottom side (smooth side) under each spreader. Transducers were used to monitor the vertical displacement.

The results are shown in Figures 6 and 7. The panels were basically designed for a load level of 6623 N and the test shows that this value is much lower than the actual failure load obtained experimentally, which is around 35 to 45 kN (depending on the loading conditions). This means that the design load was around 20% of the ultimate value. One important feature of the curves obtained is that they exhibit a large linear portion up to quite high deflections.

With the three point loading test, the stresses were concentrated essentially at the corners, and the panel had a tendency to move downwards in a rigid manner. Failure occurred at the corners. Cracks appeared little by little at each corner and suddenly grew more rapidly at one of the corners; then an unstable state took place, and the panel collapsed.

In the case of the single central point loading, the deflection was more concentrated below the single spreader, and the curvature of the panel was greater. Fracture did not occur at the corners but along the width of the panel under the spreader, indicating that the stresses were greatest at this point. However, cracks appeared almost equally at the four corners all around the steel supporting plates.

It is also interesting to see that during the three point loading test, the deflection remained symmetrical with respect to the transverse axis of symmetry of the panel till fracture occurred. In the case of the single point loading, the deflection became non-symmetrical before fracture and this accelerated the process of crack growth.

5.2 Pull out tests on the corners of the glass FRP panel

In order to determine the load capacity of each corner, local tests, i.e. pull-out tests, were carried out on the corners of one panel. Because of the particular way the corners had been built up, it was very difficult to find an adequate clamping device to prevent movement of the corner during the test. A particular device was however applied to constrain the corner not to move.

The panel was turned upside down and simply supported near the corner, as shown in Figure 8. The steel insert plate and its surrounding area were left completely free. To counterbalance the torque which was produced when the rod was pushing down on the bracket, dead weights were put on to the top side of the panel. The load was applied continuously and the load deflection curve was recorded autographically. One transducer measured the vertical displacement of the tip of the bracket.

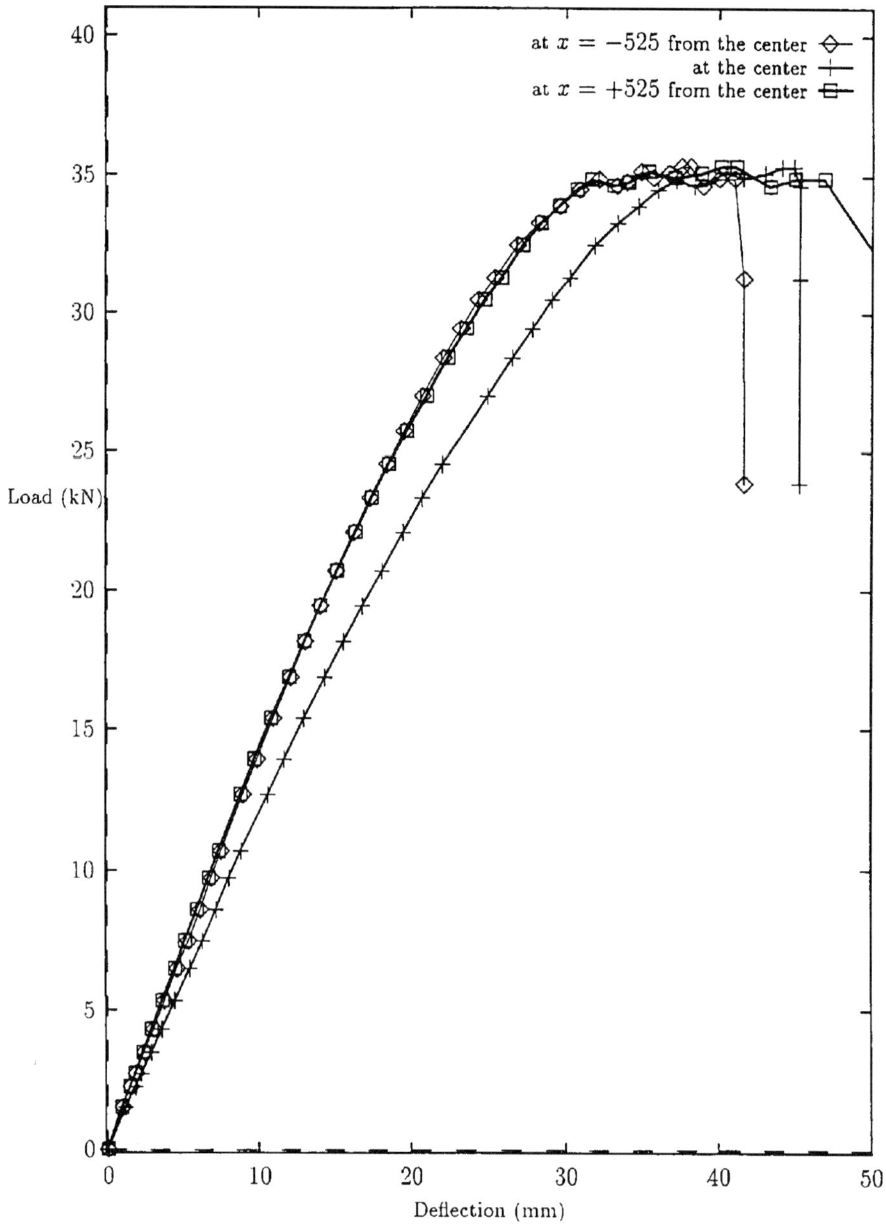

Figure 6 Overall test on a glass FRP panel: 3 point loading.

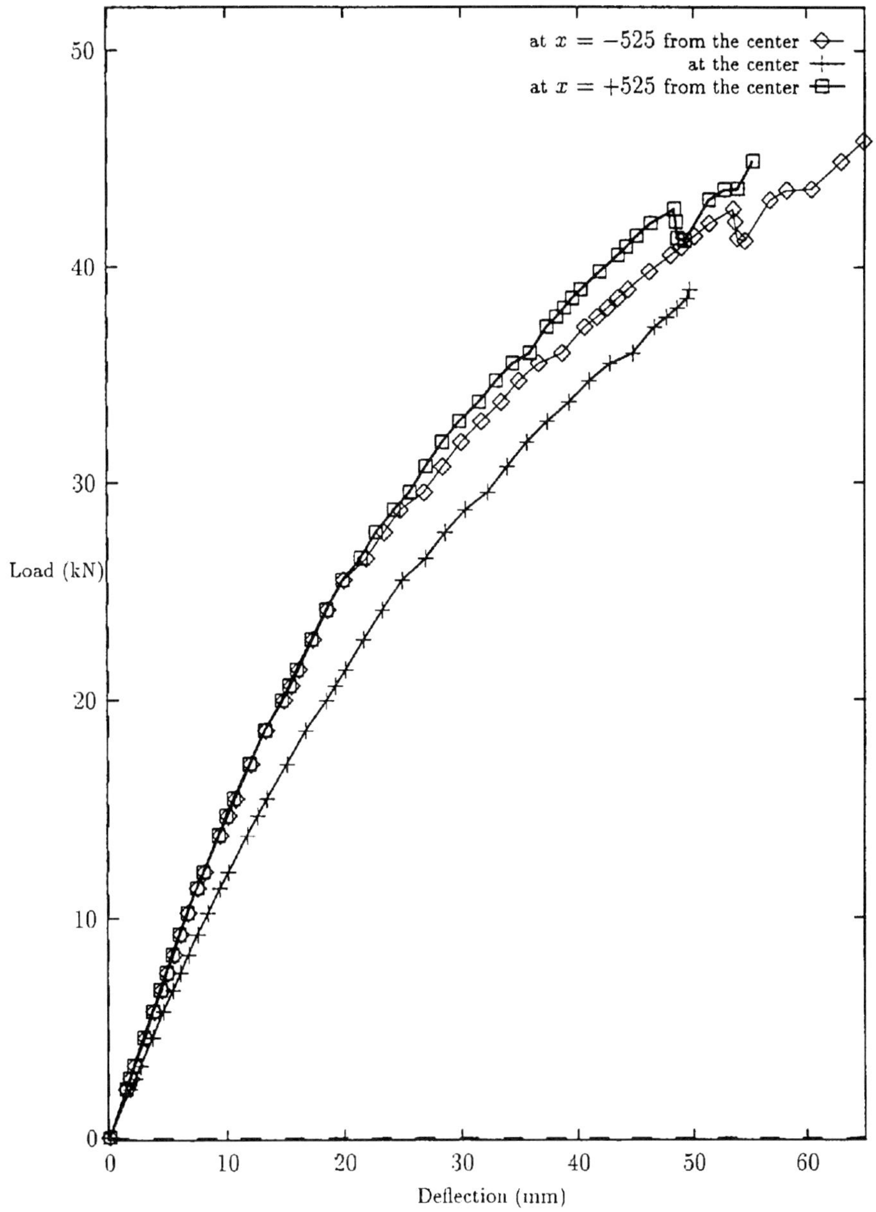

Figure 7 Overall test on a glass FRP panel: 1 point loading.

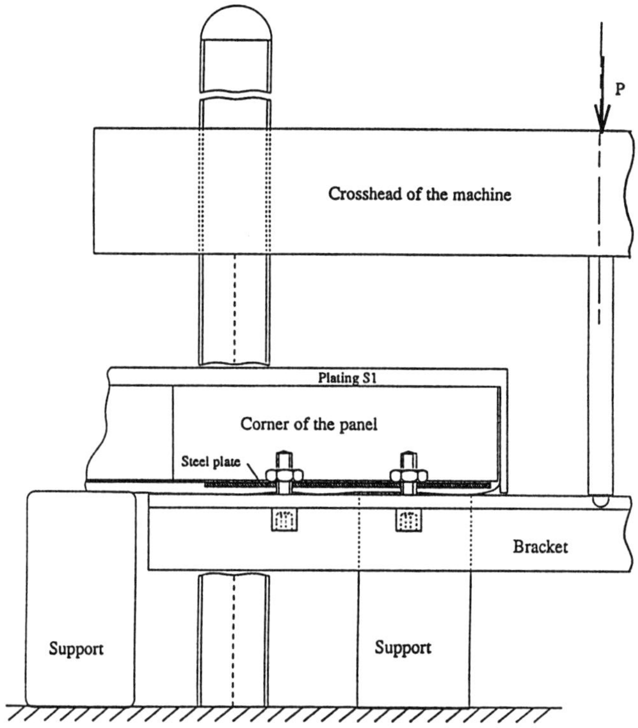

Figure 8 Corner test.

The results are shown in Figure 9. The curves obtained show that the average maximum load sustained by the corners is around 15.9 kN, with a scatter of ±2 kN, which can be attributed partly to the intrinsic structure of the corner and therefore its manufacturing, and partly to the conditions under which the panels were tested. The method of restraining the corner was not ideal.

Furthermore, this value is not far away from the value which could be deduced from the overall bending test, assuming each corner would take 1/4 of the total load. In fact as the panel fractured at around 35-45 kN, the maximum load capacity of the corner which could be expected would be approximately 8.75-11.25 kN. These values are lower than the values obtained in the individual tests. This can be explained by the fact that in the overall test, once one corner began to crack it induced in the component additional moments and forces which had the effect of accelerating the failure mechanism and thus reducing the ultimate load. This was not the case in the corner test because the corner was "isolated", even if it was still attached to the rest of the panel.

The variation of the curve is linear up to around 12.3 kN for the mean, which is 22% below the ultimate values.

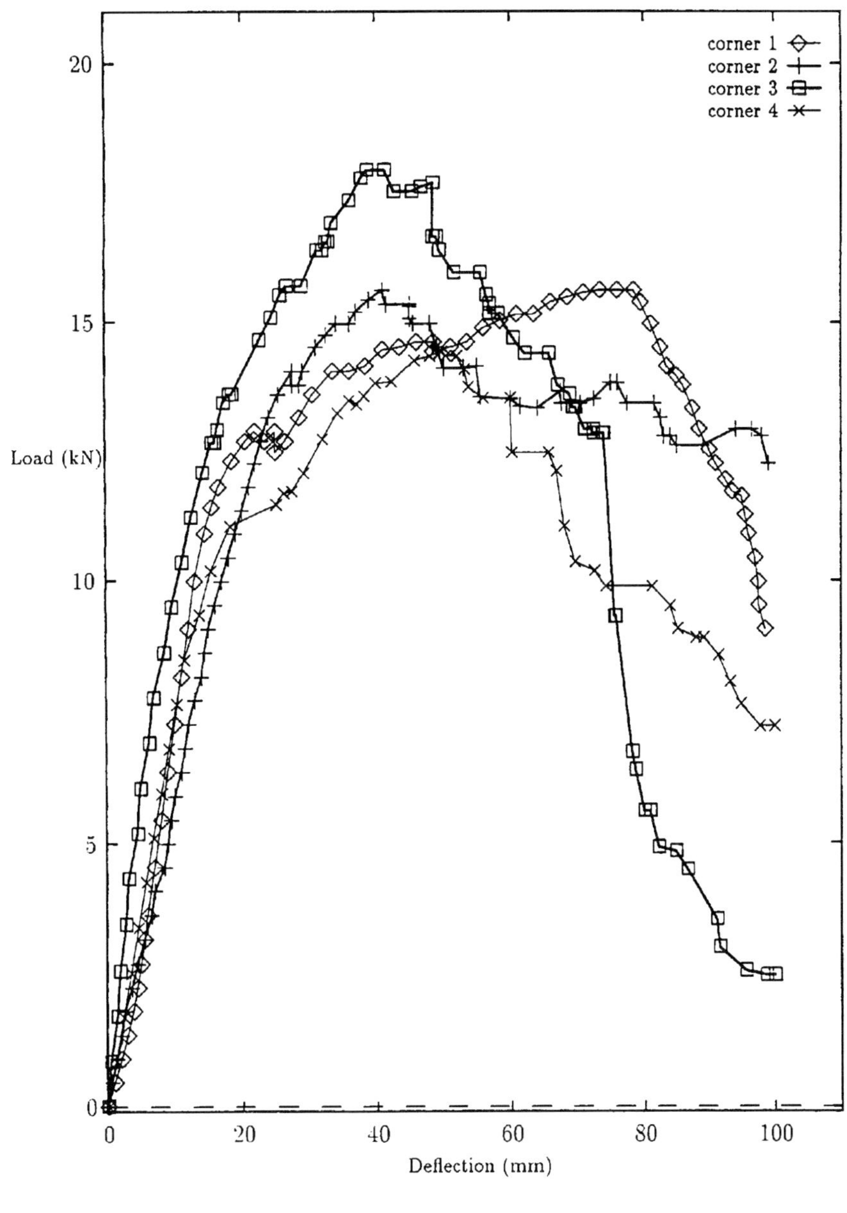

Figure 9 Test on the corners of a glass FRP panel.

5.3 Compression test on the central longitudinal stiffener

In this test a compressive in-plane load was applied to the top and bottom edges of the central longitudinal stiffener, as shown in Figure 10. The load induced at the edges of the stiffener is an incremental load from zero to a maximum value at which the structural element fails. The incremental loading was applied to monitor the instability behaviour of the stiffener until failure occurred.

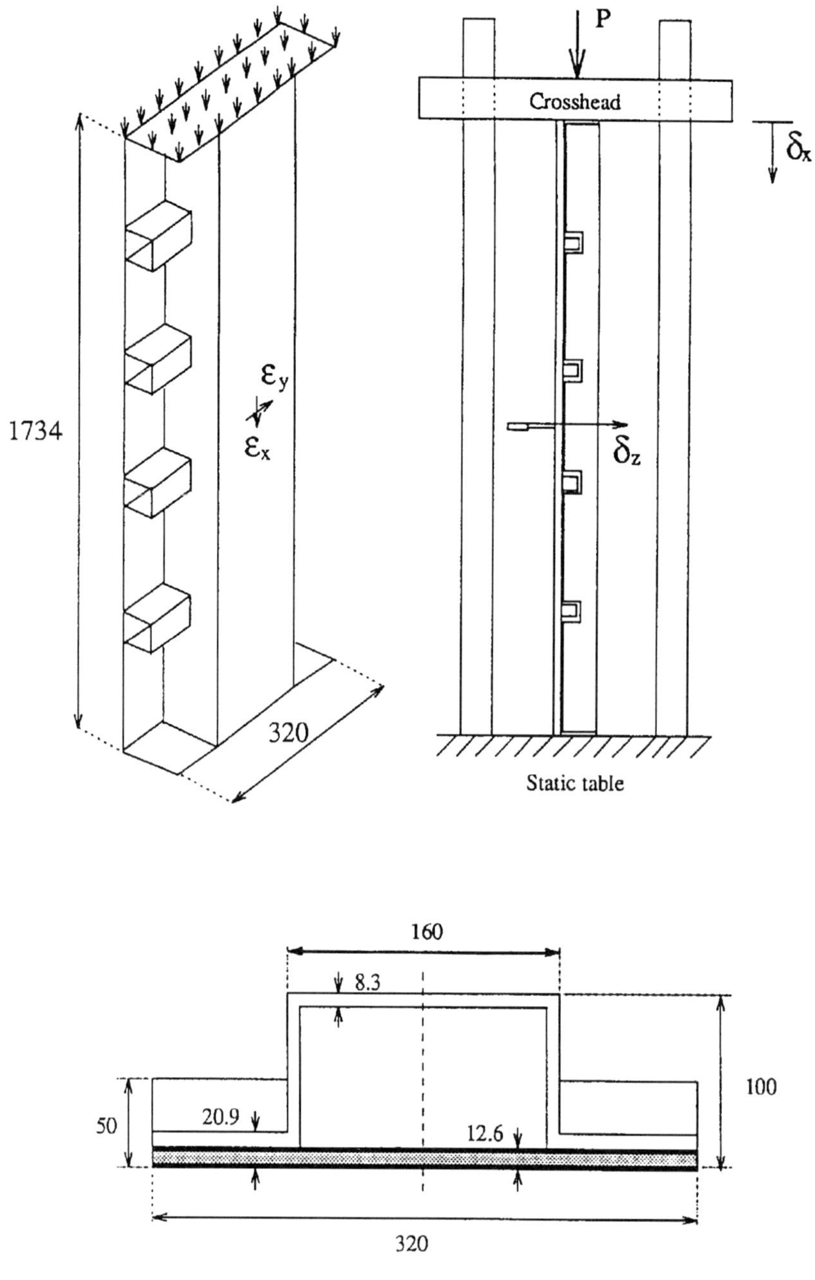

Figure 10 Compression test on the central longitudinal stiffener.

The compressive load is applied to the ends of the specimen by the crosshead of a testing machine. The bottom part of the testing machine remains static and provides a reaction force to the bottom end of the stiffener. The longitudinal displacement is defined as the displacement of the crosshead towards the static table and was measured with a transducer. The out of plane displacement was also measured with a transducer, which permitted the buckling shape of the plating side of the stiffener at mid-length to be monitored.

Two plotters were installed for this experiment. The first one recorded the buckling shape of the plating side of the stiffener versus the applied load. The second one recorded the in-plane displacement of the stiffener versus the applied load.

In addition two biaxial strain gauge rosettes were fixed on to the outer surfaces of the plating side and stiffener side at the mid-length of the stiffener. They were orientated so that the x-axis of the rosette coincided with the longitudinal axis of the specimen, and the y-axis with its transverse axis.

A typical load out of plane displacement curve is shown in Figure 11. This indicates that the fracture load is well above the theoretical critical load to cause buckling.

6 FINITE ELEMENT ANALYSIS

A finite element analysis was performed for the overall behaviour in bending under actual conditions of support of the panel and for the compression test on the central longitudinal stiffener. Shell 63 elements (4 nodes, 6 dof's per node) were used to model the panel, and Beam 4 elements (2 nodes, 6 dof's per node) were used to model the brackets at the corners. The material properties used were those computed from the theory, taking into account the orthotropic features of the panel. The load applied for the simulation of the bending test was a pressure distributed on elements corresponding to the area in contact with the spreaders during the experiments.

The load applied for the simulation of the compression test was a force applied on one node, the other nodes being coupled with this node.

The results depend strongly on the boundary conditions, which were taken at the tip of the brackets for the bending test, and on whether the central longitudinal stiffener is considered clamped or not.

In fact, it is very difficult to model the boundary conditions so that they match the real case. They are a compromise between extremes.

In the case of the compression test the ends were assumed clamped. The other boundary conditions are given in the relevant tables.

An outline of comparative results is given in Tables 4 to 7.

Table 8 gives a summary of the structural performance of the panel.

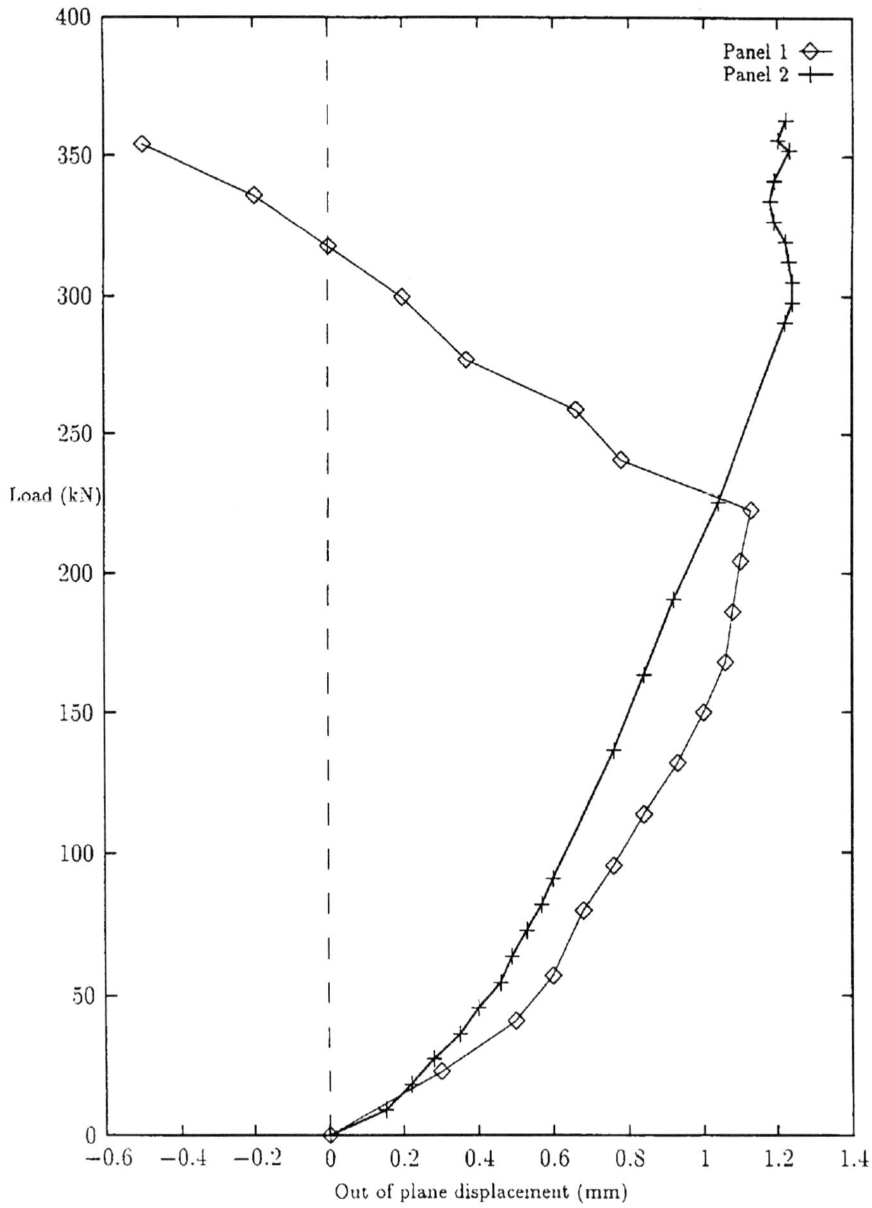

Figure 11 Buckling test: load versus out of plane displacements.

Table 4 Bending of the panel transversally supported under three equidistant static loads of 2207.5 N.

	Macaulay's method	*ANSYS*	*Bending test*
Deflection at the centre δ (mm)	2.37	2.60	2.72

Table 5 Bending of the panel supported at its four corners under a load of 15.36 kN distributed on three equidistant spreaders across the width of the panel.

	ANSYS (u_x=u_y=u_z=0)	*ANSYS* (u_y=u_z=0)	*Bending test*
Deflection at the centre δ (mm)	8.44	13.54	12.92

Table 6 Bending of the panel supported at its four corners under a load of 14.68 kN distributed on one central spreader across the width of the panel.

	ANSYS (u_x=u_y=u_z=0)	*ANSYS* (u_y=u_z=0)	*Bending test*
Deflection at the centre δ (mm)	9.22	16.49	12.6

Table 7 Compression of the central longitudinal stiffener (P= 200 kN).

	Experiment	*ANSYS*
In plane vertical displacement (mm)	7	5.3
Out of plane horizontal displacement (mm)	0.48	0.421
$\mu\varepsilon_x$ on the stiffener side	-4254	-7548
$\mu\varepsilon_x$ on the plating side	-2142	-7548
$\mu\varepsilon_y$ on the stiffener side	726	2106
$\mu\varepsilon_y$ on the plating side	556	1812

Table 8 Structural performance of the panel.

Maximum load carried by the panel under real conditions of support with three-point loading in bending	35 kN
Maximum load carried by the panel under real conditions of support with one-point loading in bending	45 kN
Average maximum load carried by one corner of the panel (pull-out test)	15.9 kN
Maximum load carried by the central longitudinal stiffener in compression	345 kN
Theoretical Euler buckling load	315 kN

7 CONCLUSIONS

The report discusses tests on a practical structural element for a bridge enclosure. The analytical approach uses the concepts developed in the EUROCOMP Design Code and EUROCOMP Handbook. Results of the test programme indicate that provided care is taken in the experimental approach the design methods proposed give realistic predictions of loads on actual structures as well as accurate estimates of mechanical properties.

BONDED JOINT TESTS FOR EUROCOMP

by

Timo Brander

Helsinki University of Technology
Laboratory of Lightweight Structures
Puumiehenkuja 5 A
FIN - 02150 Espoo

CONTENTS

BONDED JOINT TESTS

1 INTRODUCTION

The scope of this test programme was to verify certain design assumptions made in the connection design part of the EUROCOMP Design Code and in general to verify, albeit to a limited extent, the design procedures for bonded and combined joints. Bonded joints included adhesively bonded joints and laminated joints, whereas combined joints included joints based on adhesive bonding and mechanical fastening.

The test programme was necessary because of lack of a reliable joint data. For example, there are very limited data on the behaviour of laminated joints. Also, adhesive manufacturers are generally not able to provide those adhesive mechanical properties required in the design process.

The most widely used surface treatments were included in the test programme to find the minimum surface treatment requirements for a proper and durable joint.

Only the most common plate-to-plate joint configurations using one or two joint geometries could be included in the test programme, owing to limited resources. The coverage of this test programme is very limited in the field of bonded joints. However, the most important and most commonly utilized bonded joints were included to get at least some experimental data on the applicability of the EUROCOMP design procedures.

2 TEST PROGRAMME

The test programme was divided into two parts: testing of material properties and verification testing of the design assumptions and procedures. Material property tests were conducted to measure:

- mechanical properties of adhesives, and
- tensile properties of laminates.

The verification tests comprised the following types of tests:

- tests of adhesively bonded joints
- tests of laminated joints
- tests to verify material compatibility
- tests to verify minimum surface treatment requirements.

The verification tests were performed in order to verify the following assumptions made in the EUROCOMP Design Code:

- the design approach is valid for bonded joints
- single-lap and single-strap joints can be treated identically
- double-lap and double-strap joints can be treated identically
- adhesively bonded and laminated joints can be treated identically
- the tested materials are compatible

In addition the tests considered:

- minimum surface treatment requirements for composites
- effect of laminate surface layer on the bonded joint strength.

3 TEST MATERIALS AND TEST SPECIMENS

3.1 Test materials

The test materials included five different types of glass fibre reinforced laminates and three adhesives. The laminates were mainly manufactured using the hand (wet) lay-up technique, which is the predominant method within the construction industry. One of the laminate types tested was manufactured using the RTM (resin transfer moulding) method. The resins used in the laminates were two polyesters and two vinyl esters. The most widely used resin in the test programme, Neste A300, is tailored for the RTM manufacturing process, but it was also used in the hand lay-up process. The equivalent Neste resin for the hand lay-up process is G300 (see Table 3.2, EUROCOMP Design Code). Materials and stacking sequences of the laminates are given in Table 1. The amount of reinforcements in the main directions is given in Table 2.

Two epoxy adhesives and one polyurethane adhesive were included in the test programme, namely:

- Araldit 2015 (epoxy) by Ciba-Geigy
- DP-460 (epoxy) by 3M
- FOSS THAN 2K 1897 (polyurethane) by Casco Nobel.

3.2 Test specimens

The tensile properties of laminates were determined using ISO 3268-1978 Type II Rectangular Specimens. The adhesive shear properties were determined using typical thick adherend lap shear test (TALST) specimens (Figure 1). Bonded joint tests were conducted using the following joint configurations (Figure 2):

- single-lap joints
- single-strap joints
- double-lap joints
- double-strap joints.

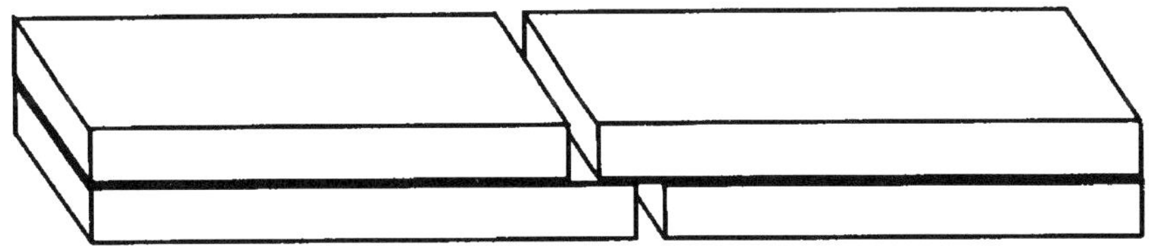

Figure 1 Test specimen for adhesive shear testing.

Table 1 Test programme laminates.

Resin	*Mfg method*	*Code*
Neste A300, UP orthophthalic	RTM	A300 RTM
Neste A300, UP orthophthalic	Hand lay-up	A300 HL
Neste K530, UP isophthalic (corr. resistant)	Hand lay-up	K530
Dow Derakane 411, VE bisphenol-A	Hand lay-up	411
Dow Derakane 470, VE novolac	Hand lay-up	470
Neste A300, UP orthophthalic	Hand lay-up	A300 WR
Neste A300, UP orthophthalic	Hand lay-up	A300 UD1
Neste A300, UP orthophthalic	Hand lay-up	A300 UD2

Reinforcements:

1.	CSM	Owens Corning MK 22-450-125 450 g/m^2
2.	CSM	Ahlstrom M113 150 g/m^2
3.	WR	Ahlstrom R12-600 9650 600 g/m^2
4.	CFM	Vetrotex U-812 375 g/m^2
5.	NCMA	Ahlstrom 2 x 300 g/m^2
6.	UD	Ahlstrom R24-600-L 9267 600 g/m^2

Stacking sequences:

RTM laminate
[CSM150/NCMA600(0°)/CFM375/NCMA600(0°)/CFM375/NCMA600 (0°)/CFM375/NCMA600(0°)/CSM150]

Hand lay-up laminates
A300, K530, 411, 470:
[CSM450/WR600(0°)/CSM450/WR600(±45°)/CSM450/WR600(0°)/ CSM450]
A300 WR:
[WR600(0°)/CSM450/CSM450/WR600(±45°)/CSM450/CSM450/WR 600(0°)]
UD1:
[UD(0°)/WR600(0°)/CSM450/WR600(0°)/CSM450/WR600(0°)/UD(0 °)]
UD2:
[UD(0°)/CSM450/WR600(0°)/CSM450/UD(0°)]

NCMA = non-crimp multi-axial

Table 2 Amount of reinforcements in the main directions with respect to the loading direction (0°).

Laminate	0° (g/m^2)	90° (g/m^2)	±45° (g/m^2)	CSM (g/m^2)	CFM (g/m^2)	Total (g/m^2)
A300 RTM	1200	1200		300	1125	3825
A300 HL[a]	600	600	600	1800		3600
A300 WR	600	600	600	1800		3600
A300 UD1	2100	900		900		3900
A300 UD2	1500	300		900		2700

a) also K530, 411, and 470

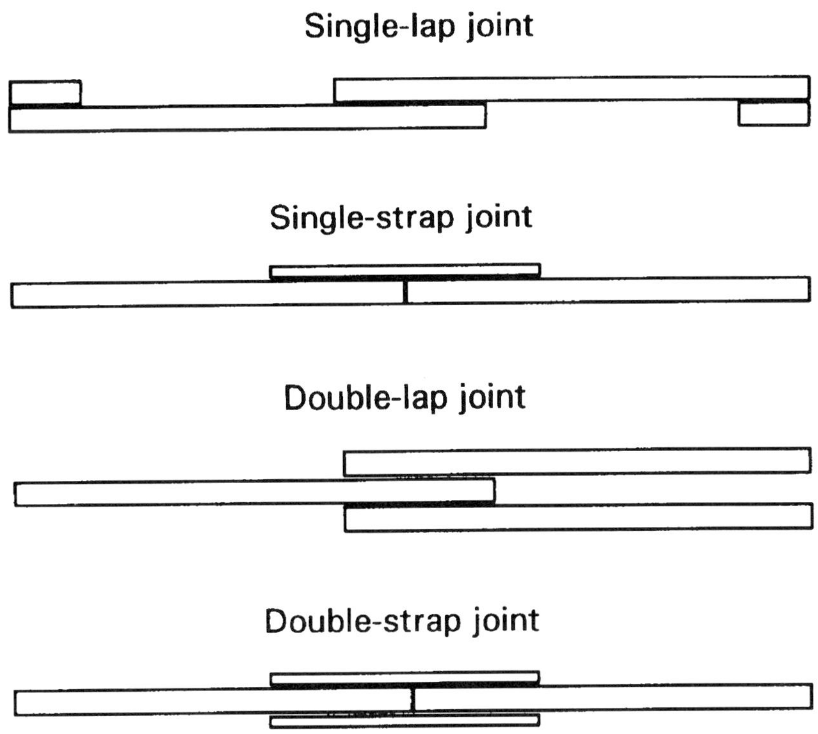

Figure 2 Joint configurations tested.

3.3 Manufacturing of laminates and test specimens

Laminates were manufactured mainly by Neste subsidiary Plastilon Oy. The laminates A300 WR, A300 UD1 and A300 UD2 were manufactured by the Helsinki University of Technology, Laboratory of Lightweight Structures (HUT/LLS). Adhesively bonded joints and specimens for the mechanical characterisation of the adhesives were made by HUT/LLS. Laminated joints were made by Neste and HUT/LLS. All laminates, laminated joints, and adhesives were cured at room temperature for 24 hours minimum and then postcured according to the procedures given in Table 3.

Table 3 Cure and postcure procedures for polymeric test materials.

Resin or adhesive/mfg method	*Postcure*
Neste A300/RTM	24h 50°C + 3h 80°C
Neste A300	5h 50°C + 3h 80°C
Neste K530	5h 50°C + 3h 80°C
Dow Derakane 411	8h 80°C
Dow Derakane 470	8h 80°C
Araldit 2015	30min 80°C
DP-460	2h 60°C
FOSS THAN 2K 1897	2h 60°C

For thick adherend lap shear test specimens a surface treatment was performed. The aluminium adherends were 10 mm thick and 25 mm wide. After the specimens were cured and postcured, the adherends were cut to form a test area with an 8 mm lap length.

Plates of 200 mm width were cut to the required lengths from the cured laminates. Areas to be bonded were treated according to specified surface treatment procedures prior to bonding.

In adhesively bonded joints copper wires with a diameter of 0.34 mm were used in between the laminates to control the adhesive layer thickness. The copper wires were placed in areas to be discarded. Adhesive fillets with a 45° angle were formed at the joint ends. The adherend ends were not shaped. Test specimens with 25 mm width were cut from the bonded laminates. Typically five specimens from each batch were tested.

In laminated joints, the other adherend or the strap or straps were laminated onto the treated laminate surface or surfaces. Samples 25 mm wide were cut from the cured laminates. Typically five specimens from each batch were tested.

4 TESTING

The testing was performed in the Laboratory of Lightweight Structures at the Helsinki University of Technology. Tests were conducted under laboratory conditions with a nominal temperature of 23°C and relative humidity of 40%. A Dartec testing system was used.

Laminate material tests were based on the ISO 3268-1978 standard. Type II rectangular specimens were used. Young's modulus was first determined using an extensometer. The same specimen was then loaded to failure without the extensometer.

Thick adherend lap shear tests (TALST) were not based on any standard.

However, standard practice used in the tests was followed. The displacement rate used was 0.3 mm/min. The adhesive layer displacement was measured using two in-house built KGR-1 type extensometers.

The joint tests were based on the ISO 4587-1979 standard when applicable. The constant displacement rate was chosen so that the failure occurred within 65±20 seconds. Typically this led to displacement rates of 1.5 or 3.0 mm/min depending on the lap length. The load and loading head displacement were recorded.

5 RESULTS

5.1 Mechanical properties of laminates

The results of laminate material tests are presented in Table 4. Mechanical properties of A300 WR were not tested as it had the same materials, ply orientations, and number of plies as the A300 HL laminate. The fibre volume fraction presented in Table 4 was calculated from the nominal mass per unit area of the reinforcements and from the laminate average thickness using the value of 2.6 g/m^3 for the specific gravity of glass.

Table 4 Laminate material properties.

Laminate	E (kN/mm^2)	σ_{ult} (N/mm^2)	N_{ult} (N/mm)	$t_{average}$ (mm)	v_f [a]
A300 RTM	14.9	268	1347	5.0	0.29
A300 HL	11.3	159	924	5.8	0.24
K 530	11.8	178	1048	5.9	0.23
411	11.3	149	960	6.4	0.22
470	12.0	164	932	5.7	0.24
A300 UD1	16.4	224	1211	5.3	0.28
A300 UD2	15.7	244	905	3.6	0.29

[a] Calculated value

5.2 Mechanical properties of adhesives

The TALST system and the KGR-1 type extensometers were used for the first time at HUT/LLS in actual testing. The system did not function correctly in every respect. The tests were performed in two sets. The second set of tests was performed about four months after the first one to fill the gaps and to verify the results of the first set.

In the first set no reliable values were obtained for the shear modulus. Also the tests on polyurethane were rather unsuccessful. In the second set more reliable results were obtained. However, since the number of specimens in each test series was limited, one should consider the results in Table 5 only as indicative and to save time they should not be used as definitive material data.

Table 5 Mechanical properties of the adhesives.

	n	G (N/mm^2)	τ_e (N/mm^2)	τ_{max} (N/mm^2)]	$\gamma_{failure}$	t_a (mm)
Set 1						
2015	3	N/A	11.2	20.9	0.597	0.41
DP460	3	N/A	18.4	31.6	0.791	0.41
1897	1	N/A	N/A	8.8	0.867	0.43
Set 2						
2015	5	670	14.2	21.0	0.709	0.37
DP460[a)]	3	705	28.1	32.9	0.478	0.39
1897	4	110	1.6	11.8	1.099	0.41

n = number of specimens tested
G = adhesive shear modulus
τ_e = adhesive elastic/plastic shear stress
τ_{max} = adhesive maximum shear stress
$\gamma_{failure}$ = adhesive failure shear strain
t_a = adhesive layer thickness

[a] Partly adhesive failure.

5.3 Bonded joints

The results of the bonded joint tests are presented in Tables 6 to 9. The results of material compatibility tests are presented in Table 10 and the results of surface treatment tests in Table 11. The following abbreviations and notation are used:

Geo. joint geometry
SL single-lap joint
DL double-lap joint
SS single-strap joint
DS double-strap joint
SC scarfed joint
/C compressive loading

L/t ratio of overlap length to adherend thickness
n number of tested specimens
N loading density
SDEV standard deviation
t adherend thickness

t_a	adhesive layer thickness
t_s	strap or the other adherend thickness
ϕ	joint efficiency: the ratio of joint strength to adherend strength

Table 6 Compilation of adhesively bonded joint tests in which a mat layer was used on laminate surfaces, RTM manufactured laminates.

Laminate	*Geo.*	L/t	N (N/mm)	*SDEV*	ϕ	t (mm)	t_a (mm)	n
Araldit 2015								
A300 RTM	SL	5	218	4.7%	0.16	4.9	0.4	5
A300 RTM	SL	25	455	1.6%	0.34	5.0	0.5	7
A300 RTM	DL	12.5	657	4.9%	0.49	5.0	0.5	5
A300 RTM	DL/C	12.5	1304	7.0%	0.97	5.1	0.5	3
A300 RTM	SS	25	404	2.4%	0.30	5.1	0.6	6
A300 RTM	DS	12.5	645	3.8%	0.48	5.0	0.8	5
A300 RTM	SC	20	282	1.9%	0.21	5.1	N/A	4
3M DP460								
A300 RTM	SL	5	232	2.9%	0.17	4.9	0.25	5
A300 RTM	SL	25	452	2.1%	0.34	5.1	0.25	6
Casco Nobel FOSS THAN 2K 1897								
A300 RTM	SL	5	196	2.4%	0.15	5.0	0.2	5
A300 RTM	SL	25	746	3.9%	0.55	5.0	0.5	5

Table 7 Compilation of adhesively bonded joint tests in which a woven roving or uni-directional layer was used on laminate surfaces, hand lay-up laminates. Araldit 2015.

Laminate	*Geo.*	L/t	N ([N/mm)	*SDEV*	ϕ	t (mm)	t_a (mm)	n
A300 WR	SL	5	243	7.2%	0.26	5.9	1.2	5
A300 WR	SS	25	482	7.4%	0.52	6.0	1.4	5
A300 WR	DS	12.5	599	12.2%	0.65	6.4	1.1	5
A300 UD1	DS	12.5	840	8.9%	0.69	5.7	1.3	5
A300 UD2	DS	12.5	478	17.7%	0.53	3.9	1.4	5

Table 8 Compilation of laminated joint tests in which a mat layer was used on laminate surfaces, hand lay-up laminates.

Laminate	*Geo.*	L/t	N (N/mm)	*SDEV*	ϕ	t (mm)	t_s (mm)	n
Neste A300								
	SL	5	244	15.2%	0.26	6.1	5.6	5
	SL	25	470	3.9%	0.51	5.9	6.2	5
	SS	5	176	11.0%	0.19	6.0	3.1	5
	SS	5	191	3.8%	0.21	5.5	5.9	5
	SS	25	424	2.6%	0.46	6.0	3.5	5
	SS	25	482	2.7%	0.52	5.9	5.0	6
	DS	12.5	463	1.3%	0.50	6.3	2.6	5
	DS	12.5	573	3.8%	0.62	5.8	2.7	5
Neste K530								
	SS	5	196	10.2%	0.19	5.9	3.55	5
	SS	25	465	1.2%	0.44	5.8	3.05	5
Derakane 411								
	SS	5	272	6.7%	0.28	6.2	2.4	5
	SS	25	464	6.7%	0.28	6.2	2.4	5
	DS	12.5	788	7.2%	0.82	5.8	3.0	5
Derakane 470								
	SS	5	234	6.0%	0.25	5.9	2.9	5
	SS	25	478	4.2%	0.51	5.9	3.1	5

Table 9 Compilation of laminated joint tests in which a woven roving layer was used on laminate surfaces, hand lay-up laminates (Neste A300).

Laminate	*Geo.*	L/t	N (N/mm)	*SDEV*	ϕ	t (mm)	t_s (mm)	n
	SS	5	229	6.7%	0.25	7.0	4.4	5
	SS	25	414	7.0%	0.45	5.9	6.6	5
	DS	12.5	653	5.3%	0.71	6.0	5.5	5

Table 10 Compilation of material compatibility tests (Dow Derakane 441 laminates).

Adhesive	*Geo.*	L/t	N (N/mm)	*SDEV*	ϕ	t (mm)	t_a (mm)	n
2015	SL	5	323	3.0%	0.34	6.3	0.1	5
1897	SL	5	229	3.0%	0.24	6.2	0.2	5

Table 11 Compilation of surface treatment tests (Neste A300 hand lay-up laminates).

Adhesive or resin	*Tre.*	*Geo.*	L/t	N (N/mm)	*SDEV*	ϕ	t (mm)	t_a or t (mm)	n
2015	PP	SL	5	268	1.3%	0.29	6.8	0.3	5
1897	PP	SL	5	275	6.4%	0.30	6.6	0.7	6
A300	PP	SL	5	305	4.6%	0.33	6.4	5.1	5
2015	MA	SL	5	256	3.8%	0.28	6.7	0.1	5
1897	MA	SL	5	223	4.0%	0.24	6.3	0.4	5
A300	MA	SL	5	335	6.8%	0.36	6.1	5.5	5

Tre. = surface treatment
PP = peel ply
MA = machine abrasion

6 DISCUSSION

6.1 Laminates

The measured laminate stiffnesses and strengths are typical for these types of laminates. It can be seen that the manufacturing method has a considerable effect on the mechanical properties. For example, the RTM manufactured laminates have higher ultimate loading density than the hand lay-up manufactured laminate A300 UD1, while the nominal mass per unit area of the aligned fibres in the 0° direction, i.e. in the loading direction, are 1200 g/m^2 and 2100 g/m^2 respectively. However, the A300 UD1 laminate has a higher Young's modulus.

The hand lay-up manufactured laminates A300 HL, K530, 411 and 470, which have identical reinforcement layers and stacking sequences, have slight differences in Young's modulus and strength. Table 12 shows these values when the nominal thickness of 6.0 mm is used for each laminate. The bisphenol-A vinyl ester laminate (411) has the highest Young's modulus and the isophthalic polyester laminate (K530) has the highest strength.

Table 12 Mechanical properties of hand lay-up manufactured laminates as measured and referred to the thickness of 6.0 mm.

Laminate	$t_{MEASURED}$ [mm]	$E_{MEASURED}$ (kN/mm^2)	$E_{6.0\ mm}$ (kN/mm^2)	$\sigma_{MEASURED}$ (N/mm^2)	$\sigma_{6.0\ mm}$ (N/mm^2)
A300	5.8	11.3	10.9	159	154
K530	5.9	11.8	11.6	178	175
411	6.4	11.3	12.1	149	159
470	5.7	12.0	11.4	164	156

6.2 Adhesives

The arrangements used for testing mechanical properties of the adhesives had been constructed recently. The determined values of the shear modulus were unreliable in the first set but it the second set the scatter of results was reasonable. However, the shear modulus values should be used realising their indicative nature while maximum shear stress values may be considered reliable. Typical shear behaviour of the tested adhesives is presented in Figure 3.

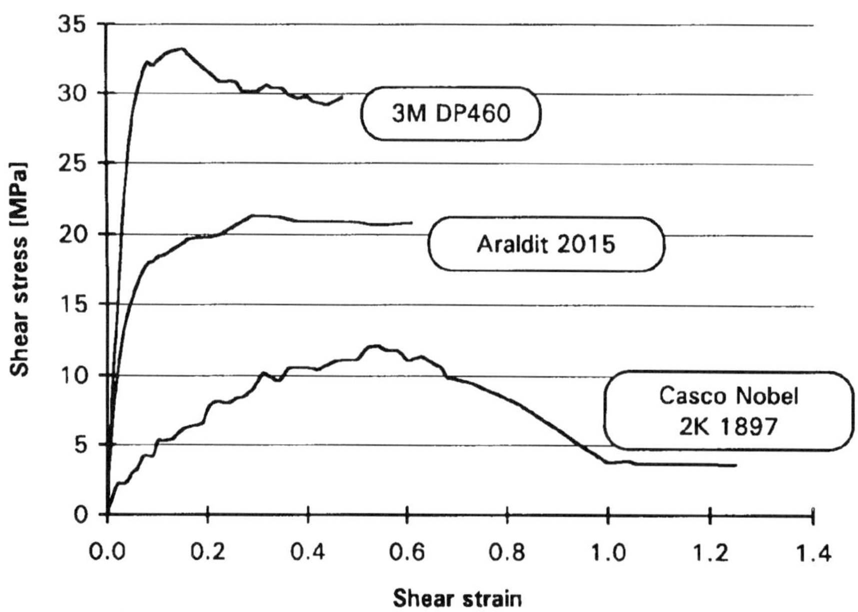

Figure 3 Typical shear behaviour of tested adhesives.

6.3 Bonded joints

The predominant failure mode in bonded joints was cohesive failure of the adherend. Only in those adhesively bonded joints with the polyurethane adhesive (1897) was a cohesive failure of the adhesive encountered. No adhesive failures were detected, indicating that the surface treatments performed had been sufficient. The quality of adhesive layers was in general satisfactory. However, in joints of laminates with a woven roving or a uni-directional surface layer (Table 7) the adhesive layers were thick and of poor quality, having large voids. With laminates of a uni-directional surface layer, this resulted in a modest (about 10%) decrease in joint strength. However, it may be concluded that the adhesive used (Araldit 2015) tolerated extensive defects in the adhesive layer well, without significant decrease in the joint strength.

The highest joint efficiency was encountered in the compression loaded double-lap joint, where an efficiency of 0.97 was achieved. The lowest efficiency, 0.15, was measured for a single-lap joint with the polyurethane adhesive.

It may be noticed that adhesively bonded joints and laminated joints appear to have similar maximum loading densities, i.e. load per unit width, as indicated in Figure 4, while the joint efficiencies of laminated joints are typically higher than those of adhesively bonded joints (Figures 5 to 7). One possible explanation of this behaviour is that in laminated strap joints the bond surface of the strap had a woven roving surface layer, while in adhesively bonded strap joints a laminate having a mat surface layer was used as straps.

According to these results it seems justifiable to use the design procedures of adhesively bonded joints also for laminated joints.

Figure 4 also shows that maximum loading densities of single-lap and single-strap joints are very close to each other. This is also true for double-lap and double-strap joints, as can be seen in Table 6. These results indicate that lap and strap joints may be treated using similar design principles and design procedures.

The increase in the overlap length from the L/t value of 5 to 25 typically doubled the joint efficiency, as shown in Figures 6 and 7. It should be realised, however, that this occurs only within this specific range of the L/t ratios. A similar increase from 25 to 125 would produce practically no increase in the joint strength.

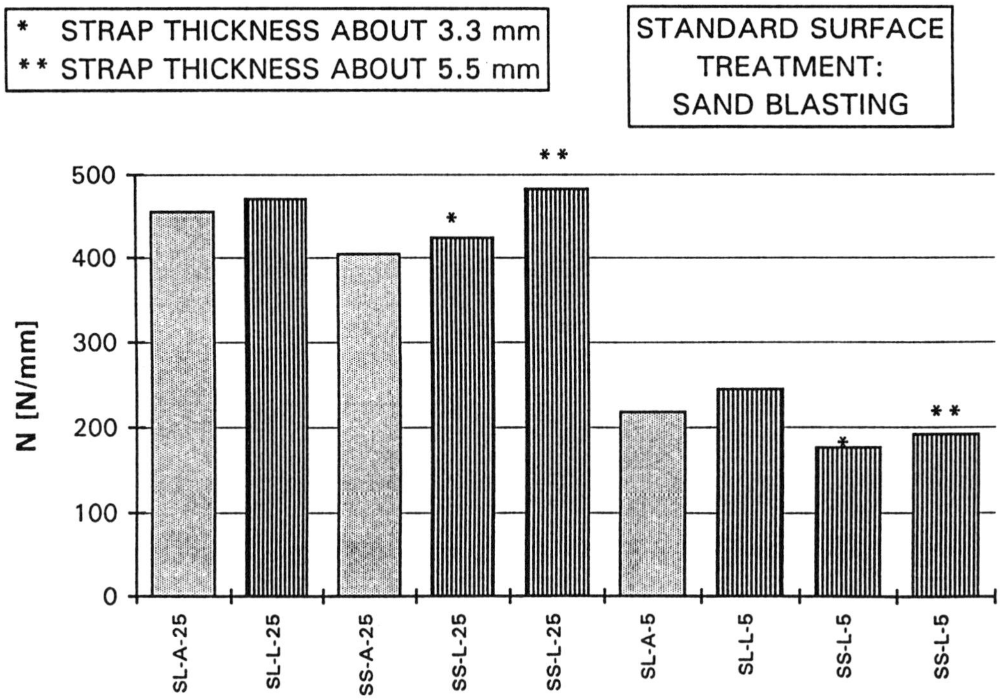

Figure 4 Adhesively bonded and laminated single-lap and single-strap joints.

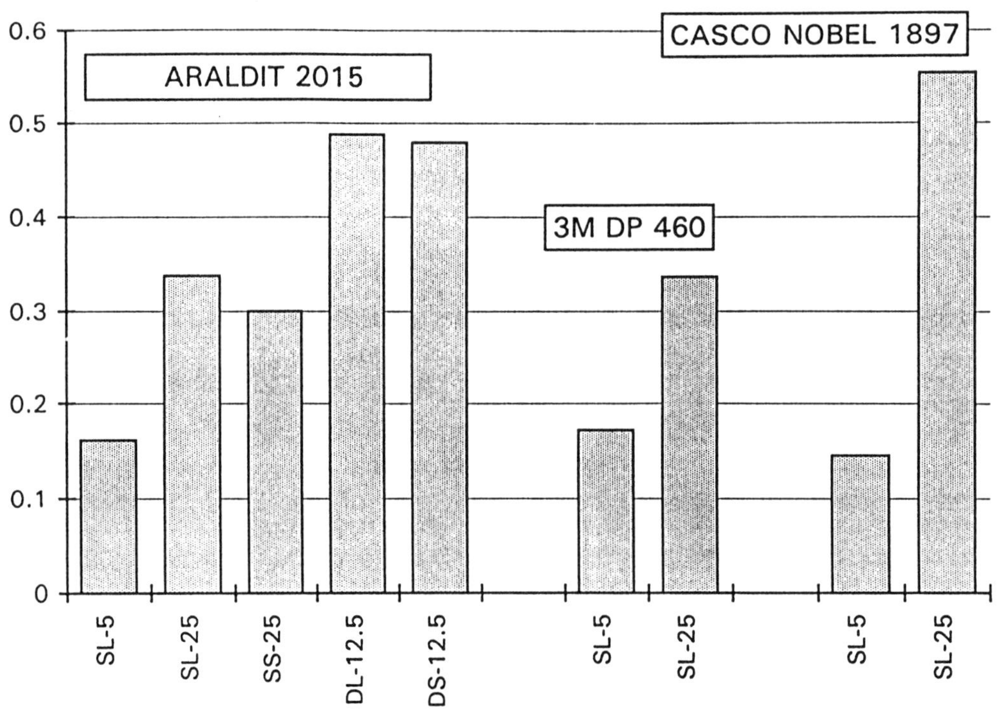

Figure 5 Joint efficiencies of adhesively bonded joints with A300 RTM adherents.

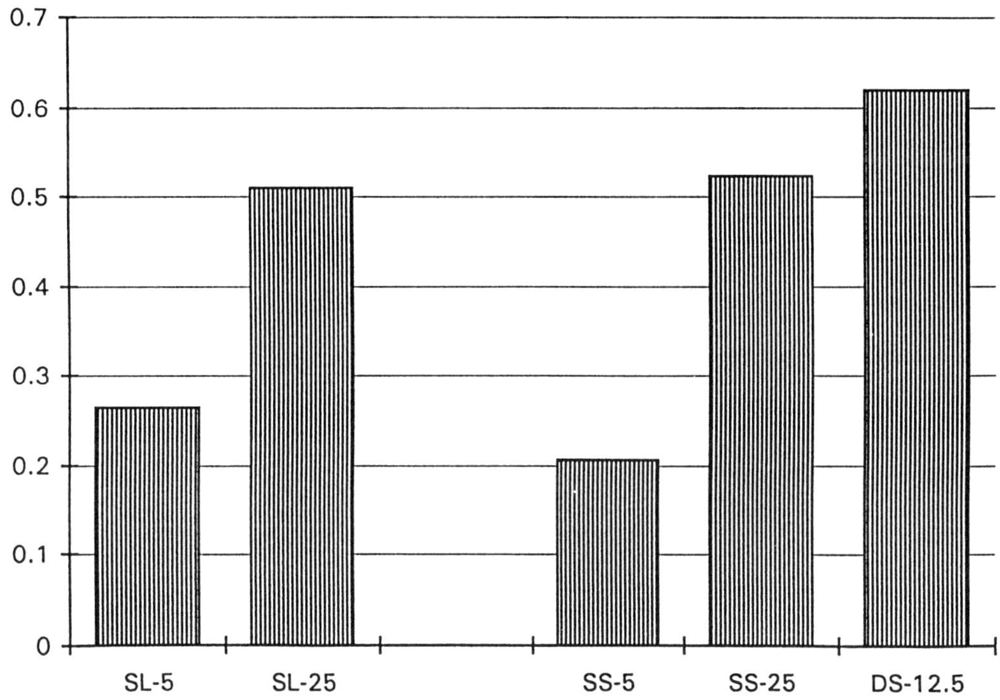

Figure 6 Joint efficiencies of laminated joints with A300 HL adherents.

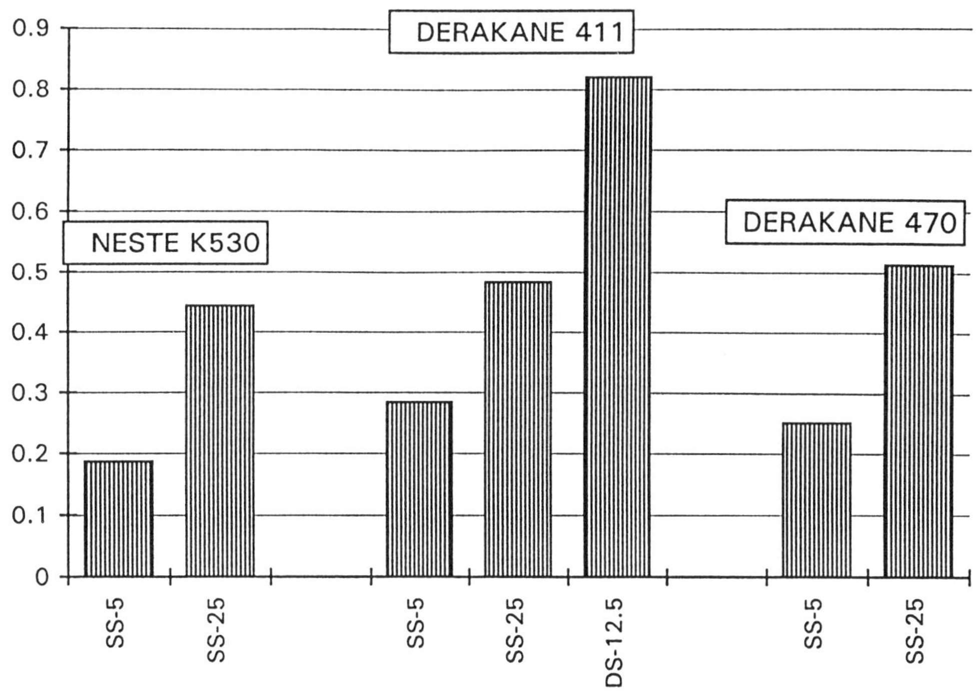

Figure 7 Joint efficiencies of laminated joints with various hand lay-up adherents.

The type of reinforcement on the bond surface has a significant effect on the joint efficiency achievable, as shown in Figure 8. The absolute magnitude of the improvement cannot be defined as the laminates having a mat surface layer or a woven roving or UD layer were not identical in respect to the manufacturing method, laminate thickness, and fibre volume fraction. However, the use of woven roving, or uni-directional reinforcement as a surface layer increases the joint strength compared to that of laminates with mat (CSM) surface layers. Naturally the fibre orientation of the surface layer has to coincide with the principal loading direction of the joint.

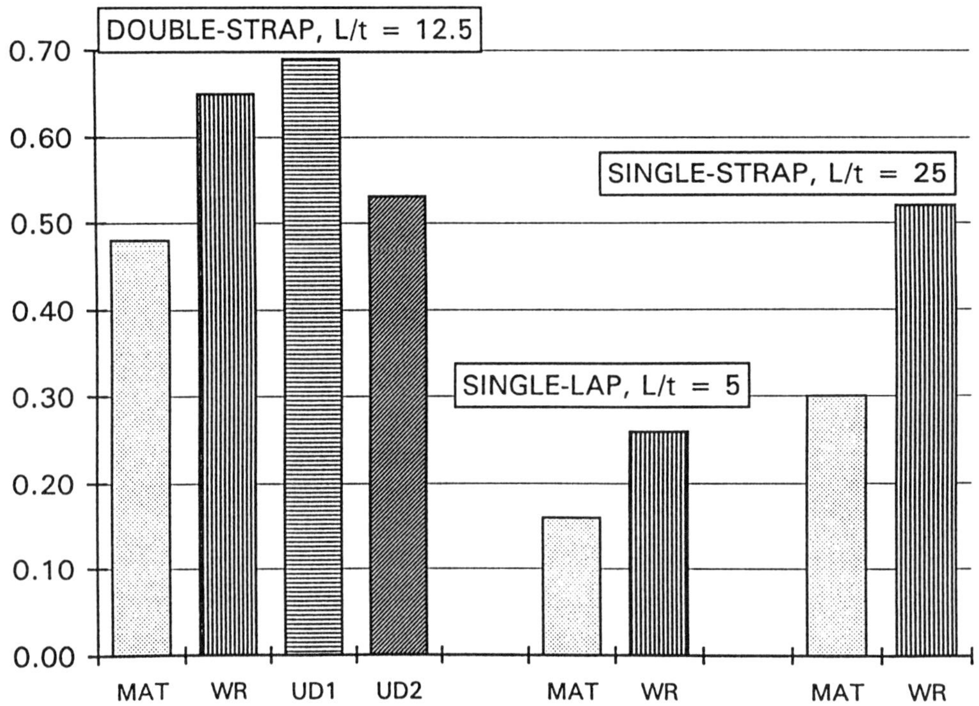

Figure 8 Effect of surface ply on the joint efficiency.

The effect of the adhesives and resins on the joint efficiency is shown in Figure 9. A possible explanation for the better performance of laminated joints compared to adhesively bonded joints is the one given above concerning the strap lay-up sequence. However, that does not explain the behaviour of the laminated single-lap joint.

Only one scarf joint geometry was tested. It had the maximum allowable scarf angle and therefore the joint was not very effective. However, it provided a maximum loading density of 282 N/mm^2, compared to 218 N/mm^2 in the single-lap joint, while the bond lengths of the scarf joint and the single-lap joint were 14 and 25 mm, respectively.

In material compatibility tests no critical material combinations in respect of adhesion were detected.

The effect of material surface treatments is presented in Figure 10. Again it should be noted that those adhesively bonded joints with sand blasting were RTM laminates, while others were hand lay-up laminates.

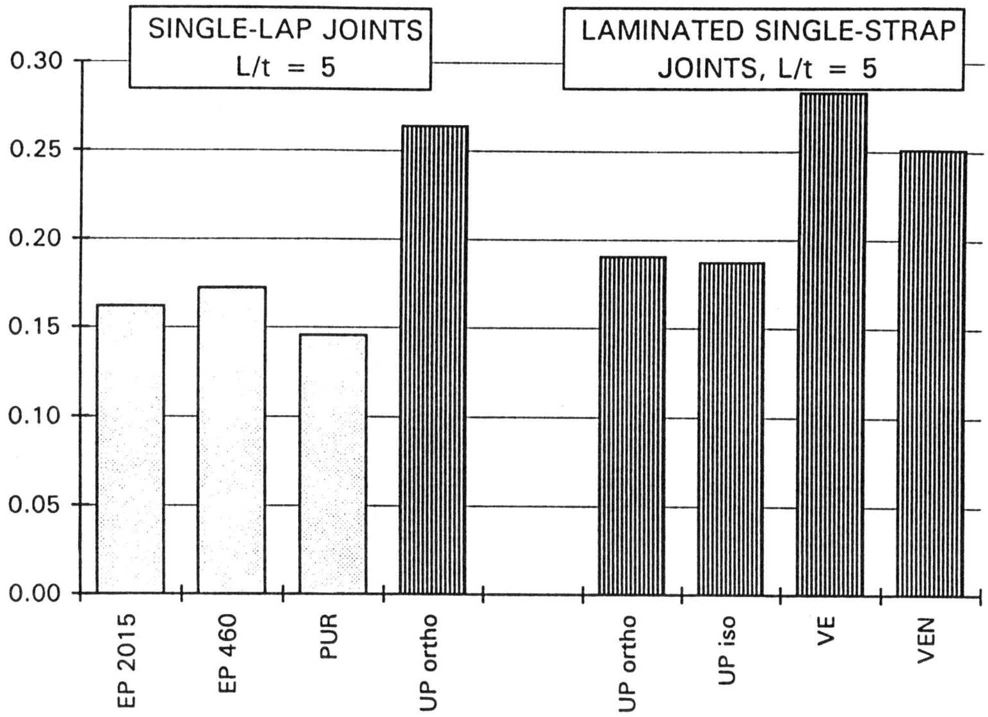

Figure 9 Effect of adhesive resin on the joint efficiency.

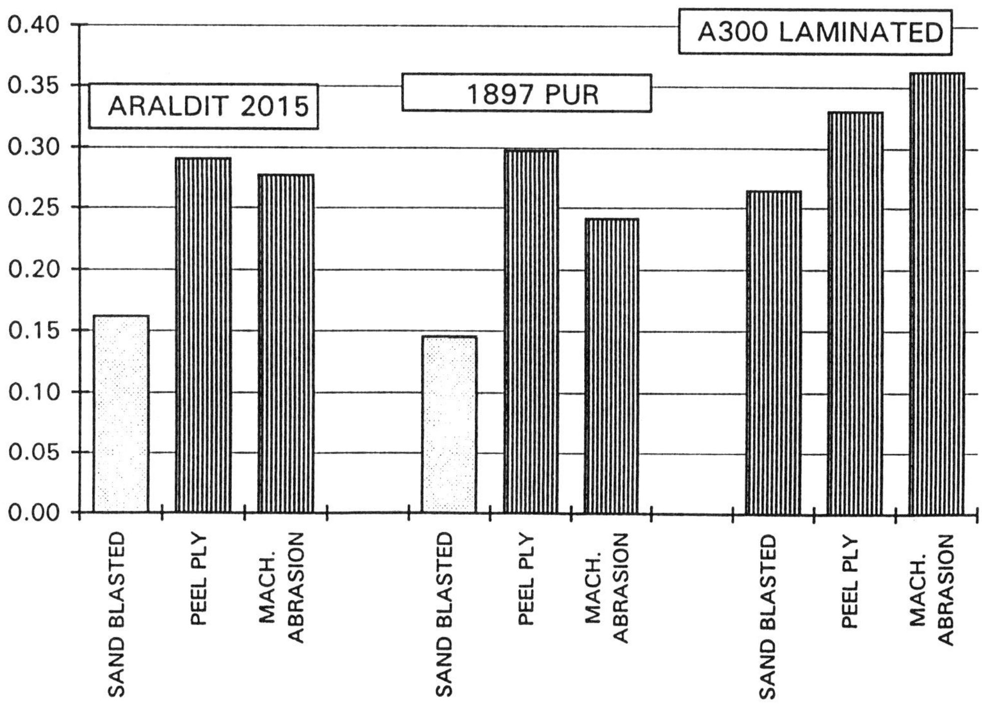

Figure 10 Effect of surface ply on the joint efficiency, single-lap joints, L/t = 5.

The load-displacement data provides some information on the behaviour of bonded joints. Especially interesting and informative is the behaviour of single-strap joints. Figure 11 shows a typical behaviour of a single-lap joint with the L/t ratio of 5. The first knee in the load-displacement curve is due to the separation of the butt ends of the adherents. The second minor knee just before the final failure is due to peeling of the strap ends. Figure 12 shows a corresponding behaviour of a single-strap joint with an L/t ratio of 25. Again the two failure modes are detected, although the second knee (with the displacement value of 2.4 mm) is relatively small.

6.4 Correlation with the EUROCOMP Design Code procedures

One of the main purposes of the test programme was to verify, to a limited extent, the design procedures presented in the EUROCOMP Design Code. Table 13 contains the measured and calculated maximum joint tensile loadings for certain adhesively bonded joints. Laminated joints could not be analysed as some basic material properties were not known. In Table 13 only adhesive shear loading is considered. Peel stresses were also calculated but as the adhesive tensile strength and the adherend through-thickness strength were unknown and had to be estimated the results are not presented here. However, when using typical values for those strengths it may be concluded that peel loading did not become critical in any of the joint configurations tested where adhesive bonding was used.

The calculations for Table 13 were performed according to the rigorous procedures for single- and double-lap joints presented in the EUROCOMP Design Code. The following values for partial safety factors were used:

γ_f = 1.35
$\gamma_{m,\ ADHESIVE}$ = 1.56
$\gamma_{m,\ ADHEREND}$ = 1.65

The calculated values, N_{calc}, indicate the maximum allowable loading determined according to the EUROCOMP design procedure using material values presented in this report and partial safety factors given above. Measured failure loads are indicated as N_{ult}.

The effect of a single specific parameter on the calculated maximum loadings is evident if one compares the loadings of A300 RTM/Araldit 2015 double-lap and double-strap joints. For both joints the maximum allowable loadings are calculated using exactly the same equation. As their joint geometries are practically the same, one would expect to obtain about the same maximum loadings. However, owing to different thicknesses of adhesive layers (Table 6), the maximum loadings are different.

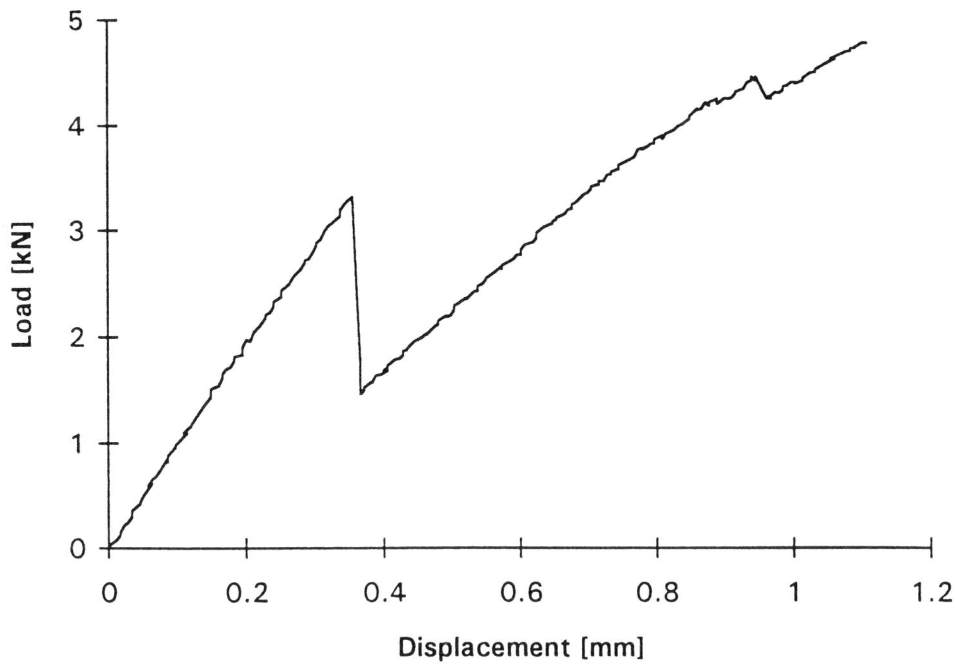

Figure 11 A typical load vs displacement curve of a single-strap joint, L/t = 5.

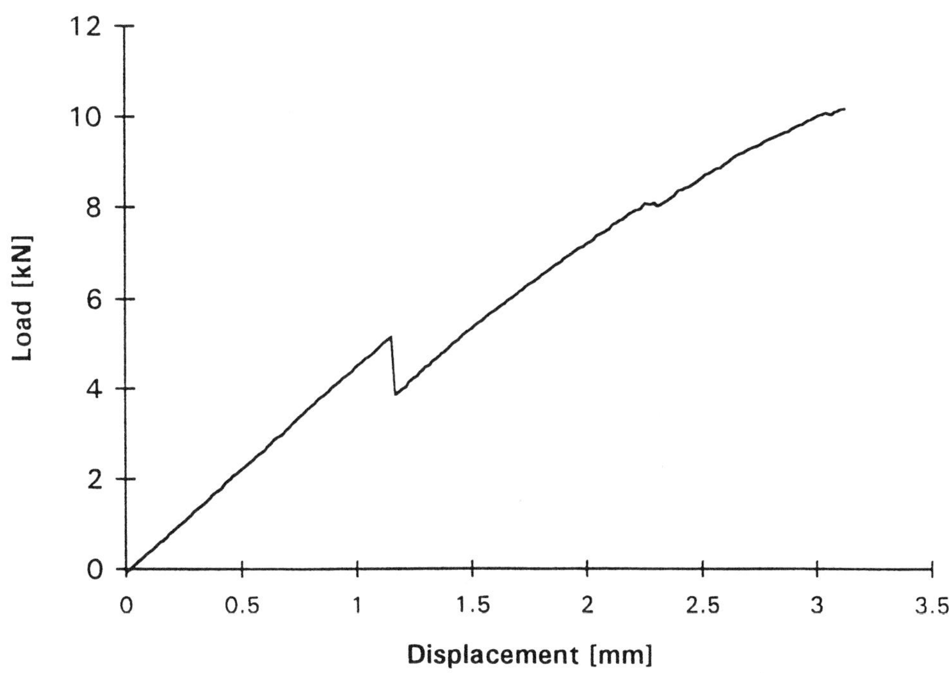

Figure 12 A typical load vs displacement curve of a single-strap joint, L/t = 25.

In the EUROCOMP design procedure the maximum calculated loading is assumed to be proportional to the square root of the adhesive layer thickness. Thus, when the adhesive layer thickness is increased from 0.5 mm to 0.8 mm (60%) the maximum calculated loading is increased from 87 kN/m to 110 kN/m (26%). However, the actual joint strength, i.e. the measured ultimate joint loads, does not increase. This indicates that the EUROCOMP design procedure does not take into account the effect of the adhesive layer thickness realistically. It should also be acknowledged that the applicability limits of the theory presented in the EUROCOMP Design Code are exceeded when thick adhesive layers are analysed.

Table 13 shows that the design procedures for bonded joints in the EUROCOMP Design Code seem to provide a reasonable conservatism in the joint design when short term static loadings are considered. However, owing to the relatively thick adhesive layers in most of the test joints, the applicability limits of the current theory may have been exceeded. This might have caused some inaccuracy in the calculated joint strengths.

Based on the test programme the EUROCOMP Design Code provides acceptable procedures for the design of bonded joints. However, the test programme provided only a limited amount of information on the behaviour of bonded joints. Excessive generalisations on the basis of this test programme should be avoided.

7 SUMMARY

Based on the programme the following conclusions can be made:

1. The methods presented in the EUROCOMP Design Code may be used to design bonded joints.
2. Single-lap and single-strap joints can be treated identically.
3. Double-lap and double-strap joints can be treated identically.
4. Adhesively bonded and laminated joints can be treated identically.
5. The materials used in the test programme had no compatibility problems in respect of adhesion or to any other property.
6. All three surface treatment procedures used in the test programme provided adequate adhesion.
7. Mat layers on the laminate bond surfaces should be avoided as they reduce the load-bearing capacity of the joint compared with that of the laminates with a woven roving or uni-directional surface layer.

Table 13 Calculated and measured maximum tensile joint loadings.

Laminate	*Adhesive*	*Geo.*	L/t	N_{ult} (kN/m)	N_{calc} (kN/m)	N_{calc}/N_{ult}
A300 RTM	2015	SL	5	218	42.5	19.5%
A300 RTM	2015	SL	25	455	61.5	13.5%
A300 RTM	2015	SS	25	404	68.5	17.0%
A300 WR	2015	SL	5	243	71.0	29.2%
A300 WR	2015	SS	25	482	106.0	22.0%
A300 RTM	DP 460	SL	5	232	65.5	28.2%
A300 RTM	DP 460	SL	25	452	88.0	19.5%
A300 RTM	2015	DL	12.5	657	87.0	13.2%
A300 RTM	2015	DS	12.5	645	110.0	17.1%
A300 WR	2015	DS	12.5	599	127.0	21.2%
A300 UD1	2015	DS	12.5	840	157.0	18.7%
A300 UD2	2015	DS	12.5	478	132.0	27.6%

TESTS ON TUBULAR MEMBERS FOR THE THE EUROCOMP PROJECT

by

P Garcin

Commissariat a L'Energie Atomique CEA/ CADARACHE
BP No. 1
13115 Saint Paul Les Durance
FRANCE

and

B Foure

Centre Experimental de Recherches et D'Etudes
du Batiment et des Travaux Publics [CEBTP]
Domaine de Saint Paul
BP 37
78470 Saint Remy Les Chevreuse
FRANCE

CONTENTS

TESTS ON TUBULAR MEMBERS

1 INTRODUCTION

The structural members studied by the French group were trusses made of polyester reinforced by long uni-directional glass fibres. These trusses are used as masts, either cantilevered or guyed. This latter situation, in which they are subjected to eccentric compression, was chosen and was simulated by compression tests in a hydraulic press.

The main vertical members are continuous; they are connected by means of glued nodal pieces to diagonal and horizontal bracing elements.

The test programme consisted of the following topics:

- basic mechanical properties of the material: basic data obtained from the manufacturer
- compressive behaviour of 18 mm diameter and 36 mm diameter tubes: axial compressive failure tests on stubs (length 1, 2, 3 and 5 diameters) and buckling tests on bars (main members and bracing members) with length equal to the distance between nodes of the truss
- loading tests on truss in the elastic range to estimate the effect of geometrical imperfections and the heterogeneity of the material
- four failure tests on trusses, with two main parameters, load position and eccentricity (in a plane of symmetry or in a perpendicular plane) and height of the truss, to study the behaviour of the trusses, including the overall second order effects.

The trusses were fabricated by LERC (a subsidiary of COFLEXIP. Tests on elements were performed by CEA and tests on both elements and trusses by CEBTP, with the financial support of EDF.

The purpose of the tests was to provide data which might be used to check the EUROCOMP Design Code's approach to specific topics such as:

- effect of partial safety factors for materials and geometrical uncertainties of cross-section
- effect of local and overall imperfection
- buckling calculation.

Full details are given in two reports (references 1 and 2). The present report is a brief overview of the programme.

2 TESTING OF ELEMENTS BY CEA

The programme consisted of testing short tubes forming parts of the model tested by CEBTP. It was necessary to design a testing procedure and measurement equipment.

The truss tested by CEBTP is based on two types of tube (the aim of the CEA testing being to determine the behaviour of these two kinds of tube in axial compression)

- 18 mm external diameter plain tube
- 36 mm external diameter tube wall, thickness 5.5 mm.

These tubes are made by pultrusion with E-glass fibres and polyester resin. According to LERC the glass volume fraction is 50%, the Young's modulus is more than 20 kN/mm^2 and the strength is more than 50 N/mm^2.

Three test programmes were performed:

2.1 Test programme N°1

36 mm diameter tubes with a length of 1, 2, 3 and 5 diameters were tested. The purpose of this programme was, for each length:

- to determine the strain ε as a function of the stress σ, so $\varepsilon = f(\sigma)$
- to measure the axial displacement of the tube.

2.2 Test programme N°2

36 mm diameter tubes with a length of 710 mm (this being the length of this kind of tube in the truss) were tested. The purpose of this programme was:

- to determine the strain ε as a function of the stress σ, so $\varepsilon = f(\sigma)$
- to measure the axial displacement of the tube
- to measure the deflection of the middle section of the tube.

2.3 Test programme N°3

18 mm diameter tubes with a length of 840 mm were tested. The purpose of this programme was the same as that of the second one.

The aim of these three programmes was to obtain the axial Young's modulus of each kind of tube and the deflection of the long tubes.

2.4 Test method

The maximum loading was 50 kN or the destruction of the tube. Three tests of each kind were performed, so that the results could be compared. On each tube, three uni-directional strain gauges were bonded to the middle section of the tube at 120°, see Figure 1.

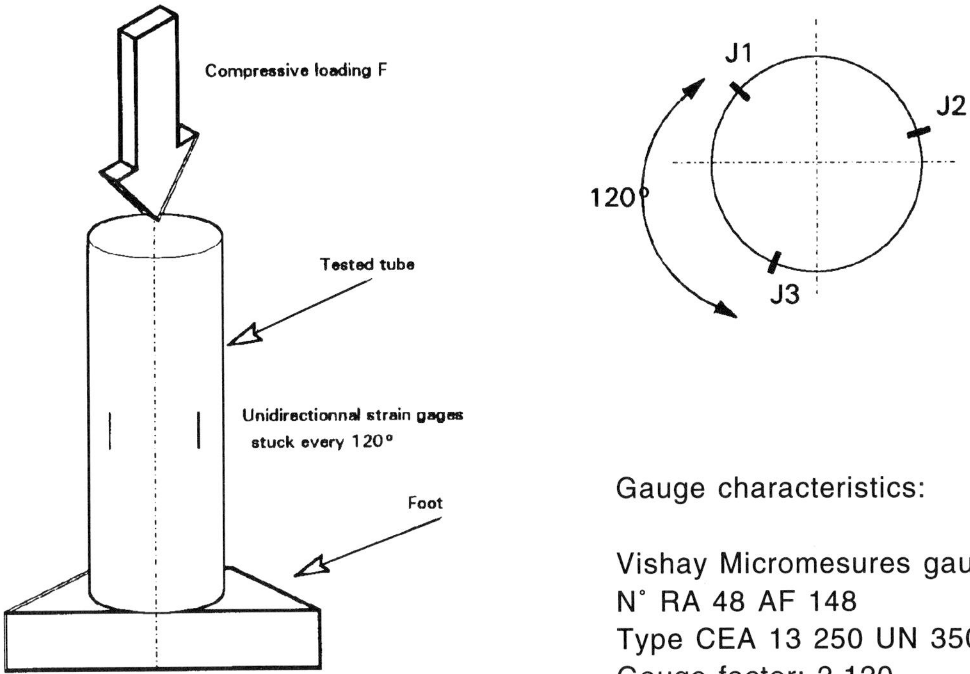

Gauge characteristics:

Vishay Micromesures gauges
N° RA 48 AF 148
Type CEA 13 250 UN 350
Gauge factor: 2.120
Maximum: 5%

Figure 1 Loading and instrumentation.

For the short tubes, the loading speed was 50N/min.

For the long tubes (36 mm diameter with length = 710 mm and 18 mm diameter with length = 840 mm) the loading speed was 250N/min. The deflection of the middle section of the tube was measured with the equipment shown in Figure 2:

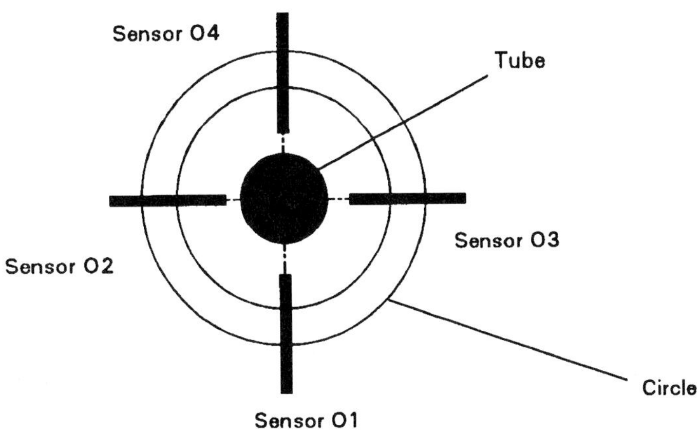

Figure 2 Measuring equipment.

The results which were recorded on a HP 3497 are:

Type of tube	axial stress as function of load		displacement as function of load		
6 dia length 36	ε = -85.59094	-0.04940182 * F	d = 0.01629547	+ 0.4398	10^{-5} * F
36 dia length 72	ε = -38.54541	-0.05244364 * F	d = 0.00939545	+ 0.5233636	10^{-5} * F
36 dia length 108	ε = -46.36365	-0.04943636 * F	d = 0.03424089	+ 0.7643455	10^{-5} * F
36 dia length 180	ε = -0.1364746	-0.04918727 * F	d = 0.04135	+ 0.1127727	10^{-4} * F
36 dia length 710	ε = -52.181	-0.05061818 * F	d = 0.09654	+ 0.377118	10^{-4} * F
18 dia length 840	ε = -28	-0.95 * F	d = 0.0021907	+ 0.102368	10^{-3} * F

ε is in microstrain, d is in mm and F is in N

The deflection of the middle section was about 1.70 mm at 50 kN.

The programme on 18 mm diameter tubes showed that:

- the tube deflected linearly until 8 kN
- between 8 kN and 12 kN, deflection was too complex to measure with the system used
- before 16 kN was reached, the tube was destroyed by deflection (stress was around 15000 microstrain and axial displacement was 30 mm before destruction).

2.5 Conclusions

The tubes which were made by pultrusion (which guarantees good mechanical characteristics), were of good quality (external aspect, lack of defects, regular dimensions, etc) and this led to very good test results:

- for the 36 mm diameter tubes of length 36, 72 and 710 mm and 18 mm tube tested at length 840 mm, the scatter of results was often less than 10% on stress and axial displacement
- for the 36 mm diameter tube tested at length 108 and 180 mm, the scatter was very low - around 2% on stress and 5% on axial displacement

In each test, the stress and axial displacement graph is perfectly linear with respect to the load. Thus the axial Young's modulus was determined as 35 kN/mm^2.

3 TESTING OF TRUSSES AT CEBTP

3.1 Check on dimension of the trusses

Only simple measurements were done to check the external diameter and the out of straightness of the main members.

3.2 Loading conditions and measurements

These tests were carried out in a hydraulic machine provided with perfectly hinged platens. Oil pressure was servo-controlled by measurement of the overall vertical displacement.

The rate of loading was approximately 0.1 $kN/mm^2.s$ in the most compressed member.

The axial deformation of the members was measured by means of strain gauges and their horizontal displacement at mid-height by means of LVDTs.

One of the trusses underwent four successive loadings in the elastic range under axial load or slightly eccentric load in each of the planes of symmetry.

Three short trusses (consisting of three modules) were loaded to failure under eccentric compression applied either in a plane of symmetry, or in a perpendicular plane (Figure 3). A fourth "long" truss (consisting of seven modules) was loaded to failure with the same load position as one of the short trusses, to study the effect of overall buckling.

(a) e = 0 or 35 mm (b) e = 80, 90 or 155 mm

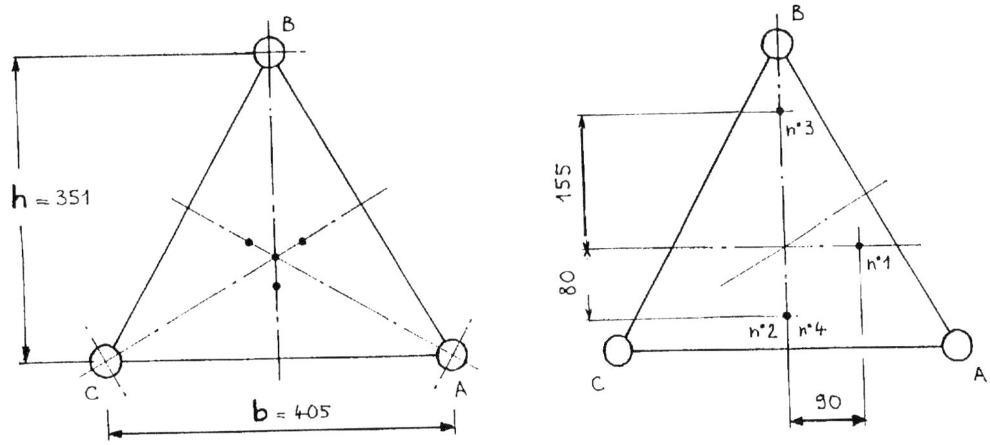

(a) Loading in the elastic condition. **(b) Test to failure**

Figure 3 Truss loading arrangement.

3.3 General behaviour

The behaviour remained linear until a high value of the load was reached. Then the onset of bending of the most compressed member could be observed and the maximum load was obtained when this member buckled rather suddenly. After that the displacement could be increased under slowly decreasing load. Increasing rotations led some nodes to burst out and the truss to collapse.

3.4 Results from strain measurements

The strain measurements gave the axial force N_i in each of the main members. The comparison between the sum ΣN_i and the total applied force F allowed an estimate of the mean modulus E to be made for each truss, which differs slightly from the value measured on test samples. However, the overall mean value for the four trusses showed no significant difference.

The comparison between the measured N values and the theoretical ones gave an indication of the geometrical imperfections of the trusses plus the inaccuracy of centring on the machine (see Figure 4). Their order of magnitude was estimated to be ±2 mm.

3.5 Evaluation of buckling

The ultimate axial force N_i in the most compressed member was determined from the strain measurements. The corresponding mean stress was far lower than the strength f_r, which indicated an elastic buckling phenomenon. Thus, the ultimate load N_n could be compared with the critical Euler's value, based on the buckling length l_c which is related to the distance l between two successive nodes: $l_c \approx 0.78$ l for the short trusses; $l_c \approx 0.91$ l for the long truss.

Comparing the behaviour with the theoretical estimation of the buckling length (Figure 5) two conclusions may be drawn:

- the rotational restraint at the nodes is negligible
- the effect of sway is not negligible for the short trusses and is rather important for the long one.

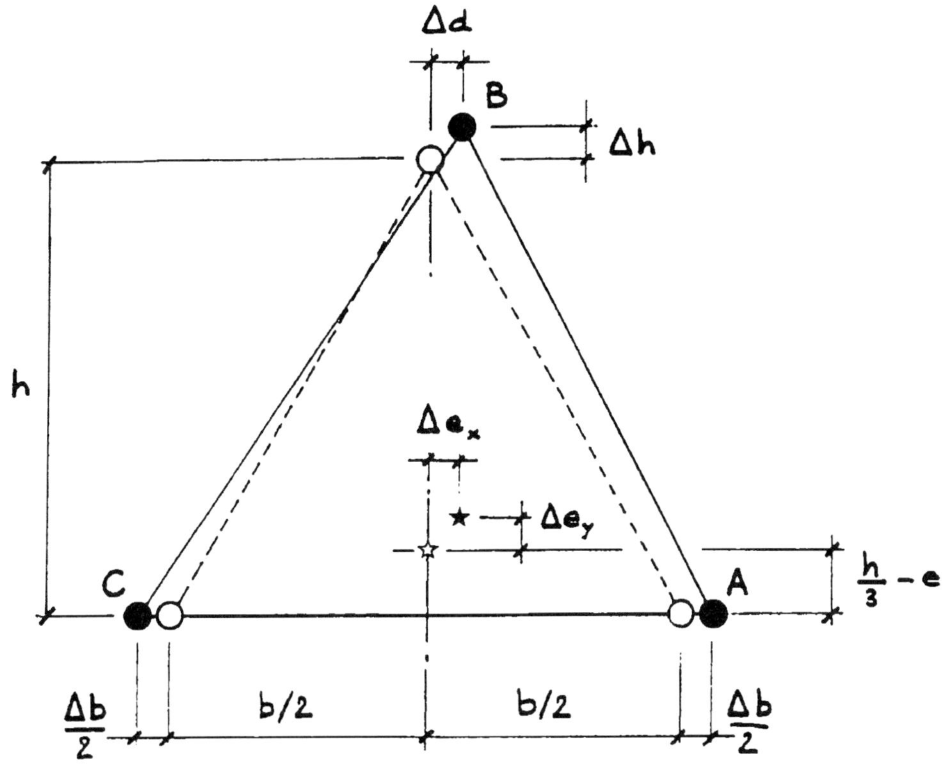

$$N_B = \frac{\frac{h}{3} - e + \Delta e_y}{h + \Delta h}\, F$$

$$N_A = \frac{\left(1 - \frac{N_B}{F}\right)\left(\frac{b + \Delta b}{2} + \Delta d\right) + \Delta e_x - \Delta d}{b + \Delta b}\, F$$

$$N_C = F - N_A - N_B$$

Figure 4 Effect of eccentricity.

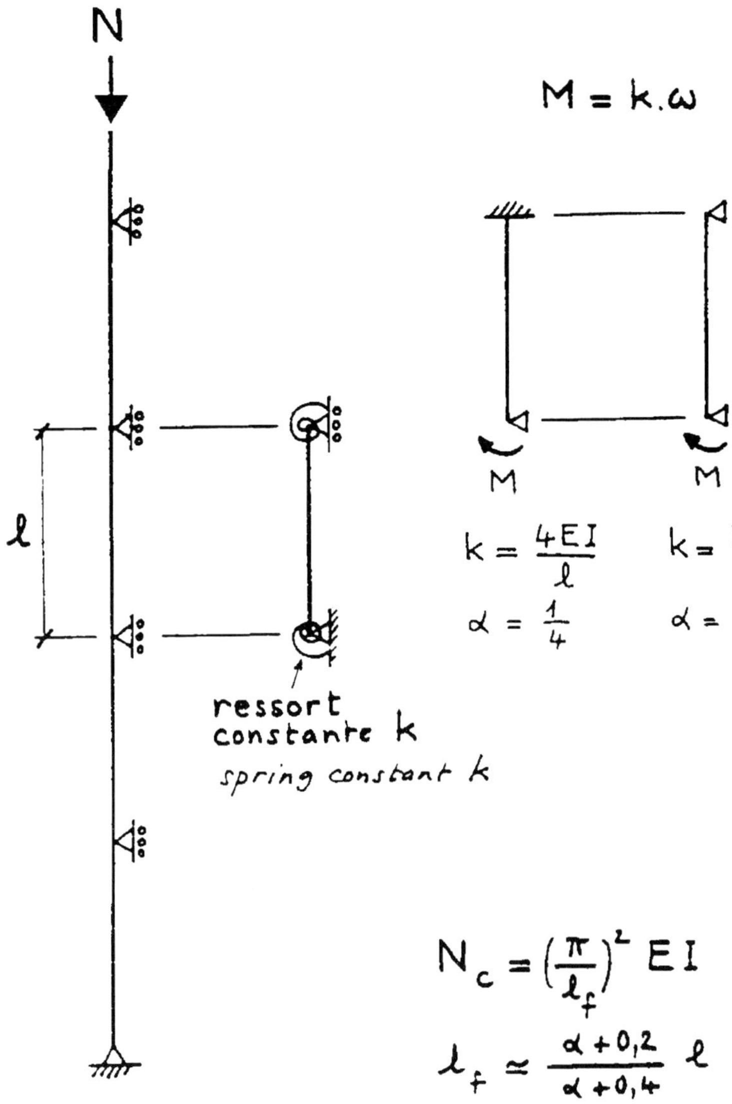

Figure 5 Buckling of truss.

3.6 Computer calculation

The geometrically perfect truss was analysed as a structure composed of beam-like elements (Navier-Bernoulli's hypothesis) with elastic-brittle behaviour. The second orders effects due to displacements were included.

The results showed that the bending moments in the main members were not negligible. But the maximum stress remained well below the strength f_r, which confirmed the elastic buckling phenomenon.

All the diagonal bars were subjected to tensile force and to a very small bending moment.

The horizontal bracing bars were subjected to a rather high bending moment. Their amplification beyond the ultimate buckling state explains the bursting out of the nodes.

3.7 Conclusion

The main results are:

- The tubes are pultruded with a polyester resin reinforced with long uni-directional fibres of E-glass. The volume fraction is 50%, the axial Young's modulus is greater than 25 000 N/mm^2 and the strength is greater than 200 N/mm^2
- The composite material exhibits a quasi-linear behaviour until failure in compression; its modulus is about 37000 N/mm^2 and its strength 475 N/mm^2. These composite material characteristics are reliable; they have been measured on 18 and 36 mm diameter tubes as well
- From the comparison between the theoretical and measured distribution of forces in the members, it appears that the geometrical defects of the trusses have an order of magnitude of 2 mm. Strain measurements show also that the mean modulus in the truss can deviate ±3% from the modulus measured on samples
- Failure of the truss is due to buckling of the more compressed member, in each *storey* of the truss at a level of stress much lower than the strength. After the occurrence of buckling, the displacement can be increased under decreasing load, until some nodes burst out and the truss totally collapses. The behaviour can be considered as quasi- linearly elastic, with very little permanent deformation even after buckling has occurred.

  Evaluation of the failure load must include the overall second order effect due to the horizontal displacement of the nodes, even for the shorter trusses.

In addition to the results given previously, these tests gave data which may be used to check specific topics in the EUROCOMP Design Code such as:

- effect of partial safety factors for materials and geometrical uncertainties of cross-section
- effect of local and overall imperfection
- buckling calculation.

REFERENCES

1 Garcin, P. and Letiec, P., CEA Laboratoire d'Etudes du Comportement des Circuits (Cadarache),
Compte rendu des essais sur tubes composites (contrat EUROCOMP), Avril 1994.

2 Foure, B., Thomas, P., Durand,J.P. and Menezes, N., CEBTP Service d'Etude des structures (Saint Rémy lès Chevreuse),
Projet EUROCOMP - Résultats d'essais sur des mâts à treillis en matériau composite,
Juin 1994.

TESTS ON NOMINALLY PINNED CONNECTIONS FOR PULTRUDED FRAMES FOR THE EUROCOMP PROJECT

by

J.T. Mottram

Engineering Department
University of Warwick
Gibbet Hill Road
Coventry CV4 7AL
UK

CONTENTS

TESTS ON NOMINALLY PINNED CONNECTIONS FOR PULTRUDED FRAMES

1 SUMMARY

Five tests on frame sub-assemblies have been conducted to study the behaviour of web cleated connections made of pultruded section. Full-size beam-to-column connections were tested to failure. Four of them were made by bolting but in one specimen bonding alone was used. The test method gave indicative results on the moment-rotation response of each connection in a frame. Results are used to provide simple design guidance for nominally pinned connections.

Research is showing that the semi-rigid action of connections can have a beneficial effect on the design of frames. Although there is not scope for an in-depth appraisal of this research here a number of references are given for the interested reader.

2 INTRODUCTION

Pultruded material manufacturers such as Morrison Molding Glass Fiber Co. and Creative Pultrusion Inc. in the USA provide a range of standard sections (e.g., I, leg-angle, box, flat sheet) for the construction of load-bearing structures. As these sections resemble their steel counterparts in appearance, it has been the practice to use knowledge available from the behaviour of steelwork. Such an approach has produced the design guidance in reference 1 for braced frames having simply supported connections. These web cleat connections are made of standard leg-angle section with jointing a combination of bonding and bolting. A number of American structures have been designed and erected in this way.

Reliance on copying steel practice has been recognized by Bank et al (reference 2) and Mottram (reference 3) to have deficiencies. For instance, differences in material properties ensure that identically shaped members will behave differently. Furthermore, the design of members in pultruded frames will often be based on preventing instabilities and limiting deflection and not strength as is often the case with steel. Deformation and instabilities of beams can be reduced by the end restraint (e.g. semi-rigid action) provided by flange cleated connections and this is one reason why such behaviour is now being studied (references 2-10). Knowledge of how to implement this connection behaviour will eventually allow the designer to increase current allowable loads (reference 1), making frames more economical and attractive.

No experimental data are available on the behaviour of recommended nominally pinned connections (reference 1). To rectify this situation the results of five such beam-to-column connection tests are presented. The test set-up, as shown in Figure 1, represents beams connected to an internal column (reference 7,8,10). The test method and salient results of each test are presented in what follows. Further details of the test progamme and an in-depth analysis of the results are to be found in reference 10.

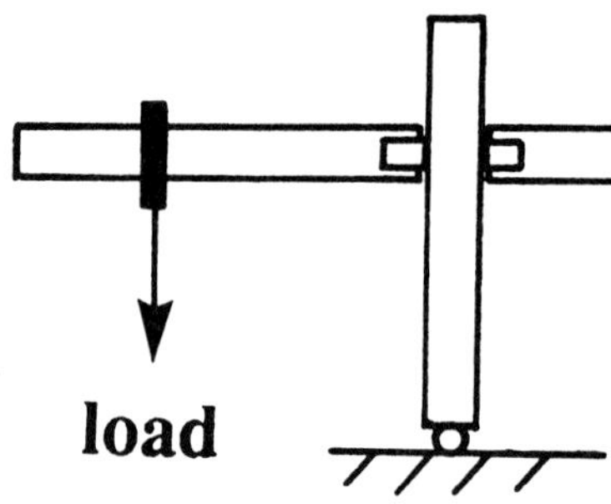

Figure 1 Test configuration.

3 MATERIAL SPECIFICATION AND CONNECTION DETAILS

Specimens were fabricated from standard members in the MMFG EXTREN 525 series (reference 1). Column and beam members were of 8 x 8 x 3/8 inch (203 x 203 x 9.5 mm) wide flange I-section. Figure 2 shows the geometry for the pinned connections and Figure 3 the nominal dimensions for the web cleats. Pieces that comprised the cleats were cut from 6 x 6 x 1/2 inch (152 x 152 x 12.7 mm) equal-leg angle. No gap was allowed between the end of a beam and the column in accordance with Figure 2.

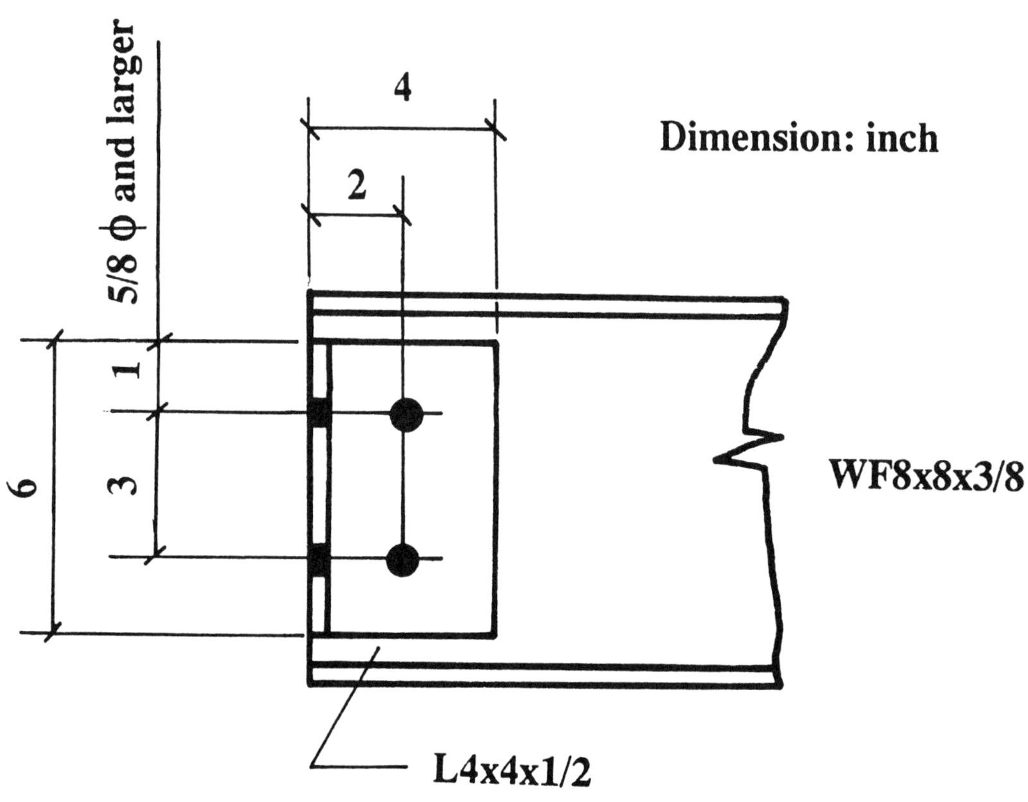

Figure 2 MMFG design manual connection (reference 1).

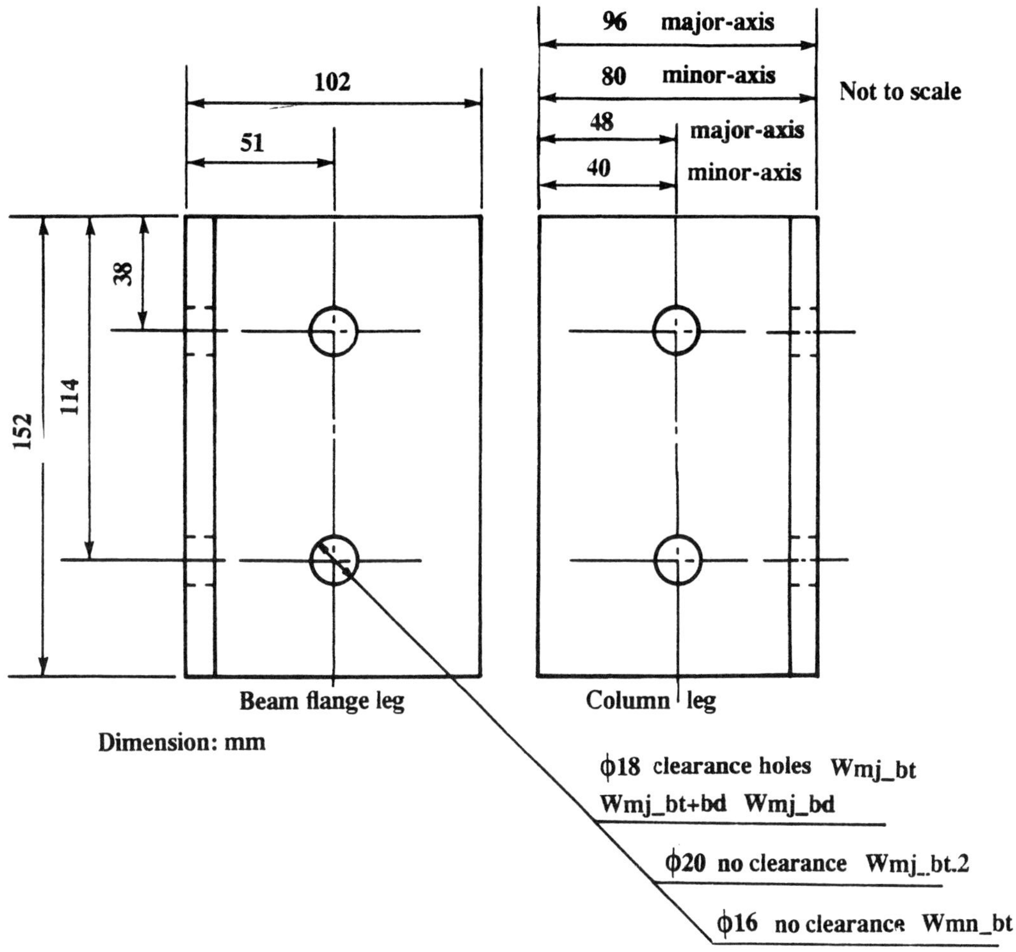

Figure 3 Nominal dimensions of web cleats.

Table 1 presents the labelling system for the specimens (reference 10).

Table 1 Summary of connection specimens.

Label	*Connection details*	*Column axis*	*Jointing*
Wmj_bt	Web cleats	major	bolting (16ϕ)
Wmj_bt.2	Web cleats	major	bolting (20ϕ)
Wmj_bt+bd	Web cleats	major	bolting (16ϕ) + bonding
Wmj_bd	Web cleats	major	bonding
Wmn_bt	Web cleats	minor	bolting(16ϕ)

Adhesive bonding in connections Wmj_bt+bd and Wmj_bd was with the toughened epoxy Araldite 2015 (Ciba Geigy, UK), a two-part cold-cure adhesive.

All mating surfaces were bonded following the procedure given by Bass (reference 7).

Three connections (Table 1) had mechanical fastening by M16 grade 8.8 steel bolts with 30 mm diameter standard size washers. These were tightened to a torque of 23.8 N m, as recommended in reference 1 for FRP bolts. To aid buildability in connections Wmj_bt, Wmj_bt+bd and Wmj_bd, bolt holes had a clearance of 2 mm. After test Wmj_bt was conducted, the specimen was modified such that the holes in the members and cleats (Figure 3) were re-drilled to provide a "tight" fit for M20 steel bolts. This additional test, Wmj_bt.2, was conducted for a number of reasons (reference 10).

To prepare the bonded connection Wmj_bd the specimen was fabricated to the specifications in Figures 2 and 3. M16 steel bolts, tightened to a torque of 23.8 N m, were used to compress the mating surfaces together while the adhesive fully cured. Before testing commenced the bolts were removed.

No clearance holes were present in the single minor-axis connection Wmn_bt. Note that this practice makes assembly difficult because the inherent flexibility and shape of the members does not allow holes to be drilled with the percision required for "tight" fitting bolts. Hole locations for jointing to the column web were assumed following those given by MMFG in Figure 2. Choosing, therefore, to place the two bolts along the centreline of the leg also made assembly difficult. To improve buildability, it is recommended that the "minor-axis" dimension of 40 mm in Figure 3 be changed to 30 mm.

4 TEST METHOD

Figure 1 shows the test configuration having the form of two back-to-back cantilever beams with a central column. Load was applied at a distance of 1 m from the connection and measured by a 9 kN load cell connected in series with a manual hydraulic jack. Rotations of the column and connections were recorded using three electronic clinometers with a resolution of 0.02 mrad (linear to $\pm 1\%$ over a 10° range). One clinometer was located along the centre-line of each beam and close to the end with the connection, while the third clinometer was located on the column above the joint. The two connection rotations were determined from the clinometer readings.

The connection moment, M, was applied in equal increments until the connection behaviour became non-linear; control was then transferred to the connection rotation ϕ. At a rotation above 12.5 mrad the specimen was unloaded and then reloaded. The purpose of this procedure was to determine the extent of permanent deformation and change in connection stiffness. A time interval of 5 to 10 minutes elapsed between each load increment to carry out visual inspection for failure and to take measurements. Moment/rotation values were recorded immediately an increment was applied and again just before the next one was applied. After the ultimate moment had been reached, each connection was taken to either ultimate failure or until rotation was excessive. Test duration was between 1 and 3 hours. Raw test data are presented in appendix A of reference 10.

5 RESULTS AND DISCUSSION

Piece-wise linear moment-rotation curves for the five connections in Table 1 are shown in Figures 4 to 8. The results plotted are for the connection in each test that failed first. The term "Left" or "Right" is used to distinguish between the two connections in each specimen (reference 10). In the figures the response plotted is that recorded at least 5 minutes after the application of the moment or rotation increment. For a moment, M, below its maximum value there was generally < 6% relaxation between the initial moment and those presented in the figures (reference 10). Defining "first failure" as the point at which visible material failure was first observed (reference 10), Table 2 presents connection properties at first failure for secant stiffness S_j, moment M_j, and rotation ϕ_j. To show when failure was first observed, its position on each curve is labelled. Table 2 also gives the ultimate moment M_{ult} and maximum rotation ϕ_{max} obtained later.

Table 2 Connection properties.

connection	$S_j = M_j/\phi_j$ (kNm/rad)	M_j (kNm)	ϕ_j (mrad)	M_{ult} (kNm)	ϕ_{max} (mrad)
Wmj_bt left	52(68)	1.64	32(24)	1.72	39
Wmj_bt right	57(76)	1.64	35(22)	1.72	49
Wmj_bt.2 left	59(74)	1.76	31(24)	2.00	45
Wmj_bt.2 right	76(86)	1.64	23(20)	2.00	30
Wmj_bt+bd left	85	1.24	7	2.07	89
Wmj_bt+bd right	172	1.24	7	2.07	33
Wmj_bd left	369	1.18	3	1.06	3
Wmj_bd right	385	1.04	3	1.18	10
Wmn_bt left	163	1.43	9	2.62	17
Wmn_bt right	234	2.06	9	2.62	44

Note: Values in parentheses are following unloading.

M-ϕ behaviour for connection Wmj_bt in Figure 4 was non-linear when M was > 0.5 kN m, in part due to differential connection slip. Such slip (horizontal and to a lesser extent vertical) occurred because of the following connection details; no adhesive bonding and 2 mm clearance holes, low frictional resistance between bearing surfaces, non-contact between beam end and column flange and an insufficient bolt torque (23.8 N m).

Significant prying action (reference 7) at the web cleats was seen when ϕ > 20 mrad. Connection resistance and first failure occurred when ϕ exceeded 30 mrad. The mode of failure took the form of tensile cracking at the column flange-web interface, initiating at the position where column bowing due to flange flexibility was largest (references 7,10).

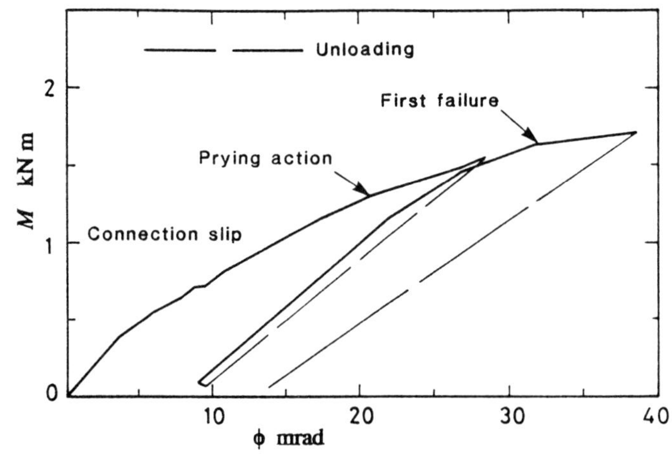

Figure 4 Moment-rotation curve of left connection Wmj_bt.

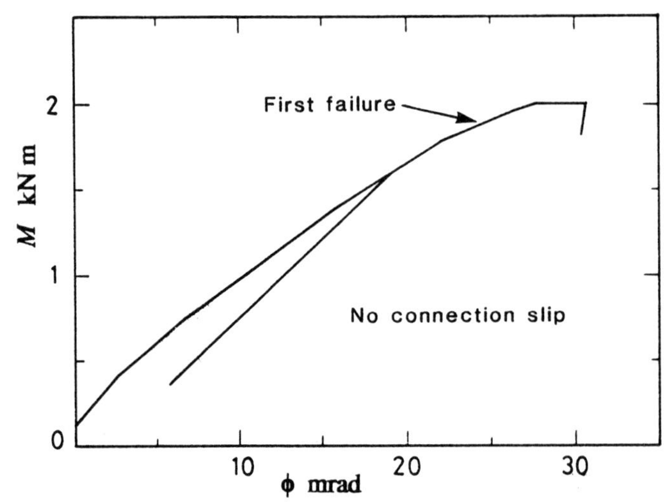

Figure 5 Moment-rotation curve of right connection Wmj_bt.2.

Connection Wmj_bt.2 (Figure 5), having 20 mm bolts and no clearance holes, gave a very different behaviour. The tight fitting bolts eliminated slip. Despite no slip and larger bolts, S_j (Table 2) was only slightly increased. However, non-linear behaviour occurred at a much higher moment.

Unloading at 25 mrad showed that permanent rotation was low and such a result suggested that the permanent rotation in test Wmj_bt (Figure 4) was due to connection slip and not permanent deformation. Delamination cracks, at the top of the web cleats, were detected when M was 1.8 kN m. The left connection collapsed due to extensive delaminations at the top of the cleats.

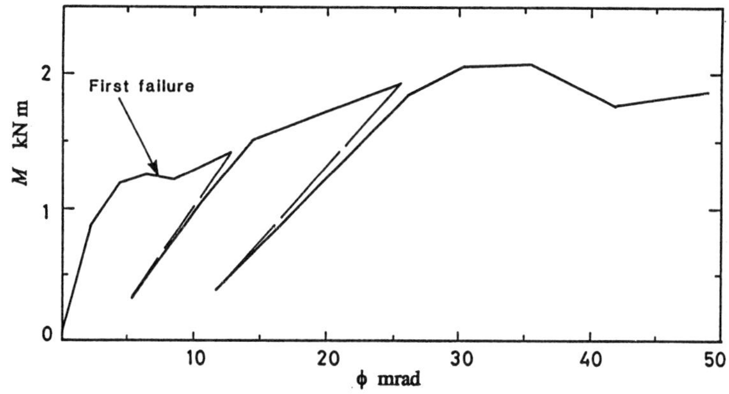

Figure 6 Moment-rotation curve of left connection Wmj_bt+bd.

The behaviour of the left and right connections was the same up to 30 mrad for Wmj_bt+bd (Figure 6). Jointing by bonding had eliminated connection slip. S_j was significantly higher than that for the bolted connections (Table 2), as bonding became the primary load path through the joint. Non-linear behaviour resulted in a noticeable knee in the M-ϕ response (c.f. Figures 4 and 6), owing to cleavage (or peel) failure of the adhesive at the web cleat/column flange bond. First failure was observed when ϕ was only 7 mrad. The presence of the bolts allowed the connection to restiffen as deformation continued (Figure 6). Despite the top bolt row acting as a "crack stopper" the debond area continued to grow with increased ϕ. First failure of a member was seen when ϕ was > 20 mrad and it took the form of a barely visible crack at the web-flange junction in the column, level with the top bolt row of the connections.

The resistance of connection Wmj_bt+bd (Table 2) had been increased by 16% by having a joint with combined bolting and bonding. The mode of ultimate failure was gross delamination just below the surface of the inside face of the left column flange.

The M-ϕ curve for the right connection Wmj_bd is presented in Figure 7. The initial part of the curve gave a behaviour similar to Wmj_bt+bd, but with a higher S_j (Table 2), showing again that when bonding was present it governed the behaviour of the connection. Adhesive debond at the web-column flange interfaces was the first and ultimate mode of failure. Such failure of the right connection was observed when ϕ was about 2.5 mrad. The area of debond continued to grow as deformation increased and ultimate failure occurred when ϕ was only 10 mrad.

One minor-axis connection test (Wmn_bt) was conducted and a M-ϕ curve is given in Figure 8. Results from major-axis tests had clearly showed that, except to prevent slip, bonding had not improved connection performance sufficiently to make such a costly practice a strong design recommendation. Here the bolts were tight fitting to minimise slip (Figure 3). S_j (Table 2) was much higher than the equivalent major-axis connection Wmj_bt, because the column web represented a rigid support. Splitting between the reinforcement layers, at the top of the left web cleats, was the first visual mode of failure. This failure grew progressively until ultimate failure.

6 DESIGN CONSIDERATIONS

It can readily be shown (reference 10,) that the support rotation capacity of a uniformly loaded simply supported beam needs to be 13 mrad to meet the serviceability limit state of a maximum deflection 1/250th of the beam span. Note that this rotation assumes the beam is attached to rigid columns and that it is therefore not necessarily the situation at external connections or at internal connections subjected to non-symmetric deformation.

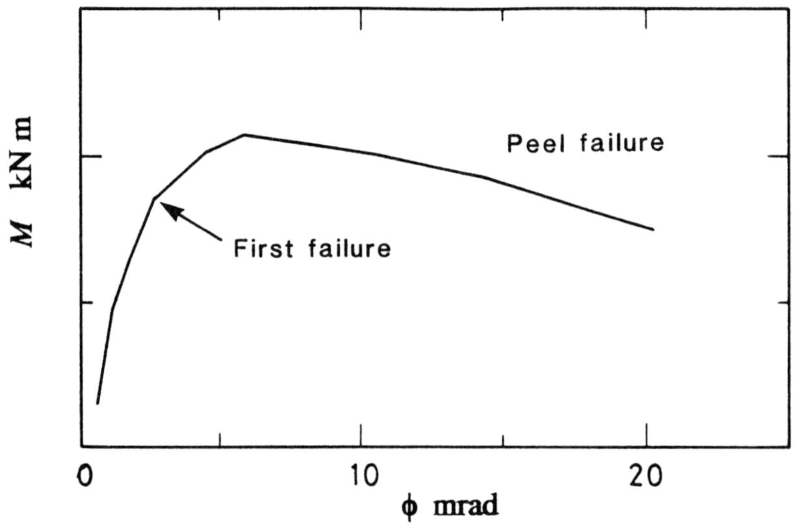

Figure 7 Moment - rotation curve of left connection Wmj_bd.

Table 2 summarises connection properties. The most important connection property for the purpose of design is its rotation at first failure ϕ_j, as this rotation may be used to determine the beam's maximum deflection. First failure is associated with deformation of the web cleats, owing to the moment transmitted by the connection (Figures 4 to 8). Any relative slip between beam and columns is not considered as failure. Here it is assumed that once material failure has occurred the web cleat connection may have its long-term performance impaired. The consequences of failure on the durability, creep and fatigue performance of the connection are not available and thus it seems prudent to be conservative when presenting design guidance.

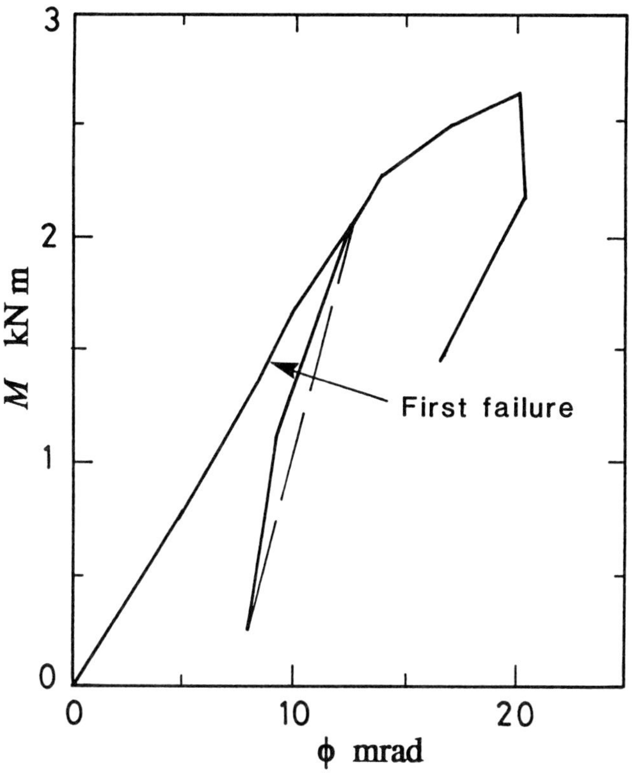

Figure 8 Moment - rotation curve of left connection Wmn_bt.

The three major-axis connections with bolting had considerable flexibility, while the bonded major-axis and minor-axis connections were significantly stiffer. In order of ascending values of ϕ_j (Table 2) the tests gave 3 mrad for Wmj_bd, 7 for Wmj_bt+bd, 9 mrad for Wmn_bt, 23(20) mrad for Wmj_bt.2 and 32(22) mrad for Wmj_bt. (Note values in parentheses are estimated capacities available when the connection was reloaded (reference 10).) Connections Wmj_bd and Wmn_bt did not achieve a rotation of 13 mrad without extensive failure. Ultimate failure of Wmj_bd actually occurred when ϕ was << 13 mrad, showing that bonding alone is unsafe and is not to be recommended.

A rotation of 13 mrad was not obtained, without adhesive failure, with the MMFG recommended connection Wmj_bt+bd. However, bond failure was restricted to the top of the web cleats by the restraining effect of the top bolt row and thus this connection has adequate performance to meet the serviceability limit state. Both bolted connections Wmj_bt and Wmj_bt.2 achieved the required ϕ of 13 mrad, but, along with connection Wmj_bt+bd, all provided insufficient rotation capacity (32.5 mrad) for the design to be considered safe if the MMFG maximum deflection limit of 1/100th of the beam span is used (reference 1]).

The lowest ϕ_j for a bolted major-axis connection was 20 mrad. However, it is seen in Figures 4 to 6 that the rotation can be higher than this without a connection failing. To ensure that a serviceability limit state is reached before an ultimate one, the ultimate connection rotation can be taken to be

25 mrad. This corresponds to a deflection limit of 1/128th of the span for a simply supported beam with uniform loading. The argument for this proposal is developed in reference 10.

The situation is further complicated because the data here is for short-term loading and do not provide information on a joint's creep behaviour. Both members (references 11,12) and the connections (reference 10) are known to creep, such that in time there will be increased beam deflection and therefore an increase in the required rotation capacity of the connection. Rigorous procedures to account for creep in the design process are not available. Designers are advised to use creep moduli (i.e. reduced values from short-term values) when determining deflections of beam members in a braced frame (reference 1). It may be assumed that the a web cleat connection does not adversely affect this design approach.

Another issue that needs consideration is whether or not clearance holes are necessary. Clause (6) in 6.4.1 of the EUROCOMP Design Code recommends that holes be formed on site by clamping together the parts to be joined and drilling through both laminates. Such practice to provide tight fitting bolts is acceptable for plate-to-plate connections but would not be practical when erecting pultruded frames on site. Drilling of frame members could perhaps take place in the works and the frame members be transported to site as sub-assemblies for final erection. However, the difference in geometries between frame connections and plate-to-plate connections means that clamping of frame members for precision drilling would be likely to cause problems that are difficult to solve. It is therefore recommended that bolted connections for frames have clearance holes and that these are of the same size as is standard with steelwork (2 mm).

Table 2 summarises the ultimate connection moment, M_{ult}. The values are relatively low. For example, it can be shown (reference 10) that if a 8 x 8 x 3/8 inch WF is assumed to be simply supported over a span 4.3 m, with uniform loading and a deflection 1/250th of span, then the calculated rotation corresponds to a maximum mid-span moment of 8kNm. This beam moment for the serviceability limit state is significantly below that required for the ultimate limit states of lateral torsional buckling (lateral restraint provided), local compression flange buckling and flexural rupture, so it governs design.

7 DESIGN GUIDANCE

The participants of the EUROCOMP project present the following as simple design guidance for nominally pinned connections. It may only be valid for the MMFG connection in Figure 2, as the other connections (reference 1) may behave differently.

Recommendations (1) to (6) are for major-axis and (7) and (8) are for minor-axis connections. The beam is assumed to have rigid supports and vertical uniform loading only.

(1) The web cleat connection design, as shown in Figures 2 and 3, can be used safely with both short- and long-term permanent loadings (i.e. without visible material failure), providing the maximum deflection is ≤ 1/250th of the beam span.

(2) For (1) to be valid, the connection must have bolting. If slip at the connection can be tolerated, it is not necessary to have bonding as well.

(3) To aid buildability it is recommended that 2 mm clearance holes be used in the erection of frames.

(4) A connection jointed by bolting alone will be susceptible to slip when there are clearance holes, as the current value of bolt torque (23.8 N m) does not provide sufficient frictional resistance at mating surfaces. This situation might be improved by the use of oversize washers of diameter 2.5 times the bolt diameter, with rounded edges, with the whole assembly tightened so that the bearing pressure is no higher than one-third of the through- thickness crushing strength of the material and in no event higher than 68 N/mm^2.

(5) To eliminate slip, jointing should be by combined bolting and bonding.

(6) It is recommended that jointing by bonding alone should not be made, as such a connection is liable to a sudden ultimate failure when the maximum deflection is > 1/1000th of the beam span.

(7) The minor-axis web cleat connection in Figure 3 with dimension 40 mm changed to 30 mm should be used.

(8) It is recommended that the connection in (7) have combined bonding and bolting only when slip is not allowed. The beam must be lightly loaded, as there will be failure of the cleats when the maximum deflection is > 1/360th of the beam span.

(9) Until relevant test data are available it should not be assumed that recommendations (1) to (8) are valid for the other MMFG connections (reference 1).

(10) To increase the rotation at first failure of the bolted connection there should be bolt clearance holes of 2 mm and a gap of 6 to 12 mm between the beam end and column face, as recommended in the 1995 addendum to the MMFG Design Manual (reference 1).

REFERENCES

1 *EXTREN Fiberglass structural shapes, Design Manual,* Morrison Molded Glass Fiber Company, Bristol, VA., 1989.

2 Bank, L.C., Mosallam, A.S. and McCoy, G.T., Design and performance of connections for pultruded frame structures, *Proc. 47th Annual SPI Conf.*, Composite Institute, Society for the Plastics Industry, Cincinnati, OH., Session 2-B, pp. 1-8, 1992.

3 Mottram, J.T., Recommendations for the optimum design of pultruded frameworks, *Mechanics of Composite Materials,* 29, 4, 1993, 675-682.

4 Bank, L.C, Mosallam A.S and Gonsior H.E, Beam-to-column connections for pultruded FRP structures, in serviceability and durability of construction materials (ed. B Suprenant.), *Proc. ASCE First Materials Engineering Congress*, Denver, CO., ASCE, 1990, 804-813.

5 Mosallam, A.S, Short and long-term behavior of a pultruded fiber reinforced plastic frame, PhD thesis, The Catholic University of America, Washington DC., Oct. 1990.

6 Bank, L.C. and Mosallam, A.S, Design and performance of connections for pultruded fiber- reinforced plastic beams, *Journal of Reinforced Plastics and Composites,* 13, March 1994, 199-212.

7 Bass, A.J., Behaviour of polymeric composite connections for pultruded frames, MSc thesis, University of Warwick, April 1994.

8 Bass, A.J. and Mottram, J.T., Behaviour of connections in frames of fibre reinforced polymer section, *The Structural Engineer,* 72 17, 1994, 280-285.

9 Mosallam, A.S., Abdelhamid, M.K. and Conway, J.H., Performance of pultruded FRP connections under static and dynamic loads, *Journal of Reinforced Plastics and Composites,* 13, May 1994, 386-407.

10 Mottram, J.T., Connection tests for pultruded frames, Civil Engineering Group, Research Report CE47, Dept. of Engineering, University. of Warwick, July 1994.

11 Bank, L.C. and Mosallam A.S, Creep and failure of a full-size fiber-reinforced plastic pultruded frame, *Composites Engineering,* 2 3, 1992, 213-227.

12 Mottram, J.T, Short and long-term structural properties of pultruded beam assemblies fabricated using adhesive bonding, *Composite Structures,* 25, 1-4, 1993, 387-395.

TESTING OF A PULTRUDED GRP PINNED BASE RECTANGULAR PORTAL FRAME FOR THE EUROCOMP PROJECT

by

G.J. Turvey

Engineering Department
Lancaster University
Bailrigg
Lancaster
LA1 4YR
UK

CONTENTS

TESTING OF A PULTRUDED GRP PORTAL FRAME

1 INTRODUCTION

The frame test work undertaken for the EUROCOMP Project is first set in context via a brief review of the scant literature on testing of GRP frame structures. Thereafter, important aspects of the frame and associated test work are described. For convenience, these are presented under the following sub-headings: Coupon Tests on WF-section Material, WF-section Beam Tests, and Portal Frame Tests. Some analytical work was also undertaken to develop closed-form expressions for deformations at particular locations of interest in pinned base, shear-deformable, semi-rigid rectangular portal frames. An example of these expressions is given and their utility in predicting the measured flexural and shear mode frame responses is assessed. It is concluded that classical rigid-jointed frame analysis does not provide an accurate means of assessing frame deformations, especially for the shear mode response. Full details of the work may be found in reference 5.

NOTATION

a	:	Frame column height
A	:	Cross-sectional area
b	:	Beam span (between column centres)
E	:	Young's modulus
E_L^t, E_L^c	:	Longitudinal tension and compression moduli
G	:	Shear modulus
G_{LT}	:	In-plane shear modulus
I	:	Second moment of area
I_{xx}, I_{yy}	:	Major and minor axis second moments of area
L	:	Span of simply supported beam
s_L^t, s_L^c	:	Longitudinal tension and compression strengths
s_{LT}^L	:	Longitudinal in-plane shear strength
w_c	:	Mid-span deflection
W	:	Point load
α (= EI/GA)	:	Shear stiffness ratio
β (= EI/K)	:	Semi-rigidity ratio (where K is the joint stiffness)
ν_{LT}^t, ν_{LT}^c	:	Longitudinal tension and compression Poisson's ratios

2 PULTRUDED GRP PORTAL FRAME TESTS - A BRIEF REVIEW

During the past ten years increasing numbers of pultruded GRP portal frame structures have been erected. Most of the rectangular portal frames are small (column height typically 1.5-2 m and beam span typically 2-3 m) and are used to support raised platforms or walkways. A few, much larger, pitched roof portal frames have also been erected. Little has appeared in the literature on any proof load or other tests carried out on these frames.

The first tests of a research nature on pultruded GRP rectangular portal frames were carried out by Mosallam (reference 1) in 1990. He tested two frames loaded statically through the section centroid by equal vertical point loads acting at the third points of the beam span. One was tested to destruction under short-term loading and deflections and strains were monitored throughout the test. The second was subjected to the same loading arrangement, but with the load maintained at about 25% of the initial failure load (70 kN) of the first frame, in order to monitor the long-term behaviour over a period of 1.15 years (10 000 hours).

The beam-column connections used in these frames consisted of equal leg pultruded GRP angles connected to the flanges and webs of the beam and column by means of fibre bolts. Because of the low stiffness and strength of their resin threads, fibre bolts are unable to support significant torques. A conventional fibre bolted angle to base plate connection was adopted for the column feet.

In 1992 a portal frame was fabricated in the Lancaster University Engineering Department (reference 2) as part of a final year undergraduate project. The overall dimensions (1829 mm column height and 2743 mm beam span) were similar to those of the frames tested by Mosallam. However, the I-section beam and columns were formed from pultruded glass FRP channel section connected back to back. The beam-column connection used mild steel bolts and did not employ pultruded glass FRP angles. Unlike Mosallam's frames, the column base arrangement used in the Lancaster frame allowed both pinned and fixed end conditions to be simulated. A further difference was in the loading arrangement. In two separate series of tests, the Lancaster frame was subjected to short-term single point loading applied normal to the top flange of the beam at its mid-span and along the beam centre-line at the top of one of the columns, i.e. the frame was tested statically in both flexural and sway modes. Two low load-unload cycles were applied in each mode and deflections were recorded at various locations throughout the frame. These measurements demonstrated that the frame response was repeatable and linear elastic. On completion of the low load tests, the frame was loaded to failure with the point load at mid-span. The failure mode was local buckling of the beam flange followed by tearing of the flange from the web close to the point of application of the load. There was also some evidence of failure initiation in the beam-column connection.

Deflections derived from classical rigid-jointed frame analysis were compared with measured frame deflections, but did not show good agreement.

3 COUPON TESTS ON WF-SECTION MATERIAL

EXTREN™ 500 Series pultruded GRP WF and equal leg angle sections were used to construct the portal frame. The nominal dimensions of the WF and angle sections were 203 x 203 x 9.5 and 76 x 76 x 9.5 mm respectively. The fibre reinforcement was in two forms, namely E-glass rovings (discrete bundles of uni-directional fibres) and continuous filament mat (CFM). The reinforcement was stabilised by, and enclosed in, a polyester resin matrix and the resulting fibre volume fraction was of the order of 40%. The outer skin of the WF and angle sections was formed by a resin rich polyester veil.

In order to predict the deformations in a simple statically loaded portal frame by means of frame analysis, it is necessary to know the stiffness of the beam and columns. These stiffnesses may be determined by measuring the material properties of coupons cut from the WF-section and combining them with nominal values of the section cross-sectional area and second moment of area. Accordingly, a series of tension, compression and shear tests on coupons cut from the flanges and web of the WF-section material was carried out.

The coupon tests, from which the modulus and strength values were obtained, were not carried out strictly in accordance with the procedures prescribed in ASTM etc. publications. For example, coupons, cut from WF-sections, have to be large in order to reduce results scatter owing to the inhomogeneous nature of the reinforcement architecture (e.g. the roving spacing may be quite large - between 10 and 20 mm) and they are usually of large thickness. These and other factors dictated the need to make adjustments to the coupon geometry and test procedures. Details of these tests are given below.

3.1 Tension coupon tests

Six parallel-sided rectangular cross-section tensile coupons were cut (length parallel to pultrusion axis) from the web and flanges of the WF-section. The nominal coupon dimensions were 275 x 25 x 9.5 mm. They were tested without end tabs, using a 50 mm grip length. A biaxial strain gauge (5 mm gauge length) was bonded to the centre of one coupon face and a uniaxial gauge (10 mm gauge length) was bonded to the centre of the other face of each coupon. The gauges recorded the longitudinal and transverse strains as the load was increased up to 20 kN and subsequently unloaded. The longitudinal tension modulus and Poisson's ratio values obtained for each coupon are listed in Table 1. The average tension modulus was 21.4 kN/mm^2 and the average Poisson's ratio was 0.31.

A further six coupons with the same overall dimensions were prepared for ultimate tension strength tests. These coupons, which were not strain gauged, were loaded continuously to failure. The tensile strengths of each coupon are listed in Table 2. It is evident that the average strength of the web coupons (285.1 N/mm^2) is significantly greater than the average strength of the flange coupons (214.8 N/mm^2).

Table 1 Longitudinal tension and compression moduli and Poisson's ratios of coupons cut from 203 x 203 x 9.5 mm WF-section.

Coupon label	*Coupon location in WF-section*	E_L^t (kN/mm^2)	E_L^c (kN/mm^2)	ν_{LT}^t	ν_{LT}^c
W2	Web	19.85	23.37	0.298	0.310
W5	Web	24.35	20.32	0.334	0.289
UF2	Upper flange	22.86	23.74	0.295	0.333
UF5	Upper flange	20.77	21.58	0.302	0.280
LF2	Lower flange	19.99	22.23	0.320	0.303
LF5	Lower flange	20.27	20.51	0.328	0.345
Minimum values (from reference 4)	-	17.24	17.24	0.33	-

Table 2 Longitudinal tension and compression strengths of coupons cut from 203 x 203 x 9.5 mm WF-section.

Coupon label	*Coupon location in WF-section*	s_L^t (N/mm^2)	s_L^c (N/mm^2)
W1	Web	286.8	342.3
W4	Web	283.4	323.8
UF1	Upper flange	212.1	358.4
UF4	Upper flange	224.4	360.0
LF1	Lower flange	202.1	340.1
LF3	Lower flange	-	330.6
LF4	Lower flange	220.5	336.5
Minimum values (reference from 4)	-	206.8	206.8

3.2 Compression coupon tests

Six coupons (nominally 230 x 25 x 9.5 mm) were cut from the WF-section with their lengths parallel to the pultrusion axis. These coupons were strain gauged in a similar manner to the tension coupons and were used to measure the longitudinal compression modulus and Poisson's ratio. The coupons were tested in an ITRII type double-wedge fixture (reference 3) without end tabs. The coupon grip and gauge lengths were each 50 mm and the 80 mm excess coupon length protruded beyond the fixture ends. Because the coupons were thick, antibuckling guides were not used. The compression coupons were loaded incrementally up to 20 kN and then unloaded. Strains were recorded for each load increment/decrement. The longitudinal compression modulus and Poisson's ratios determined for each coupon are given in Table 1. It is evident that the average compression modulus (22.0 kN/mm^2) is slightly higher than the tension modulus (21.4 kN/mm^2), but the Poisson's ratios appear to be unaffected by the type of loading.

A further seven coupons with the same nominal dimensions were tested to determine the longitudinal compression strength of the WF-section. These coupons were not strain gauged. They were tested with a 50 mm gauge length and bonded aluminium tabs over the 50 mm grip lengths. The coupons were loaded continuously to failure. The individual coupon strengths are listed in Table 2. Unlike the tensile strengths, no significant difference between the web and flange compression strengths was observed. The average compression strength (341.6 N/mm^2) appears to be about 37% higher than the corresponding tension value (249.9 N/mm^2).

3.3 Shear coupon tests

An Iosipescu type fixture was used to establish the longitudinal shear moduli and strengths of rectangular coupons with a pair of vee-notches (nominally 8.5 mm deep and 90° included angle) cut in the coupon edges at mid-length. The coupons, nominally 150 x 40 x 9.5 mm, were cut with their lengths parallel to the pultrusion axis. A biaxial strain gauge (identical to those used on the tension/compression coupons) was bonded on to one face of each coupon at its centre. The sensitive directions of the gauge were oriented at ±45° to the coupon's axes of symmetry. In order to inhibit local bearing failure on the edge of the coupon during testing, steel saddles were introduced between the longitudinal edges of the coupon and the supports of the test fixture. The shear load was applied incrementally up to a maximum load of 6 kN whereupon the coupon was unloaded following a similar pattern. The longitudinal shear modulus of each coupon is listed in Table 3. The average shear modulus is 4.77 kN/mm^2 and is significantly higher than the manufacturer's minimum value of 2.93 kN/mm^2 (reference 4).

These six coupons were subsequently loaded continously to failure. The ultimate longitudinal shear strengths obtained for each coupon are listed in Table 3. The average longitudinal shear strength is 88.83 N/mm^2.

[a] This value has been derived from a full-section test.

Further details of all of the coupon tests (tension, compression and shear), including load-strain plots are given in reference 5.

4 WF-SECTION BEAM TESTS

Instead of calculating member stiffnesses from the coupon moduli and the section geometry, they may be determined from three-point bending tests on simply supported beams. Tests of this type on pultruded GRP sections have been reported elsewhere (references 6-8).

Three-point bending tests were carried out on the WF-section beam and column members of the frame. Each WF-section member was simply supported on 25 mm diameter steel rollers (only one of which was free to roll) and loaded by a vertical line load at mid-span. Both major and minor

Table 3 Longitudinal shear moduli and strengths of coupons cut from 203 x 203 x 9.5 mm WF-section.

Coupon label	*Coupon location in WF-section*	G_{LT} (kN/mm^2)	s^L_{LT} (N/mm^2)
W1	Web	4.97	92.84
W2	Web	4.49	100.5
UF1a	Upper flange	4.92	86.58
UF2b	Upper flange	4.56	89.92
LF1a	Lower flange	4.69	74.98
LF2b	Lower flange	5.02	88.11
Minimum value (from reference 4)	-	2.93[a]	-

axis bending tests were undertaken and mid-span deflections were recorded. In the major axis tests on the beam a 10 kN load was applied in two increments followed by unloading in one step for each of five spans (L = 1.5-2.75 m). The same procedure was followed for the column major axis tests, except that because of their shorter length only three spans (L = 1.5-2.1 m) were used. Similar span ranges were used for the minor axis tests, but the loads were reduced to 2 kN and 3 kN.

It may be shown (reference 8) that the mid-span deflection (w_c) of a simply supported shear deformable beam carrying a point load (W) at its centre may be expressed as:

$$W_C = \frac{WL^3}{48EI} + \frac{WL}{4GA} \tag{1}$$

where $I = I_{xx}$ or I_{yy} according to the plane of bending. Equation (1) may be rearranged as:

$$\frac{w_c}{WL} = \frac{L^2}{48EI} + \frac{1}{4GA} \tag{2}$$

and Equation (2) may be plotted as (w_c/WL) versus L^2 to give a straight line with a slope of (1/48EI) and an intercept on the (w_c/WL)-axis of (1/4GA). Hence, plotting the data (see reference 5 for details) from the three-point bending tests on the beam and columns according to Equation (2) allows the slope and intercept values to be determined, from which the section flexural (EI) and shear (GA) stiffnesses may be calculated. Then making use of the nominal values for I and A given in reference 4, the section longitudinal elastic (E) and shear (G) moduli may be calculated. These section values are listed in Table 4.

5 PORTAL FRAME TESTS

5.1 Frame geometry and joint details

A rectangular, pinned base portal frame was fabricated from 203 x 203 x 9.5 mm WF-section material. The beam, of length 2540 mm, was connected at its ends to the two columns, each of length 2286 mm, by means of a pair of 76 x 76 x 9.5 mm pultruded glass FRP angles. The layout and dimensions of the frame are shown in Figure 1(a) and details of the bolted beam-column connection are shown in Figure 2. The connection arrangement is similar to typical details illustrated in reference 4, except that no adhesive bonding was employed between the angle legs and the beam web-column flanges. Instead, the bolts were tightened to a torque of 100 Nm. Connections of this type are assumed to provide pinned supports. The pinned base of each column was provided by means of a 25.4 mm diameter mild steel rod passing through the column web. The area around the pin was reinforced by EXTREN 500™ Series sheet 6.4 mm, thick bonded on both sides of the column web.

5.2 Frame instrumentation and test arrangement

The portal frame was reasonably comprehensively instrumented. Back-to-back rosette strain gauges were bonded to the beam neutral axis at mid-span and a further 24 uniaxial strain gauges were bonded to the frame at the locations shown in Figure 1(a). Nine dial gauges, and four LVDTs were positioned as shown on Figure 1(b) to enable deflections and rotations to be recorded.

The frame was tested in the horizontal position. The pinned bases provided vertical support at the column feet. Vertical support to the beam was provided by two 300 mm^2 plywood sheets each supported underneath at their centres by a flat, 150 mm diameter circular steel pad. The underside of each pad housed three ball-bearings which were free to roll on a horizontal steel plate. It was estimated that each steel pad supported a mass of only 25 kg in this configuration. A simple test demonstrated that the frictional resistance produced by this set-up was negligible. The point load was applied to the frame by means of a manually operated long stroke jack acting through a load cell.

Table 4 Elastic moduli and section stiffnesses calculated from simply supported three-point beam bending tests on 203 x 203 x 9.5 mm WF-section.

Frame member	*Bending axis*	E (kN/mm^2)	G (kN/mm^2)	EI (MNm^2)	GA (MN)	Loads (kN)	Spans (m)
Beam	Major	26.23	0.651	1.083	3.668	5, 10	1.524, 1.829, 2.134, 2.438, 2.743
Column 1	Major	26.74	0.623	1.104	3.532	5, 10	1.524, 1.829, 2.134
Column 2	Major	26.28	0.732	1.085	4.126	5, 10	1.524, 1.829, 2.134
Beam	Minor	19.42	1.384	0.259	7.795	2, 3	1.524, 1.829, 2.134, 2.438, 2.743
Column 1	Minor	20.97	0.939	0.280	5.288	2, 3	1.524, 1.829, 2.134
Column 2	Minor	21.39	0.713	0.285	4.016	2, 3	1.524, 1.829, 2.134

5.3 Low load flexural mode tests

Three low load tests with the point load applied at the mid-span of the beam were carried out. In the first two tests, the load was increased in 2 kN steps up to 16 kN and then unloaded in 2 kN steps. The instrumentation was read immediately after the application of each load increment/decrement. In the third test the frame was loaded up to a maximum load of 20 kN. All three tests showed good repeatability and the frame exhibited a linear elastic response. Mid-span deflection and joint rotations recorded in the second test are shown in Figure 3. Details of the load-strain plots for two of the low load and the ultimate load tests are given in reference 5.

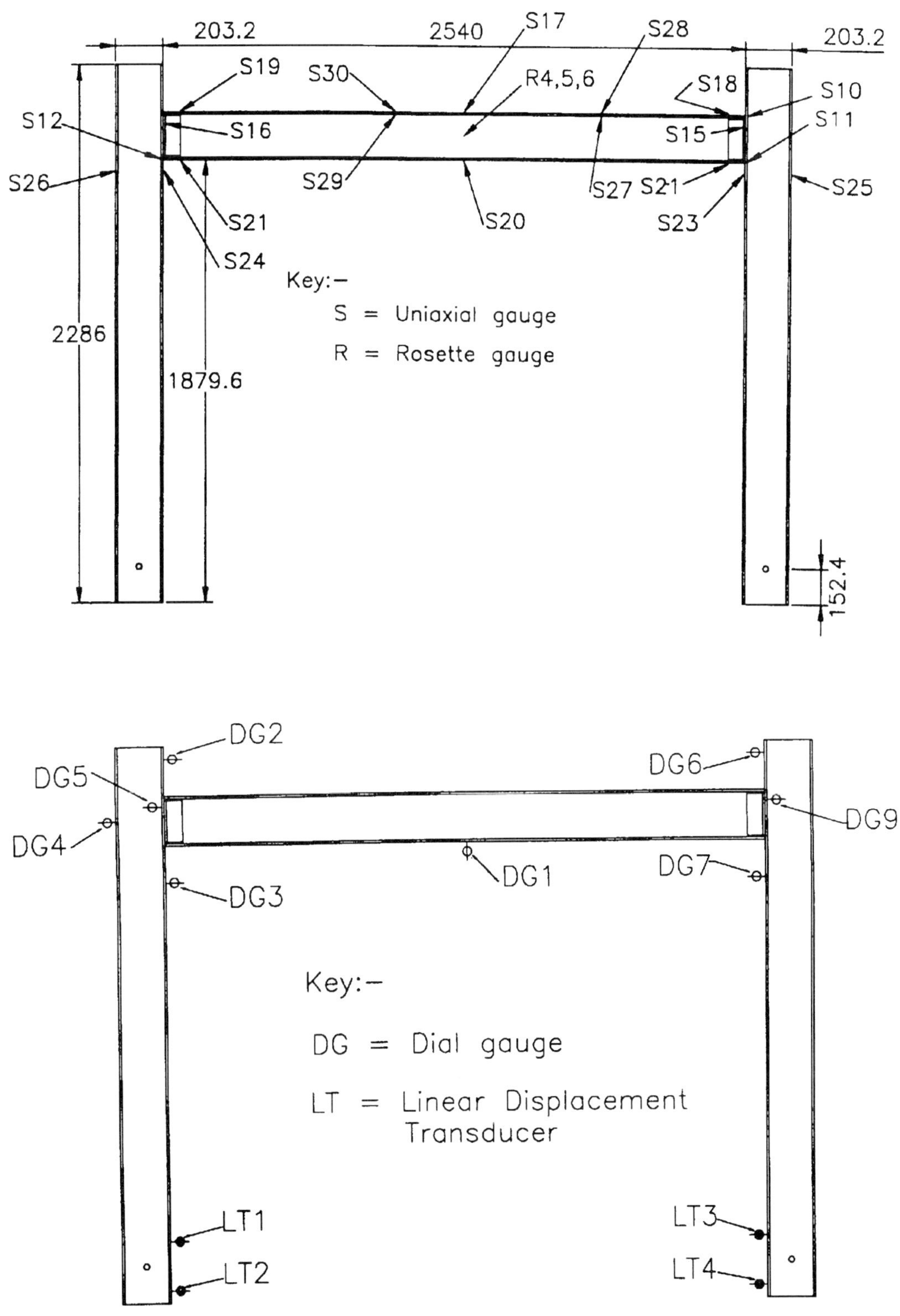

Figure 1 Pultruded GRP frame layout (dimensions in mm) showing the approximate locations of strain gauges and dial gauges and linear displacement transducers.

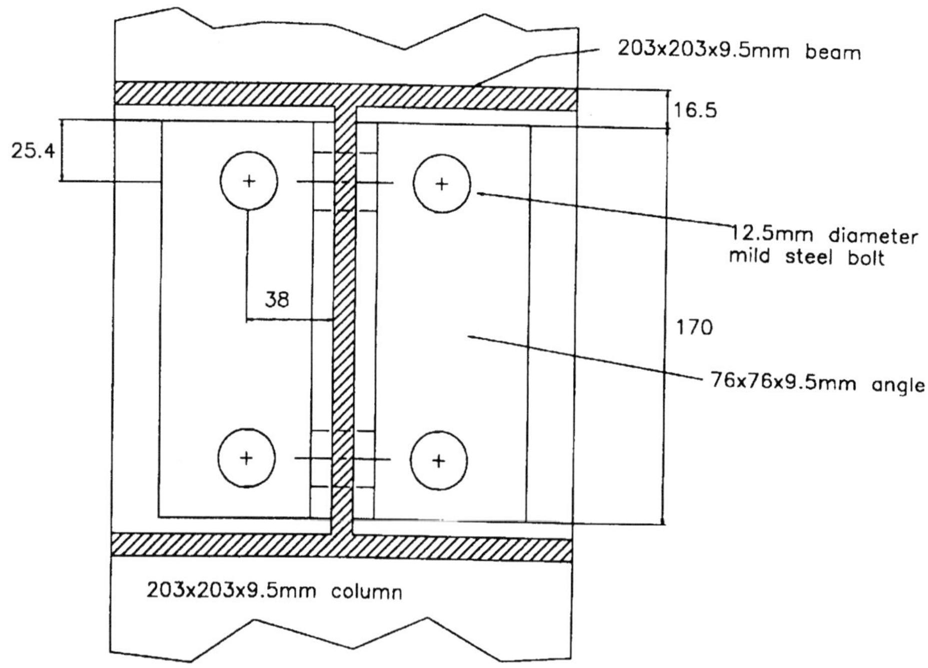

Figure 2 Section normal to the beam axis showing the beam-column connection details (dimensions in mm).

5.4 Low load sway mode frame tests

In order to carry out these two tests, the point load was transferred to a position near to the top of the right hand column so that the load acted along the beam centre-line. The load was increased in 0.2 kN increments up to 2 kN and then unloaded to zero in steps of 0.4 kN. During the first test there was evidence of acoustic emission when the load reached 1.8 kN and again at the maximum load. Examination revealed surface veil cracking in the angle cleats which extended along most of the length of the junction of the angle legs. This superficial cracking appears not to have affected the frame, as there was no evidence of acoustic emission during the second test and the results show that the response was similar in both tests. However, the frame was very flexible - the deflection at the top of the columns exceeded 20 mm and the joint rotations were of the order of 10 mrad at the maximum applied load. Sway deflection and joint rotations are shown in Figure 4.

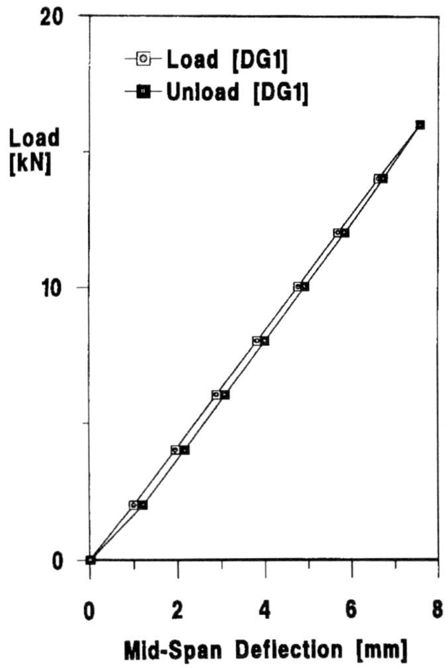

(a)

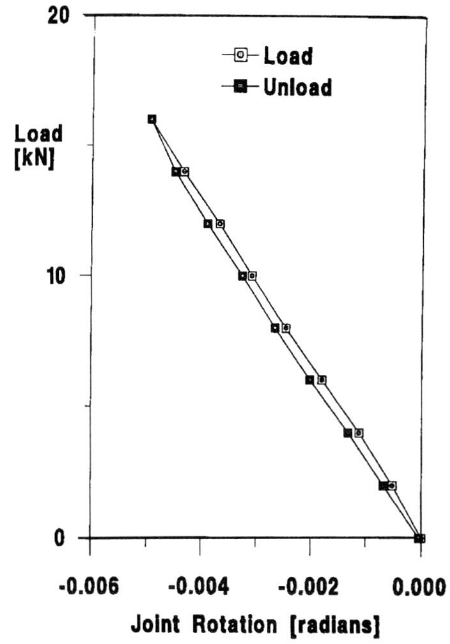

(b)

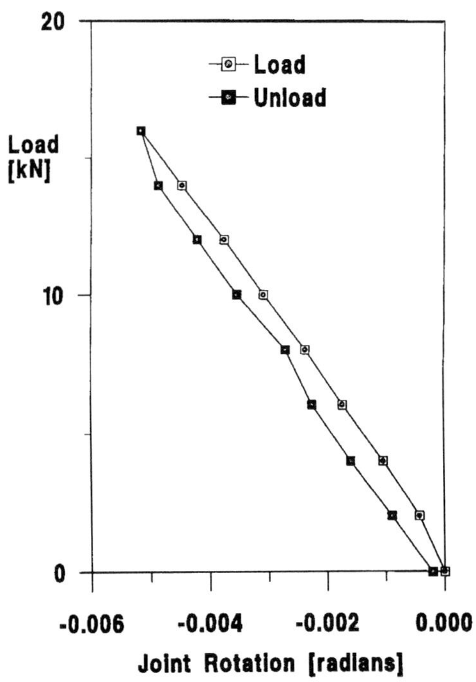

(c)

Figure 3 Second low load, flexural mode portal frame test. (a) Load versus mid-span deflection DG1 = dial gauge 1, see Figure 1(b). (b) Load versus relative rotation between the beam and column of the left-hand beam-column joint. (c) Load versus relative rotation between the beam and column of the right-hand beam-column joint (signs of rotations have been deliberately reversed).

5.5 Ultimate load flexural mode frame test

The frame failure load test was carried out with the point load applied at the mid-span of the beam. The load was increased in steps of 5 kN. Displacements and strains were recorded immediately after the application of each load increment. The frame behaved as in the previous low load tests up to a load of 20 kN. When the next and subsequent load increments were applied, creep effects began to be more significant and it became progressively more difficult to keep the point load constant during the short time it took to read the dial gauges. At a load of 30 kN there was noticeable local buckling deformation of the beam compression flange under the point load. After a delay of about 5 minutes to examine the frame, the load was increased to 35 kN. The flange local buckling deformation increased and was accompanied by considerable rotation of the flange and the loading pad. The test was again halted briefly to allow inspection of the frame. An attempt was then made to apply a further 5 kN load increment, but when the load reached 39.7 kN there was a loud cracking noise. The test was halted and the frame was examined for signs of failure A 250 mm long flexural tension crack at the flange-web junction was clearly visible under the loading pad. During the minute or so that the beam-column joints were being examined, this crack extended to 300 mm and the load dropped to 36 kN. The frame was then unloaded in one step and the deformations and strains were recorded. Further examination after unloading revealed surface cracking of the web-flange junction on the underside of the beam compression flange in two locations. The final extent of these cracks is shown in Figure 5.

The load-mid-span deflection response is shown in Figure 6(a) and the relative rotations between the tops of the columns and the ends of the beam are shown in Figures 6(b) and 6(c) respectively (the rotation signs in Figure 6(c) have been deliberately reversed). It appears that the left-hand connection has the greater stiffness and undergoes a smaller relative rotation (e.g. 11 mrads compared with 13 mrads at a load of 35 kN).

It was hoped that frame failure would initiate in the connections. However, it appears that initial imperfections (identified in a cross-section dimension survey reported in reference 5 promoted local buckling in the beam compression flange which eventually resulted in cracking of the web-flange junction at mid-span before the moment capacities of the beam-column joints were exhausted.

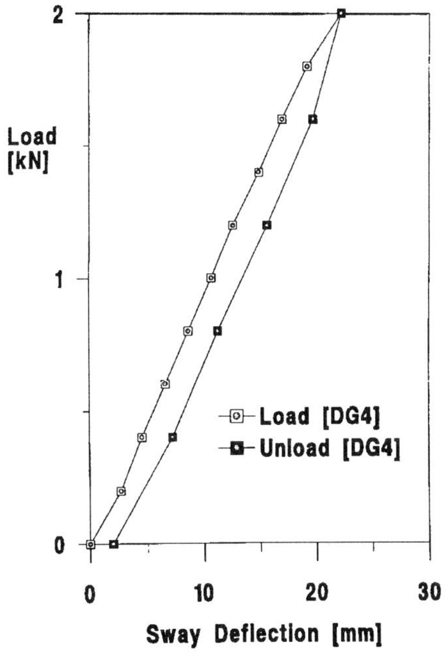

(a)

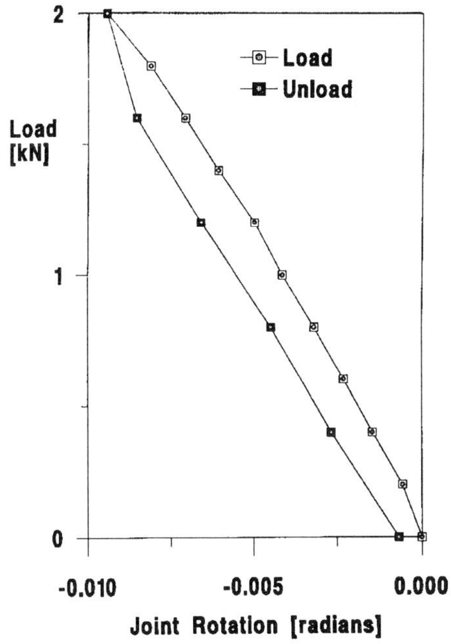

(b)

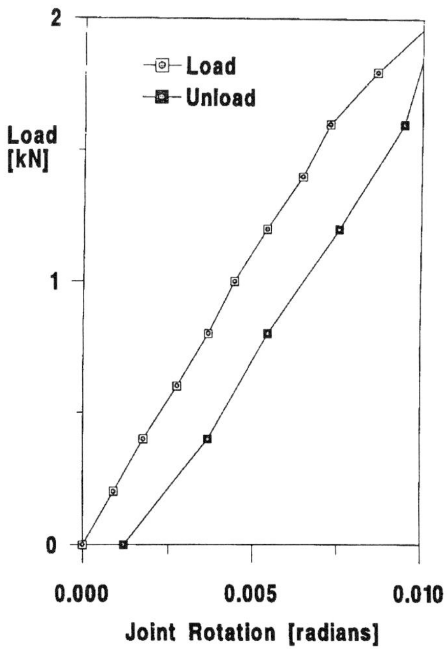

(c)

Figure 4 First low load, sway mode portal frame test. (a) Load versus deflection (DG4 = dial gauge 4, see Figure 1(b)). (b) Load versus relative rotation between the beam and column of the left-hand beam-column joint. (c) Load versus relative rotation between the beam and column of the right-hand beam- column joint (signs of rotations have been deliberately reversed).

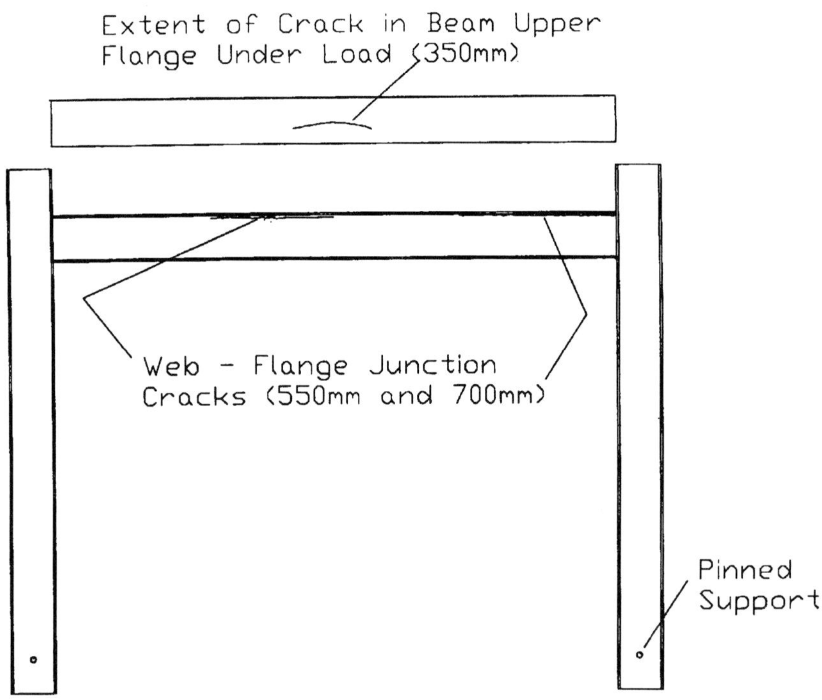

Figure 5 Extent of cracking in beam compression flange and along flange-web junction after completion of the flexural mode ultimate load portal frame test.

6 COMPARISON OF EXPERIMENTAL AND ANALYTICAL DEFORMATIONS

6.1 Shear deformable semi-rigid frame analysis

For a simple rectangular pinned base frame made of shear deformable members joined together by semi-rigid connections with a linear moment-rotation characteristic, it is possible to develop simple closed-form expressions for the deformations at points of particular interest, e.g. mid-span deflection (w_c) and joint rotations. Such expressions have been derived assuming that the beam (span = b) and columns (height = a) are made of the same sections. For example, the expression for the mid-span deflection of the beam when the frame is loaded in a flexural mode is as follows:

$$w_c = \left(\frac{Wb^3}{192EI}\right)\left[4 - \left\{\frac{9ab}{2a^2 + 3ab + 6\alpha + 6a\beta}\right\} + 48\frac{\alpha}{b^2}\right] \qquad (3)$$

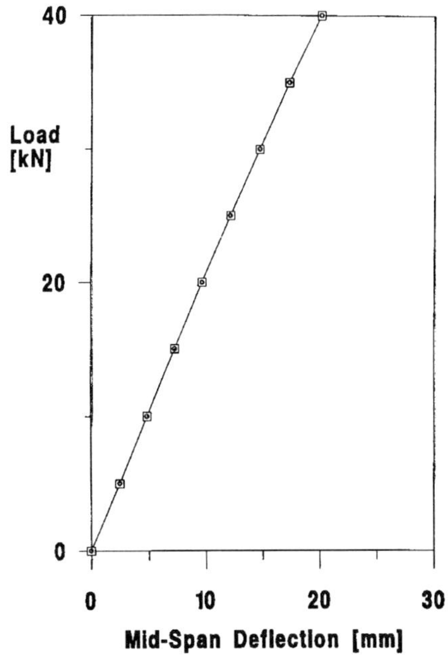

(a)

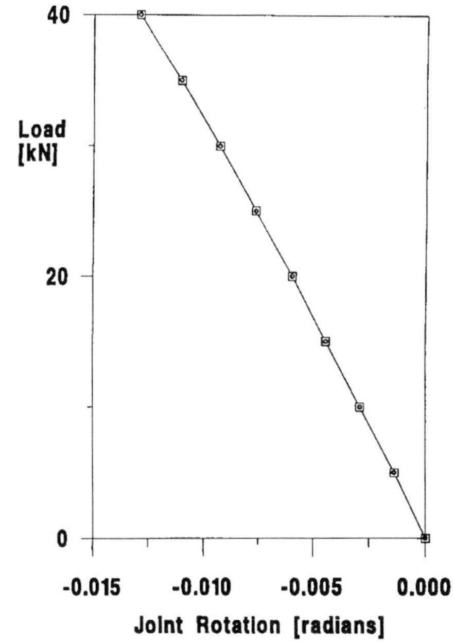

(b)

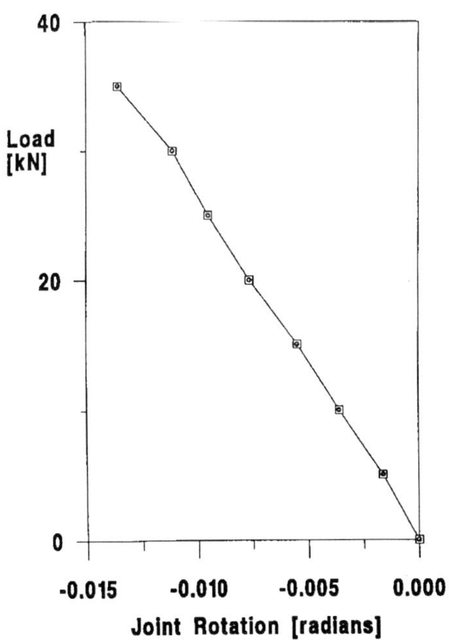

(c)

Figure 6 Ultimate load, flexural mode portal frame test. (a) Load versus mid-span deflection (DG1 = dial gauge 1, see Figure 1(b)). (b) Load versus relative rotation between the beam and column of the left-hand beam-column joint. (c) Load versus relative rotation between the beam and column of the right-hand beam- column joint.

In Equation (3) the first and third terms in the square brackets, when multiplied by the factor outside these brackets, represent respectively the flexural and shear deflections of a simply supported beam carrying a point load at its centre. The second term accounts for the fact that the beam is not simply supported but forms part of a frame with identical semi-rigid beam-column joints. Also in Equation (3), the α (= EI/GA) and β (= EI/K, where K is the joint stiffness) parameters reflect the contributions of shear deformation and semi-rigidity respectively to the mid-span deflection. Similar expressions have been derived for the beam translation under sway mode loading and the corresponding rotations at the beam-column joints and pinned bases. Some of these expressions have been used to predict the frame deformations described in the next section.

6.2 Measured and calculated deflections

For the flexural mode case, beam mid-span deflections were calculated using various beam and frame models with the point load at mid-span. In these calculations, the value of α was determined from the EI and GA values measured in the three-point beam bending tests. The beam and frame models include: shear rigid and shear deformable simply supported and clamped beams and shear rigid and shear deformable, rigid jointed, pinned base frames. Unfortunately, the semi-rigidity of the beam-column connection was not known a priori and, therefore, the semi-rigid frame analysis mid-span deflections could not be calculated directly. Details of the analytical and experimental deflections for each model are given Table 5. It is clear that the calculated deflections vary from as little as 21% (shear rigid clamped beam model) to as much as 123% (shear deformable simply supported beam model) of the measured deflection.

An attempt was made to use back analysis to evaluate the joint semi-rigidity by setting the measured mid-span deflection equal to w_c in Equation (3) and then solving for the semi-rigidity factor, β. However, when a similar back analysis was carried out for the sway mode a different value of β was obtained.

A comparison of the measured and calculated sway mode deflections is presented for different frame models. In the semi-rigid model the β-value determined from the flexural mode back analysis is used. Table 6 shows that the predicted deflections are all much less than the measured values - ranging from 16% to 45% of the actual value.

Table 5 Pultruded glass FRP portal frame tested in flexural mode (W = 20 kN) - comparison of measured and calculated deflections at the centre of the beam.

Analysis model	*Deflection at beam centre (mm)*	*Percentage of experimental mid-span deflection*
Shear-flexible, rigid-jointed frame	7.87	82.9
Shear-rigid, rigid-jointed frame	3.82	40.2
Shear-flexible, simply supported beam	11. 7	123.1
Shear-rigid, simply supported beam	7.94	83.7
Shear-flexible, clamped beam	5.73	60.3
Shear-rigid, clamped beam	1.99	20.9
Experimental value	9.49	100.0

7 CONCLUSIONS

A series of coupon and three-point simply supported beam bending tests have been carried out on pultruded glass FRP WF-sections and material cut therefrom. In addition, several low load tests on a simple rectangular pinned base portal frame fabricated from WF-sections and subjected to single point loading have been carried out. Deformations and strains were recorded in each test. An ultimate load frame test was also carried out for the point load applied at the beam mid-span. Instead of failure arising in the bolted beam web column flange connections, flexural tension cracking failure of the beam compression flange occurred in the vicinity of the point load when the load was almost 40 kN and the mid-span deflection was just over 20 mm (about 1.5 times the serviceability limit on the deflection of (span/200)). Full section stiffness data derived from the three-point flexure tests were used in a number of beam and frame models in order to establish their accuracy in predicting frame deflections. All of the models grossly under-estimate the sway mode deflection at the top of the columns and all but one under-estimate the flexural mode beam mid-span deflection. The material property data derived from the coupon tests appear to be greater than (and are, therefore, consistent with) the minimum values quoted by the manufacturer (reference 4).

Table 6 Pultruded glass FRP portal frame tested in sway mode (W = 2 kN) - comparison of measured and calculated deflections.

Analysis model	*Deflection at top of column (mm)*	*Percentage of top of column experimental deflection*
Shear-flexible, rigid-jointed frame	4.46	21.5
Shear-rigid, rigid-jointed frame	3.30	15.9
Shear-flexible, semi-rigid jointed frame	9.35	45.2
Experimental value	20.7	100.0

8 ACKNOWLEDGEMENTS

The author wishes to record his appreciation to Mr Mark Salisbury (Laboratory Technician, Engineering Department) who carried out all of the fabrication work, as well as the majority of the experimental tasks associated with this project.

The author wishes also to acknowledge the financial support received from the EC's EUREKA Programme (Project EU468) and the Engineering Department, Lancaster University, without which this project could not have been undertaken.

REFERENCES

1. Mosallam, A.S., Short and long-term behavior of a pultruded fiber reinforced plastic frame, PhD Thesis, Catholic University of America, Washington, DC, October 1990.

2. Bell, S., GRP structure, Final Year Undergraduate Project Report, pp. 92, May 1992.

3. Turvey, G.J., Tension/compression strength of unnotched/notched pultruded GRP plate. Published in *Composites Testing and Standardisation,* (eds P.J. Hogg, G.D. Sims, F.L. Matthews, A.R. Bunsell and A. Massiah), European Association for Composite Materials, pp.167-176 1992.

4. *EXTREN fiberglass structural shapes - Design Manual,* Morrison Molded Fiberglass Company Inc., Bristol, Virginia, 1989.

5. Turvey, G.J. Testing of a pultruded GRP pinned base rectangular portal frame for the EUROCOMP Project. Internal Report, Engineering Department, Lancaster University, pp.65 May 1994 (Revised July 1994).

6. Sims, G.D., Johnson, A.F. and Hill, R.D. Mechanical and structural properties of a GRP pultruded section, *Composite Structures,* **8**, No.3, pp.173-187, 1987.

7. Bank, L.C. Flexural and shear moduli of full-section Fiber reinforced plastic (FRP) pultruded beams, *ASTM Journal of Testing and Evaluation,* **17**, No.1, pp.40-45, 1989.

8. Mottram, J.T. Lateral-torsional buckling of a pultruded I-beam, *Composites,* **23**, No.2, 1992, pp.81-92.

9. Timoshenko, S.P. and Gere, J.M., *Mechanics of materials,* Brooks/Cole Engineering Division, Monterey, California, 2nd Edition, pp.407-414, 1984.

INDEX

Page references in **bold** refer to figures and page references in *italic* refer to tables.

Index

Index

Index